The Exercise Effect on Mental Health

The Exercise Effect on Mental Health contains the most recent and thorough overview of the links between exercise and mental health, and the underlying mechanisms of the brain. The text will enhance interested clinicians' and researchers' understanding of the neurobiological effect of exercise on mental health. Editors Budde and Wegner have compiled a comprehensive review of the ways in which physical activity impacts the neurobiological mechanisms of the most common psychological and psychiatric disorders, including depression, anxiety, bipolar disorder, and schizophrenia. This text presents a rigorously evidence-based case for exercise as an inexpensive, time-saving, and highly effective treatment for those suffering from mental illness and distress.

Henning Budde, Ph.D., is Professor of Sports Science and Research Methodology at the MSH Medical School in Hamburg and an affiliating professor at the Lithuanian Sport University and Reykjavík University in Iceland. Henning finished his "Habilitation," the highest academic degree in Germany, at the Humboldt-Universität zu Berlin, and also received an honorary doctorate from the Lithuanian Sport University (LSU). His main research interest is exercise neuroscience and how these findings can be implemented in school settings. He is funded by the DFG (German Research Foundation) BU 1837/5-1 and BU 1837/5-2.

Mirko Wegner, Ph.D., is currently a Professor of Sport Psychology at Humboldt-Universität zu Berlin in Germany. His research interests focus on physiological and neurobiological responses to physical and psychological stress, and their affective, health-related, and cognitive consequences. He is an expert in motivation and self-regulation research and specialized in implicit motivational processes and their behavioral physiological associations. He applies his knowledge to the areas of health, physical education in school, and expert sport performance.

The Exercise Effect on Mental Health

Neurobiological Mechanisms

Edited by
Henning Budde and
Mirko Wegner

NEW YORK AND LONDON

First edition published 2018
by Routledge
711 Third Avenue, New York, NY 10017

and by Routledge
2 Park Square, Milton Park, Abingdon, Oxon, OX14 4RN

Routledge is an imprint of the Taylor & Francis Group, an informa business

Library of Congress Cataloging-in-Publication Data
Names: Budde, Henning, editor. | Wegner, Mirko, editor.
Title: The exercise effect on mental health : neurobiological mechanisms / edited by Henning Budde and Mirko Wegner.
Description: First edition. | New York, NY : Routledge, 2018. | Includes bibliographical references.
Identifiers: LCCN 2017061586 (print) | LCCN 2017061851 (ebook) | ISBN 9781315113906 (ebk) | ISBN 9781498739511 (hbk) | ISBN 9780815348863 (pbk)
Subjects: | MESH: Exercise–psychology | Mental Disorders–therapy | Mental Health | Psychophysiology–methods | Neurobiology–methods
Classification: LCC RC489.E9 (ebook) | LCC RC489.E9 (print) | NLM WM 405 | DDC 616.89/13–dc23
LC record available at https://lccn.loc.gov/2017061586

ISBN: 978-1-498-73951-1 (hbk)
ISBN: 978-0-815-34886-3 (pbk)
ISBN: 978-1-315-11390-6 (ebk)

Typeset in Galliard
by Out of House Publishing

Contents

List of Figures

List of Tables

Notes on Contributors

Aderbal Silva Aguiar, Jr., Ph.D., is a professor of kinesiology in the department of health sciences at Universidade Federal de Santa Catarina in Brazil. She researches in the field of exercise physiology and the neurobiology of physical exercise.

Brandon L. Alderman, Ph.D., is an Associate Professor and Vice Chair of Kinesiology and Health at Rutgers, The State University of New Jersey. He received his Ph.D. in Interdisciplinary Exercise Science in 2004 from Arizona State University. He has established a patient-oriented research program to study how exercise and other behavioral interventions can be used to enhance physiological, neurocognitive, and psychological resilience.

Sophie Bischoff, is a psychologist at the Clinic for Psychiatry at the Campus Charite Mitte Berlin.

Inna Bragina, studied Sports Psychology and Sports Medical Training at University of Bielefeld and Goethe University Frankfurt. She is research associate at Chemnitz University of Technology, Germany since 2015. Her research interests focus on the effect of physical activity on psychological health and on the exercise effect on proprioception and postural control under single and dual task conditions.

Annmarie Chizewski, is a Ph.D. student in Kinesiology at the University of Illinois Urbana-Champaign. Her research areas of interest are exercise promotion among firefighters and first responders and the effects exercise has on firefighters' emotional health with a focus on psychophysiology.

Jo Corbett, is a Principal Lecturer in Exercise Physiology and Associate Head in the Department of Sport and Exercise Science at the University of Portsmouth. Jo works in the Extreme Environments Laboratory where his research interests encompass a range of themes but are primarily related to the role of environmental stressors (heat, cold, and hypoxia) in performance, health, and disease.

Eco de Geus, Ph.D., is head of the department of Biological Psychology at the Vrije Universiteit Amsterdam that maintains the Netherlands Twin

Register (NTR). NTR is a population-based cohort of over 100,000 participants with longitudinal information that extends from genotype to biomarkers, gene expression to rich behavioral information including biennial reports on personality, mental health, and lifestyle. Within the NTR de Geus leads the genetic epidemiology research on the association between exercise behaviors and mental and cardiovascular disease.

Marleen De Moor, Ph.D., is Assistant Professor at the sections of Clinical Child and Family Studies and Methods at the Vrije Universiteit Amsterdam. She investigates the relationship between parenting, genetic susceptibility and child development (in particular motor development and physical activity). She also teaches statistics in the bachelors of Psychology and Educational and Family Studies. Marleen obtained her Ph.D. on the genetics of exercise behavior and mental health in 2009 in the department of Biological Psychology at the Vrije Universiteit Amsterdam.

Zsolt Demetrovics, Ph.D., is professor of psychology at the Eötvös Loránd University (ELTE), Budapest, Hungary. He is clinical psychologist (specialized in addiction), cultural anthropologist, and has a Ph.D. in clinical and health psychology (addictive behaviors) and a D.Sc. in psychology. He is currently the dean of the Faculty of Education and Psychology at ELTE. His primary interest is focused on the characteristics of substance use disorders and behavioral addictions.

Guy Faulkner, Ph.D., is a Professor and CIHR-PHAC Chair in Applied Public Health at the University of British Columbia. He received his Ph.D. in exercise psychology in 2001 from Loughborough University, UK. His research focuses on the development and evaluation of physical activity interventions.

Viviane Grassmann, Ph.D., is currently an independent researcher based in Toronto, Canada. She received her Ph.D. in Psychobiology in 2015 from the Universidade Federal de São Paulo. Her studies focus on the effects of exercise on cognitive function and mental health.

Daniel R. Greene, Ph.D., an Assistant Professor of Kinesiology at Augusta University in Augusta, Georgia. He received his Ph.D. in Kinesiology in 2017 from The University of Illinois at Urbana-Champaign. His current research agenda is focused on the psychological and physiological adaptations to high-intensity body weight/interval training, especially within populations living with full to subsyndromal post-traumatic stress.

Tina A. Greenlee, Ph.D., is a Program Manager for the Geneva Foundation, currently based out of the Center for the Intrepid (San Antonio, TX) and focused primarily on physical therapy research in the military population. She received her Ph.D. in Kinesiology in 2016 from the University of Illinois at Urbana-Champaign. Her research interests include the influence of stress on cognition and affect, exercise behavior, pain, injury prevention and prediction, and prediction of outcomes following treatment for musculoskeletal injuries.

Mark D. Griffiths, Ph.D., is a Chartered Psychologist and Distinguished Professor of Behavioral Addiction at the Nottingham Trent University, and Director of the International Gaming Research Unit. He has spent 30 years in the field and is internationally known for his work into gambling, gaming, and behavioral addictions (including online addictions).

Megan Herting, Ph.D., is an Assistant Professor in Preventive Medicine at The University of Southern California. She received her Ph.D. in Behavioral Neuroscience in 2012 from Oregon Health and Science University. Her research aims to address how environmental factors, including aerobic exercise, influence the developing brain and behavior.

Robert Kohn, M.D., is a Professor of Psychiatry and Human Behavior at the Warren Alpert Medical School of Brown University in Providence, Rhode Island, U.S.A. He is the Director of the Brown University World Health Organization Collaborating Center for Research on Psychiatric Epidemiology and Mental Health.

Alexandra Latini, Ph.D., is an Associate Professor of Biochemistry in the Biochemistry Department at the Center for Biological Sciences, Universidade Federal de Santa Catarina. She has a Doctorate degree in Chemical Sciences and Biochemistry and is a researcher in the field of mitochondrial physiology and dynamics, and metabolism.

Stephen Lawrie, M.D., is a Professor and Head of Psychiatry at the University of Edinburgh. He received his M.D. (Hons) in Psychiatry in 2007 from the same university. His research aims to identify schizophrenia at the earliest possible stage and to intervene with a view to secondary and potentially even primary prevention.

Nico Lehmann, Ph.D., is a professor of motor skill learning, neuroplasticity, and neuromodulation at the Max Planck Institute for Human Cognitive and Brain Sciences in Leipzig, Germany, where he also received his Ph.D. in 2014. He is currently a research assistant at the University of Fribourg in Switzerland.

Terry McMorris, Ph.D., is Emeritus Professor of the Cognitive Neuroscience of Exercise and Sport at the University of Chichester and Visiting Professor at the University of Portsmouth. He also teaches part-time at Northumbria University, Newcastle-upon-Tyne.

Lorna McWilliams, Ph.D., is a research fellow at The Christie NHS Foundation Trust and is affiliated with the Manchester Centre for Health Psychology at the University of Manchester. She received her Ph.D. in Psychiatry and Applied Psychology in 2013 from the University of

Nottingham. Her research focuses on behavior change for a range of health behaviors in public health and clinical settings.

George Mammen, Ph.D., is a Post-Doctoral Fellow at the Centre for Addiction and Mental Health (Toronto, Canada) within the Institute for Mental Health Policy Research. He received his Ph.D. in Exercise Science in 2015 from the University of Toronto, which partially entailed studying physical activity's preventive effect on depression. Dr. Mammen's current research revolves around studying cannabis' medical application to mood and anxiety disorders, along with the public health implications of legalizing recreational cannabis.

Gillian Mead, Ph.D., is Professor of Stroke and Elderly Care at University of Edinburgh. She received her M.D. from the University of Cambridge. She has major research interests in the role of exercise for health and for recovery after stroke

David Moreau, Ph.D., is a cognitive neuroscientist at the University of Auckland. He is interested in the plasticity of the brain, that is, the capacity for the brain to change throughout the lifespan, a line of research that he started at Princeton University and is now expanding to neurodevelopmental disorders. He is currently leading the MovinCog Initiative, a program of research investigating cognitive malleability in children with learning disorders, based on ongoing research on brain plasticity.

Jennifer Mumm, M.Sc., is a psychologist at the Clinic for Psychiatry at the Campus Charite Mitte Berlin.

Nanette Mutrie, Ph.D., is Chair of Physical Activity for Health at the University of Edinburgh. She received her Ph.D. in Exercise Science in 1985 from the Pennsylvania State University and was awarded an MBE in 2015 for services to physical activity promotion. Nanette directs the Physical Activity for Health Research Centre which develops and tests interventions to encourage people to sit less and move more.

Claudia Niemann, studied Biology and Physical Education at Free University of Berlin and Humboldt University of Berlin. She received her Ph.D. from Jacobs University Bremen in the field of movement and cognitive science in 2015.

Ryan L. Olson, Ph.D., is an assistant professor in the Department of Kinesiology, Health Promotion, and Recreation at the University of North Texas, Denton. He received his doctorate from Rutgers University.

Steven J. Petruzzello, Ph.D., is a Professor in the Department of Kinesiology and Community Health at the University of Illinois Urbana-Champaign and Director of the Exercise Psychophysiology Laboratory. He received his Ph.D. in Exercise Science from Arizona State University in 1991. His research examines the psychological impact of exercise from a multi-level perspective captured within three major themes: (1) the influence of exercise intensity on affective responses; (2) psychophysiological aspects of exercise in extreme environments; and (3) individual differences in exercise-related affective and cognitive responses.

Michel Probst, Ph.D., is a professor in the Department of Rehabilitation Sciences at the Univesrity of Leuven.

Katie Richards, is a Research Assistant at the University of Edinburgh, Scotland.

Simon Rosenbaum, Ph.D., is a professor in the Department of Exercise Physiology in the School of Medical Sciences at the University of South Wales, Sydney, Australia.

Kathryn Rougeau, is an Assistant Professor of Wellness and Health Promotion at Oakland University in Rochester, Michigan. She received her Ph.D. in Kinesiology in 2017 from the University of Illinois at Urbana-Champaign. Her research involves the exploration of the impact of wellness and health promotion in individuals with disabilities.

Andreas Ströhle, Ph.D., is a specialist for psychiatry and psychoterhapy. Since 2002 he has been a senior physician at the Clinic for Psychiatry at the Campus Charité Mitte Berlin.

Brendon Stubbs, Ph.D., is a researcher in the Physiotherapy Department at the South London and Maudsley NHS Foundation Trust in London as well as a professor in the Health Service and Population Research Department at the Institute of Psychiatry at King's College, London.

Attila Szabo, Ph.D., is professor of psychology at the Institute of Health Promotion and Sport Sciences at the Eötvös Loránd University (ELTE) in Budapest, Hungary. He completed his university studies up to and including the Ph.D. in Montreal, Canada, then worked as Senior Lecturer and subsequently as Reader in Nottingham, United Kingdom for nearly nine years, and in 2007 he joined the ELTE team in Budapest. His research interests falls within the area of health psychology with a special focus on behavioral addictions, affective effects of exercise, and the role of expectations in human behavior.

Marco Taubert, Ph.D., studied Sport Science at Martin-Luther University Halle-Wittenberg. He received his Ph.D. from University Leipzig for his work on "Plasticity in the Sensorimotor System" in 2012. In recent years, he authored and co-authored several articles in international journals on the effect of motor learning and physical activity on human brain structure and function.

Adrian Taylor, has worked in the field of exercise and mental health since 1990, and co-founded the international journal, *Mental Health and Physical*

Activity in 2008. He has led many studies to improve our understanding of how single sessions of exercise impact on different dimensions of mood and affect, and other psychological constructs (e.g. nicotine, alcohol and food cravings and withdrawal symptoms, cognitive functioning, attentional bias to substances) associated with addictions and behaviors which are often poorly self-regulated. He has also led numerous randomized trial on the effectiveness of physical activity interventions for people with mental health conditions.

Tom Thompson, Ph.D., is a Research Fellow within Plymouth University Peninsula Schools of Medicine and Dentistry, supported by grants from several National Institute of Health Research grants and he is currently the Principal Investigator on a Research for Patient Benefit funded systematic review of physical activity and the prevention, reduction, and treatment of alcohol and substance use across the lifespan.

Davy Vancampfort, Ph.D., is a professor in the Department of Rehabilitation Sciences at the University of Leuven. He received his doctorate in biomedical sciences.

Claudia Voelcker-Rehage, Ph.D., is Full Professor of Sports Psychology (with a focus on Prevention and Rehabilitation) at Chemnitz University of Technology, Germany. Her research interests include the neurocognition and control of movement, learning, and plasticity and the role of physical activity for cognitive development and health. She examines both children and adolescents as well as, in particular, older adults.

Preface

The Exercise Effect on Mental Health

An editorial welcome to this new inter-disciplinary book. Whether you are a novice or a well-established researcher in the field, we hope it will appeal to readers from a wide range of backgrounds. Exercise or physical activity (PA) has been recommended by scientists for many years to prevent physical health problems, such as obesity and hypertension and it is also well-known in treating a variety of other diseases such as Diabetes Mellitus or Peripheral Arterial Disease once they have occurred. In the past 30 years we have witnessed that exercise or PA can also prevent and treat mental diseases, so the time is right to bring the related areas such as psychology, sport science, medicine, and neuroscience together in one publication. This book has one primary aim: to enhance our understanding of the relationship between body and mind for the benefit of those who are interested in the phenomena of mental health problems.

The body–mind connection has recaptured scientific interest in recent years with more than a dozen academic books, several special issues in journals (e.g. *Neural Plasticity*, *CNS and Neurological Disorders – Drug Targets*) and even one international impact-factored journal (*Mental Health and Physical Activity*, first issue 2008) entirely dedicated to this topic. The number of manuscripts published on the topic in peer-reviewed journals has increased from 10 in 1990, to about 60 in 2000, to over 700 in 2017, based on a raw search on *PubMed* using the words "exercise" and "mental health". This reflects the growth in interest available through just this one search engine. It is agreed that mental health is not just the absence of mental disorders. *Mental health* means mental functioning and has a physiological base. It is interconnected with physical and social functioning as well as with health outcomes (World Health Organization 2001). This definition of mental health also includes concepts like subjective well-being, self-efficacy, autonomy, competence, intergenerational dependence, and being able to utilize one's intellectual (e.g. cognitive functioning) and emotional potential (e.g. absence of depression and anxiety), and ensures social functioning (World Health Organization, 2001, 2004) and life satisfaction (Gauvin & Spence 1996). This book focuses on several aspects included in mental health. *Physical activity* refers to body movement that leads to energy expenditure and is initiated by skeletal muscles

(Caspersen, Powell, & Christenson 1985; Budde et al. 2016). *Exercise* has been previously defined as a disturbance of homeostasis through muscle activity resulting in movement and increased energy expenditure (Scheuer & Tipton 1977). The critical difference between both terms refers to the planned and structured nature of exercise (Caspersen et al. 1985).

The pandemic nature of mental health problems like anxiety and depression (see Chapter 1 by Kohn in this book) in combination with the costs and time associated with cognitive psychotherapy, as well as the potential side effects of various drugs used in the treatment of such diseases, stipulate the search for a potent cure. Provided with positive attributes, exercise or PA could be the treatment of choice if it could be found to be effective and applicable in treating as well as preventing mental diseases. Previous research has strongly suggested that exercise or PA benefit different areas of mental health (Taylor, Sallis, & Needle 1985; Hughes 1984) including depression, anxiety, cognitive functioning, and psychological well-being in adults as well as in the elderly (Gauvin & Spence 1996; Hillman, Erickson, & Kramer 2008; Wegner, Helmich, Machado Arias-Carrión, and Budde 2014) and in children and adolescents (Lagerberg, 2005; Donaldson and Ronan 2006; Biddle and Asare 2011; Sibley and Etnier 2003). However, the effectiveness of exercise or PA as a mental health intervention is not universally acknowledged nor perfectly understood. Still little is known about the exercise–mental health relationship regarding physiological mechanisms. Future studies are strongly needed that more closely investigate the underlying neurobiological mechanisms. However, for various reasons, researchers and practitioners in medicine, psychology, and public health remain either uninformed or unconvinced about the potential of this intervention to promote mental health (Ekkekakis 2013). Some of these reasons may be due to concerns about the methodological accuracy of the studies. Distinguishing between good and poor methodology paired with suggestions for future studies is a topic that runs like a continuous thread through the book. Additionally, mostly due to inexperience, practitioners find it hard to “subscribe” exercise (e.g. in an in-patient setting for majorly depressive individuals) as an effective treatment. To this end, the present book is an attempt to give a state-of-the-art overview of the topic and the editors hope that it will encourage more collaboration between researchers and practitioners to further utilize exercise and PA as effective means to support mental health.

Along with Biddle, Fox, Boutcher, and Faulkner (2000) there are five reasons why physical activity may be a very effective strategy to promote mental health. One is that physical activity is potentially cost effective – it is relatively inexpensive to deliver as an intervention and to participate in. Second and third, in contrast to pharmacological interventions, physical activity is associated with minimal adverse side-effects but it has the potential to simultaneously improve physical health and well-being. Fourth, physical activity can be indefinitely sustained by the individual unlike pharmacological and psychotherapeutic treatments, which often have a specified end-point. Finally, many other non-drug treatments (such as cognitive behavioral therapy) are expensive and therefore often in short supply.

The idea behind this book is to foster the understanding of the link between mental health and exercise or physical activity in a potentially broad field from neuroscience across many different dimensions of mental health and psychological well-being. In this book we have drawn together researchers from different disciplines, e.g. brain science, cognitive psychology, neuroscience, exercise science as well as psychophysiology to present a state-of-the-art summary of what is known about exercise/PA and mental health. However, the explaining mechanisms underlying the effects of regular physical activity and exercise on the different facets of mental health are not completely understood up to now.

In Section 1, the authors discuss the mechanisms behind the benefits of exercise as a theoretical introduction. Robert Kohn starts this section by giving a detailed epidemiological overview of the occurrence of common mental disorders including, for example, the prevalence and incidence of mental disorders, sociodemographic risk factors, the comorbidity and the financial burden for the society. Following, Terry McMorris and Jo Corbett present existing evidence for neurobiological mechanisms behind the benefits of exercise for mental health. The purpose of their chapter is to outline the neurobiological changes induced by acute and chronic exercise, which are thought to influence depression, anxiety, and psychological well-being. In their chapter entitled "Causality in the associations between exercise, personality, and mental health," Marleen De Moor and Eco de Geus show that part of the same set of genetic factors that influence whether people participate in regular exercise also affect mental health outcomes. Based on a review of published evidence they further conclude that, in the population at large, the association between exercise participation and higher levels of life satisfaction and happiness and lower levels of anxiety and depression is not causal. The authors include evidence that immunomodulation is a biological mechanism responsible for the anti-inflammatory effects of regular exercise. Aderbal Silva Aguiar, Jr. and Alexandra Latini are taking a very close look at an immune perspective when talking about treating depression with exercise.

The focus of Section 2 are age-related effects of exercise on mental health which will start out with a chapter on the younger age group by Mirko Wegner and Henning Budde. This chapter aims to illustrate how physical activity and exercise benefit different areas of mental health in children and adolescents including their general well-being, reductions in depression and anxiety, and benefits to cognitive functioning. Exercise may positively affect well-being through different neurobiological mechanisms which are highlighted for this age group of children and adolescents. The second chapter of this section is a contribution by Inna Bragina, Claudia Niemann, and Claudia Voelcker-Rehage and focuses on the exercise effect on mental health in older adults. The authors provide insights on how PA and exercise interact with cognitive functioning and psychological well-being in older adults. They include research on healthy aging of cognitive functions and the brain, on cognitive and neuronal plasticity, dementia and mild cognitive impairment, on how physical activity can prevent or postpone mild cognitive impairment (MCI) and dementia, and support psychological well-being, self-efficacy, and self-esteem. They further show how diseases like dementia, depression, and anxiety interact and explain the underlying neurobiological mechanisms.

Section 3 of this book is dedicated to the exercise effects on cognition and motor learning. It starts with David Moreau, who stresses a possible impact of physical exercise on cognitive enhancement with a summary of recent work in this field. He presents new experimental evidence for integrated activities combining physical and cognitive demands with applications in the classroom. Nico Lehmann and Marco Taubert then talk about exercise-induced improvement in motor learning. They review physical (endurance) exercise as a new approach to enhance motor skill learning through facilitation of the underlying neurobiological processes. Megan Herting finishes this third section with a review of research on exercise and learning. In her chapter she reviews some basic concepts of the ways in which we learn (and remember) information before she dives into the established associations between exercise and these cognitive abilities as well as the underlying neurobiology.

The effects of sport and exercise on emotions and psychological diseases is the topic in Section 4. The first chapter by Adrian Taylor and Tom Thompson focuses on exercise in the prevention, treatment, and management of addictive substance use. The chapter begins with identifying the prevalence of addictions, and the implications for health and society. They then briefly identify the scope and extent of evidence for the effectiveness of current treatment options. Following, they turn to summarizing the evidence (and quality/risk of bias) of the chronic and acute effects of physical activity on different addictive behaviors. Finally, they consider the possible mechanisms for how PA has an impact on addictions and they further identify future research questions and discuss the practical implications. Attila Szabo, Zsolt Demetrovics and Mark D. Griffiths in their chapter more closely examine morbid exercise behavior and discuss exaggerated exercise behavior from the perspective of behavioral addictions. The chapter is followed by a chapter by Davy Vancampfort, Simon Rosenbaum, Michel Probst and Brendon Stubbs who provide new insights to the benefits of aerobic exercise for people with schizophrenia. Their chapter has two aims, the first is to provide a systematic overview of intervention characteristics, exercise outcomes and motivational skills used in recent (i.e. last decade) randomized controlled trials investigating the effect of aerobic exercise in schizophrenia. The second aim is to provide evidence-based clinical and research recommendations regarding the prescription of exercise in schizophrenia. Jennifer Mumm, Sophie Bischoff, and Andreas Ströhle review the links between exercise on anxiety disorders. They focus on the effect of different kinds of exercise or PA typically investigated in research on various forms of anxiety. They also discuss psychological and biological mechanisms of exercise on anxiety. The chapter by Lorna McWilliams is entitled "Exercise and ADHD: Implications for treatment." The author outlines the research surrounding the promising use of physical activity as a treatment for the disorder, particularly in the aid of development of executive functions. Nanette Mutrie, Katie Richards, Stephen Lawrie, and Gillian Mead's topic is: Can physical activity prevent or treat clinical depression? In this chapter the authors explore the role of physical activity for the prevention and treatment of depression. They discuss in depth the possible neurobiological mechanisms by which exercise might benefit depression. Finally, they point out various guidelines about the role of physical activity that mental health professionals might find useful.

The final section of the book binds chapters dealing with research implications for the health sector and for schools. Brandon L. Alderman and Ryan L. Olson begin the section by summarizing the available evidence on mode and dose-response relationships of exercise on anxiety, depression, and cognitive functioning. Steven J. Petruzzello, Dan Greene, Annmarie Chizewski, Kathryn Rougeau, and Tina Greenlee provide an overview of acute and chronic effects of exercise on important mental health outcomes. In addition, they provide important implications for health care practitioners. Viviane Grassmann, George Mammen, and Guy Faulkner raise the question: Can physical activity prevent mental illness? The focus of their section is to explore whether physical activity may also serve a role in preventing mental disorders – in particular dementia and depression.

The inter-disciplinary research on the link between exercise and mental health represented in this edition is an exciting and vibrant field of research in which scholars from different areas meet and produce findings that have strong effects on different domains of society. Our hope is that the interested reader finds the most recent and in-depth overviews on different aspects of the exercise-mental health links including references to underlying neurobiological mechanisms. We are looking forward to seeing more researchers and practitioners working cooperatively to utilize exercise and physical activity in the applied setting. Finally, we hope that the contributions to this edition encourages researchers to further explore the field and contribute health and well-being for future generations.

Henning Budde and Mirko Wegner
Am Wriezener Bahnhof Berlin, Janaury 2018

References

Biddle, S. J. H., & M. Asare. 2011. "Physical activity and mental health in children and adolescents: A review of reviews." *British Journal of Sports Medicine 45*: 886–95. doi:10.1136/886 bjsports-2011–090185

Biddle, S. J. H., K. R. Fox, S. H. Boutcher, & G. Faulkner. 2000. "The way forward for physical activity and the promotion of psychological well-being." In S. J. H. Biddle, Fox, K. R. & S. H. Boutcher (Eds.), *Physical activity and psychological well-being* (pp. 154–168). London: Routledge.

Budde, H., R. Schwarz, B. Velasques, P. Ribeiro, M. Holzweg, S. Machado,…M. Wegner. 2016. "The need for differentiating between exercise, physical activity, and training." *Autoimmunity Reviews 15*: 110–11. doi:10.1016/j.autrev.2015.09.004

Caspersen, C. J., K. E. Powell, & G. M. Christenson. 1985. "Physical activity, exercise, and physical fitness: Definitions and distinctions for health-related research." *Public Health Reports 100* (2): 131.

Donaldson, S. J., & K. R. Ronan. 2006. "The effects of sports participation on young adolescents' emotional well-being." *Adolescence 41* (162): 369–89.

Ekkekakis, P. 2013. *Routledge handbook of physical activity and mental health.* New York: Routledge.

Gauvin, L., & J. C. Spence. 1996. "Physical activity and psychological well-being: Knowledge base, current issues, and caveats." *Nutrition Reviews 54* (4): S53–S67.

Hillman, C. H., K. I. Erickson, & A. F. Kramer. 2008. "Be smart, exercise your heart: Exercise effects on brain and cognition." *Nature Reviews Neuroscience 9* (1): 58–65.

Hughes, J. R. 1984. "Psychological effects of habitual aerobic exercise: A critical review." *Preventive Medicine 13*: 66–78.

Lagerberg, D. 2005. "Physical activity and mental health in schoolchildren: A complicated relationship." *Acta Paediatrica 94*: 1699–701.

Scheuer, J., & C. M. Tipton. 1977. "Cardiovascular adapations to physical training." *Annual Review of Physiology 39*: 221–51.

Sibley, B. A., & J. L. Etnier. 2003. "The relationship between physical activity and cognition in children: A meta analysis." *Pediatric Exercise Science 15*: 243–56.

Taylor, C. B., J. F. Sallis, & R. Needle. 1985. "The relation of physical activity and exercise to mental health." *Public Health Reports 100* (2): 195–202.

Wegner, M., I. Helmich, S. Machado, O. Arias-Carrión, & H. Budde. 2014. "Effects of exercise on anxiety and depression disorders: Review of meta-analyses and neurobiological mechanisms." *CNS and Neurological Disorders – Drug Targets 13* (6): 1002–14. doi:10.2174/1871527313666140612102841

World Health Organization. 2001. *The World Health Report 2001: Mental health: New understanding, new hope.* Geneva: World Health Organization.

World Health Organization. 2004. "Prevalence, severity, and unmet need for treatment of mental disorders in the World Health Organization: World Mental Health Survey." *Journal of the American Medical Association 291*: 2581–90.

Section 1

The Benefits of Exercise – a Theoretical Introduction (Mechanisms)

1 Epidemiology of Common Mental Disorders

Robert Kohn

Prevalence and Incidence

In DSM-5 (American Psychiatric Association 2013) major depressive disorder is defined as having five or more symptoms during the same two-week period, which cause clinically significant distress or impairment, and at least one should be a depressed mood or loss of interest or pleasure. The other symptoms include: change in weight; sleep disturbance; psychomotor agitation or retardation; fatigue or loss of energy; worthlessness or guilt; impaired concentration; and thoughts of death. The current DSM-5 criteria are similar to the DSM-IV (American Psychiatric Association 2000) diagnostic criteria used in most of the psychiatric epidemiological surveys conducted since the 1990s.

The anxiety disorders have been redefined from DSM-IV to DSM-5. In DSM-IV the anxiety disorders included panic disorder, agoraphobia without a history of panic disorder, specific phobia, social phobia, post-traumatic stress disorder, acute stress disorder, generalized anxiety disorder, and obsessive-compulsive disorder. The anxiety disorders in DSM-5 have been narrowed with post-traumatic stress disorder and acute stress disorder being moved into a category entitled Trauma – and Stressor- Related Disorders and obsessive-compulsive disorder being placed in a category entitled Obsessive-Compulsive and Related Disorders. Current psychiatric epidemiological studies are primarily based on the DSM-IV definition of anxiety disorders. Acute stress disorder is not usually included among general population psychiatric epidemiological surveys and the inclusion of obsessive-compulsive disorder may be variable.

Prevalence and incidence studies are based on representative community household surveys that utilize a structured diagnostic instrument, such as the Composite International Diagnostic Interview (CIDI) (Robins et al. 1988). The prevalence of major depressive disorder varies widely cross nationally. For example, the 12-month prevalence ranges from 0.3% in Vietnam to 10.2% in Iran. Across studies based on DSM-IV using the CIDI the estimated median 12-month prevalence is 4.7% (n = 42 studies) (see Table 1.1). Similarly, there is a broad range of lifetime prevalence rates from 3.2% in Nigeria to 20.4% in France. The median lifetime prevalence is estimated at 9.9% (n = 39 studies). Major depressive disorder has an early age of onset; the median age in epidemiological studies is approximately 27.6 years of age (n = 23 studies).

Data on the one-year incidence of major depression is based on a small number of prospective studies. The Epidemiological Catchment Area (ECA) survey reported an incidence rate of 1.6 per 100 person years at risk, 2.0 females and 1.1 males (Eaton et al. 1989). The Baltimore site of the ECA, 1981–1993, had an incidence rate of 0.3 (Eaton, Kalaydjian, Scharfstein, Mezuk, & Ding 2007). During the 1993–2004 period the incidence rate was 0.2. However, the difference between the two time periods, and incidence rates also found in shorter follow-up studies, may be due to attrition in the sample and not that the incidence of major depression is declining. For both periods, female gender and age of 30–44 had the highest incidence rates. A study based in Edmonton, Canada had an incidence rate of 2.8, with a rate for females of 3.7 and males 2.0 (Newman & Bland 1998). The incidence rate in the Canadian study was highest for age 45–64. The Netherlands Mental Health Survey and Incidence Study (NEMESIS) reported an incidence rate of 2.7 for major depression, 3.9 females and 1.7 males. The incidence rate was highest for males for ages 25–34 and for females for age 35–44 (Bijl, de Graaf, Ravelli, Smit, & Vollebergh 2002). The NEMISIS study examined risk factors associated with the incidence of a mood disorder; female gender, having a negative life event in the past 12 months, ongoing difficulties in the past 12 months, high neuroticism and sleep problems were all associated in an adjusted logistic regression model (de Graaf, Bijl, Ravelli, Smit, & Vollebergh 2002). In the National Epidemiologic Survey on Alcohol and Related Conditions (NESARC) risk

Table 1.1 Twelve-month and lifetime DSM-III-R and DSM-IV prevalence rates and age of onset of major depressive episodes and anxiety disorders from selected Composite International Diagnostic Interview studies

Country	*Major Depression*			*Anxiety Disorders*		
	12M	*LT*	*Onset*	*12M*	*LT*	*DX*
Australia	4.8	12.8		11.8	20.0	1–5,7
Belgium	5.2	14.1	29.4	8.4	13.1	1,2,4,5,7,8
São Paulo, Brazil	10.1	18.0	24.3	19.9	28.1	1–8
Bulgaria	3.0	6.7		5.6		1,2,4,7,8
Beijing and Shanghai, China	2.0	3.8	30.3	3.0	4.8	1,2,4,5,7,8
Shenzhen, China	3.6	6.8	18.8			
Canada	4.8	12.2		4.7		1,4,7
Chile*	5.7	9.2		9.9	16.2	1,2,4,5,7,8
Colombia	5.3	11.8	23.5	14.4	25.3	1,2,4,5,7,8
Medellin, Colombia	3.8	9.9				
Czech Republic*	2.0	7.8				
Finland	7.4					
France	5.6	20.4	28.4	13.7	22.3	1,2,4,5,7,8
Germany	3.1	10.3	27.6	8.3	14.6	**1,2,4,5,7,8**
Guatemala	0.8	3.2		2.3	5.2	1–5,7
Pondicherry, India	4.5	9.0	31.9			
Iran	10.2			15.6		1–5,7
Iraq	3.9	7.2	46.0	10.4	13.8	1,2,4,5,7,8
Israel	5.9	9.8	25.5	3.6	5.2	1,2,4,5,7,8
Italy	2.9	9.7	27.7	6.5	11.0	1,2,4,5,7,8
Japan	2.4	6.8	30.1	4.2	6.9	1,2,4,5,7,8
Lebanon	4.9	10.3	23.8	12.2	16.7	1,2,4,5,7,8
Mexico	3.7	7.6	23.5	8.4	14.3	1,2,4,5,7,8
Netherlands	4.9	18.0	27.2	8.9	15.93	1,2,4,5,7,8
New Zealand	5.7	15.8	24.2	15.0	24.6	1,2,4,5,7,8
Nigeria	1.1	3.2	29.2	4.2	6.5	1,2,4,5,7,8
Northern Ireland	8.8	17.7		12.3		1,2,4,7,8
Norway*	7.3	17.8				
Peru	2.7	6.4	38.0	7.9	14.9	1,2,4,5,7,8
Poland	1.6	3.8				
Portugal	7.0	17.4		13.7		1,2,4,7,8
Romania	1.5	2.9		4.2		1,2,4,7,8
Singapore	2.2	5.8	26.0	0.4	0.9	2
South Africa	4.9	10.4	22.3	8.2	15.8	1,2,4,5,7,8
South Korea	3.1	6.7		6.8	8.7	1,2,4,5,7,8
Spain	3.8	10.6	30.0	6.6	9.9	1,2,4,5,7,8
Murcia, Spain	6.0	13.8		9.7	15.0	1–8
Bangkok, Thailand		19.9			10.2	9
Turkey*	3.5	6.3		5.8	7.4	1,2,4,7,8

(continued)

Table 1.1 (cont.)

Country	*Major Depression*			*Anxiety Disorders*		
	12M	*LT*	*Onset*	*12M*	*LT*	*DX*
Izmir, Turkey	8.2					
United States	8.3	19.2	22.7	19.0	31.0	1,2,4,5,7,8
Ukraine	8.4	14.6	27.8	6.8	10.9	1,2,4,5,7,8
Mekong Delta, Vietnam	0.3			0.4		1–5,7

* DSM-III-R diagnoses others are DSM-IV

12M = 12-month prevalence; LT = lifetime prevalence; Onset = median age of onset for depressive disorder; DX = Type of anxiety disorder included in study: agoraphobia without panic disorder = 1, generalized anxiety disorder = 2, obsessive-compulsive disorder = 3, panic disorder = 4, post-traumatic stress disorder = 5, separation anxiety disorder = 6, social phobia = 7, specific phobia = 8, not specified = 9

Alhasnawi et al. 2009; Andrade et al. 2003; Andrade et al. 2012; Bromet et al. 2011; Cho et al. 2015; Chong et al. 2012; Chong, Vaingankar, Abdin, and Subramaniam 2012; Fiestas and Piazza 2014; Gureje, Uwakwe, Oladeji, Makanjuola, and Esan 2010; Kessler et al. 2009; Kessler et al. 2015; Kohn 2013; Kringlen, Torgersen, and Cramer 2001; Lee et al. 2009; Markkula et al. 2015; McDowell et al. 2014; McEvoy, Grove, and Slade 2011; Navaro-Mateu et al. 2015; Patten et al. 2015; Piazza and Fiestas 2014; Sharifi et al. 2015; Steel et al. 2009; Thavichachart et al. 2001; Topuzoğlu et al. 2015; Vaingankar et al. 2013; Viana and Andrade 2012; Vicente et al. 2006

factors for incidence of major depression included female gender, age 18–55, low income and being unmarried. The incidence rate for the NESARC study was 1.5, similar to the ECA study (Grant et al. 2009). The NESARC study also examined the relationship of incidence with comorbidity. Significantly higher incident rates for major depression were found among those who had dysthymia, an anxiety disorder (panic disorder, social phobia, specific phobia, generalized anxiety disorder, post-traumatic stress disorder) and certain personality disorders (paranoid, schizotypal, borderline, narcissistic).

Sociodemographic Risk Factors

Gender

Traditionally, a 2:1 female to male gender ratio has been associated with major depression (Kohn, Dohrenwend, & Mirotznik 1998). There are exceptions to this nearly universal finding. In the ECA Jewish men had a similar rate of major depression as Jewish women, although while non-Jews had the expected 2:1 female to male ratio (Levav, Kohn, Golding, & Weissman 1997). No gender differences in the rate of major depression were found in a study of the Old-Order Amish (Egeland & Hostetter 1983). A lack of a gender association was also found in the Chinese American Psychiatric Epidemiological Study (Takeuchi et al. 1998); however, the gender difference began to appear as Chinese Americans acculturate. Interestingly, communality between these studies exists, a low rate of alcohol use disorders in men and a near 1:1 gender ratio for major depression between men and woman. However, a prevalence study of Beijing and Shanghai is not consistent with this hypothesis, as men had a significantly higher rate of alcohol use disorders with a 0.8:1 female to male gender ratio for mood disorders (Shen et al. 2006). Psychiatric epidemiological surveys conducted in Muslim countries where the rate of alcohol use disorders are very low have found a less robust female to male gender difference in rates of mood disorders, 1.4:1 in Iraq (Alhasnawi et al. 2009) and 1.5 in Iran (Sharifi et al. 2015); in both studies the 12-month prevalence rate of alcohol use disorders was 0.9%. In the World Mental Health (WMH) surveys an analysis of 18 countries found that the female to male ratio in low- to middle-income countries was 2.1 and in high-income countries 1.8 (Bromet et al. 2011).

A number of explanatory models have been proposed to explain the gender difference in the rate of major depression. The sex role hypothesis suggests that certain gender-based role experiences shape the development of self leading to a propensity to depression (Gove & Tudor 1973). For example, girls have a propensity to internalize, they are more passive and are not as mentally tough as boys. The social factor hypothesis argues that women are exposed to a greater number of stressful life events or are more vulnerable to them (Tennant, Bebbington, & Hurry 1982). If these two hypotheses could explain the gender difference, then the gender difference would already be seen in childhood and not emerge starting in adolescence. In addition, arguing against these two hypotheses are that studies adjusting for social factors and trauma are unable to eliminate the gender difference, and recurrent episodes of depression are not found at greater rates in women than in men (Parker & Brotchie 2010).

The biological hypothesis currently has the most support (Dohrenwend & Dohrenwend 1976). Support for the biological hypothesis is found in the rates of childhood major depression where the boys have a statistically higher rate than girls (Douglas & Scott 2014); however, in adolescence the incidence of depression increases dramatically in girls (Naninck, Lucassen, & Bakker 2011). This has led to view that the female to male ratio change may be a result of gonadal steroid changes and subsequent sexual differentiation in the brain, which may impact on hippocampal anatomy and its structural plasticity, as well as the limbic system and subsequent interactions with the neurotransmitter system (Naninck et al. 2011).

Anxiety disorders also have a strong female to male gender ratio (Kohn et al. 1998). In the WMH surveys post-traumatic stress disorder had the strongest (Seedat et al. 2009) female to male odds ratio 2.6 and social phobia the weakest 1.3. The theories that have been put forth for the gender difference in major depression are also used to explain the finding for anxiety disorders. The notion that the higher rates of major depression and anxiety disorders are due to women and men responding differently to stress has limited support, as most studies have not found such an association (Hill & Needham 2013).

Age

There is an inverse relationship with the prevalence of major depression and anxiety disorders and age. This finding is not limited to major depression and anxiety disorders, but also most other psychiatric disorders studied in community household surveys. Methodological issues may in part explain this consistent finding, such as bias due to older individuals denying symptoms when subjected to complex interview protocols (O'Connor & Parslow 2010). Studies do show, however, that the elderly are less likely to report one-year incident cases of major depression than younger age groups (Grant et al. 2009). In addition, there is evidence to support that younger age groups do have an increasing risk for major depression across time. Cross sectional studies of the same population have shown increasing rates of major depression, a birth cohort effect (Hasin et al. 2011).

Late onset depression, after the age 50, is considered to possibly be distinct from major depression with onset earlier in life. The vascular hypothesis for late-onset depression is based on the hypothesis that subclinical cerebrovascular disease can cause depression (Alexopoulos et al. 1997). Risk factors such as age, hypertension, hyperlipidemia, smoking and diabetes lead to atherosclerosis and subsequently deep white matter lesions in the brain making it more vulnerable to psychosocial risk factors resulting in a "vascular depression" and associated executive dysfunction (Alexopoulos et al. 1997; Sneed & Culang-Reinlieb 2011). Although there is strong evidence for this hypothesis based on clinical research; epidemiological studies using community samples are less robust. The Rotterdam Study, as an example, found that lower blood flow velocity as measured by transcranial doppler ultrasonography predicted depressive disorders (Direk et al. 2012). However, the same longitudinal cohort study could not find an association between atherosclerosis and increased risk for late-onset depression (Newson et al. 2010). In another analysis by this same group, moderate support was found as diabetes mellitus, stroke risk, current smoking, and antihypertensive drug use were associated with late-onset depression. No relation with depression was noted for cholesterol, diastolic and systolic blood pressure, history of cardiovascular disease, atrial fibrillation, left ventricular hypertrophy or the use of statins and anticoagulants (Luijendijk, Stricker, Hofman, Witteman, & Tiemeier 2008). As in other studies examining this issue diagnosis of depression was based on a screening instrument and not a diagnostic interview schedule.

Marital Status

Married persons in most studies have a significantly lower risk of major depression. In the WMH surveys the relationship between major depressive episodes and marital status differed significantly between high and low-middle income countries (Kessler & Bromet 2013). In high-income countries there was a stronger association with being separated and never married with depression;

and in low-middle income countries the association was strongest with being divorced and widowed. Being divorced has the strongest association with having an anxiety disorder (McDowell et al. 2014). Longitudinal incidence studies can provide insight into the causal direction, which cross-sectional studies cannot. The NESARC study found that among those who were separated, widowed or divorced there was an elevated risk of developing major depression compared to those who were married; however, individuals who were never married did not have such an increased risk (Grant et al. 2009). Incident panic disorder, social phobia and specific phobia were not associated with marital status; however, being separated or divorced increased the risk of developing generalized anxiety disorder.

Social Economic Status

Commonly used measures of social economic status (SES) may include income, education, occupation and employment. In two early reviews individuals in the lower SES had a higher risk of reporting depression, 1.81 (Lorant et al. 2003) and 2.4 (Kohn et al. 1998). As for anxiety disorders an inverse relationship existed with SES, the risk was markedly higher for the lower SES: panic disorders 5.6, phobic disorders 2.5, and generalized anxiety disorder 1.7 (Kohn et al. 1998). In the WMH survey there was an inverse association with income and having a major depressive episode in about half of the high-income countries; while in low-middle income countries this association was noted only in one country (Bromet et al. 2011). For anxiety disorders specific phobia and agoraphobia have the most robust inverse relationship with income (Kawakami et al. 2012). Early onset major depression and social phobia, and to a less degree specific phobia and agoraphobia, in the WMH surveys was associated with substantial decrements in household income (Kawakami et al. 2012). The association between lower education and major depressive episode was weak in the WMH survey (Bromet et al. 2011). For both major depression and anxiety disorders early termination of secondary education was associated with an increased risk for major depression and anxiety disorders in high-income but not low-middle income countries (Lee et al. 2009). Subjective social status also has been shown to be inversely associated with major depression and anxiety disorders (Scott et al. 2014). An inverse association with social class may be due to either social causation or social selection; tests of these two competing hypothesis have suggested that social causation has a larger role, a finding that is stronger among women (Dohrenwend et al. 1992; Ritsher, Warner, Johnson, & Dohrenwend 2001; Simmons, Braun, Charnigo, Havens, & Wright 2008). Although there are numerous measure of social class including education, income, poverty, employment status, occupation, occupational prestige, family size, and subjective social status which measure has the strongest relationship with depression or anxiety disorders varies widely across countries and studies.

Life Events

Stressful life events in adulthood have been shown to be predictive of major depression, as well as recurrence of depression. This relationship is reciprocal as major depression can elicit or exacerbate certain stressful life events (Kessler 1997).

The National Comorbidity Survey – Replication (NCS-R) survey examined the role of childhood adversity and first onset mood disorders (Green et al. 2010). Maladaptive family functioning which includes parent mental illness, parent substance abuse, parent criminality, family violence, physical abuse, sexual abuse and neglect were all significantly associated with both mood and anxiety disorders. However, parental death, parental divorce, serious physical illness, and family economic adversity were not associated with first onset mood or anxiety disorders in adults.

Studies of the relationship of life events have numerous methodological pitfalls. Most studies use checklists, which do not account for the contextual meaning of the life event. Whether or not the event is independent or fateful is frequently not assessed.

Stigma and Discrimination

Stigma is frequently defined as an attribute that is deeply discrediting and that reduces the bearer from a whole and usual person to a tainted, discounted one (Goffman 1963). Link and Phelan (2001) describe stigma as occurring when elements of labeling, stereotyping, separation, status loss, and discrimination co-occur in a power situation that allows the components of stigma to unfold. Perceived stigma is strongly related to an increased prevalence of both depression and anxiety disorders in cross-national prevalence studies (Alonso et al. 2008). Perceived discrimination also is a risk factor for depression and anxiety disorders (Capezza, Zlotnick, Kohn, Vicente, & Saldivia 2012; Moomal et al. 2009). Interestingly, the South African Stress and Health Study did not find a relationship for racial discrimination and mental illness, but only non-racial discrimination (Moomal et al. 2009).

Comorbidity

Psychiatric Comorbidity

Major depression is highly comorbid with anxiety and substance use disorders (Kessler et al. 2010). The greater the number of comorbid disorders the stronger the association with major depression. The NCS-R found that three-fourths of those with lifetime major depressive disorder met criteria for at least one other lifetime mental disorder; 59.0% anxiety disorder, 31.9% impulse dyscontrol disorder, and 24.0% substance use disorder (Kessler, Merikangas, & Wang 2007). Two-thirds of those with 12-month prevalent major depressive disorder were comorbid for another 12-month prevalent mental disorder;

57.5% anxiety disorder, 20.8% impulse dyscontrol disorder, and 8.5% substance use disorder. Major depression has been shown to temporally have a later age of onset than the comorbid disorders.

The National Comorbidity Survey (NCS) ten-year follow-up provided temporal information on the relationship between major depression and other mental disorders. The longitudinal nature of this study permitted evaluation of the strength of the association of major depression with incidence of other mental disorders. Major depression was significantly associated with the incidence of nicotine dependence, but not alcohol or illicit drug use disorders (Swendsen et al. 2010). One-third of the comorbid cases of major depression and generalized anxiety disorder occur within the same year (Kessler et al. 2008). Major depression has not been shown to be a predictor for persistence of generalized anxiety disorder; however, generalized anxiety disorder has been shown to be a predictor of persistence of major depression. The NESARC 3-year follow-up study found that lifetime anxiety disorders and alcohol use disorders were predictive of developing first onset major depressive disorder (Peters, Shankman, Deckersbach, & West 2015).

Physical Comorbidity

Physical disorders are highly comorbid with both major depression and anxiety disorders as found in the WMH survey. The rate of comorbidity for 12-month prevalence of major depression with back neck pain was 10.0%; headache was 12.8%; arthritis was 8.0%; any pain was 8.6%; diabetes was 6.5%; COPD or asthma was 9.6%; heart disease was 9.1%; ulcers was 9.7%; and chronic physical disorder was 8.6% (Gureje 2009). The rate of comorbidity for 12-month prevalence of anxiety disorders with back neck pain was 11.4%; headache was 14.7%; arthritis was 9.9%; any pain was 10.4%; diabetes was 8.3%; COPD or asthma was 11.9%; heart disease was 9.8%; ulcers was 12.1%; and chronic physical disorder was 10.3%.

Twelve-month prevalent major depression and anxiety disorders are associated with lifetime COPD (Rapsey et al. 2015). The 12-month prevalence of major depression in those with COPD was 8.7%, and 20.8% for lifetime prevalence of major depression for adult onset asthma (Alonso et al. 2014). Twelve-month and lifetime prevalence of anxiety disorders among those with COPD varies markedly based on the specific anxiety disorder. The 12-month prevalence comorbidity rate for specific anxiety disorders is: panic disorder 2.2%, generalized anxiety disorder 6.4%, social phobia 7.0%, specific phobia 8.5%, agoraphobia with panic disorder 1.4% and post-traumatic stress disorder 3.2%. The lifetime comorbidity rate for specific anxiety disorders is: panic disorder 4.2%, generalized anxiety disorder 9.5%, social phobia 7.1%, specific phobia 11.2%, agoraphobia with panic disorder 1.2%, and post-traumatic stress disorder 8.3%.

An association with subsequent diabetes, hypertension, heart disease, ulcers, arthritis, back and neck pain, and chronic headaches was established

(Demyttenaere et al. 2007; He et al. 2008; Scott et al. 2013; Stein et al. 2014) for major depression both with and without comorbity with anxiety disorders. The positive relationship between subsequent onset of heart disease and major depression although statistically significant was weaker than that for most anxiety disorders, impulse-control disorders and substance disorders (Scott et al. 2013). A positive association with subsequent onset of non-fatal stroke has been noted for major depression and among the anxiety disorder for panic disorder, social phobia, specific phobia, and post-traumatic stress disorder (Swain et al. 2015). Among individuals with adult onset-diabetes, 14.8% have a lifetime prevalence of major depression, and 11.3% predate the onset of the diabetes (de Jonge et al. 2014). Although anxiety disorders are associated with diabetes in unadjusted analyses, this finding no longer remains significant when comorbidity for other mental disorders is accounted for. Depression is associated with self-reported cancer, in particular with earlier onset depression, as is post-traumatic stress disorder (O'Neill et al. 2014). The association for cancer remained significant for panic disorder and specific phobia after adjusting for comorbid mental disorders. A modest association was noted between major depression, anxiety disorders and obesity. This association was stronger among women and those with severe obesity (Scott et al. 2008). The relationship with obesity and major depression is inconsistent across studies; for example the NESEARC study found a positive relationship with atypical depression but not "classical depression" (Levitan et al. 2012).

Physical Activity

A number of representative population surveys have examined the incidence of major depression and anxiety disorders and their relationship to physical activity (see Table 1.2). A cohort of 14–24 year olds from Munich, Germany, was followed for 4 years. No association was found with incidence of major depression and physical activity (Ströhle et al. 2007). The baseline study, a cross-sectional study, found a relationship between major depression and non-regular physical activity, but not between no physical activity and regular physical activity. The relationship with anxiety disorders and physical activity was somewhat different. At baseline anxiety disorders were less prevalent among those with regular physical activity compared to no physical activity, but not non-regular physical activity. The incidence of anxiety disorders was also less frequent among those with regular physical activity as compared to those with no activity; for specific anxiety disorders, however, this could only be shown for agoraphobia and specific phobia. Similarly, a Dutch study with a 3-year follow-up found an association with exercise and major depression and anxiety disorder in the baseline prevalence study, and in the incidence analysis an inverse relationship with major depression and anxiety disorder was noted in those who exercised 1–3 hours but not greater than 4 hours (ten Have, de Graaf, & Monshouwer 2011). Recovery from major depression was not associated with exercise; but recovery from anxiety disorders were, but not

Table 1.2 Studies of major depression and anxiety disorders based on the CIDI and the relationship with physical activity in representative community-based household surveys

First Author	*Location*	*Sample*	*Age*	*Type Study*	*Diagnosis*	*Results*
Beard, Heathcote, Brooks, Earnest, and Kelly 2007	New South Wales, Australia 1997	968	18–85	2-year LI	ICD-10	Negative
Chazelle et al. 2011	Ireland SLAN 2007	9978	18+	CS	DSM-IV	Mixed
Goodwin 2003	USA NCS 1990–1992	5877	15–54	CS	DSM-III-R	Positive
Haarasilta, Marttunen, Kaprio, and Aro 2004	Finland FHCS 1996	942	15–24	1-year LI	DSM-III-R	Mixed
Joutsenniemi et al. 2013	Finland 2000–2001	3658	30–64	CS	DSM-IV	Mixed
Ströhle et al. 2007	Munich, Germany EDSP 1994	2548	14–24	4-year LI	DSM-IV	Mixed
ten Have 2011	Netherlands NEMESIS 1996	7076	18–64	3-year LI	DSM-III-R	Mixed

CS = Cross-sectional study; LI = Longitudinal incidence study
EDSP = Early Developmental Stages of Psychopathology; FHCS = Finnish Health Care Survey; NCS = National Comorbidity Survey; NEMESIS = Netherlands Mental Health Survey and Incidence Study; NoRMHS = Northern Rivers Mental Health Study; SLAN = Survey of Lifestyle, Attitudes and Nutrition

for any specific disorder. In a one-year longitudinal Finnish study lack of exercise was not associated with increased incidence of major depression among 15–19 year olds; however, an association existed among 20–24 year olds (Haarasilta, Marttunen, Kaprio, & Aro 2004). An Australian 2-year longitudinal survey found no relationship with major depression or anxiety disorders and physical activity (Beard, Heathcote, Brooks, Earnest, & Kelly, 2007).

In addition several large-scale cross-sectional national population surveys have examined the role of physical activity and major depression. The National Comorbidity Survey from the United States, ages 15–54, found not only an inverse relationship between physical activity and one-year prevalence of major depression and anxiety disorders, but also a dose response (Goodwin 2003). A subsequent Finish population based study among 30–64 year olds examined the role of physical activity on major depression and cognition. This study found that regular physical activity was associated with better performance in reaction time tests and better verbal fluency among men with a major depressive episode (Joutsenniemi et al. 2013). An Irish cross-sectional survey found an association between low physical activity and major depression, as well as generalized anxiety disorder, but not moderate physical activity (Chazelle et al. 2011).

These studies are household surveys that have all used a variation of the CIDI. Other than the four incidence studies, the others are cross-sectional and can only demonstrate an association between major depression and physical activity. Each of these studies, however, relies on self-report of physical activity. The findings from these psychiatric epidemiological surveys provide a mixed picture for an association between physical activity and major depression.

Course of Major Depression

The course of major depression differs in community and primary care populations compared to clinical samples. In community-based and primary care studies a sizable proportion of individuals after a single depressive episode have a stable prolonged recovery, between 35% and 60% with a mean of 48% (Steinert, Hofman, Kruse, & Leichsenring 2014). Approximately, 10% to 17% have a chronic course of depression. The average length of a depressive episode to recovery varies widely across studies from 20 weeks to 20 months. On average woman has 3.6 years until a recurrence and males 4.6 years. Lack of recovery is associated with being abused in childhood, family history of depression, younger age of onset, lower education, daily smoking, comorbid borderline personality disorder, comorbid substance use disorder, severity of depressive episode and a prior episode of depression. Comorbidity with panic disorder and generalized anxiety disorder also are associated with lack of recovery (Agosti 2014).

The NEMESIS study examined recurrence of major depression (Hardeveld, Spijker, de Graaf, Nolen, & Beekman 2013). In multivariate analysis younger age, a higher number of previous episodes, a severe last depressive episode, negative youth experiences and ongoing difficulties were significant. In the longitudinal Canadian National Population Health Survey the risk for recurrence of depression progressively increased with the number of prior episodes; in addition the risk of recurrence was greater in younger people and women (Bulloch, Williams, Lavorato, & Patten 2014).

The NEMISIS study found long-term debilitating effects of depression even three years after recovery (Rhebergen et al. 2010). The duration of the depressive episode is associated with impaired functioning after recovery.

Child and Adolescent Studies

Representative community-based prevalence studies of adolescents and children remain few (see Table 1.3). Unlike studies of adults, there is little consistency in what is considered the gold-standard instrument to obtain the

Table 1.3 Prevalence of depression and anxiety disorders in selected child and adolescent community-based studies

Country	*Age*	*Diagnostic*	*Depression*			*Anxiety Disorders*		
	Range	*Instrument*	*Boys*	*Girls*	*Total*	*Boys*	*Girls*	*Total*
Chile	4–18	DISC	3.1	7.0	5.1	5.8	11.0	8.3
China	6–17	DAWBA			1.3			6.1
Colombia	13–17	CIDI-A	3.5	6.5	5.0	3.5	7.5	5.5
Goa, India	12–16	DAWBA			0.5			1.0
Israel	14–17	DAWBA	1.9	4.7	3.3	3.1	9.2	6.1
Mexico City, Mexico	12–17	CIDI-A	2.0	7.6	4.8	24.6	35.0	29.8
Puerto Rico	4–16	DISC			3.6			9.5
Seoul, South Korea	6–12	DISC	0	0.1	0.1	10.3	9.5	11.1
United Kingdom	5–15	DAWBA	0.7	0.7	0.7	3.5	4.0	3.8
USA	13–18	CIDI-A	4.6	10.7	8.2			24.9
USA	8–15	DISC	1.8	3.7	2.7	0.4	0.9	0.7

Avenevoli et al. 2015; Benjet et al. 2009; Canino et al. 2004; Farbstein et al. 2010; Ford, Goodman, and Meltzer 2003; Merikangas et al. 2010; Park et al. 2015; Pillai et al. 2008; Torres de Galvis et al. 2012; Vicente et al. 2012; Xiaoli et al. 2014

DAWBA = Development and Well-Being Assessment inventory (Goodman, Ford, Richards, Meltzer, & Gatward 2000)

DISC = Diagnostic Interview Schedule for Children (Shaffer, Fischer, Lucas, Dulcan, & Schwab-Stone 2000)

CIDI-A = Composite International Diagnostic Interview Schedule – Adolescent Supplement (Merikangas, Avenevoli, Costello, Koretz, & Kessler 2009)

prevalence of mental illness. This can be noted by the marked difference in the two national studies of the United States, the NHANES study had a prevalence rate for depression of 3.2% for the age range of 12–19 (Merikangas, Mendola, Pastor, Reuben, & Cleary 2012) and 2.7% for the age range of 8–15 (Merikangas et al. 2010); whereas the adolescent CIDI study had a prevalence of 8.2% for age range 13–18 (Avenevoli, Swendsen, He, Burstein, & Merikangas 2015). Other methodological issues such as the parent interview versus the child-only interview and use of impairment criteria remain open discussions.

Even fewer studies have examined risk factors for depression in children and adolescents. Parents' education, urban-rural differences, SES and marital status of the parents are not associated with major depression in children and adolescents (Avenevoli et al. 2015; Merikangas et al. 2012; Canino et al. 2004; Vicente et al. 2012). Obesity is also not associated with depression (Merikangas et al. 2012). Parental psychopathology, not living with both parents, and to a less degree, in bivariate but not multivariate analysis, impaired family functioning are risk factors for depression (Vicente et al. 2012). Comorbidity between depression and other mental disorders is common, 63.7% had comorbidity primarily with anxiety and behavioral disorders (Merikangas et al. 2012).

Anxiety disorders as a group are more prevalent than major depression. Rates across studies are difficult to compare as the specific anxiety disorders include vary. Risk factors for anxiety disorders include female gender, lower SES, poorer family functioning and being in a single-parent household (Vicente et al. 2012).

Societal Cost

Major depression results in 34.4 mean days out of role per year. The number of mean days out of role is highest in lower-income countries 35.8, compared to medium income countries 34.8 and higher-income countries 33.7 (Alonso et al. 2011). Major depression is not the psychiatric disorder with the most days out of role; bipolar disorder 41.2, panic disorder 42.9, specific phobia 33.8, social phobia 39.3, generalized anxiety disorder 39.8, drug abuse 36.6, and post-traumatic stress disorder 42.7, all have higher days out of role. However, depression and anxiety disorders have higher days out of role than "any physical disorder" 24.5, including cancer which has 31.9 days out of role. As for partial disability, having to cut down on what you did or not get as much done as usual depression ranked 5th among mental and physical disorders. Approximately half of the impact of major depression on role functioning was mediated by problems with cognition and by feelings of embarrassment or shame (Buist-Bouwman et al. 2008). Depression was the second most common reason to cut back on the quality of what one did, and experience extreme effort to perform as usual; only post-traumatic stress disorder resulted in more disability for the mental and physical disorders examined (Bruffaerts et al. 2012).

Major depression is a significant predictor of increased teen childbearing (Kessler et al. 1997). Depression also has an indirect effect on cognition, mobility, self-care, family burden, and stigma (Alonso et al. 2013). In the United States alone the human capital loss of major depression is estimated to between $30.1 billion to $51.5 billion dollars (Kessler 2012).

Global Burden of Disease

The Global Burden of Disease uses two measures of disability. Disability Adjusted Life Years (DALY) is the potential years of life lost due to premature death to include equivalent years of healthy life lost by virtue of individuals being in states of poor health or disability. Years lived with disability (YLD) is defined as years of healthy life lost as a result of disability (Murray & Lopez 1996). Mental disorders account for a disproportionate amount of the burden of disease (7.1% DALY; 21.2% YLD), with depressive disorders (major depression and dysthymia) accounting for 36.4% DALY and 38.3% YLD caused by mental disorders in 2013. Across all ages depressive disorders

Table 1.4 Global percentage and ranking of DALY and YLD accounted by depressive disorders out of 163 diseases and injuries by gender, developed country, developing country and age group, 2013

		Both Genders		*Male*		*Female*		*Developed*		*Developing*	
		%	*Rank*	*%*	*Rank*	*%*	*Rank*	*%*	*Rank*	*%*	*Rank*
DALY	All Ages	2.51	11	1.77	18	3.39	5	3.17	4	2.38	13
	5–14	2.98	11	2.28	14	3.78	7	5.29	5	2.81	11
	15–49	4.80	4	3.41	7	6.54	2	6.09	2	4.60	4
	50–69	2.39	8	1.55	14	3.52	7	2.64	9	2.32	9
	70 ≥	1.17	17	0.75	29	1.69	13	1.21	19	1.15	19
YLD	All Ages	8.07	2	6.55	3	9.42	2	7.06	2	8.36	2
	5–14	5.92	3	4.62	5	7.33	3	6.32	5	5.86	3
	15–49	10.49	2	8.75	2	12.03	2	10.05	2	10.58	2
	50–69	7.10	3	5.56	5	8.50	3	6.39	4	7.38	3
	70 ≥	4.23	8	3.16	10	5.00	6	3.57	8	4.75	5

Source: Institute for Health Metrics and Evaluation (IHME). GBD Compare. Seattle, WA: IHME, University of Washington, 2015. Available from http://vizhub.healthdata.org/gbd-compare.

account for 2.5% of all DALY and 8.1% of all YLD. Nearly 1 in 12 of all years lived with disability of all disorders is accounted by depressive disorders (see Table 1.4). Depressive disorder is the 11th ranked disorder globally contributing to DALY, while it is the second highest ranked disorder for YLD. It is the second ranked disorder for YLD across all ages among females and third among males. The DALY and YLD contribution for developing countries for depressive disorders (843.19 per 100,000) is lower than developed countries (953.76 per 100,000).

Examining the DALY and YLD contribution more specifically for major depression, it accounts for 2.1% of DALYs, ranking 12th among 301 diseases and injuries and accounts for 6.3% of YLD ranking second (see Table 1.5). It is the number one cause of YLD among females age 15–49. Major depression accounts for 31.2% DALY and 32.9% YLD caused by mental disorders in 2013. The DALY and YLD contribution for developing countries for major depression (762.59 per 100,000) is lower than developed countries (713.77 per 100,000). Anxiety disorders account for 1% of DALYs and 3.2% of YLDs. Anxiety ranks ninth among diseases contributing to YLD (see Table 1.6). Among persons age 15–49 anxiety disorders are the fifth leading cause of YLD.

The reasons that depressive disorders and anxiety disorders account for such a high percentage of the global burden of disease is its high prevalence, young age of onset, and chronicity. One of the reasons for the higher rate of DALY and YLD in developed countries for depressive and anxiety disorders compared to developing countries is that lower respiratory diseases, diarrheal diseases, malaria, HIV, and preterm birth complications are all disorders that

Table 1.5 Global percentage and ranking of DALY and YLD accounted by major depression out of 301 diseases and injuries by gender, developed country, developing country and age group, 2013

		Both Genders		*Male*		*Female*		*Developed*		*Developing*	
		%	*Rank*	*%*	*Rank*	*%*	*Rank*	*%*	*Rank*	*%*	*Rank*
DALY	All Ages	2.11	12	1.48	17	2.85	6	2.58	8	2.02	13
	5–14	2.88	7	2.20	10	3.65	5	5.08	4	2.72	8
	15–49	4.06	3	2.88	7	5.53	2	4.96	3	3.92	3
	50–69	1.93	10	1.24	20	2.85	7	2.12	11	1.87	10
	70 ≥	0.93	21	0.59	-	1.26	17	0.97	23	0.91	20
YLD	All Ages	6.23	2	5.51	2	7.92	2	5.76	2	7.09	2
	5–14	5.72	2	4.46	3	7.09	3	6.08	4	5.67	2
	15–49	8.87	2	7.40	2	10.18	1	8.20	2	9.01	2
	50–69	5.71	4	4.44	6	6.85	3	5.13	5	5.94	4
	70 ≥	2.12	8	2.50	11	4.01	9	2.86	11	3.77	7

Source: Institute for Health Metrics and Evaluation (IHME). GBD Compare. Seattle, WA: IHME, University of Washington, 2015. Available from http://vizhub.healthdata.org/gbd-compare.

Table 1.6 Global percentage and ranking of DALY and YLD accounted by anxiety out of 301 diseases and injuries by gender, developed country, developing country and age group, 2013

		Both Genders		*Male*		*Female*		*Developed*		*Developing*	
		%	*Rank*	*%*	*Rank*	*%*	*Rank*	*%*	*Rank*	*%*	*Rank*
DALY	All Ages	0.99	-	0.62	-	1.43	19	1.48	18	0.90	-
	5–14	1.99	11	1.31	16	2.76	7	5.49	3	1.73	13
	15–49	1.97	11	1.25	21	2.87	7	2.96	7	1.82	14
	50–69	0.76	-	0.43	-	1.19	-	1.10	-	0.65	-
	70 ≥	0.30	-	0.17	-	0.42	-	0.39	-	0.24	-
YLD	All Ages	3.19	9	2.32	9	3.97	7	3.31	9	3.16	9
	5–14	5.72	4	4.46	6	7.09	3	6.08	3	5.67	5
	15–49	4.30	5	3.20	6	5.28	5	4.89	5	4.18	5
	50–69	2.25	10	1.55	16	2.88	10	2.67	11	2.08	11
	70 ≥	1.07	19	0.70	-	1.34	17	1.16	19	1.00	19

Source: Institute for Health Metrics and Evaluation (IHME). GBD Compare. Seattle, WA: IHME, University of Washington, 2015. Available from http://vizhub.healthdata.org/gbd-compare.

result in early mortality or years life lost in developing countries, whereas depressive disorders result in relatively few deaths. As the epidemiological transition evolves in developing countries the contribution of depressive and anxiety disorders to DALY and YLD will increase. This is already evident because from 1990 to 2013 there was a 17.4% increase in DALY attributed to depressive disorders in developing countries, with only a 4.8% increase in developed countries.

Treatment Gap

Most people who suffer from depression as well as anxiety disorders (as primary disorder or comorbidity) are often seen for the first time at the primary care level. However, these disorders are frequently not recognized by the health care team, and when they are, they are not always managed appropriately. When depression, in particular, is not treated properly and in a timely manner, it often leads to recurrent symptoms or becomes chronic, resulting in severe disability, death by suicide, or the prolonged suffering of the patient and his or her family. Depression and other manifestations of the disease that are masked by physical symptoms or diverse complaints, lead to repeated medical consultations, tests, inappropriate treatment or unproductive consultations among physicians that do not improve the patient's condition and increase the cost of medical care. As a result of these factors and other barriers to care, such as believing the problems will resolve on their own or one can resolve the problems themselves; financial burden; fear of being diagnosed; having no time to seek treatment; lack of confidence in the health care system or that treatment will not help; and stigma has resulted in a wide treatment gap for both depression and anxiety disorders that needs to be bridged (Vicente, Kohn, Saldivia, Rioseco, & Torres 2005).

The treatment gap represents the absolute difference between the true prevalence of a disorder and the treated proportion of individuals affected by the disorder (Kohn, Saxena, Levav, & Saraceno 2004). Treatment gap may be expressed as the percentage of individuals who require care but do not receive treatment. In 2004 the treatment gap for major depression globally was estimated to be 56% and varied widely across regions of the world. In Africa and Arab countries it was estimated at 67% and 70% respectively, while in Europe it was 45% compared to 57% in the Americas. The most recent estimate of the treatment gap for major depression in the Americas is between 43.2% and 54.8% (Kohn 2013). The treatment gap in the United States is 43.2% (Wang et al. 2005). The World Mental Health surveys provided estimates for all mood disorders. A number of countries had treatment gaps above 80% (e.g. Colombia, Lebanon, Nigeria, South Africa, Ukraine) (Kohn 2014). European countries reported a treatment gap of 69.1% for mood disorders (Alonso et al. 2004). The treatment gap may be underestimated as no data is available for most low-income countries.

The treatment gap for anxiety disorders is even higher than for major depression. The treatment gap for anxiety disorders varies from 50.1% to 66.9% in the Americas, with panic disorder having the lowest treatment gap and generalized anxiety disorder the highest (Kohn 2013). Among European countries the treatment gap for anxiety disorders was 73.8% (Alonso et al. 2004).

Treatment lag may be as high as 4 years with only 35.4% of those who receive treatment doing so in the first year in the United States (Wang et al. 2005). The proportions of lifetime cases with mood disorders that make treatment contact in the first year ranges worldwide from 6.0% to 52.1%

(Wang et al. 2007). The median duration of delay is shortest in some western European countries, China, and Japan, approximately 1 year with longest median delay being 14.0 years in Mexico among World Mental Health survey countries. Earlier onset of disorder, older age, and male gender are associated with greater delay in treatment.

A number of recommendations have been proposed to address the treatment gap for depressive disorders as well as other mental disorders. The World Health Organization (2001) has outlined ten recommendations to implement at both the national and community levels in order to reduce the treatment gap in mental health: make mental health treatment accessible in primary care; psychotropic drugs need to be readily available; shift care away from institutions toward community care; educate the public; involve family, communities, and consumers; establish national mental health programs; increase and improve training of mental health professionals; increase links with other governmental and non-governmental institutions; provide monitoring of the mental health system with quality indicators; and support more research. To close the treatment gap 25 grand challenges were identified by the National Institute of Mental Health as research priorities for the next ten years that need to be addressed (Collins et al. 2011).

Conclusion

Major depression is a highly prevalent disorder worldwide with an early age of onset. It is frequently a chronic and remitting disorder. It is more prevalent in women, in younger individuals, and in those who are not married. Life events including those associated with maladaptive family functioning is a risk factor. Major depression is highly comorbid with anxiety disorders and substance use disorders, as well as physical disorders. The role of physical activity in major depression remains an open question based on psychiatric epidemiological surveys. Major depression has a significant societal cost and accounts for one of the highest burdens of disease of any disorder.

References

Agosti, V. 2014. "Predictors of remission from chronic depression: A prospective study in a nationally representative sample." *Compr Psychiatry 55* (3): 463–7.

Alexopoulos, G. S., B. S. Meyers, R. C. Young, S. Campbell, D. Silbersweig, & M. Charlson. 1997. "'Vascular depression' hypothesis." *Arch Gen Psychiatry 54* (10): 915–22.

Alhasnawi, S., S. Sadik, M. Rasheed, A. Baban, M. M. Al-Alak, A. Y. Othman,…Iraq Mental Health Survey Study Group. 2009. "The prevalence and correlates of DSM-IV disorders in the Iraq Mental Health Survey (IMHS)." *World Psychiatry 8* (2): 97–109.

Alonso, J., M. C. Angermeyer, S. Bernert, R. Bruffaerts, T. S. Brugha, H. Bryson,… ESEMeD/MHEDEA 2000 Investigators, European Study of the Epidemiology of Mental Disorders (ESEMeD) Project. 2004. "Use of mental health services in

Europe: Results from the European Study of the Epidemiology of Mental Disorders (ESEMeD) project." *Acta Psychiatr Scand Suppl 420*: 47–54.

Alonso, J., A. Buron, R. Bruffaerts, Y. He, J. Posada-Villa, J. P. Lepine,...World Mental Health Consortium. 2008. "Association of perceived stigma and mood and anxiety disorders: Results from the World Mental Health Surveys." *Acta Psychiatr Scand 118* (4): 305–14.

Alonso, J., P. de Jonge, C. C. Lim, S. Aguilar-Gaxiola, R. Bruffaerts, J. M. Caldas-de-Almeida,...K. M. Scott. 2014. "Association between mental disorders and subsequent adult onset asthma." *J Psychiatr Res 59*: 179–88.

Alonso, J., M. Petukhova, G. Vilagut, S. Chatterji, S. Heeringa, T. B. Üstün,... R. C. Kessler. 2011. "Days out of role due to common physical and mental conditions: Results from the WHO World Mental Health surveys." *Mol Psychiatry 16* (12): 1234–46.

Alonso, J., G. Vilagut, N. D. Adroher, S. Chatterji, Y. He, L. H. Andrade,...R. C. Kessler. 2013. "Disability mediates the impact of common conditions on perceived health." PLoS One. *8* (6): e65858.

American Psychiatric Association. 2000. *Diagnostic and Statistical Manual of Mental Disorders, Fourth Edition, Text Revised*. Washington, DC: American Psychiatric Association.

American Psychiatric Association. 2013. *Diagnostic and Statistical Manual of Mental Disorders, Fifth Edition*. Arlington, VA: American Psychiatric Association.

Andrade, L., J. J. Caraveo-Anduaga, P. Berglund, R. Bijl, R. de Graaf, W. Vollebergh,... H. U. Wittchen. 2003. "The epidemiology of major depressive episodes: Results from the International Consortium of Psychiatric Epidemiology (ICPE) surveys." *Int J Methods Psychiatr Res 12* (1): 3–21.

Andrade, L. H., Y. P. Wang, S. Andreoni, C. M. Silveira, C. Alexandrino-Silva, E. R. Siu,...M. C. Viana. 2012. "Mental disorders in megacities: Findings from the São Paulo megacity mental health survey, Brazil." *PLoS One 7* (2): e31879.

Avenevoli, S., J. Swendsen, J. P. He, M. Burstein, & K. R. Merikangas. 2015. "Major depression in the national comorbidity survey-adolescent supplement: Prevalence, correlates, and treatment." *J Am Acad Child Adolesc Psychiatry 54* (1): 37–44.

Beard, J. R., K. Heathcote, R. Brooks, A. Earnest, & B. Kelly. 2007. "Predictors of mental disorders and their outcome in a community based cohort." *Soc Psychiatry Psychiatr Epidemiol 42* (8): 623–30.

Benjet, C., G. Borges, M. E. Medina-Mora, E. Mendez, C. Feliz, & C. Cruz. 2009. "Diferencias de sexo en la prevalencia y severidad de trastornos psiquiátricos en adolescentes de la Ciudad de México." *Salud Mental Mexico 32* (2):155–63.

Bijl, R. V., R. de Graaf, A. Ravelli, F. Smit, & W. A. M. Vollebergh. 2002. "Gender and age-specific first incidence of DSM-III-R psychiatric disorders in the general population: Results from the Netherlands Mental Health Survey and Incidence Study (NEMESIS)." *Soc Psychiatry Psychiatr Epidemiol 37* (8): 372–9.

Bromet, E., L. H. Andrade, I. Hwang, N. A. Sampson, J. Alonso, G. de Girolamo,...R. C. Kessler. 2011. "Cross-national epidemiology of DSM-IV major depressive episode." *BMC Med 9*: 90.

Bruffaerts, R., G. Vilagut, K. Demyttenaere, J. Alonso, A. Alhamzawi, L. H. Andrade,... R. C. Kessler. 2012. "Role of common mental and physical disorders in partial disability around the world." *Br J Psychiatry 200* (6): 454–61.

Buist-Bouwman, M. A., J. Ormel, R. de Graaf, P. de Jonge, E. van Sonderen, J. Alonso,...ESEMeD/MHEDEA 2000 investigators. 2008. "Mediators of the association between depression and role functioning." *Acta Psychiatr Scand 118* (6): 451–8.

Bulloch, A., J. Williams, D. Lavorato, & S. Patten. 2014. "Recurrence of major depressive episodes is strongly dependent on the number of previous episodes." *Depress Anxiety 31* (1): 72–6.

Canino, G., P. E. Shrout, M. Rubio-Stipec, H. R. Bird, M. Bravo, R. Ramirez,...A. Martinez-Taboas. 2004. "The DSM-IV rates of child and adolescent disorders in Puerto Rico: Prevalence, correlates, service use, and the effects of impairment." *Arch Gen Psychiatry 61* (1): 85–93.

Capezza, N. M., C. Zlotnick, R. Kohn, B. Vicente, & S. Saldivia. 2012. "Perceived discrimination is a potential contributing factor to substance use and mental health problems among primary care patients in Chile." *J Addict Med 6* (4): 297–303.

Chazelle, E., C. Lemogne, K. Morgan, C. C. Kelleher, J. F. Chastang, & I. Niedhammer. 2011. "Explanations of educational differences in major depression and generalised anxiety disorder in the Irish population." *J Affect Disord 134* (1–3): 304–14.

Cho, M. J., S. J. Seong, J. E. Park, I. W. Chung, Y. M. Lee, A. Bae,...J. P. Hong. 2015. "Prevalence and correlates of DSM-IV mental disorders in South Korean adults: The Korean epidemiologic Catchment Area Study 2011." *Psychiatry Investig 12* (2): 164–70.

Chong, S. A., E. Abdin, J. A. Vaingankar, D. Heng, C. Sherbourne, M. Yap,...M. Subramaniam. 2012. "A population-based survey of mental disorders in Singapore." *Ann Acad Med Singapore 41* (2): 49–66.

Chong, S. A., J. Vaingankar, E. Abdin, & M. Subramaniam. 2012. "The prevalence and impact of major depressive disorder among Chinese, Malays and Indians in an Asian multi-racial population." *J Affect Disord 138* (1–2): 128–36.

Collins, P. Y., V. Patel, S. S. Joestl, D. March, T. R. Insel, A. S. Daar, A. S., ... D. J. Stein. 2011. "Grand challenges in global mental health." *Nature 475* (7354): 27–30.

de Graaf, R., R. V. Bijl, A. Ravelli, F. Smit, & W. A. Vollebergh. 2002. "Predictors of first incidence of DSM-III-R psychiatric disorders in the general population: Findings from the Netherlands Mental Health Survey and Incidence Study." *Acta Psychiatr Scand 106* (4): 303–13.

de Jonge, P., J. Alonso, D. J. Stein, A. Kiejna, S. Aguilar-Gaxiola, M. C. Viana,... K. M. Scott. 2014. "Associations between DSM-IV mental disorders and diabetes mellitus: A role for impulse control disorders and depression." *Diabetologia 57* (4): 699–709.

Demyttenaere, K., R. Bruffaerts, S. Lee, J. Posada-Villa, V. Kovess, M. C. Angermeyer,... M. von Korff. 2007. "Mental disorders among persons with chronic back or neck pain: Results from the World Mental Health Surveys." *Pain 129* (3): 332–42.

Direk, N., P. J. Koudstaal, A. Hofman, M. A. Ikram, W. J. Hoogendijk, & H. Tiemeier. 2012. "Cerebral hemodynamics and incident depression: The Rotterdam Study." *Biol Psychiatry 72* (4): 318–23.

Dohrenwend, B. P., & B. S. Dohrenwend. 1976. "Sex differences and psychiatric disorders." *Am J Sociol 81* (6): 1447–54.

Dohrenwend, B. P., I. Levav, P. E. Shrout, S. Schwartz, G. Naveh, B. G. Link,...A. Stueve. 1992. "Socioeconomic status and psychiatric disorders: The causation-selection issue." *Science 255* (5047): 946–52.

Douglas, J., & J. Scott. 2014. "A systematic review of gender-specific rates of unipolar and bipolar disorders in community studies of pre-pubertal children." *Bipolar Disord 16* (1): 5–15.

Eaton, W. W., A. Kalaydjian, D. O. Scharfstein, B. Mezuk, & Y. Ding. 2007. "Prevalence and incidence of depressive disorder: The Baltimore ECA follow-up, 1981–2004." *Acta Psychiatr Scand 116* (3): 182–8.

Eaton, W. W., M. Kramer, J. C. Anthony, A. Dryman, S. Shapiro, & B. Z. Locke. 1989. "The incidence of specific DIS/DSM-III mental disorders: Data from the NIMH Epidemiologic Catchment Area Program." *Acta Psychiatr Scand 79* (2): 163–78.

Egeland, J. A., & A. M. Hostetter. 1983. "Amish Study, I: Affective disorders among the Amish, 1976–1980." *Am J Psychiatry 140* (1): 56–61.

Farbstein, I., I. Mansbach-Kleinfeld, D. Levinson, R. Goodman, I. Levav, I. Vograft,...Apter, A. 2010. "Prevalence and correlates of mental disorders in Israeli adolescents: Results from a national mental health survey." *J Child Psychol Psychiatry 51* (5): 630–9.

Fiestas, F., & M. Piazza. 2014. "Lifetime prevalence and age of onset of mental disorders in Peru: Results of the World Mental Health Study, 2005." *Rev Peru Med Exp Salud Publica 31* (1): 39–47.

Ford, T., R. Goodman, & H. Meltzer. 2003. "The British Child and Adolescent Mental Health Survey 1999: The prevalence of DSM-IV disorders." *J Am Acad Child Adolesc Psychiatry 42* (10): 1203–11.

Goffman, E. 1963. *Stigma: Notes on the Management of Spoiled Identity*. Englewood Cliffs, NJ: Prentice Hall.

Goodman, R., T. Ford, H. Richards, H. Meltzer, & R. Gatward. 2000. "The Development and Well-being Assessment: Description and initial validation of an integrated assessment of child and adolescent psychopathology." *J Child Psychol Psychiatry 41* (5): 645–57.

Goodwin, R. D. 2003. "Association between physical activity and mental disorders among adults in the United States." *Prev Med 36* (6): 698–703.

Gove, W. R., & J. Tudor. 1973. "Adult sex roles and mental illness." *Am J Sociol 78* (4): 812–35.

Grant, B. F., R. B. Goldstein, S. P. Chou, B. Huang, F. S. Stinson, D. A. Dawson,...W. M. Compton. 2009. "Sociodemographic and psychopathologic predictors of first incidence of DSM-IV substance use, mood and anxiety disorders: Results from the Wave 2 National Epidemiologic Survey on Alcohol and Related Conditions." *Mol Psychiatry 14* (11): 1051–66.

Green, J. G., K. A. McLaughlin, P. A. Berglund, M. J. Gruber, N. A. Sampson, A. M. Zaslavsky, & R. C. Kessler. 2010. "Childhood adversities and adult psychiatric disorders in the National Comorbidity Survey Replication I: Associations with first onset of DSM-IV disorders." *Arch Gen Psychiatry 67* (2): 113–23.

Gureje, O. 2009. "The pattern and nature of mental-physical comorbidity: Specific or general?" In M. R. von Korff, K. M. Scott, & O. Gureje (Eds.), *Global Perspectives on Mental-Physical Comorbidity in the WHO World Mental Health Surveys* (pp. 51–83). New York: Cambridge University Press.

Gureje, O., R. Uwakwe, B. Oladeji, V. O. Makanjuola, & O. Esan. 2010. "Depression in adult Nigerians: Results from the Nigerian Survey of Mental Health and Well-being." *J Affect Disord 2120* (1–3): 158–64.

Haarasilta, L. M., M. J. Marttunen, J. A. Kaprio, & H. M. Aro. 2004. "Correlates of depression in a representative nationwide sample of adolescents (15–19 years) and young adults (20–24 years)." *Eur J Public Health 14* (3): 280–5.

Hardeveld, F., J. Spijker, R. de Graaf, W. A. Nolen, & A. T. Beekman. 2013. "Recurrence of major depressive disorder and its predictors in the general population: Results from the Netherlands Mental Health Survey and Incidence Study (NEMESIS)." *Psychol Med 43* (1): 39–48.

Hasin, D. S., M. C. Fenton, & M. M. Weissman. 2011. "Epidemiology of depressive disorders." In M. T. Tsuang, M. Tohen, & P. Jones, *Textbook of psychiatric epidemiology* (pp. 289–309). West Sussex, UK: John Wiley & Sons.

He, Y., M. Zhang, E. H. Lin, R. Bruffaerts, J. Posada-Villa, M. C. Angermeyer,...R. Kessler. 2008. "Mental disorders among persons with arthritis: Results from the World Mental Health Surveys." *Psychol Med 38* (11): 1639–50.

Hill, T. D., & B. L. Needham. 2013. "Rethinking gender and mental health: A critical analysis of three propositions." *Soc Sci Med 92*: 83–91.

Joutsenniemi, K., A. Tuulio-Henriksson, M. Elovainio, T. Härkänen, P. Sainio, S. Koskinen,...T. Partonen. 2013. "Depressive symptoms, major depressive episodes and cognitive test performance – what is the role of physical activity?" *Nord J Psychiatry 67* (4): 265–73.

Kawakami, N., E. A. Abdulghani, J. Alonso, E. J. Bromet, R. Bruffaerts, J. M. Caldas-de-Almeida,...R. C. Kessler. 2012. "Early-life mental disorders and adult household income in the World Mental Health Surveys." *Biol Psychiatry 72* (3): 228–37.

Kessler, R. C. 1997. "The effects of stressful life events on depression." *Annu Rev Psychol 48*: 191–214.

Kessler, R. C. 2012. "The costs of depression." *Psychiatr Clin North Am 35* (1): 1–14.

Kessler, R. C., S. Aguilar-Gaxiola, J. Alonso, S. Chatterji, S. Lee, J. Ormel,...P. S. Wang. 2009. "The global burden of mental disorders: An update from the WHO World Mental Health (WMH) surveys." *Epidemiol Psichiatr Soc 18* (1): 23–33.

Kessler, R. C., P. A. Berglund, C. L. Foster, W. B. Saunders, P. E. Stang, & E. E. Walters. 1997. "Social consequences of psychiatric disorders, II: Teenage parenthood." *Am J Psychiatry 154* (10): 1405–11.

Kessler, R. C., H. G. Birnbaum, V. Shahly, E. Bromet, I. Hwang, K. McLaughlin,...D. J. Stein. 2010. "Age differences in the prevalence and co-morbidity of DSM-IV major depressive episodes: Results from the WHO World Mental Health Survey Initiative." *Depress Anxiety 27* (4): 351–64.

Kessler, R. C., & E. J. Bromet. 2013. "The epidemiology of depression across cultures." *Annu Rev Public Health 34*: 119–38.

Kessler, R.C., M. Gruber, J. M. Hettema, I. Hwang, N. Sampson, & K. A. Yonkers. 2008. "Co-morbid major depression and generalized anxiety disorders in the National Comorbidity Survey follow-up." *Psychol Med 38* (3): 365–74.

Kessler, R. C., K. R. Merikangas, & P. S. Wang. 2007. "Prevalence, comorbidity, and service utilization for mood disorders in the United States at the beginning of the twenty-first century." *Annu Rev Clin Psychol 3*: 137–58.

Kessler, R. C., N. A. Sampson, P. Berglund, M. J. Gruber, A. Al-Hamzawi, L. Andrade,...M. A. Wilcox. 2015. "Anxious and non-anxious major depressive disorder in the World Health Organization World Mental Health Surveys." *Epidemiol Psychiatr Sci 24* (3): 210–26.

Kohn, R. 2013. *Treatment Gap in the Americas.* Washington, DC: Pan American Health Organization.

Kohn, R. 2014. "Trends and gaps in mental health disparities." In S. O. Okpaku (Ed.), *Global mental health: Essential concepts* (pp. 27–38). New York: Cambridge University Press.

Kohn, R., B. P. Dohrenwend, & J. Mirotznik. 1998. "Epidemiologic findings on selected psychiatric disorders in the general population." In B. P. Dohrenwend

(Ed.), *Adversity, Stress, and Psychopathology* (pp. 235–84). New York: Oxford University Press.

Kohn, R., S. Saxena, I. Levav, & B. Saraceno. 2004. "The treatment gap in mental health care." *Bull World Health Organ 82* (11): 858–66.

Kringlen, E., S. Torgersen, & V. Cramer. 2001. "A Norwegian psychiatric epidemiological study." *Am J Psychiatry 158* (7): 1091–8.

Lee, S., A. Tsang, J. Breslau, S. Aguilar-Gaxiola, M. Angermeyer, G. Borges,...R. C. Kessler. 2009. "Mental disorders and termination of education in high-income and low- and middle-income countries: Epidemiological study." *Br J Psychiatry 194* (5): 411–7.

Lee, S., A. Tsang, Y. Q. Huang, Y. L. He, Z. R. Liu, M. Y. Zhang,...R. C. Kessler. 2009. "The epidemiology of depression in metropolitan China." *Psychol Med 39* (5): 735–47.

Levav, I., R. Kohn, J. M. Golding, & M. M. Weissman. 1997. "Vulnerability of Jews to affective disorders." *Am J Psychiatry 154* (7): 941–7.

Levitan, R. D., C. Davis, A. S. Kaplan, T. Arenovich, D. I. Phillips, & A. V. Ravindran. 2012. "Obesity comorbidity in unipolar major depressive disorder: Refining the core phenotype." *J Clin Psychiatry 73* (8): 1119–24.

Link, B. G., & J. C. Phelan. 2001. "Conceptualizing stigma." *Annu Rev Sociol 27*: 363–85.

Lorant, V., D. Deliège, W. Eaton, A. Robert, P. Philippot, & M. Ansseau. 2003. "Socioeconomic inequalities in depression: A meta-analysis." *Am J Epidemiol 157* (2): 98–112.

Luijendijk, H. J., B. H. Stricker, A. Hofman, J. C. Witteman, & H. Tiemeier. 2008. "Cerebrovascular risk factors and incident depression in community-dwelling elderly." *Acta Psychiatr Scand 118* (2): 139–48.

Markkula, N., J. Suvisaari, S. I. Saarni, S. Pirkola, S. Peña, S. Saarni,...T. Härkänen. 2015. "Prevalence and correlates of major depressive disorder and dysthymia in an eleven-year follow-up–results from the Finnish Health 2011 Survey." *J Affect Disord 173*: 73–80.

McDowell, R. D., A. Ryan, B. P. Bunting, S. M. O'Neill, J. Alonso, R. Bruffaerts,...T. Tomov. 2014. "Mood and anxiety disorders across the adult lifespan: A European perspective." *Psychol Med 44* (4): 707–22.

McEvoy, P. M., R. Grove, & T. Slade. 2011. "Epidemiology of anxiety disorders in the Australian general population: Findings of the 2007 Australian National Survey of Mental Health and Wellbeing." *Aust N Z J Psychiatry 45* (11): 957–67.

Merikangas, K., S. Avenevoli, J. Costello, D. Koretz, & R. C. Kessler. 2009. "National comorbidity survey replication adolescent supplement (NCS-A): I. Background and measures." *J Am Acad Child Adolesc Psychiatry 48* (4): 367–9.

Merikangas, K. R., J. P. He, D. Brody, P. W. Fisher, K. Bourdon, & D. S. Koretz. 2010. "Prevalence and treatment of mental disorders among US children in the 2001–2004 NHANES." *Pediatrics 125* (1) :75–81.

Merikangas, A. K., P. Mendola, P. N. Pastor, C. A. Reuben, & S. D. Cleary. 2012. "The association between major depressive disorder and obesity in US adolescents: Results from the 2001–2004 National Health and Nutrition Examination Survey." *J Behav Med 35* (2): 149–54.

Moomal, H., P. B. Jackson, D. J. Stein, A. Herman, L. Myer, S. Seedat,...D. R. Williams. 2009. "Perceived discrimination and mental health disorders: The South African Stress and Health study." *S Afr Med J 99* (5 Pt 2): 383–9.

Murray, C. J. L. & A. D. Lopez. 1996. *The global burden of disease: A comprehensive assessment of mortality and disability from diseases, injuries, and risk factors in 1990 and projected to 2020*. Cambridge, MA: Harvard University Press.

Naninck, E. F., P. J. Lucassen, & J. Bakker. 2011. "Sex differences in adolescent depression: Do sex hormones determine vulnerability?" *J Neuroendocrinol 23* (5): 383–92.

Navarro-Mateu, F., M. J. Tormo, D. Salmerón, G. Vilagut, C. Navarro, G. Ruíz-Merino,...J. Alonso. 2015. "Prevalence of mental disorders in the south-east of Spain, one of the European regions most affected by the economic crisis: The Cross-Sectional PEGASUS-Murcia Project." *PloS One 10* (9): e0137293.

Newman, S. C., & R. C. Bland. 1998. "Incidence of mental disorders in Edmonton: Estimates of rates and methodological issues." *J Psychiatr Res 32* (5): 273–82.

Newson, R. S., K. Hek, H. J. Luijendijk, A. Hofman, J. C. Witteman, & H. Tiemeier. 2010. "Atherosclerosis and incident depression in late life." *Arch Gen Psychiatry 67* (11): 1144–51.

O'Connor, D. W., & R. A. Parslow. 2010. "Differences in older people's responses to CIDI's depression screening and diagnostic questions may point to age-related bias." *J Affect Disord 125* (1–3): 361–4.

O'Neill, S., J. Posada-Villa, M. E. Medina-Mora, A. O. Al-Hamzawi, M. Piazza, H. Tachimori,...K. M. Scott. 2014. "Associations between DSM-IV mental disorders and subsequent self-reported diagnosis of cancer." *J Psychosom Res 76* (3): 207–12.

Park, S., B. N. Kim, S. Cho, J. W. Kim, M. S. Shin, & H. J. Yoo. 2015. "Prevalence, correlates, and comorbidities of DSM-IV psychiatric disorders in children in Seoul, Korea." *Asia Pac J Public Health 27* (2): NP1942–51.

Parker, G., & H. Brotchie. 2010. "Gender differences in depression." *Int Rev Psychiatry 22* (5): 429–36.

Patten, S. B., J. V. Williams, D. H. Lavorato, J. L. Wang, K. McDonald, & A. G. Bulloch. 2015. "Descriptive epidemiology of major depressive disorder in Canada in 2012." *Can J Psychiatry 60* (1): 23–30.

Peters, A. T., S. A. Shankman, T. Deckersbach, & A. E. West. 2015. "Predictors of first-episode unipolar major depression in individuals with and without sub-threshold depressive symptoms: A prospective, population-based study." *Psychiatry Res 230* (2): 150–6.

Piazza, M., & F. Fiestas. 2014. "Annual prevalence of mental disorders and use of mental health services in Peru: Results of the World Mental Health Survey." 2005. *Rev Peru Med Exp Salud Publica 31* (1): 30–8.

Pillai, A., V. Patel, P. Cardozo, R. Goodman, H. A. Weiss, & G. Andrew. 2008. "Non-traditional lifestyles and prevalence of mental disorders in adolescents in Goa, India." *Br J Psychiatry 192* (1): 45–51.

Rapsey, C. M., C. C. Lim, A. Al-Hamzawi, J. Alonso, R. Bruffaerts, J. M. Caldas-de-Almeida,...K. M. Scott. 2015. Associations between DSM-IV mental disorders and subsequent COPD diagnosis. *J Psychosom Res 79* (5): 333–9.

Rhebergen, D., A. T. Beekman, R. de Graaf, W. A., Nolen, J. Spijker, W. J. Hoogendijk, & B. W. Penninx. 2010. "Trajectories of recovery of social and physical functioning in major depression, dysthymic disorder and double depression: A 3-year follow-up." *J Affect Disord 124* (1–2): 148–56.

Ritsher, J. E., V. Warner, J. G. Johnson, & B. P. Dohrenwend. 2001. "Inter-generational longitudinal study of social class and depression: A test of social causation and social selection models." *Br J Psychiatry Suppl 40*: s84–90.

Robins, L. N., J. Wing, H. U. Wittchen, J. E. Helzer, T. F. Babor, J. Burke,...L. H. Towle. 1988. "The Composite International Diagnostic Interview. An epidemiologic Instrument suitable for use in conjunction with different diagnostic systems and in different cultures." *Arch Gen Psychiatry 45* (12): 1069–77.

Scott, K. M., A. O. Al-Hamzawi, L. H. Andrade, G. Borges, J. M. Caldas-de-Almeida, F. Fiestas,...R. C. Kessler. 2014. "Associations between subjective social status and DSM-IV mental disorders: Results from the World Mental Health surveys." *JAMA Psychiatry 71* (12): 1400–8.

Scott, K. M., J. Alonso, P. de Jonge, M. C. Viana, Z. Liu, S. O'Neill,...R. C. Kessler. 2013. "Associations between DSM-IV mental disorders and onset of self-reported peptic ulcer in the World Mental Health Surveys." *J Psychosom Res 75* (2): 121–7.

Scott, K. M., R. Bruffaerts, G. E. Simon, J. Alonso, M. Angermeyer, G. de Girolamo,... M. von Korff. 2008. "Obesity and mental disorders in the general population: Results from the world mental health surveys." *Int J Obes* (Lond) *32* (1): 192–200.

Scott, K. M., P. de Jonge, J. Alonso, M. C. Viana, Z. Liu, S. O'Neill,...R. C. Kessler. 2013. "Associations between DSM-IV mental disorders and subsequent heart disease onset: Beyond depression." *Int J Cardiol 168* (6): 5293–9.

Seedat, S., K. M. Scott, M. C. Angermeyer, P. Berglund, E. J. Bromet, T. S. Brugha,... R. C. Kessler. 2009. "Cross-national associations between gender and mental disorders in the World Health Organization World Mental Health Surveys." *Arch Gen Psychiatry 66* (7): 785–95.

Shaffer, D., P. Fischer, C. P. Lucas, M. K. Dulcan, & M. E. Schwab-Stone. 2000. "NIMH Diagnostic Interview Schedule for Children Version IV (NIMH DISC-IV): Description, differences from previous versions, and reliability of some common diagnoses." *J Am Acad Child Adolesc Psychiatry 39* (1): 28–38.

Sharifi, V., M. Amin-Esmaeili, A. Hajebi, A. Motevalian, R. Radgoodarzi, M. Hefazi, & A. Rahimi-Movaghar. 2015. "Twelve-month prevalence and correlates of psychiatric disorders in Iran: The Iranian Mental Health Survey, 2011." *Arch Iran Med 18* (2): 76–84.

Shen, Y. C., M. Y. Zhang, Y. Q. Huang, Y. L. He, Z. R. Liu, H. Cheng,...R. C. Kessler. 2006. "Twelve-month prevalence, severity, and unmet need for treatment of mental disorders in metropolitan China." *Psychol Med 36* (2): 257–67.

Simmons, L. A., B. Braun, R. Charnigo, J. R. Havens, & D.W. Wright. 2008. "Depression and poverty among rural women: A relationship of social causation or social selection?" *J Rural Health 24* (3): 292–8.

Sneed, J. R., & M. E. Culang-Reinlieb. 2011. "The vascular depression hypothesis: An update." *Am J Geriatr Psychiatry 19* (2): 99–103.

Steel, Z., D. Silove, N. M. Giao, T. T. Phan, T. Chey, A. Whelan,...R. A. Bryant. 2009. "International and indigenous diagnoses of mental disorder among Vietnamese living in Vietnam and Australia." *Br J Psychiatry 194* (4): 326–33.

Stein, D. J., S. Aguilar-Gaxiola, J. Alonso, R. Bruffaerts, P. de Jonge, Z. Liu,...K. M. Scott. 2014. "Associations between mental disorders and subsequent onset of hypertension." *Gen Hosp Psychiatry 36* (2): 142–9.

Steinert, C., M. Hofman, J. Kruse, & F. Leichsenring. 2014. "The prospective long-term course of adult depression in general practice and the community: A systematic literature review." *J Affect Disord 152–154*: 65–75.

Ströhle, A., M. Höfler, H. Pfister, A. G. Müller, J. Hoyer, H. U. Wittchen, & R. Lieb. 2007. "Physical activity and prevalence and incidence of mental disorders in adolescents and young adults." *Psychol Med 37* (11): 1657–66.

Swain, N. R., C. C. Lim, D. Levinson, F. Fiestas, G. de Girolamo, J. Moskalewicz,... Scott, K. M. 2015. "Associations between DSM-IV mental disorders and subsequent non-fatal, self-reported stroke." *J Psychosom Res 79* (2): 130–6.

Swendsen, J., K. P. Conway, L. Degenhardt, M. Glantz, R. Jin, K. R. Merikangas,... R. C. Kessler. 2010. "Mental disorders as risk factors for substance use, abuse and dependence: Results from the 10-year follow-up of the National Comorbidity Survey." *Addiction 105* (6): 1117–28.

Takeuchi, D. T., R. C. Chung, K. M. Lin, H. Shen, K. Kurasaki, C. A. Chun, & S. Sue. 1998. "Lifetime and twelve-month prevalence rates of major depressive episodes and dysthymia among Chinese Americans in Los Angeles." *Am J Psychiatry 155* (10): 1407–14.

ten Have, M., R. de Graaf, & K. Monshouwer. 2011. "Physical exercise in adults and mental health status findings from the Netherlands mental health survey and incidence study (NEMESIS)." *J Psychosom Res 71* (5): 342–8.

Tennant, C., P. Bebbington, & J. Hurry. 1982. "Female vulnerability to neurosis: The influence of social roles." *Aust N Z J Psychiatry 16* (3): 135–40.

Thavichachart, N., P. Intoh, T. Thavichachart, O. Meksupa, S. Tangwongchai, A. Sughondhabirom, & P. Worakul. 2001. "Epidemiological survey of mental disorders and knowledge attitude practice upon mental health among people in Bangkok Metropolis." *J Med Assoc Thai 84* Suppl 1: S118–26.

Topuzoğlu, A., T. Binbay, H. Ulaş, H. Elbi, F. A. Tanık, N. Zağlı, & K. Alptekin. 2015. "The epidemiology of major depressive disorder and subthreshold depression in Izmir, Turkey: Prevalence, socioeconomic differences, impairment and help-seeking." *J Affect Disord* 181: 78–86.

Torres de Galvis, Y., J. Posada Villa, J. Bareño Silva, D. Y. Berbesi Fernández, M. Sierra Hincapie, L. P. Montoya Velez, & R. M. Montoya. 2012. "Salud mental en adolescentes en Colombia." *Revista del Observatorio Nacional de Salud Mental 1* (1): 17–27.

Vaingankar, J. A., G. Rekhi, M. Subramaniam, E. Abdin, & S. A. Chong. 2013. "Age of onset of life-time mental disorders and treatment contact." *Soc Psychiatry Psychiatr Epidemiol 48* (5): 835–43.

Viana, M. C., & L. H. Andrade. 2012. "Lifetime Prevalence, age and gender distribution and age-of-onset of psychiatric disorders in the São Paulo Metropolitan Area, Brazil: Results from the São Paulo Megacity Mental Health Survey." *Rev Bras Psiquiatr 34* (3): 249–60.

Vicente, B., R. Kohn, P. Rioseco, S. Saldivia, I. Levav, & S. Torres. 2006. "Lifetime and 12-month prevalence of DSM-III-R disorders in the Chile psychiatric prevalence study." *Am J Psychiatry 163* (8): 1362–70.

Vicente, B., R. Kohn, S. Saldivia, P. Rioseco, & S. Torres. 2005. "Service use patterns among adults with mental health problems in Chile." *Rev Panam Salud Publica 18* (4–5): 263–70.

Vicente, B., S. Saldivia, F. de la Barra, R. Kohn, R. Pihan, M. Valdivia,...R. Melipillan. 2012. "Prevalence of child and adolescent mental disorders in Chile: a community epidemiological study." *J Child Psychol Psychiatry 53* (10): 1026–35.

Wang, P. S., P. Berglund, M. Olfson, H. A. Pincus, K. B. Wells, & R. C. Kessler. 2005. "Failure and delay in initial treatment contact after first onset of mental disorders in the National Comorbidity Survey Replication." *Arch Gen Psychiatry 62* (6): 603–13.

Wang, P. S., M. Angermeyer, G. Borges, R. Bruffaerts, W. Tat Chiu, G. de Girolamo,… T. B. Ustün. 2007. "Delay and failure in treatment seeking after first onset of mental disorders in the World Health Organization's World Mental Health Survey Initiative." *World Psychiatry* 6 (3): 177–85.

World Health Organization. 2001. *The World Health Report 2001 Mental Health: New Understanding New Hope*. Geneva: WHO.

Xiaoli, Y., J. Chao, P. Wen, X. Wenming, L. Fang, L. Ning,…P. Guowei. 2014. "Prevalence of psychiatric disorders among children and adolescents in northeast China." *PLoS One* 9 (10): e111223.

2 Neurobiological Changes as an Explanation of Benefits of Exercise

Terry McMorris and Jo Corbett

Introduction

The purpose of this chapter is to outline the neurobiological changes induced by acute and chronic exercise, which are thought to influence depression, anxiety and psychological well-being. In this section, we briefly outline the hypothesized causes of depression and anxiety, which we believe exercise-induced neurobiological changes can affect. In the next section, we describe the physiological changes that occur as a result of acute and chronic exercise, as it is the neurobiological responses to these changes which ultimately affect depression and anxiety, and indeed psychological well-being. In later sections, we discuss the hypothesized neurobiological changes that are claimed to result from exercise, and the evidence that they do, in fact, occur. In the penultimate section, we look for definitive, empirical evidence for an exercise-neurobiology-psychological well-being interaction and finally we make our conclusions.

Neurochemical Hypotheses for Depression and Anxiety

In this sub-section, we will outline the main hypotheses that have been posited in order to explain depression and anxiety. We are, however, in agreement with Hasler (2010), who stated that our current knowledge does not allow for a unified hypothesis.

Catecholamines Hypothesis

Basically the catecholamines hypothesis for depression states that depression is due to deficiencies in the synthesis and release of the brain catecholamines neurotransmitters, ***noradrenaline*** (NA) and ***dopamine*** (DA) (Schildkraut 1965). In a healthy and efficient brain, these neurotransmitters activate the prefrontal cortex, which controls emotion and behavior, through projections to subcortical regions like the hypothalamus, amygdala, and brainstem nuclei (Price, Carmichael, & Drevets 1996). In optimal conditions, NA activates the high affinity α_{2A}-adrenoceptors, which increase the strength of neural signaling in the preferred direction by inhibiting cyclic adenosine monophosphate (cAMP) activation (Roth, Tam, Ida, Yang, & Deutch 1988). Similarly, the high affinity D_1- receptors are activated by DA, which dampens the "noise" by inhibiting firing to non-preferred stimuli (Finlay, Zigmond, & Abercrombie 1995). So DA and NA, working together, improve the signal to "noise" ratio and the prefrontal cortex can operate efficiently. In depression, when catecholamines concentrations are low, the appropriate sequence of neuronal activation cannot be obtained as a result of neurons being at such a low level of excitation that they cannot be stimulated to an adequate level of summation. Hence prefrontal cortex activity is inefficient and input to the limbic system is disrupted. Furthermore, the limbic system itself is negatively affected by low levels of activity, as it too depends on catecholamines for activation. Low NA neurotransmission leads to decreased alertness, low energy, problems of inattention, concentration, and cognitive ability, while limited DA activity is implicated in problems of motivation, pleasure, and reward (Moret & Briley 2011).

Prefrontal cortex efficiency and input to the limbic system can also be disrupted in other ways. When stress levels are high, as during periods of anxiety, NA and DA concentrations become excessive. In the prefrontal cortex, the excess NA activates the lower affinity α_1- and β-adrenoceptors (Roth et al. 1988). Activation of α_1-adrenoceptors results in reduced neuronal firing by phosphatidylinositol-protein kinase C intracellular signaling pathway activation, while excessive stimulation of D_1-receptors and β-adrenoceptors induces excess activity of the secondary messenger cAMP, which dampens all neuronal activity, thus weakening the signal to "noise" ratio (Arnsten 2009). Thus, the prefrontal cortex is unable to inhibit unwanted limbic system activity. In fact, when stress levels are very high, the prefrontal cortex may be inactivated (Arnsten 2009). Although catecholamines concentrations are very high during the manic stage of bipolar disorder, it is actually a little too simplistic to state that the mania is due to prefrontal cortex inefficiency. It appears that the situation is more complex (Mann, Currier, Quiroz, & Manji 2012). The same can be said of the catecholamines hypothesis itself, as other neurochemicals are involved, not to mention morphological factors.

Monoamine Hypothesis

The monoamine hypothesis (Coppen 1967), or the monoamine deficiency hypothesis, incorporates the actions of the indoleamine neurotransmitter *5-hydroxytryptamine* (5-HT), also known as serotonin, with those of NA and DA. Basically, the hypothesis states that depression and anxiety are the result of low concentrations of 5-HT and catecholamines in the brain (see Hasler 2010; Moret & Briley 2011). According to Mann et al. (2012), low concentrations of 5-HT may be due to low levels of release, fewer receptors or impaired serotonin receptor-mediated signal transduction.

Of particular importance is the fact that low concentrations of 5-HT lead to lower activation of the serotonergic receptors 5-HT_{1A} and 5-HT_{2A}. The former are found in high density in the hippocampus, entorhinal cortex, septum, amygdala, frontal cortex, and the dorsal and median raphe nuclei. The latter are found in the frontal cortex, amygdala, hippocampus, and basal ganglia. 5-HT_{1A} is responsible for regulating serotonergic function and is involved in the modulation of excitatory glutamatergic neurotransmission. 5-HT_{2A} receptors play roles in both excitatory glutamatergic and inhibitory γ-amino butyric acid (GABA)-ergic activity. Furthermore, the 5-HT system induces activation of the *hypothalamic-pituitary-adrenal* (HPA) *axis* system, which is also seen as a strong contender for control of depression and anxiety (Hasler 2010; Moret & Briley 2011). Given these actions, one can see that failure to activate the serotonergic neurotransmitters will result in inefficient working of the limbic system. This will be added to by dysfunction of the noradrenergic and dopaminergic pathways, hence depression and/or anxiety.

Hypothalamic-Pituitary-Adrenal Axis Hormones, and Depression and Anxiety

Hyperactivity of the HPA axis has for some time been linked to depression and anxiety (Carroll, Curtis, & Mendels 1976). When the individual is stressed, feedback to the hypothalamus initiates the synthesis and release of *corticotropin releasing factor* (CRF), *adrenocorticotropin hormone* (ACTH) and *cortisol*. However, the activity of the HPA axis can be maintained in homeostasis.

Crucial to this process are the roles of the corticosteroid receptors, ***mineralo-corticoid receptors*** (MRs) and ***glucocorticoid receptors*** (GRs). Both GRs and MRs are found in most parts of the brain, with MRs showing high density in dentate gyrus, CA2 and CA1 regions of the hippocampus, lateral septum and central amygdala. GRs demonstrate high density in the hippocampus, neocortex and hypothalamic nuclei, such as the paraventricular nucleus (PVN), and supraoptic nucleus (De Kloet et al. 2000). MRs have a high affinity with cortisol and as a result, even under baseline early morning conditions, show a high occupancy. To the contrary, GRs are hardly occupied under baseline conditions. In times of stress, however, both demonstrate a high degree of occupancy by cortisol (Reul & De Kloet 1985). As a result of these findings, Reul and De Kloet (1985) argued that hippocampal MRs have a tonic, inhibitory influence on circulating glucocorticoids that restrains baseline HPA axis activity, thus maintaining homeostasis. Reul et al. (2015) claimed that the stress-CRF-MR mechanism prevents glucocorticoid hypersecretion and that, in depressed individuals, this mechanism "may be failing" (p. 46). This results in increased CRF concentrations, as there is no increase in MRs and hence less inhibition of CRF activity. This in turn results in increased ACTH and cortisol concentrations, which interact with NA and 5-HT, resulting in negative effects on depression and anxiety.

As well as hyperactivity of the HPA axis hormones inducing depression, hypocortisolism has also been shown to play a role (Vreeburg, Hoogendijk, & DeRijk 2013). According to Heim, Ehlert, and Hellhammer (2000), hypocortisolism may be due to the down-regulation of CRF receptors in the pituitary due to chronic and/or recurrent stress-induced hypothalamic CRF secretion, which results in lower ACTH synthesis and release, and hence lower cortisol concentrations. Heim et al. also pointed to increased negative feedback sensitivity of the HPA-axis, at the level of the pituitary, as a possible reason for hypocortisolism.

Brain-Derived Neurotrophic Factor Hypothesis and Morphological Considerations

Several authors have expressed dissatisfaction with the neurochemical hypotheses outlined above (e.g. Duman, Heninger, & Nestler 1997), particularly due to limitations in the success rates of anti-depressant treatments based on these hypotheses and the fact that there is a 2–3 week latency before the anti-depressant effects are demonstrated, despite the fact that the drugs have rapid synaptic effects (Autry & Monteggia 2012). This led Duman et al. (1997) to suggest that depression was due to neural dysfunction. This was supported by observations that ***depressed individuals demonstrated reduced hippocampal volume*** (Bremner et al. 2000). Furthermore, rodent studies showed decreases in hippocampal levels of ***brain-derived neurotrophic factor*** (BDNF) messenger ribonucleic acid (mRNA), with greatest reductions occurring in the dentate gyrus, and CA1 and CA3 pyramidal cell layers of the hippocampus (Ueyama et al. 1997). Findings such as these supported the notion that neural dysfunction was an important factor in depression and anxiety, and presented the likelihood of a major role for BDNF.

BDNF is a protein and a member of the neurotrophic family. It plays major roles in neurogenesis and neuroprotection, and synaptic transmission, and deficiencies are seen as negatively affecting these activities (Binder and Scharfman 2004). It was, therefore, no surprise to find that BDNF levels were decreased in regions of the hippocampus in postmortem tissue taken from suicide victims or patients with major depressive disorder (MDD). Humans with MDD also demonstrated decreased volume in the prefrontal cortex and this correlated with decreased BDNF levels (Dwivedi et al. 2003). Also concentrations of BDNF were low in the serum of patients with MDD (Castrén, Võikar, & Rantamäki 2007; Castrén & Rantamäki 2010). This led to an initial acceptance of the hypothesis that low levels of BDNF resulted in depression. The veracity of the hypothesis has, however, been questioned (Autry & Monteggia 2012).

While hypotrophy of the prefrontal cortex and hippocampus, and depletion of hippocampal BDNF mRNA are consistent with the hypothesis, findings for other brain regions are not. In fact, they are contradictory. Depressed patients have shown increased BDNF protein in the nucleus acumbens (NAc) tissue (Krishnan et al. 2007). Moreover, hypertrophy of the amygdala has been demonstrated in MDD patients (Frodl et al. 2002). These results show that depression is related to different BDNF functional properties in the ventral tegmental area-NAc pathway compared to those in the hippocampus (Siuciak, Lewis, Wiegand, & Lindsay 1997), i.e. signaling is increased in the amygdala and NAc, and decreased in the hippocampus and prefrontal cortex. These findings strongly suggest that the roles of BDNF in depression and anxiety are very complex and brain region specific.

Physiological Changes Caused by Acute and Chronic Exercise

In this section, we outline the main peripheral, physiological responses to exercise. The neurobiological responses in the brain begin with the decision to exercise, made by the higher centers of the brain, which stimulate the premotor cortex and supplementary motor area (SMA) to initiate neural and physiological feedforward in order to undertake the exercise. Feedback from the periphery, as a result of the peripheral, physiological changes, stimulates neurobiological responses in the brain. So, the exercise-induced neurobiological changes in the brain are the result of the peripheral physiological demands.

The physiological changes occurring with exercise can be studied in terms of those changes occurring with *acute exercise*, i.e. a single bout of exercise, or *chronic* changes, which are induced by multiple bouts of exercise over a prolonged period of time, i.e. weeks or months. The physiological changes induced by acute exercise are characterized as responses, defined as *sudden temporary changes in function(s)*, whereby the disruption to function is restored shortly after exercise is completed. In contrast, chronic exercise induces *adaptations, which can be defined as a longer-term change in structure and/or function in response to regular exercise.* In each instance, the requirement to maintain the internal environment of the body within a relatively narrow range (homeostasis), whilst at the same time meeting the elevated energetic demand of exercise, is a key driving force underpinning the physiological changes that occur.

The transition from rest to exercise induces a coordinated array of physiological responses. Indeed, it would be hard to argue that there is any body-system that is not affected by exercise to some extent. Adenosine triphosphate (ATP) is the body's "energy currency" and it is the liberation of free energy from the phosphate bonds in the ATP molecule that is the predominant mechanism for powering biological processes, including muscular contraction. However, the human body contains only a very limited store of ATP (~80 g). During very intense exercise it would be possible to deplete this store in a few seconds, whereas during more prolonged exercise, such as a marathon run, ~80 kg of ATP would be required. Consequently, the body has a coordinated spectrum of energy systems for resynthesizing ATP in order to meet the increased demand for ATP imposed by exercise, the relative contribution of which depends upon the nature of the exercise task as influenced by the exercise duration and intensity.

Phosphocreatine (PCr), stored within the skeletal muscle at an approximately three-fold greater concentration than ATP, acts as an initial energy buffer, being broken down to release energy, creatine, and an inorganic phosphate molecule, thereby enabling the resynthesize of ATP from adenosine diphosphate (ADP) and inorganic phosphate (P_i). The adenylate kinase reaction can also resynthesize ATP, using two ADP molecules to produce one ATP and one adenosine monophosphate (AMP) molecule. However, although these reactions enable the body to respond to large, rapid changes in energy demand, the capacity for energy provision is limited; the energy available from the PCr stores will be depleted within a matter of seconds of intense exercise. Nonetheless, these reactions also influence the AMP, P_i and ADP concentrations, which provide the cellular signals to activate the glycolytic and the oxidative pathway.

The glycolytic pathway (glycolysis) uses glucose, derived from the foodstuffs consumed, and from the plasma as well as muscle and liver glycogen stores, to resynthesize ATP. Glycolysis is a series of fermentation reactions taking place in the cytoplasm of the cell. Each glucose molecule produces a net gain of two ATP molecules by substrate-level phosphorylation and results in the production of pyruvate. Because of a difference in the initial reaction to produce glucose-6-phosphate, a net gain of three ATP molecules is produced from each glucosyl unit liberated from glycogen. Glycolysis can resynthesize ATP relatively quickly and is the primary energy system for all-out bouts of exercise lasting for between ~30 s and ~ 2 minutes, although it generates only about 5% of the ATP available from the complete degradation of a glucose molecule. It also produces hydrogen atoms which pass their electron to nicotinamide adenine dinucleotide (NAD^+) to form NADH, which is used to shuttle electrons into the mitochondria where they can be used to resynthesize further ATP by oxidative phosphorylation. However, during intense exercise, the rate at which NADH is processed by the respiratory chain may be insufficient; this could lead to insufficient NAD^+ and cessation of glycolysis, but this is prevented when hydrogens from NADH combine with pyruvate to form lactate, catalyzed by lactate dehydrogenase. This process frees up NAD^+ to accept hydrogen generated in glycolysis, thereby enabling glycolysis to continue. Some lactate is always being produced, even at rest, but this is oxidized in muscles fibers with high oxidative capacity. Nonetheless, during intense exercise the rate of lactate production can exceed the rate at which it is removed, resulting in lactate accumulation; the work rate, or exercise intensity at which lactate begins to accumulate in the blood, is termed the *lactate threshold* (LT). This point is associated with the development of metabolic acidosis, and, in conjunction with other intra-cellular changes occurring with anaerobic glycolysis, is associated with inhibited glycolytic enzyme function, impaired muscular contraction, and fatigue.

Thus far, the mechanisms described for the resynthesize of ATP are anaerobic, i.e. they do not require oxygen. At lower work rates, the pyruvate that is produced can be converted into acetyl-CoA and can enter the Krebs, or tricarboxylic acid, cycle within the cell's mitochondria. Importantly, acetyl-CoA can also be formed from the catabolism of fats and proteins consumed in foodstuffs, but the relative contribution of protein to energy expenditure is generally small in the healthy and fed state. In the Krebs cycle, acetyl-CoA is degraded to carbon dioxide and further hydrogen atoms are released to NAD^+ and flavin adenine dinucleotide (FAD), forming NADH and $FADH_2$; one ATP molecule is regenerated directly by substrate-level phosphorylation. In turn, the NADH and $FADH_2$ are oxidized in the electron transport chain, whereby the hydrogen atoms are split into protons and electrons, and pairs of electrons flow down a series of reactions in the inner mitochondrial membrane that are linked to ATP resynthesis by a process called chemiosmotic coupling. Importantly, oxygen acts as the final electron acceptor in this process, and is reduced to water. This oxygen requiring (aerobic) energy process is termed oxidative phosphorylation and predominates in maximal exercise beyond ~ 2 minutes. The balance of fat and carbohydrate use is influenced by diet as well as exercise intensity and duration; because fat oxidation cannot resynthesize ATP as quickly as carbohydrate utilization, fat oxidation is decreased at high intensities, whereas with prolonged exercise decreases in carbohydrate availability will result in an increased utilization of fats, although this may necessitate a reduced exercise work-rate. Nevertheless, if insufficient oxygen is available, electrons cannot flow down the respiratory chain and this process cannot function. Consequently, as work-rate, and therefore energy demand, increases, the rate of oxygen uptake (VO_2) will increase in proportion, in an approximately linear manner. Thus, many of the physiological responses occurring during exercise are directly related to the necessity to efficiently deliver oxygen from the atmosphere to the working tissues, as well as the removal of carbon dioxide and other metabolic waste products.

To supply the body with the oxygen that is required during exercise, an individual must ventilate the lungs with ambient air. Thereafter, the oxygen contained within the ambient air must be transported across the lung's alveolar membrane into the capillary network perfusing the lung, where it will bind to hemoglobin within the blood. Subsequently, the oxygen-rich blood must be pumped by the heart, though the vasculature, to the capillaries perfusing the working muscles, where the oxygen will dissociate from the hemoglobin before being used in the mitochondria contained within the muscle cells. Conversely, the resultant carbon dioxide that is produced must be excreted and is transported to the lung; about one-fifth of the carbon dioxide is transported bound to hemoglobin (carbaminohemoglobin), with about 10% transported dissolved in blood plasma, and the remainder as bicarbonate ions. Thus, when exercise is initiated the respiratory and cardiovascular systems must respond accordingly.

Stimulation of the respiratory centers in the brain stem, from the motor cortex and from proprioceptive feedback in the muscles, increases pulmonary ventilation. At low work rates, pulmonary ventilation is primarily increased through increases in the tidal volumes as this is more efficient, whereas at higher work rates the respiratory rate is also increased, with an approximately 20-fold increase in ventilation possible from rest to maximal exercise. Pulmonary ventilation is also influenced by changes in PCO_2, H^+ and temperature, although the importance of these factors tends to be greater during prolonged or intense exercise. Indeed, during intense exercise, buffering of the protons that dissociate from lactic acid by the bicarbonate buffering system results in the production of additional carbon dioxide, which stimulates an increase in ventilation to blow-off this "excess" carbon dioxide at the lung. As a consequence, a ***ventilatory threshold*** (VT) is typically observed at approximately the same work rate as the LT and, therefore, the VT is often used as a surrogate measure of the LT. Inside the lung, gaseous exchange is driven by the pressure differential of the gases that exist across the alveolar membrane, and the solubility of the gas; oxygen diffuses from the alveolar air into the alveolar blood, whereas carbon dioxide moves from the blood and into the alveolar air.

To deliver the oxygen-rich blood to the working muscles, the cardiac output, i.e. the amount of blood pumped from the left ventricle of the heart per minute, also increases during exercise, in proportion to the intensity of exercise. The cardiac output is determined by the product of the heart rate and the stroke volume, and is typically in the region of about 5 L at rest. Heart rate increases across the work rate spectrum, whereas stroke volume may plateau at ~50% of individual's maximum rate of oxygen consumption (VO_{2MAX}), although there is evidence to suggest that stroke volume may increase throughout the work rate spectrum in well-trained endurance athletes. As a consequence of their greater stroke volumes, the maximal cardiac outputs of trained endurance athletes may exceed 35 $L{\cdot}min^{-1}$, whereas values of ~ 25 $L{\cdot}min^{-1}$ are typical in a healthy male during maximal exercise. Most lines of evidence suggest that, under most circumstances, an individual's VO_{2MAX}, which represents the upper limit of aerobic ATP resynthesis and is a key indicator of "aerobic fitness," is limited by the supply of oxygen rich blood to the working muscles, as determined by the oxygen carrying capacity of the blood and the cardiac output, rather than the uptake of oxygen across the lung, or the utilization of oxygen within the working muscles.

As well as an increase in cardiac output, the distribution of blood flow is also altered when moving from rest to exercise. Sympathetic nervous system activity, as well as local chemical changes, directs the blood from non-essential areas such as the kidneys, liver, stomach, and intestines to those with increased activity, such as the working muscle. The working muscles may only receive ~20% of the resting cardiac output, but may receive as much as 80% of the cardiac output during maximal exercise. When exercising in the heat more blood may be directed to the skin to help lose heat to the environment, but this can compromise muscle blood flow and partially explains the decrement in endurance performance that is often seen in hot environments. Together, the exercise-induced changes in cardiac output and blood flow distribution affect blood pressure, with the systolic pressure increasing as a function of the exercise-induced increases in cardiac output, whereas the diastolic pressure remains comparatively unchanged due to a decreased peripheral resistance to flow as a consequence of vasodilation on the active muscles. The resultant increase in mean arterial pressure helps to maintain perfusion. In the main, the cardio-vascular and respiratory responses to resistance exercise are similar to those occurring with endurance exercise, albeit generally at the lower end of the spectrum of changes due to the typically lower oxidative demands. However, the blood pressure response differs because the high forces within muscles during resistance exercise can lead to compression in smaller arteries and a consequent increase in the total peripheral resistance.

Once the blood has been delivered to the relevant tissue it must be utilized. The arterial-mixed venous oxygen difference, or a- $\bar{v}$ O_2 difference, refers to the difference in the oxygen content of arterial and mixed-venous blood, and primarily results from oxygen extraction from the arterial blood as it passes through the exercising muscle, diffusing across the capillary membrane, through the tissue fluid and into the cells. At rest, the a-$\bar{v}$ O_2 difference is ~5 mL of oxygen per deciliter of blood, but may increase to ~15 mL of oxygen per deciliter of blood when the tissue demand for blood is elevated, such as during maximal exercise. This is mainly due to a decrease in venous O_2 content, because in most instances the arterial O_2 content changes little with exercise intensity. The a-$\bar{v}$ O_2 difference is primarily influenced by the tissue PO_2, which becomes reduced with increasing exercise intensity, thereby promoting the offloading of oxygen from the blood. The offloading of O_2 from hemoglobin at the tissues is further facilitated by exercise-induced changes in PCO_2, pH and temperature. Myoglobin within the muscle cells provide a further important O_2 store, particularly when there are substantial declines in muscle PO_2, such as at the onset of exercise and during intense exercise.

The endocrine system also plays an important role in maintaining the stability of the internal environment during a bout of exercise, with hormones produced by endocrine glands exerting a wide array of influences on target organs and cells. During exercise, changes are typically observed in the concentration of insulin, glucagon, growth hormone, testosterone, ACTH and thyrotropin, to list but a few. Although space does not permit a full description of the multitude of hormones exerting effects during exercise, the main neurochemically relevant hormones are the catecholamines, adrenaline and NA, which are secreted from the adrenal medulla and exert their effects in many target tissues. The effects of exercise on these neurotransmitters are discussed in some detail in the next section.

When repeated, the physiological perturbations induced by bouts of exercise serve as the stimulus for the adaptations seen with chronic exercise. The nature of the adaptation is dependent upon the adaptation stimulus, i.e. training specificity, and the exercise must be at a level that is greater than that to which the tissue or system is accustomed for adaptation to take place, i.e. overload. If the training stimulus is substantially diminished, a loss of these physiological adaptations will occur over time, i.e. reversibility. Nonetheless, regardless of the type of training undertaken, the magnitude of change will be influenced by the baseline value and will incorporate a substantial genetic component.

With regular prolonged (aerobic) exercise, the training-induced changes are largely characterized by adaptations improving the delivery and utilization of oxygen, and removal of carbon dioxide and metabolic waste products, but these may be preceded by the rapidly observed reductions in the plasma catecholamines responses to exercise. Although pulmonary ventilation is unchanged, or possibly even reduced at rest and during exercise at a given submaximal work-rate, maximal pulmonary ventilation is increased as a consequence of increases in tidal volume and respiratory rate. Likewise, pulmonary diffusion rates are unchanged at rest and at a given submaximal work rate, but maximal pulmonary diffusion rates are increased, likely as a consequence of increased maximal pulmonary ventilation and increased perfusion of the lung with blood. Blood volume is increased, due to increases in both plasma volume and the red blood cell volume, although the expansion in plasma volume is often greater than the increase in red blood cell volume, particularly during the initial phases of an exercise program. Regardless, the enhanced blood volume, in combination with morphological changes to the heart, results in an increase in the stroke volume of the heart. Consequently, a given cardiac output can be achieved for a lower heart rate, and resting, as well as submaximal exercise heart rate, is, therefore, reduced. Although changes in maximal heart rate may be negligible, the larger stroke volume precipitates an increased maximal cardiac output, thereby increasing the oxygen delivery to the working muscle during intense exercise. Similarly, arterial pressure will be lower at a given submaximal work rate, whilst at maximal work rates the systolic blood pressure will be increased and the diastolic pressure decreased. Within the exercising muscle there is an increase in capillary density, which increases the capillary to muscle fiber ratio, and a greater opening of the capillaries and more efficient blood flow redistribution to the working muscles during exercise. The decreased blood viscosity, which occurs as a consequence of the greater plasma volume expansion, relative to red cell expansion, may also help to improve oxygen delivery through the narrow capillaries. Aerobic training also increases the cross-sectional area of those muscle fibers in which aerobic metabolism predominates (slow twitch or type I muscle fibers). Within the muscle fibers themselves, there is an increase in the mitochondrial density and higher oxidative enzyme activity. Finally, the more efficient distribution of blood flow to the working muscle, in combination with the increased oxidative capacity of the muscle, results in a reduction in the oxygen content of mixed venous blood and an increase in the a-$\bar{v}$ O_2 difference. Together, these adaptations may shift the balance of fat and carbohydrate usage to spare carbohydrate until higher work rates, increase the work rate at the LT and VT, which is linked to the ability to sustain a high percentage of the $\dot{V}O_{2MAX,}$ and increase the catecholamines threshold (see following section) and $\dot{V}O_{2MAX}$. Thus, following an effective aerobic exercise training intervention, a given absolute work rate will elicit a reduced physiological "stress," whereas, conversely, the same physiological "stress" will elicit a higher work rate; the ability to perform prolonged (aerobic) work will also be improved.

With high-intensity "anaerobic" and resistance-type exercise, the adaptations relating to oxygen delivery are much less pronounced, and adaptations associated with the predominant energy systems involved in this type of activity are evident. Muscle biopsy data have demonstrated increases in the concentration of key substrates involved in anaerobic energy supply, such as ATP, PCr and glycogen, and there is also some evidence for changes in the activity of some of the key enzymes regulating anaerobic metabolic pathways. However, the impact of these changes on the capacity to produce energy from anaerobic metabolism is not entirely clear. In contrast, anaerobic training, and particularly resistance training, clearly increases the cross sectional area of muscle fibers (hypertrophy) and in particular the cross sectional area of fast twitch (type 2) muscle fibers; there may also be an increase in the percentage of type 2 muscle fibers, with a concomitant decrease in the percentage of type 1 muscle fibers. Neural adaptations are also evident with resistance exercise. These typically precede the hypertrophic responses and are characterized by an improved ability to recruit motor units and, possibly, better synchronization of motor unit recruitment. Together these adaptations combine to increase muscle force output following a successful period of resistance training.

Neurochemicals, Exercise and Depression, Anxiety and Well-Being

Catecholamines

Synthesis and Release

The synthesis of catecholamines takes place both centrally, within the brain, and peripherally. The precursor of catecholamines synthesis is the aromatic amino acid tyrosine, which is either taken directly from food or is formed in the liver by the hydroxylation of phenylalanine. Thus it is readily available peripherally and is transported across the blood–brain barrier by the facilitative transporter L1 (Hawkins, O'Kane, Simpson, & Viña 2006). In both the brain and peripherally, tyrosine is broken down into the metabolite 3, 4 dihydroxy–L-phenylalanine (L-DOPA), under the influence of *tyrosine hydroxylase* (TH) with the cofactor tetrahydrobiopterin, both of which are found in all cells that synthesize catecholamines. L-DOPA is then catalyzed by aromatic amino acid decarboxylase (AADC), with the cofactor pyridoxal phosphate, and DA is formed. In neurons that use DA as a neurotransmitter, no further action occurs and the DA is stored in vesicles. In neurons that use NA as neurotransmitters, DA is further synthesized into NA. This takes place, with the aid of DA-β-hydroxylase (DBH) and its cofactor, ascorbate. The majority of NA is stored in vesicles in these neurons and there is no further processing. In the periphery, DA is stored in some neurons in the pulmonary artery and kidney. The majority of peripheral DA, however, is further synthesized into NA. This takes place in the granules of cells in the adrenal medulla. In about 15% of the granules the process terminates and NA is stored. The rest of the NA diffuses back into the cytoplasm where it is *N*-methylated by phenylethanolamine-*N*-methyltransferase (PNMT), with *S*-adenosylmethionine as the methyl donor, and adrenaline is synthesized. Adrenaline is then transported back into chromaffin granules for storage. In the brain, as in the periphery, the further synthesis of NA into adrenaline requires the presence of PNMT. This is present only in a few neurons in the pons and medulla. Some NA is *N*-methylated by PNMT in these neurons, thus a small amount of adrenaline is synthesized and stored in the brain. It should be noted that TH is the rate-limiting enzyme in the whole process (Fernstrom & Fernstrom 2007).

Catecholamines During Exercise

During and even immediately before exercise, the hypothalamus and brainstem initiate action of the *sympathoadrenal system* (SAS), which is part of the Autonomic Nervous System (ANS). This results in the release of DA and NA at the postganglionic cells of those neurons that require activating or inhibiting. Within the brain these neurotransmitters activate the motor cortex, SMA, basal ganglia and cerebellum, which collectively control movement.

SAS synthesized DA only plays a role centrally. As exercise increases in intensity, there is also release of ***adrenaline*** and, to a lesser extent NA, into the blood from the adrenal medulla. The catecholamines play important roles in the organism's ability to undertake exercise. Peripherally, during low-intensity exercise, NA and adrenaline aid lipolysis by activating receptors in adipose tissue, resulting in the mobilization of free fatty acids as fuel; stimulate receptors in muscle, causing utilization of free fatty acids as the fuel source; activate receptors in the pancreas to suppress insulin release; and stimulate secretion of the hormones glucagon, growth hormone and cortisol (Arakawa et al. 1995).

As exercise intensity increases hypoglycemia occurs, which results in large increases in plasma NA concentrations. Adrenaline concentrations rapidly increase when there is a decline in hepatic glucose concentrations. At this stage, adrenaline stimulates glycogenolysis and hepatic glucose release in the liver, and glycogenolysis in muscle. Adrenaline and, to a lesser extent, NA also act on the cardiovascular system by activating receptors responsible for increasing heart rate and contractile force, while stimulating arteriolar constriction in renal, splanchnic and cutaneous vascular beds. The points at which the sudden rise in NA and adrenaline plasma concentrations are demonstrated are known as the NA and adrenaline thresholds. They generally show moderate to high correlations (Podolin, Munger, & Mazzeo 1991), therefore they are often referred to as the ***catecholamines thresholds*** (CT). It is generally thought that intensity needs to be ~ 75% VO_{2MAX} (Podolin et al. 1991), which according to Arts and Kuipers (1994) equates to ~ 65% maximum power output ($\dot{W}_{MAX}$), but there are large inter-individual variations (Urhausen, Weiler, Coen, & Kindermann 1994). Moreover, blood lactate concentrations follow a similar exponential profile and the LT also shows moderate to high correlations with CT (Podolin et al. 1991). If exercise intensity increases beyond the threshold, plasma catecholamines concentrations continue to rise and soon reach very high levels.

The effect of increasing exercise intensity on plasma catecholamines concentrations is best seen during incremental exercise (i.e. when exercise intensity is gradually increased in increments) to exhaustion. However, steady state moderate intensity exercise ($\geq$ 40% but < 80% VO_{2MAX}) also demonstrates an exponential increase in plasma catecholamines concentrations but it can take at least as long as 30 mins (Hodgetts, Coppack, Frayn, & Hockaday 1991). The importance of the circulating NA and adrenaline is that they activate β-adrenoceptors on the ***afferent vagus nerve***, which runs from the abdomen through the chest, neck and head, and terminates in the ***nucleus tractus solitarii*** (NTS) within the blood–brain barrier. The excitatory neurotransmitter glutamate mediates synaptic communication between the vagal afferents and the NTS, allowing noradrenergic cells in the NTS, which project into the ***locus coeruleus***, to stimulate NA synthesis and release to other parts of the brain (Miyashita & Williams 2006). Also, Grenhoff and Svensson (1993) have shown that stimulation of α_1-adrenoceptors potentiates the firing of DA neurons in the ventral tegmental area, probably due to α_1-adrenoceptor activation inducing enhanced glutamate release, which affects the excitability of DA neurons (Velásquez-Martinez, Vázquez-Torres, & Jiménez-Riveira 2012). Thus, ***acute-exercise induced increases in peripheral adrenaline and NA concentrations can directly affect brain synthesis and release of the catecholamines brain neurotransmitters, DA and NA***. Moreover, indirectly, feedback to the hypothalamus via the thalamus, reticular activation system and limbic system, concerning stress on the cardiorespiratory system, pain and glycogen depletion may well induce central release of catecholamines. According to Mason (1975a, 1975b), if the individual perceives the situation as being unpredictable and/or one in which he/she is not in control, central synthesis and release of DA and NA will be initiated.

Evidence for Acute Exercise-Induced Increases in Brain Concentrations of Catecholamines

Research into the effect of exercise on brain concentrations of catecholamines has been largely in animal studies. The reliance on animal studies is unavoidable due to the presence of the blood–brain barrier. Catecholamines do not readily cross the blood–brain barrier (Cornford, Braun, Oldendorf, & Hill 1982), therefore plasma concentrations of catecholamines are almost entirely the result of peripheral activity only. At best they are only capable of indicating that similar changes in concentrations may have taken place in the brain. Obviously it is not possible to directly test brain concentrations in humans. However, in animals, microdialysis has been widely used and provides good data (see Meeusen, Piacentini, & De Meirleir 2001, for a description of the microdialysis process).

It is necessary for some substances to cross the blood–brain barrier. Crossing is mostly by diffusion and depends on the lipid solubility of the substance. Although catecholamines do not readily cross the blood–brain barrier, the precursor of catecholamines synthesis, the aromatic amino acid tyrosine, does. Tyrosine is transported across the blood–brain barrier by the facilitative carrier L1. However, it must compete with other amino acids, such as leucine, valine, methionine, histidine, isoleucine, tryptophan. phenylalanine and thronine for transportation by L1. During exercise, the branched-chain amino acids, leucine, valine and isoleucine, are taken up by muscle and so plasma concentrations fall. It was thought that this may allow greater quantities of tyrosine to be transported by L1, as tyrosine was thought not to be taken up or metabolized by skeletal muscle. Research, however, has shown that tyrosine concentrations in muscle do increase during exercise, possibly due to protein degradation (Blomstrand & Newsholme 1992). Moreover, brain microdialysis studies have shown no significant increase in brain concentrations of tyrosine during or following exercise (Foley & Fleshner 2008).

Animal studies have also shown that acute exercise has little positive effect on whole brain concentrations of NA. Meeusen and colleagues (see Meeusen &

De Meirleir 1995; Meeusen et al. 2001), in very thorough systematic reviews, found either a decrease in NA concentrations or no significant effect. However, research has shown increased DA concentrations, particularly in the brainstem and hypothalamus, during and immediately following acute exercise (see Meeusen et al. 2001; Meeusen & De Meirleir 1995, for reviews). Despite this limited support for acute exercise-induced increases in brain concentrations of NA, animal studies have shown increases in brain concentrations of the NA metabolite 3-methoxy 4-hydroxyphenylglycol (MHPG) in most brain regions (Meeusen et al. 2001). Similarly, increased concentrations of the DA metabolites 3, 4-dihydroxyphenylacetic acid (DOPAC) and 4-hydroxy 3-methoxyphenylacetic acid, also known as homovanillic acid (HVA), have also been shown, particularly in the brainstem and hypothalamus (Meeusen et al. 1997). This is evidence of acute exercise-induced turnover of NA and DA in the brain.

Only two studies have, so far, been attempted with humans. Wang et al. (2000) examined the effect of treadmill running on striatal DA release in the human brain. Pre- and post-exercising on a treadmill at an intensity > 85% estimated maximum heart rate, participants underwent an intravenous injection of the radiotracer [^{11}C]raclopride and positron emission tomography scans of the putamen and cerebellum were made. The authors expected to see a small but significant decrease in [^{11}C]raclopride binding following exercise, as raclopride's low affinity for DA D_2 receptors means that increases in synaptic DA release affect [^{11}C]raclopride binding. No significant effect was shown. The authors argued that this was probably due to the positron emission tomography protocol not being robust enough to highlight small changes in [^{11}C]raclopride binding.

Dalsgaard et al. (2004) examined NA and adrenaline concentrations in cerebrospinal fluid (CSF) by lumbar puncture and the arterial to internal jugular venous difference (*a-v* diff) following exercise to exhaustion. Arterial concentrations of NA and adrenaline showed large increases but the *a-v* diff was not affected significantly. Exercise increased the CSF concentrations of NA only. The authors argued that the CSF increase in NA concentrations was probably due to activity in the locus coeruleus. This study provides some support for acute exercise inducing increased brain concentrations of NA in humans.

Long-Term Effects of Exercise-Induced Changes in Brain Catecholamines Concentrations

The effects outlined in the previous sub-section are transient. The half-life of catecholamines in the periphery is only ~ 3 mins, while the half-life of brain catecholamines is in the range of 8–12 hours (Eisenhofer et al. 2004). Despite this, there is evidence from animal studies to show chronic exercise-induced changes in brain concentrations of NA and DA. Meeusen and colleagues (Meeusen & De Meirleir 1995; Meeusen et al. 2001) reported that, overall, chronic exercise resulted in increases in whole brain NA concentrations. They stated that results for DA concentrations tended to be region specific, with increases in hypothalamus and midbrain but decreases in prefrontal cortex, hippocampus, and striatum.

The most interesting recent research, however, has examined the effect of chronic exercise on TH activity in the brain. A number of studies have measured the effect of chronic exercise on TH mRNA expression in animals. Although O'Neal, Hoomissen, Holmes, & Dishman (2001) showed no significant effect, others (Foley & Fleshner 2008; Gavrilović et al. 2012; Tümer et al. 2001) found positive results. Tümer et al.'s findings were affected by age and brain region. Young rats (6 months) showed increased brain TH mRNA expression in the locus coeruleus and ventral tegmental areas but not in the substantia nigra, while old rats (2 years) demonstrated increased TH mRNA expression in the substantia nigra only. Foley and Fleshner examined the effect of chronic exercise on TH mRNA expression in the substantia nigra pars compacta (SNpc) and found a significant increase in the mid-SNpc and an increase approaching significance ($p = 0.07$) in the caudal SNpc. Gavrilović et al. showed increased TH mRNA expression in stellate ganglia following 12 weeks of 20 mins per day running. In normal circumstances, TH synthesis is modulated by end-product inhibition, as it competes with free intra-neuronal catecholamines (Alousi & Weiner 1966). It would appear that prolonged activation of TH heightens its affinity for the biopterin cofactor and it becomes less sensitive to end-product inhibition (Zigmond, Schwarzschild, & Rittenhouse 1989). As well as increased brain TH mRNA expression, a positive effect of chronic exercise on the density of DBH in the PVN has been demonstrated (Higa-Tanaguchi, Silva, Silva, Michelini, & Stern 2007), as well as increased DBH mRNA in stellate ganglia (Gavrilović et al. 2012). DBH is essential for the synthesis of NA from DA.

5-Hydroxytryptamine

Synthesis and Release

The amino acid tryptophan, the main source of which is dietary protein, is the precursor of 5-HT. It is transported across the blood–brain barrier by the facilitative transporter L1. Having crossed the blood–brain barrier, tryptophan is hydroxylated to 5-hydroxytryptophan (5-HTP), under the influence of tryptophan hydroxylase. It is further broken down by AADC into 5-HT. This process takes place mainly in the raphe nuclei of the brain. This is the only place that 5-HTP is found. 5-HT is stored in vesicles, mainly the parafollicular cells of the thyroid (Lefebvre et al. 2001). Tryptophan hydroxylase is the rate-limiting enzyme for 5-HT synthesis and is not fully saturated under normal conditions, therefore increases in brain concentrations of tryptophan will facilitate 5-HT synthesis. As we saw when discussing the monoamine hypothesis, the serotonergic system innervates most of the brain, particularly the hippocampus, septum, hypothalamus, striatum, frontal cortex, amygdala, substantia nigra, and caudate putamen (Frazer & Hensler 1999).

5-Hydroxytryptamine and Exercise

Tryptophan is found in plasma either bound to albumin or unbound. Unbound tryptophan readily crosses the blood–brain barrier. According to Blomstrand (2006), during exercise free fatty acids displace tryptophan from binding with albumin, therefore there is an increase in unbound tryptophan. This crosses into the brain and forms 5-HT. Tryptophan is not taken up by muscle, therefore plasma concentrations rise readily during exercise and availability for transport across the blood–brain barrier increases markedly (Blomstrand 2006; Hawkins et al. 2006). Some albumin-bound tryptophan also crosses the blood–brain barrier, probably due to a dissociation mechanism that takes place at the surface of the brain capillary endothelium (Pardridge 1998). Fernstrom and Fernstrom (2006) questioned the role of free fatty acids and unbinding of tryptophan from albumin as the cause for exercise-induced increases in 5-HT concentrations but they did agree that exercise induces increased brain concentrations of 5-HT. It may be that the increase is centrally mediated in the same way as exercise-induced perceptions of stress result in increased catecholamines synthesis and release.

Given that exercise facilitates the crossing of the blood–brain barrier by tryptophan, the fact that the serotonergic system innervates most of the brain, and that tryptophan hydroxylase is not saturated, it is not surprising to find strong evidence of acute exercise-induced increases in brain concentrations in animal studies (see Meeusen & De Meirleir 1995; Meeusen et al. 2001). Moreover, Blomstrand et al. (2005) examined the brain uptake of tryptophan during prolonged exercise (3 h at 200 ± 7 W, on a cycle ergometer) in humans, by calculating the *a-v* diff multiplied by plasma flow. They found large increases in cerebral uptake. It is important to note that the exercise duration was long. The authors claimed that the increases in cerebral uptake were a direct result of the action of unbinding tryptophan from albumin as a result of the organism's use of fat as the main energy supply, thus easing the crossing of the blood–brain barrier for tryptophan. Fat rather than carbohydrates is recruited mostly in sub-maximal, long-duration exercise. In shorter intensity, heavy exercise, lactate restricts the transport of free fatty acids in the blood (Bülow, Madsen, Astrup, & Christensen 1985) as does α-adrenergic action (Gullestad, Hallén, & Sejersted 1993).

Extensive reviews by Meeusen and colleagues (Meeusen & De Meirleir 1995; Meeusen et al. 2001) have shown that chronic exercise has also generally resulted in increases in whole brain 5-HT concentrations, particularly in the striatum, hippocampus, hypothalamus, and frontal cortex. Although results are not unequivocal, other recent studies have generally supported the findings reported in these reviews (e.g. Chen et al. 2008; Chennaoui et al. 2001), although Chennaoui et al. (2001) also showed a decrease in frontal cortex concentrations.

Hypothalamic-Pituitary-Adrenal Axis Hormones

Synthesis and Release

When stress is perceived, the hypothalamus initiates the synthesis of CRF, often called Corticotropin Releasing Hormone, from the prepro-CRF gene, a process that takes place in the PVN of the hypothalamus (Vale & Rivier 1977). CRF is secreted into the hypophyseal vessels in the median eminence, where it stimulates the synthesis and release by exocytosis of ACTH from its precursor preproopiomelanocortin. The nonapeptide, arginine vasopressin (AVP), also known as antidiuretic hormone, also acts as a stimulator of ACTH synthesis and release. It is primarily released from its preprohormone in response to feedback concerning osmolality of body fluids, although it is also secreted as the result of other physiological responses. AVP is synthesized by cells in the supraoptic and PVN of the hypothalamus, then transported to hypophyseal vessels in the median eminence.

Following synthesis, ACTH passes into the zona fasciculata of the adrenal cortex where it stimulates the synthesis and secretion of cortisol in humans and corticosterone in rodents. The precursor of cortisol is cholesterol. In the adrenals, stored, esterified cholesterol is hydrolyzed and free cholesterol is transported from storage vacuoles to the outer mitochondria. Steroidogenic acute regulatory protein mediates transfer to the inner mitochondrial membrane. Cholesterol is then converted to Δ^5-pregnenolone catalyzed by the enzyme P-450_{SCC}. Within the endoplasmic reticulum, Δ^5-pregnenolone is converted to 11-deoxycortisol catalyzed by 17-hydroxylase, 3β-ol-dehydrogenase and 21-hydroxylase. 11-deoxycortisol is transferred back to the mitochondria and hydroxylated by 11-hydroxylase to cortisol, which diffuses from the cell (Brandenberger et al. 1980).

Evidence for Acute Exercise-Induced Increases in Brain Concentrations of HPA Axis Hormones

During exercise, in the periphery cortisol plays major roles in glucose production from proteins, the facilitation of fat metabolism and muscle function, and the maintenance of blood pressure (Deuster et al. 1989). ACTH, which is synthesized in the gastrointestinal tract as well as in the pituitary, also plays a role in lipolysis (Borer 2003). Research examining exercise-induced changes in plasma and salivary ACTH and cortisol concentrations strongly suggests that, although the HPA axis hormones undergo circadian rhythms, normal concentrations are sufficient for efficient performance unless the exercise is heavy or of moderate intensity but carried out for long periods. Generally speaking, exercise needs to be $\geq$ 80% VO_{2MAX} for it to induce increases in plasma and salivary concentrations of ACTH and cortisol (De Vries, Bernards, De Rooij, & Koppeschaar 2000). Exercise of < 80% VO_{2MAX} has needed to be of at least 45 mins in duration (Bridge, Weller, Rayson, & Jones 2003) to affect plasma and salivary concentrations of ACTH and cortisol.

With regard to how peripheral changes affect the brain, the need for cortisol and ACTH peripherally obviously requires increased hypothalamic PVN synthesis and release of CRF. Rodent studies have shown evidence of acute exercise-induced increases in CRF mRNA expression in the PVN (e.g. Jiang et al. 2004; Kawashima et al. 2004; Timofeeva, Huang, & Richard 2003). With regard to ACTH, although it is synthesized and released from the anterior pituitary which, although outside of the blood–brain barrier, lies within the Central Nervous System (CNS), researchers have focused on plasma ACTH concentrations. However, ACTH does not cross the blood-cerebrospinal fluid (CSF)-barrier and CSF concentrations have been shown not to correlate with plasma concentrations (Allen, Kendall, McGilvra, & Vancura 2011). While plasma ACTH concentrations may not correlate with CSF concentrations and hence are not directly indicative of CNS activity, studies examining the effect of non-physical, psychological stress on plasma ACTH concentrations have demonstrated significant increases (Weinstein et al. 2010). Thus some interaction must be taking place. The situation with plasma and salivary cortisol concentrations is different because cortisol does cross the blood–brain barrier, hence increased peripheral cortisol concentrations mean that there is increased cortisol availability in the brain.

Rodent studies examining extracellular corticosterone concentrations in the brain support the peripheral plasma results. Droste et al. (2008) demonstrated significant increases in extracellular corticosterone concentrations in the hippocampus following a severe stressor, a forced swim, and a mild stressor, being placed in a novel environment. The increase as a result of the mild stress was much less than that induced by the forced swim. Although the authors put this down to the severity of the stressor it could also have been due to the swim inducing greater physiological demands. Moreover, extracellular corticosterone concentrations in the brain peaked 20 mins after plasma concentrations but returned to normal at the same time as plasma concentrations. Thus the brain was exposed to increased corticosterone for a considerably shorter period than peripheral plasma concentrations would suggest. Under normal conditions hippocampal, extracellular corticosterone and plasma corticosterone concentrations change synchronously due to circadian rhythms.

Long-Term Effects of Exercise-Induced Changes in Brain HPA Axis Hormones

In humans, research has generally shown that chronic exercise does not alter basal plasma ACTH and cortisol concentrations (e.g. Chatzitheodorou, Kabitsis, Malliou, & Mougios 2007; Duclos, Corcuff, Pehourcq, & Tabarin 2001; Wittert, Livesey, Espiner, & Donald 1996). However, Wittert et al. found athletes to have higher plasma and salivary ACTH concentrations than controls, even though there were no significant differences in cortisol concentrations. Blaney, Sothmann, Raff, Hart, and Horn (1990) found no significant differences between the plasma concentrations of ACTH and cortisol between active and sedentary individuals, while undertaking a modified Stroop color test. In a follow-up experiment, a group of sedentary individuals undertook a 4-month training program, which resulted in an 18% improvement in VO_{2MAX}, but again ACTH and cortisol plasma concentrations did not differ from a control group, while undertaking the Stroop color test.

However, as we saw above, ACTH plasma concentrations provide only very limited information concerning brain concentrations. Fortunately brain extracellular corticosterone concentrations and CRF mRNA expression have been studied in rodents following chronic exercise. Results concerning CRF mRNA activity are somewhat equivocal. Expression has been shown to be decreased in the hypothalamic PVN following long-term exposure to voluntary exercise (Bi, Scott, Hyun, Ladenheim, & Moran 2005; Kawashima et al. 2004), but some researchers found no significant effect (Droste, Chandramohan, Hill, Linthorst, & Reul 2007; Levin & Dunn-Meynell 2004), while Park et al. (2005) demonstrated an initial increase followed by a return to original levels. Forced exercise regimens have, however, resulted in significant increases (Chennaoui, Gomez Merino, Lesage, Drogou, & Guezennec 2002; Harbuz & Lightman 1989).

Examining the concentrations of free corticosterone in the brain, Droste, Collins, Lightman, Linthorst, & Reul (2009) found that training needed to be long-term before any changes in ultradian or circadian rhythms would be shown. Following long-term exercise, they found that exercising rats showed an increased afternoon/evening pulse in hippocampal free corticosterone concentrations compared to sedentary animals. However, they demonstrated no significant differences between exercising and sedentary rats following a forced swim, although both groups demonstrated significant increases in hippocampal-free corticosterone concentrations.

Several authors have examined the effect of chronic exercise on brain corticosteroid receptor mRNA gene expression. There would appear to be little effect of chronic exercise on GR mRNA expression in the hippocampus (Chang et al. 2008; Droste et al. 2003), although Park et al. (2005) found a significant decrease in hippocampal region CA4 only, while Droste et al. (2007) found increased GR mRNA levels in hippocampal layers. Reductions in MR mRNA expression have been shown (Chang et al. 2008; Droste et al. 2003) but some authors found no significant effect (Droste et al. 2007; Park et al. 2005). Droste et al. (2007) also found no significant changes in GR mRNA expression in the hypothalamic PVN, frontal cortex or anterior pituitary. Park et al. demonstrated initial decreases in PVN and anterior pituitary GR mRNA but, as training continued, these returned to pre-training levels. Overall, the evidence for increased chronic exercise-induced changes in GR mRNA and MR mRNA expression, and CRF, ACTH and cortisol brain concentrations are very weak indeed.

Growth Factors, Exercise and Depression, Anxiety and Well-Being

Brain-Derived Neurotrophic Factor

Synthesis and Release

BDNF is widely distributed throughout the CNS but is particularly well represented in the hippocampus, neocortex, cerebellum, striatum, and amygdala (Binder and Scharfman, 2004). It is also found peripherally in sensory neurons and glial cells (Knaeppen, Goekint, Heyman, & Meeusen 2010). BDNF is initially encoded by the BDNF gene to pro-BDNF, which is either proteolytically cleaved intracellularly by pro-convertases and secreted as mature BDNF; or secreted as pro-BDNF and then cleaved by extracellular proteases to mature BDNF (see Lessmann, Gottmann, & Malcangio 2003, for a review).

The effects of BDNF on neurogenesis and synaptic transmission are triggered when it binds with one of its receptors, the high-affinity tropomyosin-related kinase-B (Trk-B). Binding to Trk-B results in receptor dimerization and trans-autophosphorylation of tyrosine residues in the cytoplasmic domains of the receptor, which initiates a number of intracellular signaling cascades, including calcium/calmodulin kinase II (CaMKII) and mitogen-activated protein kinase (MAPK), resulting in the phosphorylation of cAMP-response element binding protein (CREB) (Binder & Scharfman 2004).

Brain-Derived Neurotrophic Factor and Exercise

During exercise, increased synthesis and release of BDNF begins as part of the hypothalamic response to undertaking exercise, particularly in its role in regulating peripheral energy metabolism (Wisse & Schwartz 2003). BDNF is also involved in ANS control of cardiovascular function, probably via signaling in central autonomic nuclei of the brainstem (Neeper, Gómez-Pinilla, Choi, & Cotman 1996). Several studies have shown that bouts of acute exercise have induced significant increases in serum or plasma BDNF concentrations in humans (e.g. Ferris, Williams, & Shen 2007; Griffin et al. 2011; Rasmussen et al. 2009; Tang, Chu, Hui, Helmeste, & Law 2008; Winter et al. 2007). In a meta-analysis, Szuhany, Bugatti, and Otto (2015) demonstrated a low to moderate mean effect size ($g = 0.46$), following a single bout of exercise for previously sedentary individuals, but this rose to $g = 0.59$ in individuals following a training regimen. Several authors, however, have pointed to evidence that gender may be a moderating variable with females demonstrating less of an effect (e.g. Szuhany et al. 2015; Voss et al. 2013). No-one has yet been able to determine why this may be.

It would also appear that the increase is transient and soon returns to baseline levels. The possibility that intensity may have an effect cannot be ruled out. Ferris et al. (2007) demonstrated that 30 mins cycling at 20% below VT had no significant effect on serum BDNF concentrations, however 30 mins at 10% above VT induced increased concentrations. Winter et al. (2007) reported that 40 mins of moderate intensity running, with blood lactate concentrations < 2 mmol[1], had no significant effect on serum BDNF concentrations but 2 x 3 min sprints to exhaustion, with a 2 mins between sprints rest period, induced increased BDNF concentrations. However, observation of the results reported show that moderate intensity exercise demonstrated a borderline significant ($p = 0.05$) pre- to post-exercise effect. Ferris et al. (2007) provided further evidence for an intensity effect as they showed a significant correlation ($r = 0.57$) between Δ BDNF serum concentrations and Δ blood lactate levels.

The picture with regard to chronic exercise in humans is far less clear. Szuhany et al. (2015) found a low but significant effect size ($g = 0.27$). In a narrative review, we (McMorris 2016) have reported that findings tend to show no significant effect. We found that results were not unequivocal, however, as Zoladz et al. (2008) found a significant increase in basal plasma concentrations and increased BDNF response in plasma to an acute bout of exercise. Also, Seifert et al. (2010) showed a significant increase in basal concentrations but no change in the response to a bout of acute exercise. Griffin et al. (2011) found that 3 weeks of 60 mins cycling at 60% VO_{2MAX} blunted the serum BDNF response to a bout of exercise to exhaustion. After 5 weeks training, participants showed increased post-exercise serum BDNF concentrations compared to pre-training but only when samples were taken 30 mins following cessation of the exercise. There were no significant effects of either 3 or 5 weeks on basal concentrations.

Evidence for Exercise-Induced Increases in Brain Concentrations of BDNF

During acute exercise, it is generally accepted that BDNF crosses the blood–brain barrier (Sartorius et al. 2009), although not unequivocally (Seredinin et al. 2013). Despite this, empirical evidence for increased brain concentrations during exercise are not readily available. Rasmussen et al. (2009) examined the effect of a single bout of acute exercise on BDNF concentrations in humans, measuring *a-v* diff. They showed that both arterial and internal jugular venous BDNF concentrations increased significantly during 4 h of rowing at ~ 85%–90% LT. Comparison of the *a-v* diff at rest and after 4 h demonstrated an increase in release of BDNF from the brain. It should be noted that, after 2 h, there were no significant effects of exercise. After 1 h of recovery, release of BDNF from the brain had returned to resting levels. Moreover, Rasmussen et al. showed that ~ 70% of resting plasma and serum BDNF originates from the brain and, during exercise, there is little or no change in this percentage.

The strongest evidence for acute-exercise-induced increases in the brain comes from rodent studies. These provide very strong evidence that acute exercise induces increased BDNF and/or BDNF mRNA expression in the brain, in particular in the hippocampus (e.g. Berchtold, Castello, & Cotman 2010). These increases are transient.

The situation with chronic exercise is slightly different. We are aware of only two studies which provide any evidence for chronic exercise-induced changes in human BDNF expression in the brain. Seifert et al. (2010) showed that 3 months of aerobic training resulted in increased jugular venous BDNF concentrations at rest, but there was no significant effect of training on *a-v* diff during a subsequent acute bout of exercise. Erickson et al. (2011) had elderly individuals, aged 55–80 years, undertake a 1-year program exercising at 60–77% maximum heart rate reserve for 40 mins per week. Increases in serum BDNF concentrations from pre- to post-treatment were not significant but Δ BDNF concentrations correlated significantly with Δ hippocampal volume as measured by magnetic resonance imaging ($r = 0.36$, $p < 0.01$, for the left hippocampus and $r = 0.37$, $p < 0.01$ for the right). According to Cohen (1988), these are moderate effect sizes, however, they only account for 12.96% and 13.69% of the relationship. Given that BDNF release triggers a whole range of activity which is involved in neurogenesis, one should not be surprised at the low correlation coefficients. As with acute exercise, rodent studies provide very strong evidence that chronic exercise induces increased BDNF and/or BDNF mRNA expression in the brain, again particularly in the hippocampus (e.g. Gomez-Pinilla, Vaynman, & Ying 2008; Griesbach, Hovda, & Gomez-Pinilla 2009; Huang et al. 2006).

To summarize, we can say that evidence from human studies for exercise-induced increases in BDNF concentrations in serum and plasma is limited. However, animal studies provide strong evidence for such changes. As one might expect, evidence for exercise-induced changes in brain concentrations in humans are very limited. However, animal studies provide strong evidence (see Table 2.1).

Other Growth Factors

Although there are many growth factors which may affect the brain-exercise interaction, the two which have received the most interest, apart from BDNF, are ***insulin-like growth factor*** (IGF-1) and ***vascular endothelial growth factor*** (VEGF). In this sub-section, we outline the evidence for exercise-induced increases in brain IGF-1 and VEGF, and introduce the interactions between these growth factors and BDNF, leading into the next sub-section, in which we examine effects of exercise on brain morphology.

IGF-1 is primarily generated by the liver, under the control of growth hormone. Growth hormone binds with its hepatic receptor and stimulates expression and release of IGF-1 peptide in the circulation. This has high affinity for IGF-1 binding proteins (IGFBPs) and represents the endocrine form of IGF-1 (Delafontaine 1995). In addition to the liver, many other organs produce IGF-1 but in a form which has a lower affinity for IGFBPs, and represents the autocrine and paracrine forms of IGF-1. These organs include neurons, glia, and vascular cells within the brain, providing evidence that these hormones have an important role in brain function (Donahue, Kosik, & Shors 2006).

Table 2.1 Exercise effects on brain concentrations of catecholamines 5-hydroxytryptamine, glucocorticoids, and selected proteins

	Acute exercise	*Chronic exercise*
Catecholamines	Increased concentrations Transient	Increased tyrosine hydroxylase mRNA
5-hydroxytryptamine	Increased concentrations Transient	Increased concentrations
Corticotropin releasing factor	Increased concentrations Transient	Evidence somewhat equivocal
Adrenocorticotropin hormone	Increased concentrations Transient	Little support for any changes
Cortisol	Increased concentrations Transient	Little support for any changes
Brain-derived neurotrophic factor	Increased concentrations Transient	Increased concentrations
Insulin-like growth factor	Increased concentrations Transient	Increased concentrations
Vascular endothelial growth factor	Increased concentrations Transient	Increased concentrations Increased mRNA

Given that, peripherally, IGF-1 is essential in glucose metabolism, tissue maintenance (Schwarz, Brasel, Hintz, Mohan, & Cooper 1996) and cerebrovascular function (Sonntag et al. 2000), it is not surprising to find that both animal and human studies demonstrate exercise-induced increases in peripheral IGF-1 (Gill 2007). Moreover, circulating IGF-1 crosses the blood–brain barrier (Pan & Kastin 2000). Interestingly, this process has been shown to be activity-dependent. Glutamate release at active synapses triggers two parallel processes, vasodilation to increase local availability of serum IGF-I and increased activity of matrix metalloprotease 9 (MMP9), an IGFBP3 protease, which is released in response to neuronal activity (Michaluk et al. 2007). These actions result in increased local availability of free serum IGF-I and this is transcytosed through the blood–brain barrier by the action of low-density lipoprotein receptor related protein (LRP1) (Nishijima et al. 2010). IGF-1 also crosses the blood–CSF barrier at the choroid plexus epithelium (Marks, Porte, & Baskin 1991). Increases in systemic IGF-I, at the blood-CSF interface, facilitate transportation by the protein transporter, LRP2 (Carro, Nuñez, Busiguina, & Torres-Aleman 2000). Carro et al. argued that this is the main method by which IGF-1 enters the brain. Whatever the method, animal studies strongly support acute exercise-induced increases in brain IGF-1 concentrations (see Carro et al. 2000; Cotman, Berchtold, & Christie 2007; Delafontaine, Song, & Li 2004). Moreover, evidence from studies, comparing trained versus sedentary individuals, shows that IGF-1 levels were significantly higher in trained individuals (Pareja-Galeano et al. 2013; Voss et al. 2013).

VEGF is a 45-kDa secretable basic heparin-binding homodimeric glycoprotein (Ferrara & Henzel, 1989). It is produced by skeletal muscle cells and secreted into the circulation (Kraus, Stallings, Yeager, & Gavin 2004). In humans, acute exercise increases VEGF mRNA in skeletal muscle (Gavin et al. 2004). In contrast, skeletal muscle VEGF protein is reduced immediately after acute exercise, probably due to release into the circulation by skeletal muscle (Gavin et al. 2004; Kraus et al. 2004). Chronic exercise not only restores skeletal muscle VEGF mRNA and protein levels but actually increases them (Gustafsson et al. 2002), while acute exercise-induced increases in VEGF mRNA are attenuated (Richardson et al. 2000). VEGF plays a major role in oxygen transport and it is the organs which are essential to this process, the brain, lung, and locomotor skeletal muscles, which demonstrate exercise-induced increases in VEGF and VEGF mRNA levels (Tang, Xia, Wagner, & Breen 2010). Although VEGF does not readily cross the blood–brain barrier, animal studies show increased levels in the brain (Jiang et al. 2014; Tang et al. 2010). As Tang et al. found a positive correlation between VEGF transcription and mRNA levels, they argued that increased levels were due to VEGF transcription occurring in the brain. In the brain, this increase was in the hippocampus only.

Interactions Between Growth Factors

In the next section, we examine the morphological changes induced by the effects of the growth factors. In this sub-section, we briefly outline the roles of the growth factors and the way in which BDNF, IGF-1 and VEGF interact. The roles of the growth factors in neurogenesis, neuroprotection, synaptic transmission and neuroplasticity are well-documented (see Binder & Scharfman 2004). BDNF appears to play the largest role but interacts with IGF-1 and VEGF. Cotman et al. (2007) claimed that the effects of exercise on neuroplasticity and, indeed, depression are predominantly regulated by IGF-1 and BDNF, whereas exercise-dependent stimulation of angiogenesis and hippocampal neurogenesis seems to be regulated by IGF-1 and VEGF. As Cotman et al. pointed out, however, the mechanism by which growth factors have antidepressant effects is largely unknown. Castrén et al. (2007) suggested that neurotrophic factors themselves do not control mood, but rather they facilitate the mechanisms that are required to induce antidepressant effects. Similarly, we are proposing that the effects of growth factors on brain morphology and neuroplasticity may, in fact, be the way in which growth factors alleviate depression and anxiety. We believe that this is supported, at least to some extent, by the findings of Voss et al. (2013). They demonstrated a relationship between BDNF and functional connectivity in the brain, which they argued may be modulated by IGF-1 and VEGF. Both of these growth factors stimulate the growth of endothelial cells, which in turn express nitric oxide synthase, which is required for exercise-induced up-regulation of BDNF in the hippocampus (Chen, Ivy, & Russo-Neustadt 2006). BDNF probably improves functional connectivity by increasing synaptogenesis and dendritic spine density (Vaynman, Ying, & Gomez-Pinill 2004). Moreover, Voss et al.

pointed out that there is also evidence that neural activity is increased during aerobic exercise, based on increased c-Fos expression in the hippocampus (Carro et al. 2000; Clark, Bhattacharya, Miller, & Rhodes 2011) and cortex (Carro et al. 2000). Moreover, Erickson et al. (2011) demonstrated increases in hippocampal volume, which were related to serum BDNF concentrations, in elderly participants following a year-long training regimen.

Morphology

As we saw in the Introduction, Duman et al. (1997) hypothesized that depression was due to neural dysfunction. This was supported by observations that depressed individuals demonstrated reduced hippocampal volume (Bremner et al. 2000) and prefrontal cortex volume (Dwivedi et al. 2003). In this section, we have also seen that growth factors, in particular BDNF, trigger neurogenesis, synaptogenesis, angiogenesis, and dendritic, astrocytic and glia cell hypertrophy (Cotman et al. 2007; Kleim et al. 2007). Given that exercise induces increased growth factor levels in the brain, we would expect to see significant effects of exercise on brain morphology. In this sub-section, we outline the results from humans and animals, which generally support some changes but possibly of a more limited extent than that which many would expect.

While the evidence from human studies is less robust, reviews of animal and human studies generally provide sound evidence for chronic exercise-induced increases in hippocampal volume (van Praag 2008; Voss et al. 2013). Animal studies show that these increases are primarily due to neurogenesis. Increased volume in the prefrontal cortex has also been well-supported (Fincham, Carter, van Veen, Stenger, & Anderson 2002; van Veen, Cohen, Botvinick, Stenger, & Carter 2001; van Veen, Krug, & Carter 2008) and is particularly important with regard to depression and anxiety. The basal ganglia, also important in depression and anxiety, have been shown to increase volume in children due to chronic exercise (Chaddock et al. 2010a, 2010b). Changes in these regions are probably due to dendritic, astrocytic and glia cell hypertrophy (Kleim et al. 2007). Finally, we should note that animal studies have shown that NA plays a role in neurogenesis, via activation of β_2–adrenoceptors (Jhaveri et al. 2010; Masuda et al. 2011), and this could be affected by increased exercise-induced TH mRNA expression.

Empirical Evidence for an Exercise-Neurobiology-Depression/Anxiety/Psychological Well-Being Interaction

In the previous sections we have outlined the theoretical, neurobiological causes of depression and anxiety, and the evidence concerning how exercise can have a preventative and/or healing effect on these conditions. Although we have pointed out several times that exercise has been shown to have preventative and healing effects, we have not presented any empirical evidence to directly support the theoretical roles of exercise-induced changes in brain neurobiology. Most researchers examining the effect of exercise on depression, anxiety and psychological well-being point to the theoretical roles of neuroendocrines and/or changes in brain morphology, but without presenting any empirical evidence for an interaction between neurobiological changes and psychological state. Many point to animal studies which show changes in brain neurobiology and, as we have done so far, point out that exercise induces similar changes in humans, but they do not present any what we would call "direct evidence." By "direct evidence," we mean where a definite interaction between the exercise, the neurobiological changes and the changes in depression or anxiety can be demonstrated. The reason for this is that, in fact, there is very little such research.

Following a systematic review of the literature using Pubmed, SCOPUS, SportsDISCUS and Web of Knowledge databases, we found only three. Ströhle et al. (2010) examined the interaction between acute exercise, serum BDNF concentrations and emotional well-being (arousal and anxiety) of individuals with panic disorder. Treadmill walking for 30 min at 70% VO_{2MAX} had no significant effect on arousal and anxiety, but 30 min of aerobic exercise significantly increased serum BDNF concentrations in the panic disorder group. Wipfli, Landers, Nagoshi, and Ringenbach (2011) examined the effect of a 7-week stationary cycling intervention, completing three 30-min bouts of cycling, at 70% of age-predicted heart rate maximum, per week on anxiety and depression levels in Kinesiology students, who had no clinical disorder. Anxiety demonstrated a main effect of intervention but no difference between the exercise and control groups. However, exercise had a positive effect on depression. Moreover, 5-HT concentrations in the exercise group showed a significantly larger post-treatment decrease than those in the control group. More importantly for us, the authors carried out mediation tests on causes of the reduction in depression and found that percent change in 5-HT was determined to partially mediate the relationship between exercise and depression, but it only explained 9% of the variance in the interaction. This is somewhat surprising given that the monoamine hypothesis for depression states that problems are caused by low concentrations of 5-HT, not high. However, these participants were not clinically depressed.

Prakhinkit, Suppapitiporn, Tanaka, and Suksom (2014) examined the effects of walking meditation and normal walking exercise, compared to one another and a control group, on depression in elderly females (mean age 76.6, SD = 1.77), with mild-to-moderate depression as defined by Geriatric Depression Scale scores of 13–24. The walking groups' program was divided into two phases. Phase 1 (weeks 1–6) was conducted at mild intensity (20%–39% individually determined heart rate reserve) performed for 20 minutes, three times a week. In phase 2 (weeks 7–12), the training intensity was increased to moderate intensity (40%–50% heart rate reserve) performed for 30 minutes, three times a week. The walking meditation program included "typical Buddhist meditation practices" (p. 412). The walking meditation group demonstrated significant improvement in depression score and a significant reduction in plasma cortisol concentrations. The control group showed a significant increase in cortisol concentrations, pre- to post-experiment, but the normal walking group demonstrated no significant effect. However, observation of pre-treatment to post-treatment data for the normal walking group suggest a possible Type II error, as the effect size (Cohen's d = 1.0) is very high.

The rodent literature also shows only a handful of studies. Reduced anxiety and/or depression accompanied by increased BDNF mRNA expression in the hippocampus was shown by Aguiar et al. (2014), while Duman, Schlesinger, Russell, and Duman (2008) showed reduced anxiety in exercising mice. The mice demonstrated increases in BDNF mRNA expression in hippocampal area CA1, after 3 weeks of wheel running exercise, and in the dentate gyrus after 1 or 3 weeks of exercise. However, BDNF expression was not altered in area CA3 or in the frontal cortex. Lee, Ohno, Ohta, and Mikami (2013) showed that exercise-induced reductions in depression were affected by increases in hippocampal NA but there were no significant changes in hippocampal 5-HT concentrations.

The limited number of studies, the small sample sizes in the human studies, the limited measures of depression and anxiety, and the limited amounts of neurobiological variables measured mean that these studies tell us very little indeed.

Conclusion

The empirical literature reviewed in the previous section shows very little support for exercise-induced, neurochemical, and morphological changes being responsible for the observed improvements in psychological well-being following exercise. However, there is a lack of research, so we have to turn to the exercise biochemistry literature, which provides strong evidence from animal studies to show that both acute and chronic exercise result in increased concentrations of catecholamines, 5-HT, BDNF, IGF-1 and VEGF in the brain. There is some, albeit limited, support from human studies. Moreover, there is evidence from humans and animals for chronic exercise-induced increases in hippocampal, basal ganglia and prefrontal cortex volumes, as well as improved functional connectivity in the brain. One might argue that this is not definitive proof of cause and effect, "smoking gun" evidence. However, the fact that the neurochemicals, brain structures, and morphology effected are those responsible for depression and anxiety provides strong circumstantial evidence that it is the exercise-induced effects on these neurochemicals and structures that induce relief and protection from depression and anxiety. However, we have little knowledge of the intensities, duration or frequency of exercise that may be required for these changes to interact with depression and anxiety.

References

Aguiar, A. S., Jr., E. Stragier, D. da Luz Scheffer, A. P. Remor, P. A. Oliveira, R. D. Prediger,...L. Lanfumey. 2014. "Effects of exercise on mitochondrial function, neuroplasticity and anxio-depressive behavior of mice." *Neuroscience 271*: 56–63.

Allen, J. P., J. W. Kendall, R. McGilvra, & C. Vancura. 2011. "Immunoreactive ACTH in cerebrospinal fluid." *J Clin Endocrinol Metab 38*: 586–93.

Alousi, A., & N. Weiner. 1966. "The regulation of norepinephrine synthesis in sympathetic nerves: Effect of nerve stimulation, cocaine and catecholamine-releasing agents." *Proc Natl Acad Sci USA 56*: 1491–6.

Arakawa, K., S. Miura, M. Koga, A. Kinoshita, H. Urata, & A. Kiyonaga. 1995. "Activation of renal dopamine system by physical exercise." *Hypertens Res 18*: S73–7.

Arnsten, A. F. T. 2009. "Stress signalling pathways that impair prefrontal cortex structure and function." *Nat Rev Neurosci 10*: 410–22.

Arts, F. J. P., & H. Kuipers. 1994. "The relation between power output, oxygen uptake and heart rate in male athletes." *Int J Sports Med 15*: 228–31.

Autry, A. E., & L. M. Monteggia. 2012. "Brain-derived neurotrophic factor and neuropsychiatric disorders." *Pharmacol Rev 64*: 238–58.

Berchtold, N. C., N. Castello, & C. W. Cotman. 2010. "Exercise and time-dependent benefits to learning and memory." *Neuroscience 167*: 588–97.

Bi, S., K. A. Scott, J. Hyun, E. E. Ladenheim, & T. H. Moran. 2005. "Running wheel activity prevents hyperphagia and obesity in Otsuka Long-Evans Tokushima fatty rats: role of hypothalamic signaling." *Endocrinology 148*: 1678–85.

Binder, D.K., & H. E. Scharfman. 2004. "Brain-derived neurotrophic factor." *Growth Factors 22*: 123–31.

Blaney, J., M. Sothmann, H. Raff, B. Hart, & T. Horn. 1990. "Impact of exercise training on plasma adrenocorticotropin response to a well-learned vigilance task." *Psychoneuroendocrinology 15*: 453–62.

Blomstrand, E. 2006. "A role for branched-chain amino acids in reducing central fatigue." *J Nutr 136*: 544S–7S.

Blomstrand, E., K. Møller, N. H. Secher, & L. Nybo. 2005. "Effect of carbohydrate ingestion on brain exchange of amino acids during sustained exercise in human subjects." *Acta Physiol Scand 185*: 203–9.

Blomstrand, E., & E. A. Newsholme. 1992. "Effect of branched-chain amino acid supplementation on the exercise-induced change in aromatic amino acid concentration in human muscle." *Acta Physiol Scand 146*: 293–8.

Borer, K. T. 2003. *Exercise endocrinology*. Champaign, IL: Human Kinetics.

Brandenberger, G., M. Follenius, G. Wittersheim, P. Salame, M. Simeoni, & B. Reinhardt. 1980. "Plasma-catecholamines and pituitary-adrenal hormones related to mental task demand under quiet and noise conditions." *Biol Psychol 10*: 239–52.

Bremner, J. D., M. Narayan, E. R. Anderson, L. H. Staib, H. L. Miller, & D. S. Charney. 2000. "Hippocampal volume reduction in major depression." *Am J Psychiatry 157*: 115–8.

Bridge, M. W., A. S. Weller, M. Rayson, & D. A. Jones. 2003. "Ambient temperature and the pituitary hormone responses to exercise in humans." *Exp Physiol 88*: 627–35.

Bülow, J., J. Madsen, A. Astrup, & N. J. Christensen. 1985. "Vasoconstrictor effect of high FFA/albumin ratios in adipose tissue in vivo." *Acta Physiol Scand 125*: 661–7.

Carro, E., A. Nuñez, S. Busiguina, & I. Torres-Aleman. 2000. "Circulating insulin-like growth factor I mediates effects of exercise on the brain." *J Neurosci 20*: 2926–33.

Carroll, B. J., G. C. Curtis, & J. Mendels. 1976. "Neuroendocrine regulation in depression. I. Limbic system-adrenocortical dysfunction." *Arch Gen Psychiatry 33*: 1039–44.

Castrén, E., & T. Rantamäki. 2010. "The role of BDNF and its receptors in depression and antidepressant drug action: Reactivation of developmental plasticity." *Dev Neurobiol 70*: 289–97.

Castrén E., V. Võikar, & T. Rantamäki. 2007. "Role of neurotrophic factors in depression." *Curr Opin Pharmacol 7*: 18–21.

Chaddock, L., K. I. Erickson, R. S. Prakash, J. S. Kim, M. W. Voss, M. Vanpatter ,... A. F. Kramer. 2010a. "A neuroimaging investigation of the association between aerobic fitness, hippocampal volume and memory performance in preadolescent children." *Brain Res 1358*: 172–83.

Chaddock, L., K. I. Erickson, R. S. Prakash, M. VanPatter, M. W. Voss, M. B. Pontifex,...A. F. Kramer. 2010b. "Basal ganglia volume is associated with aerobic fitness in preadolescent children." *Dev Neurosci 32*: 249–56.

Chang, Y. T., Y. C. Chen, C. W. Wu, H. I. Chen, C. J. Jen, & Y. M. Kuo. 2008. "Glucocorticoid signaling an exercise-induced downregulation of the mineralocorticoid receptor in the induction of adult mouse dentate neurogenesis by treadmill running." *Psychoneuroendocrinology 33*: 1173–82.

Chatzitheodorou, D., C. Kabitsis, P. Malliou, & V. Mougios. 2007. "A pilot study of the effects of high-intensity aerobic exercise versus passive interventions on pain, disability, psychological strain, and serum cortisol concentrations in people with low back pain." *Phys Ther 87*: 304–12.

Chen, H. I., L. C. Lin, L. Yu, L. Y. F. Liu, Y. M. Kuo, A. M. Huang,...C. J. Jen. 2008. "Treadmill exercise enhances avoidance learning in rates: the role of down-regulated serotonin system ion the limbic system." *Neurobiol Learn Mem 89*: 489–96.

Chen, M. J., A. S. Ivy, & A. A. Russo-Neustadt. 2006. "Nitric oxide synthesis is required for exercise-induced increases in hippocampal BDNFand phosphatidylinositol 3' kinase expression." *Brain Res Bull 68*: 257–68.

Chennaoui, M., C. Drogou, D. Gomez-Merino, B. Grimaldi, G. Fillion, & C. Y. Guezennec. 2001. "Endurance training effects on 5-HT(1B) receptors mRNA expression in cerebellum, striatum, frontal cortex and hippocampus of rats." *Neurosci Lett 307*: 33–6.

Chennaoui, M., D. Gomez Merino, J. Lesage, C. Drogou, & C. Y. Guezennec. 2002. "Effects of moderate and intensive training on the hypothalamic-pituitary-adrenal axis in rats." *Acta Physiol Scand 175*: 113–21.

Clark, P. J, T. K. Bhattacharya, D. S. Miller, & J. S. Rhodes. 2011. "Induction of c-Fos, Zif268, and Arc from acute bouts of voluntary wheel running in new and pre-existing adult mouse hippocampal granule neurons." *Neuroscience 184*: 16–27.

Cohen, J. 1988. *Statistical power analysis for the behavioral sciences*, 2nd edition. Hillsdale, NJ: Lawrence Erlbaum Associates.

Coppen, A. 1967. "The biochemistry of affective disorders." *Br J Psychiatry 113*: 1237–64.

Cornford, E. M., L. D. Braun, W. H. Oldendorf, & M. A. Hill. 1982. "Comparison of lipid-related blood–brain barrier penetrability in neonates and adults." *Am J Physiol 243*: C161–8.

Cotman, C. W., N. C. Berchtold, & L. A. Christie. 2007. "Exercise builds brain health: Key roles of growth factor cascades and inflammation." *Trends Neurosci 30*: 464–72.

Dalsgaard, M. K., P. Ott, F. Dela, A. Juul, B. K. Pedersen, J. Warberg,...N. H. Secher. 2004. "The CSF and arterial to internal jugular venous hormonal differences during exercise in humans." *Exp Physiol 89*: 271–7.

De Kloet, E. R., S. A. B. E. Van Acker, R. M. Siburg, M. S. Oitzl, O. C. Meijer, K. Rahmouni, W. de Jong. 2000. "Aldosterone action in nonepithial cells." *Kidney Int 57*: 1329–36.

Delafontaine, P. 1995. "Insulin-like growth factor I and its binding proteins in the cardiovascular system." *Cardiovasc Res 30*: 825–34.

Delafontaine, P., Y. H. Song, & Y. Li. 2004. "Expression, regulation, and function of IGF-1, IGF-1R, and IGF-1 binding proteins in blood vessels." *Arterioscler Thromb Vasc Biol 24*: 435–44.

Deuster, P. A., G. P. Chrousos, A. Luger, J. E. DeBolt, L. L. Bernier, U. H. Trostmann,...D. L. Loriaux. 1989. "Hormonal and metabolic responses of untrained, moderately trained, and highly trained men to 3 exercise intensities." *Metabolism 38*: 141–8.

De Vries, W. R., N. T. M. Bernards, M. H. De Rooij, & H. P. F. Koppeschaar. 2000. "Dynamic exercise discloses different time-related responses in stress hormones." *Psychosom Med 62*: 866–72.

Donahue, C. P., K. S. Kosik, & T. J. Shors. 2006. "Growth hormone is produced within the hippocampus where it responds to age, sex, and stress." *Proc Natl Acad Sci U S A 103*: 6031–6.

Droste, S. K., Y. Chandramohan, L. E. Hill, A. C. Linthorst, & J. M. Reul. 2007. "Voluntary exercise impacts on the rat hypothalamic-pituitary-adrenocortical axis mainly at the adrenal level." *Neuroendocrinology 86*: 26–37.

Droste, S. K., A. Collins, S. L. Lightman, A. C. E. Linthorst, & J. M. H. M. Reul. 2009. "Distinct, time-dependent effects of voluntary exercise on circadian and ultradian rhythms and stress responses of free corticosterone in the rat hippocampus." *Endocrinology 150*: 4170–9.

Droste, S. K., L. de Groote, H. C. Atkinson, S. L. Lightman, J. M. Reul, & A. C. Linthorst. 2008. "Corticosterone levels in the brain show a distinct ultradian rhythm but a delayed response to forced swim stress." *Endocrinology 149*: 3244–53.

Droste, S. K., A. Gesing, S. Ulbricht, M. B. Müller, A. C. E. Linthorst, & J. M. H. M. Reul. 2003. "Effects of long-term voluntary exercise on the mouse hypothalamic-pituitary-adrenocortical axis." *Endocrinology 144*: 3012–23.

Duclos, M., J-B. Corcuff, F. Pehourcq, & A. Tabarin. 2001. "Decreased pituitary sensitivity to glucocorticoids in endurance-trained men." *Eur J Endocrinol 144*: 363–8.

Duman, R. S., G. R. Heninger, & E. J. Nestler. 1997. "A molecular and cellular theory of depression." *Arch Gen Psychiatry 54*: 597–606.

Duman, C. H., L. Schlesinger, D. S. Russell, & R. S. Duman. 2008. "Voluntary exercise produces antidepressant and anxiolytic behavioral effects in mice." *Brain Res 1199*: 148–58.

Dwivedi, Y., H. S. Rizavi, R. R. Conley, R. C. Roberts, C. A. Tamminga, & G. N. Pandey. 2003. "Altered gene expression of brain-derived neurotrophic factor and receptor tyrosine kinase B in postmortem brain of suicide subjects." *Arch Gen Psychiatry 60*: 804–15.

Erickson, K. I., M. W. Voss, R. S. Prakash, C. Basak, A. Szabo, L. Chaddock,...A. F. Kramer. 2011. "Exercise training increases size of hippocampus and improves memory." *Proc Natl Acad Sci U S A 108*: 3017–22.

Fernstrom, J. D., & M. H. Fernstrom. 2006. "Exercise, serum free tryptophan, and central fatigue." *J Nutr 136*: 553S–9S.

Fernstrom, J. D., & M. H. Fernstrom. 2007. "Tyrosine, phenylalanine, and catecholamine synthesis and function on the brain." *J Nutr 137*: 1539S–47S.

Ferrara, N., & W. J. Henzel. 1989. "Pituitary follicular cells secrete a novel heparin-binding growth factor specific for vascular endothelial cells." *Biochem Bioph Res Co 161*: 851–8.

Ferris, L. T., J. S. Williams, & C. Shen. 2007. "The effect of acute exercise on serum brain-derived neurotrophic factor levels and cognitive function." *Med Sci Sports Exerc 39*: 728–34.

Fincham, J. M., C. S. Carter, V. van Veen, V. A. Stenger, & J. R. Anderson. 2002. "Neural mechanisms of planning; a computational analysis using event-related fMRI." *Proc Natl Acad Sci U S A, 99*: 3346–51.

Finlay, J. M., M. J. Zigmond, & E. D. Abercrombie. 1995. "Increased dopamine and norepinephrine release in medial prefrontal cortex induced by acute and chronic stress: Effects of diazepam." *Neuroscience 64*: 619–28.

Foley, T. E., & M. Fleshner. 2008. "Neuroplasticity of dopamine circuits after exercise: Implications for central fatigue." *Neuromol Med 10*: 67–80.

Frazer, A., & J. G. Hensler. 1999. "Serotonin." In G. J. Siegel, B. W. Agranoff., R. W. Abers., S. K. Fisher, & M. D. Uhler (Eds.), ***Basic neurochemistry: Molecular, cellular and medical aspects, 6th edition*** (pp. 263–92). Philadelphia, PA: Lippincott, Williams, and Wilkins.

Frodl, T., E. Meisenzahl, T. Zetzsche, R. Bottlender, C. Born, C. Groll,...H. J. Möller. 2002. "Enlargement of the amygdala in patients with a first episode of major depression." *Biol Psychiatry 51*: 708–14.

Gavin, T. P., C. B. Robinson, R. C. Yeager, J. A. England, L. W. Nifong, & R. C. Hickner. 2004. "Angiogenic growth factor response to acute systemic exercise in human skeletal muscle." *J Appl Physiol 96*: 19–24.

Gavrilović, L., V. Mandusić, V. Stojilิković. J. Kasapovic, S. Stojiljkovic, S. B. Pajovic, & S. Dronjak. 2012. "Effect of chronic forced running on gene expression of catecholamine biosynthetic enzymes in stellate ganglia of rats." *J Biol Regul Homeost Agents 26*: 367–77.

Gill, J. M. 2007. "Physical activity, cardiorespiratory fitness and insulin resistance: A short update." *Curr Opin Lipidol 18*: 47–52.

Gomez-Pinilla, F., S. Vaynman, & Z. Ying. 2008. "Brain-derived neurotrophic factor functions as a metabotrophin to mediate the effects of exercise on cognition." *European Journal of Neuroscience 28*: 2278–87.

Grenhoff, J., & T. H. Svensson. 1993. "Prazosin modulates the firing pattern of dopamine neurons in rat ventral tegmental area." *Eur J Pharmacol 233*: 79–84.

Griesbach, G. S., D. A. Hovda, & F. Gomez-Pinilla. 2009. "Exercise-induced improvement in cognitive performance after traumatic brain-injury in rats is dependent on BDNF activation." *Brain Res 1288*: 105–15.

Griffin, É. W., S. Mullally, C. Foley, S. A. Warmington, S. M. O'Mara, & Á. M. Kelly. 2011. "Aerobic exercise improves hippocampal function and increases BDNF in the serum of young adult males." *Physiol Behav 104*: 934–41.

Gullestad, L., J. Hallén, & O. M. Sejersted. 1993. "Variable effects of beta-adrenoreceptor blockade on muscle blood flow during exercise." *Acta Physiol Scand 149*: 257–71.

Gustafsson, T., A. Knutsson, A. Puntschart, L. Kaijser, A. C. Nordqvist, C. J. Sundberg, & E. Jansson. 2002. "Increased expression of vascular endothelial growth factor in human skeletal muscle in response to short-term one-legged exercise training." *Pflugers Arch 444*: 752–9.

Harbuz, M. S., & S. L. Lightman. 1989. "Responses of hypothalamic and pituitary mRNA to physical and psychological stress in the rat." *J Endocrinol 122*: 705–11.

Hasler, G. 2010. "Pathophysiology of depression: Do we have any solid evidence of interest to clinicians?" *World Psychiatry 9*: 155–61.

Hawkins, R. A., R. L. O'Kane, I. A. Simpson, & J. R. Viña. 2006. "Structure of the blood–brain barrier and its role in the transport of amino acids." *J Nutr 136*: 218S–26S.

Heim, C., U. Ehlert, & D. H. Hellhammer. 2000. "The potential role of hypocortisolism in the pathophysiology of stress-related bodily disorders." *Psychoneuroendocrinology 25*: 1–35.

Higa-Tanaguchi, K. T., F. C. Silva, H. M. Silva, L. C. Michelini, & J. E. Stern. 2007. "Exercise-training induced remodeling of paraventricular nucleus (nor)adrenergic innervation in normotensive and hypertensive rats." *Am J Physiol Reg-I 292*: R1717–27.

Hodgetts, V., S. W. Coppack, K. N. Frayn, & T. D. R. Hockaday. 1991. "Factors controlling fat mobilization from human subcutaneous adipose-tissue during exercise." *J Appl Physiol 71*: 445–51.

Huang, A. M., C. J. Jen, H. F. Chen, L. Yu, Y. M. Kuo, & H. I. Chen. 2006. "Compulsive exercise acutely upregulates rat hippocampal brain-derived neurotrophic factor." *J Neural Transm 113*: 803–11.

Jhaveri, D. J., E. W. Mackay, A. S. Hamlin, S. V. Marathe, L. S. Nandam, V. A. Vaidya, & P. F. Bartlett. 2010. "Norepinephrine directly activates adult hippocampal precursors via β3 adrenergic receptors." *J Neurosci 30*: 2795–806.

Jiang, P., R. L. Dang, H. D. Li, L. H. Zhang, W. Y. Zhu, Y. Xue, & M. M. Tang. 2014. "The impacts of swimming exercise on hippocampal expression of neurotrophic factors in rats exposed to chronic unpredictable mild stress." *Evid-Based Compl Alt.* doi:10.1155/2014/729827

Jiang, Q. Y., H. Kawashima, Y. Iwasaki, K. Uchida, K. Sugimoto, & K. Itoi. 2004. "Differential effects of forced swim-stress on the corticotropin-releasing hormone and vasopressin gene transcription in the paravocellular division of the paraventricular nucleus of rat hypothalamus." *Neurosci Lett 358*: 201–4.

Kawashima, H., T. Saito, H. Yoshizato, T. Fujikawa, Y. Sato, B. S. McEwen, & H. Soya. 2004. "Endurance treadmill training in rats alters CRH activity in the hypothalamic paraventricular nucleus at rest and during acute running according to its period." *Life Sci 76*: 763–74.

Kleim, J. A., J. A. Markham, K. Vij, J. L. Freese, D. H. Ballard, & W. T. Greenough. 2007. "Motor learning induces astrocytic hypertrophy in the cerebellar cortex." *Behav Brain Res 178*: 244–9.

Knaeppen, K., M. Goekint, E. M., Heyman, & R. Meeusen. 2010. "Neuroplasticity – Exercise-induced response of peripheral brain-derived neurotrophic factor: A systematic review of experimental studies in human subjects." *Sports Med 40*: 765–801.

Kraus, R. M., H. W. Stallings, R. C. Yeager, & T. P. Gavin. 2004. "Circulating plasma VEGF response to exercise in sedentary and endurance-trained men." *J Appl Physiol 96*: 1445–50.

Krishnan, V., M. H. Han, D. L. Graham, O. Berton, W. Renthal, S. J. Russo,...E. J. Nestler. 2007. "Molecular adaptations underlying susceptibility and resistance to social defeat in brain reward regions." *Cell 131*: 391–404.

Lee, H., M. Ohno, S. Ohta, & T. Mikami. 2013. "Regular moderate or intense exercise prevents depression-like behavior without change of hippocampal tryptophan content in chronically tryptophan-deficient and stressed mice." *PLoS One 8*. doi:10.1371/journal.pone.0066996

Lefebvre, H., P. Compagnon, V. Contesse, C. Delarue, C. Thuillez, H. Vaudry, & J. M. Kuhn. 2001. "Production and metabolism of serotonin (5-HT) by the human adrenal cortex: Paracrine stimulation of aldosterone secretion by 5-HT." *J Clin Endocr Metab 86*: 5001–7.

Lessmann, V., K. Gottmann, & M. Malcangio. 2003. "Neurotrophin secretion: Current facts and future prospects." *Prog Neurobiol 69*: 341–74.

Levin, B. E., & A. Dunn-Meynell. 2004. "Chronic exercise lowers the defended body weight gain and adiposity in diet-induced obese rats." *Am J Physiol Reg-I 286*: R771–8.

McMorris, T. 2016. "Chronic exercise and cognition in humans: A review of the evidence for a neurochemical basis." In T. McMorris (Ed.), *Exercise and cognition: neuroscience perspectives* (pp. 167–186). New York: Academic Press.

Mann, J. J., D. Currier, J. A. Quiroz, & H. K. Manji. 2012. "Neurobiology of severe mood and anxiety disorders." In S. T. Brady, G. J. Siegel, R. W. Albers et al. (Eds.), *Basic neurochemistry, 8th edition* (pp. 1021–36). New York: Academic Press.

Marks, J. L., D. Porte, & D. G. Baskin. 1991. "Localization of type I insulin-like growth factor receptor messenger RNA in the adult rat brain by in situ hybridization." *Mol Endocrinol 5*: 1158–68.

Mason, J. W. 1975a. "A historical view of the stress field. Part I." *J Hum Stress 1*: 6–12.

Mason, J. W. 1975b. "A historical view of the stress field. Part II." *J Hum Stress 1*: 22–36.

Masuda, T., S. Nakagawa, & S. Boku. 2011. "Noradrenaline increases neural precursor cells derived from adult rat dentate gyrus through β2 receptor." *Prog Neuropsychopharmacol 36*: 44–51.

Meeusen, R., & K. De Meirleir. 1995. "Exercise and brain neurotransmission." *Sports Med 20*: 160–88.

Meeusen, R., M. F. Piacentini, & K. De Meirleir. 2001. "Brain microdialysis in exercise research." *Sports Med 31*: 965–83.

Meeusen, R., J. Smolders, S. Sarre, K. de Meirleir, H. Keizer, M. Serneels,...Y. Michotte. 1997. "Endurance training effects on neurotransmitter release in rat striatum: an in vivo microdialysis study." *Acta Physiol Scand 159*: 335–41.

Michaluk, P., L. Kolodziej, B. Mioduszewska, G. M. Wilczynski, J. Dzwonek, J. Jaworski,...L. Kaczmarek. 2007. "Beta-dystroglycan as a target for MMP-9, in response to enhanced neuronal activity." *J Biol Chem 282*: 1636–41.

Miyashita, T., & C. L. Williams. 2006. "Epinephrine administration increases neural impulses propagated along the vagus nerve: role of peripheral beta-adrenergic receptors." *Neurobiol Learn Mem 85*: 116–24.

Moret, C., & M. Briley. 2011. "The importance of norepinephrine in depression." *Neuropsychiatric Disease and Treatment 7* Suppl 1: 9–13.

Neeper, S. A., F. Gómez-Pinilla, J. Choi, & C. W. Cotman. 1996. "Physical activity increases mRNA for brain-derived neurotrophic factor and nerve growth factor in rat brain." *Brain Res 726*: 49–56.

Nishijima, T., J. Piriz, S. Duflot, A. M. Fernandez, G. Gaitan, U. Gomez-Pinedo,...I. Torres-Aleman. 2010. "Neuronal activity drives localized blood–brain-barrier transport of serum insulin-like growth factor-I into the CNS." *Neuron 67*: 834–46.

O'Neal, H. A., J. D. van Hoomissen, P. V. Holmes, & R. K. Dishman. 2001. "Preprogalanin messenger RNA levels are increased in rat locus coeruleus after treadmill exercise training." *Neurosci Lett 299*: 69–72.

Pan, W., & A. J. Kastin. 2000. "Interactions of IGF-1 with the blood–brain barrier in vivo and in situ." *Neuroendocrinology 72*: 171–8.

Pardridge, W. M. 1998. "Blood–brain barrier carrier-mediated transport and brain metabolism of amino acids." *Neurochem Res 23*: 635–44.

Pareja-Galeano, H., T. Brioche, F. Sanchis-Gomar, A. Montal, C. Jovaní, C. Martínez-Costa,…J. Viña. 2013. "Impact of exercise training on neuroplasticity-related growth factors in adolescents." *J Musculoskelet Neuronal Interact 13*: 368–71.

Park, E., O. Chan, Q., Li, M., Kiraly, S. G. Matthews, M. Vranic, & M. C. Riddell. 2005. "Changes in basal hypalamo-pituitary-adrenal activity during exercise are centrally mediated." *Am J Physiol Reg-I 289*: R1360–71.

Podolin, D. A., P. A. Munger, & R. S. Mazzeo. 1991. "Plasma-catecholamine and lactate response during graded-exercise with varied glycogen conditions." *J Appl Physiol 71*: 1427–33.

Prakhinkit, S., S. Suppapitiporn, H. Tanaka, & D. Suksom. 2014. "Effects of Buddhism walking meditation on depression, functional fitness, and endothelium-dependent vasodilation in depressed elderly." *J Altern Complement Med 20*: 411–16.

Price, J. L., S. T. Carmichael, & W. C. Drevets. 1996. "Networks related to the orbital and medial prefrontal cortex; a substrate for emotional behavior?" *Prog Brain Res 107*: 523–36.

Rasmussen, P., P. Brassard, H. Adser, M. V. Pedersen, L. Leick, E. Hart,…H. Pilegaard. 2009. "Evidence for release of brain-derived neurotrophic factor from the brain during exercise." *Exp Physiol 94*: 1062–9.

Reul, J. M., A. Collins, R. S. Saliba, K. R. Mifsud, S. D. Carter, M. Gutierrez-Mecinas,…A. C. E. Linthorst. 2015. "Glucocorticoids, epigenetic control and stress resilience." *Neurobiology of Stress 1*: 44–59. doi:org/10.1016/j.ynstr.2014.10.0012352-2895

Reul, J. M., & E. R. De Kloet. 1985. "Two receptor systems for corticosterone in rat brain: microdistribution and differential occupation." *Endocrinology 117*: 2505–11.

Richardson, R. S., H. Wagner, S. R. Mudaliar, E. Saucedo, R. Henry, & P. D. Wagner. 2000. "Exercise adaptation attenuates VEGF gene expression in human skeletal muscle." *Am J Physiol-Heart C 279*: H772–8.

Roth, R. H., S-Y. Tam, Y. Ida, J-X. Yang, & A. Y. Deutch. 1988. "Stress and the mesocorticolimbic dopamine systems." *Ann N Y Acad Sci 537*: 138–47.

Sartorius, A., R. Hellweg, J. Litzke, M. Vogt, C. Dormann, B. Vollmayr,…P. Gass. 2009. "Correlations and discrepancies between serum and brain levels of neurotrophins after electroconvulsive treatment in rats." *Pharmacopsychiatry 42*: 270–6.

Schildkraut, J. J. 1965. "The catecholamine hypothesis of affective disorders: A review of supporting evidence." *Am J Psychiatry 122*: 509–22.

Schwarz, A. J., J. A. Brasel, R. L. Hintz, S. Mohan, & D. M. Cooper. 1996. "Acute effect of brief low- and high-intensity exercise on circulating insulin-like growth factor (IGF) I, II, and IGF-binding protein-3 and its proteolysis in young healthy men." *J Clin Endocrinol Metab 81*: 3492–7.

Seifert, T., P. Brassard, M. Wissenberg, P. Rasmussen, P. Nordby, B. Stallknecht,…N. H. Secher. 2010. "Endurance training enhances BDNF release from the human brain." *Am J Physiol Regul-I 298*: R372–7.

Seredenin, S. B., T. A. Voronina, T. A., Gudasheva, T. L. Garibova, G. M. Molodavkin, S. A. Litvinova,…V. I. Poseva. 2013. "Antidepressant effect of dimeric dipeptide

GSB-106, an original low molecular weight mimetic of BDNF." *Acta Naturae 8*: 105–9.

Siuciak, J. A., D. R. Lewis, S. J. Wiegand, & R. Lindsay. 1997. "Antidepressant-like effect of brain-derived neurotrophic factor (BDNF)." *Pharmacol Biochem Behav 56*: 131–7.

Sonntag, W. E., C. Lynch, P. Thornton, A. Khan, S. Bennett, & R. Ingram. 2000. "The effects of growth hormone and IGF-1 deficiency on cerebrovascular and brain ageing." *J Anat 197*: 575–85.

Ströhle, A., M. Stoy, B. Graetz, M. Scheel, A. Wittmann, J. Gallinat,...R. Hellweg. 2010. "Acute exercise ameliorates reduced brain-derived neurotrophic factor in patients with panic disorder." *Psychoneuroendocrinology 35*: 384–8.

Szuhany, K. L., M. Bugatti, & M. W. Otto. 2015. "A meta-analytic review of the effects of exercise on brain-derived neurotrophic factor." *J Psychiatr Res 60*: 56–64.

Tang, S. W., E. Chu, T. Hui, D. Helmeste, & C. Law. 2008. "Influence of exercise on serum brain-derived neurotrophic factor concentrations in healthy human subjects." *Neurosci Lett 431*: 62–5.

Tang, K., F. C. Xia, P. D. Wagner, & E. C. Breen. 2010. "Exercise-induced VEGF transcriptional activation in brain, lung and skeletal muscle." *Resp Physiol Neurobiol 170*: 16–22.

Timofeeva, E., Q. Huang, & D. Richard. 2003. "Effects of treadmill running on brain activation and the corticotropin-releasing hormone system." *Neuroendocrinology 77*: 388–405.

Tümer, N., H. A. Demirel, L. Serova, E. L. Sabban, C. S. Broxson, & S. K. Powers. 2001. "Gene expression of catecholamine biosynthetic enzymes following exercise: Modulation by age." *Neuroscience 103*: 703–11.

Ueyama, T., Y. Kawai, K. Nemoto, M. Sekimoto, S. Toné, & E. Senba. 1997. "Immobilization stress reduced the expression of neurotrophins and their receptors in the rat brain." *Neurosci Res 28*: 103–10.

Urhausen, A., B. Weiler, B. Coen, & W. Kindermann. 1994. "Plasma catecholamines during endurance exercise of different intensities as related to the individual anaerobic threshold." *Eur J Appl Physiol 69*: 16–20.

Vale, W., & C. Rivier. 1977. "Substances modulating the secretion of ACTH by cultural anterior pituitary cells." *Fed Proc 36*: 2094–9.

van Praag, H. 2008. "Neurogenesis and exercise: past and future directions." *Neuromol Med.* doi:10.1007/s12017-008-8028-z

van Veen, V., J. D. Cohen, M. M. Botvinick, V. A. Stenger, & C. S. Carter. 2001. "Anterior cingulate cortex, conflict monitoring, and levels of processing." *NeuroImage 14*: 1302–8.

van Veen, V., M. K. Krug, & C. S. Carter. 2008. "The neural and computational basis of controlled speed-accuracy tradeoff during task performance." *J Cognitive Neurosci 20*: 1952–65.

Vaynman, S., Z. Ying, & F. Gomez-Pinilla. 2004. "Hippocampal BDNF mediates the efficacy of exercise on synaptic plasticity and cognition." *Eur J Neurosci 20*: 2580–90.

Velásquez-Martinez, M. C., R. Vázquez-Torres, & C. A. Jiménez-Riveira. 2012. "Activation of alpha1-adrenoreceptors enhances glutamate release onto ventral tegmental area dopamine cells." *Neuroscience 216*: 18–30.

Voss, M. W, K. I. Erickson, R. S. Prakash, L. Chaddock, J. S. Kim, H. Alves,...A. F. Kramer. 2013. "Neurobiological markers of exercise-related brain plasticity in older adults." *Brain Behav Immun 28*: 90–9.

Vreeburg, S. A., W. J. Hoogendijk, & R. H. DeRijk. 2013. "Salivary cortisol levels and the 2-year course of depressive and anxiety disorders." *Psychoneuroendocrinology 38*: 1494–502.

Wang, G-J., N. D. Volkow, J. S. Fowler, D. Franceschi, J. Logan, N. R. Pappas,...N. Netusil. 2000. "PET studies of the effects of aerobic exercise on human striatal dopamine release." *J Nucl Med 41*: 1352–6.

Weinstein, A. A., P. A. Deuster, J. L. Francis, R. W. Bonsall, R. P. Tracy, & W. Kop. 2010. "Neurohormonal and inflammatory hyper-responsiveness to acute mental stress in depression." *Biol Psychol 84*: 228–34.

Winter, B., C. Breitenstein, F. C. Mooren, K. Voelker, M. Fobker, A. Lechtermann,...S. Knecht. 2007. "High impact running improves learning." *Neurobiol Learn Mem 87*: 597–609.

Wipfli, B., D. Landers, C. Nagoshi, & S. Ringenbach. 2011. "An examination of serotonin and psychological variables in the relationship between exercise and mental health." *Scand J Med Sci Sports 21*: 474–81.

Wisse, B. E., & M. W. Schwartz. 2003. "The skinny on neurotrophins." *Nat Neurosci 6*: 655–6.

Wittert, G. A., J. H. Livesey, E. A. Espiner, & R. A. Donald. 1996. "Adaptation of the hypothalamo-pituitary adrenal axis to chronic exercise in humans." *Med Sci Sports Exerc 28*: 1015–9.

Zigmond, R. E., M. A. Schwarzschild, & A. R. Rittenhouse. 1989. "Acute regulation of tyrosine hydroxylase by nerve activity and by neurotransmitters via phosphorylation." *Annu Rev Neurosci 12*: 415–61.

Zoladz, J. A., A. Pilc, J. Majerczxak, M. Grandys, J. Zapart-Bukowska, & K. Duda. 2008. "Endurance training increases plasma brain-derived neurotrophic factor concentration in young healthy men." *J Physiol Pharmacol 59* Suppl. 7: 119–32.

3 Causality in the Associations Between Exercise, Personality, and Mental Health

Marleen De Moor and Eco de Geus

Introduction

Poor mental health is a major contributor to the global disease burden (Whiteford, Ferrari, Degenhardt, Feigin, a& Vos 2015), associated with huge losses of quality of life (Ustun, Ayuso-Mateos, Chatterji, Mathers, & Murray 2004; Pirkola et al. 2009), increased mortality (Cuijpers & Smit 2002), high service use, and enormous economic costs (Cuijpers et al. 2007; Cuijpers et al. 2010; Wittchen et al. 2011). Inspired by a robust association of regular exercise behavior in both adolescents and adults with multiple indicators of mental health, a strong belief has developed in preventive medicine that exercise interventions can be used to increase psychological well-being and attenuate or even prevent the development of anxiety and depressive disorders (see Chapters 1, 2, and 4). This belief converges with folk wisdom propagated by the popular press that exercise "makes you feel better" and can help "combat stress."

Notwithstanding the widespread endorsement of exercise as a panacea for threats to mental health, the majority of the population is effectively not engaging in regular exercise activities at, or even near the recommended levels (Martinez-Gonzalez et al. 2001; Hallal et al. 2012; Troiano et al. 2008; Heath et al. 2012). How can this be? Why do not all or most people engage regularly in activities that so clearly improve their well-being? These questions raise concern about the chain-of-causality underlying the association between exercise and mental health. Apart from a true causal effect of exercise at least two other explanations can account for the observed association between regular exercise behavior and mental health. First, we cannot rule out reverse causality such that a minimal level of well-being itself is a necessary condition to engage in regular exercise. Emotionally well-adjusted, outgoing, self-regulating, and self-confident individuals with low levels of stress could be simply more attracted to sports and exercise, and only such persons may have the necessary energy and self-discipline to maintain an exercise regime. More importantly, the exercise–mental health association can be due to underlying factors that independently influence both exercise behavior and mental health. Personality structure may be such an underlying factor. Scoring higher (or lower) on a particular personality dimension, for example scoring high on neuroticism, might have a detrimental effect on regular exercise behavior while simultaneously increasing the risk for low mental health. If such independent effects of personality on two different behavioral outcomes exist, this leads to an observed association between the two without a true causal effect being present.

A rapidly increasing body of evidence substantiates the role of personality in regular exercise behavior. Regular exercisers, defined as individuals who voluntarily seek out moderate to vigorous physical activity on a weekly or even daily basis, score lower on neuroticism and higher on extraversion, conscientiousness, and sensation seeking (Hoyt, Rhodes, Hausenblas, & Giacobbi 2009; De Moor, Beem, Stubbe, Boomsma, & de Geus 2006; Rhodes & Smith 2006; Wilson, Das, Evans, & Dishman 2015; Wilson & Dishman 2015; Wilkinson et al. 2013). The link between personality and exercise behavior makes good neurobiological sense in at least two major theories of personality (Gray & McNaughton 1983; Gray 1970; Gray 1985; Gray & McNaughton 2000; Eysenck 1967; Eysenck 1990; Eysenck & Eysenck 1985). The central personality traits in Eysenck's personality theory are neuroticism and extraversion. Neurotic individuals are hypothesized to have higher activation levels in their limbic brain structures, leading them to be prone to show emotional instability and more likely to report feelings of anxiety, guilt, and tension. Stables, on the other hand, have less activation and higher thresholds for activation of the limbic system, leading them to be emotionally stable even under moderate to high levels of stress. The extraversion-introversion dimension is thought to be regulated by the brain systems involved in regulating arousal. An individual's comfort level at any given time will depend on the interaction between their basal cortical arousal and the type of situation they are in; being under or over-aroused are both less desirable than a moderate level of arousal. Introverts have a higher base level of activation in the ascending reticular activating system and a corresponding higher cortical arousal level than extraverts. Introverts are thus prone to be overstimulated by sensory stimuli and consequently tend to withdraw from social situations, are less willing to take risks, and will be less drawn to exercise. In contrast, due to their lower base level of cortical arousal, extraverts tend to be more lively and sensation and experience seeking and will be attracted to social situations, including sports activities (Eysenck, Nias, & Cox 1982).

Extraversion and neuroticism are strongly related to the concepts of behavioral approach system (BAS) and behavioral inhibition system (BIS) from Gray's personality theory (Gray, Liebowitz, & Gelder 1987). The behavioral inhibition system (BIS) is activated by novelty and stimuli associated with punishment, that is, aversive stimuli or omission of reward. The behavioral approach system (BAS) is activated by appetitive stimuli associated with reinforcement, that is, reward or termination of punishment. This has immediate relevance for exercise behavior. To maintain regular exercise participation, the net appetitive effects of exercise activities during and shortly after exertion need to outweigh the net aversive effects (de Geus & De Moor 2008; de Geus & De Moor 2011). Individuals for whom the aversive effects of exercise are stronger than the rewarding effects will eventually cease their exercise participation. In contrast, individuals for whom the appetitive effects are stronger than the aversive effects will maintain exercise participation and become regular exercisers. The balance of appetitive and aversive effects can be influenced by many factors at different levels, e.g. activity-drive related to hypothalamic energy balance systems (Rowland 1998), social cognitions about exercise (e.g. attitudes, social support and self-perceived and peer-perceived exercise competence) and actual exercise ability (e.g. strength, endurance, motor skills). However, stable individual differences in sensitivity to stimuli associated with punishment or reinforcement are likely to play a key role (Schneider & Graham 2009; Wichers et al. 2015). As stated above, these differences and the coupled behavioral tendencies of avoidance and approach could be directly linked to personality. Specifically, individual differences in the functioning of the reward system in response to appetitive aspects of exercise are implicated in the personality traits of extraversion and novelty seeking/impulsivity (Gray & McNaughton 1983; Gray & McNaughton 2000). Individual differences in activity of the punishment system in response to aversive aspects of exercise are implicated in the personality traits of neuroticism and harm avoidance (Gray & McNaughton 1983; Gray & McNaughton 2000).

A vast literature exists showing that personality traits, in particular neuroticism/harm avoidance and extraversion, also play a key role in mental health. Extraversion is associated with higher levels of psychological well-being (Weiss, Bates, & Luciano 2008) and neuroticism is regarded to be a major intermediate risk factor for the development of anxiety and depressive disorders (Cloninger, Svrakic, & Przybeck 2006; Hettema et al. 2006; Kendler, Neale, Kessler, Heath, & Eaves 1993; Middeldorp, Cath, Beem, Willemsen, & Boomsma 2008). In short, there is abundant evidence that personality is associated with exercise behavior as well as mental health, and the directions of effects are mostly such that personality traits that are associated with increased exercise behavior are also associated with increased mental health (e.g. extraversion, conscientiousness), whereas personality traits that are associated with decreased exercise behavior are also associated with decreased mental health (e.g. neuroticism).

Most theories of personality assume it to be a stable disposition across the life span and this assumption is supported by substantial evidence (Gillespie, Evans, Wright, & Martin 2004; Johnson, McGue, & Krueger 2005; Loehlin & Martin 2001; Pedersen & Reynolds 1998). Whereas the environment is in constant flux and causes changes in an individual's behavior over time, personality is considered the constant factor that causes stability in behavior. This constancy is likely to be hard-wired in our biology, with effects on brain structure and function the obvious focus for such hard-wiring. *Heritability* estimates for a wide range of personality traits range between 40–50%, independent of the instruments used, and besides additive genetic variance, non-additive genetic variance is also at play for some traits (Vukasovic & Bratko 2015). Interestingly, voluntary exercise behavior also has been shown to be highly heritable – particularly in late adolescence where 80% of the individual differences could be explained by genetic factors (Stubbe, De Moor, Boomsma, & de Geus 2006; De Moor & de Geus 2012; Van der Aa, de Geus, van Beijsterveldt, Boomsma, & Bartels 2010; de Geus, Bartels, Kaprio, Lightfoot, & Thomis 2014). Heritability decreases in adulthood but remains well above 40% even in middle-aged samples (De Moor et al. 2011; Vink et al. 2011). Part of the heritability of exercise behavior and personality may thus reflect common genetic factors.

Below we will discuss existing and novel methodologies to test causal hypotheses in genetically informative samples and illustrate its application to the associations between exercise, personality, and mental health. Before we move to these methods we should briefly indicate how it differs from, and can improve on, the "classical" approaches to causality: the randomized controlled trial (RCT) and prospective epidemiology.

Causality in the Association Between Exercise and Mental Health

Randomized Controlled Exercise Training Studies

In keeping with the tenets of evidence-based medicine, the hypothesis that exercise can improve mental health has been tested by experimentally manipulating exercise levels in participants. Such so-called training studies must be carefully designed. To make sure that changes in mental health do not simply reflect the passage of time, one or more control conditions are needed. To prevent self-selection biases, the individuals cannot be allowed to choose between the conditions, but must be randomly assigned to either exercise training or the control condition. Several well-designed training studies have reported improved mood, increased coping behavior or reductions in depression and anxiety after a program of aerobic exercise training in comparison to control manipulations (Barbour & Blumenthal 2002; Brosse, Sheets, Lett, & Blumenthal 2002; Steptoe, Edwards, Moses, & Mathews 1989). Unfortunately, others have failed to replicate these training effects (de Geus, van Doornen, & Orlebeke 1993; King, Taylor, Haskell, & Debusk 1989). Without denying the potential of beneficial effects of exercise in subsets of individuals, many reviews on this topic express only cautious optimism about the use of exercise for the enhancement of mental health in the population at large (Lawlor & Hopker 2001; Dunn, Trivedi, Kampert, Clark, & Chambliss 2005; Daley 2008; Cooney et al. 2013).

The most promising evidence came from studies in individuals who had low initial levels of well-being at the start of the exercise program, like clinically depressed or clinically anxious patients (Rethorst, Wipfli, & Landers 2009; Wipfli, Rethorst, & Landers 2008). Various randomized controlled trials showed that regular exercise can be used as a treatment to relieve symptoms in patients diagnosed with a depressive disorder (Babyak et al. 2000; Blumenthal et al. 2007). These studies tend to report beneficial psychological effects of exercise that match or even exceed those of pharmacological treatment. Although these results were promising, it should be kept in mind that these studies examined the effects of prescribed and externally monitored exercise treatments in selected subgroups of patients. Self-selection bias is a concern even with randomization (Ekkekakis 2008). Those that are mildly affected or on the verge of spontaneous remission may be more likely to enter the study than those with more severe symptoms, or a more recent onset of the current episode. This may lead to more positive post-training outcomes than would have been achieved in a truly unselected sample. Likewise, the widespread beliefs about the efficacy of this intervention will bias most RCT samples to consist of patient-volunteers that firmly believe exercise to be a credible treatment modality for their depression or anxiety, creating potential for distortion of the self-reported well-being after the trial. Generalization from clinical trials in anxious or depressive patients willing to participate in training to *all* anxious or depressive patients is therefore hazardous. This caution was further reinforced by recent RCTs that failed to find robust antidepressant effects of exercise (Chalder et al. 2012; Krogh, Saltin, Gluud, & Nordentoft 2009; Krogh, Videbech, Thomsen, Gluud, & Nordentoft 2012).

The above leads us to suggest that methodological difficulties in training studies in clinical samples with positive findings may have led us to *overestimate* the causal effects of exercise on mental health. However, there is also good ground for the opposite concern that training studies may actually act to *underestimate* the beneficial effects of exercise, particularly in the population at large. First of all, training studies tend to be relatively short, with a duration of the exercise activities of weeks or months at best. This will not capture effects of regular exercise that require more prolonged exposure and habit

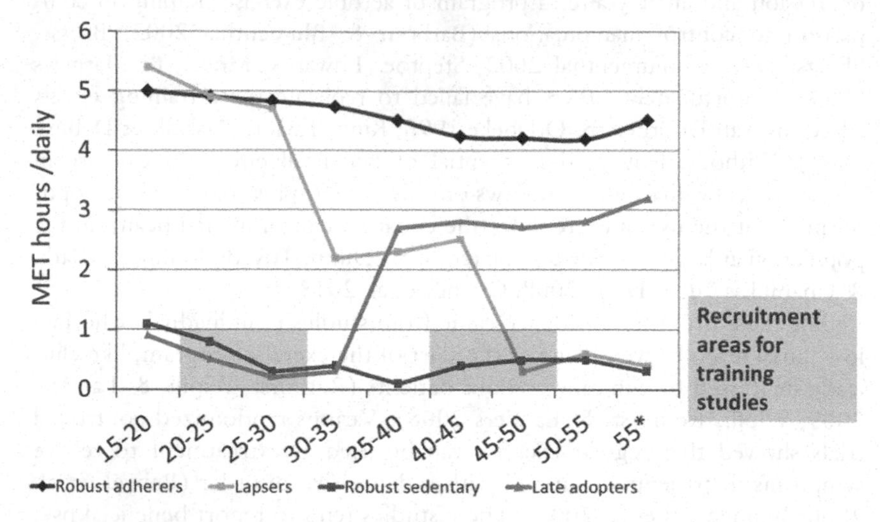

Figure 3.1 Four groups of (non)-exercisers in the population

formation. Second, whereas training studies in patient samples may suffer from a self-selection bias, training studies in population-based samples will suffer from another form of selection bias related to inclusion or exclusion criteria set by researchers.

Let us assume for a moment that the population consist of the four types of people depicted in Figure 3.1. A first trajectory describes individuals that persistently adhere to regular exercise for the largest part of their lives ("robust exercisers"). A second trajectory describes "robust sedentary" individuals that persistently abstain from voluntary exercise throughout their lifespan. The third trajectory describes individuals who never much liked exercise in their childhood, adolescence and early adulthood, perhaps turned off by its competitive nature, but begin to enjoy it as a meaningful leisure time activity at a later point in their lives ("late adopters"). The fourth trajectory of "lapsers" were fervent exercisers in their youth but gradually lost interest when reaching young adulthood, possibly reflecting a choice to focus more time on family and work-related chores. The four trajectories that have been depicted in Figure 3.1 are based on longitudinal analyses of the tracking of exercise behavior across the life span (Picavet, Wendel-vos, Vreeken, Schuit, & Verschuren 2011; Barnett, Gauvin, Craig, & Katzmarzyk 2008; Stephens, Craig, & Ferris 1986; Pinto Pereira, Li, & Power 2015). The key point is that training studies are not based on a random selection of the population at large. They tend to select individuals that are currently sedentary to maximize the effect size of the exercise manipulation. Depending on the age of the participants, this would include individuals from the robust sedentary and lapser trajectories or individuals that come from the robust sedentary and late adopter groups. Robust adherers will be almost always left out.

The choice to leave out robust exercisers is a rational choice from an experimental viewpoint. It does not seem very useful to additionally train a person who already exercises 2 to 3 hours weekly. For many variables, i.e. like aerobic fitness or cardiovascular risk factors, the benefits tend to level off and more frequent or longer exercise does not add much. By selecting sedentary individuals, a biased selection of the four groups depicted in Figure 3.1 will be made. That is unfortunate, because robust adherers or late adopters may constitute exactly the part of the population that reap psychological benefits from exercise – which in turn explains why they are currently exercising. The robust sedentary group in contrast may be enriched with individuals who did not respond well to exercise (which explains why they are sedentary). This selection bias can easily lead us to underestimate the true causal effect of exercise in the full population, when we only consider results from training studies.

An overall conclusion is that, in spite of best intentions, training studies do not truly randomize an unselected sample to an exercise intervention versus a control intervention. Yet, this is what an unbiased test of causality would require. We direly need alternative approaches to test causality that can be used in population-based samples.

Longitudinal Studies in Population-Based Samples

To help resolve causality, epidemiological studies typically address the exercise–mental health association in longitudinal designs. Most of these longitudinal studies reported that regular exercise at baseline was associated with less depression and anxiety at follow-up (Brown, Ford, Burton, Marshall, & Dobson 2005; McKercher et al. 2009; Patten, Williams, Lavorato, & Eliasziw 2009; Motl, Birnbaum, Kubik, & Dishman 2004; Sagatun, Sogaard, Bjertness, Selmer, & Heyerdahl 2007; Jerstad, Boutelle, Ness, & Stice 2010; Wise, Adams-Campbell, Palmer, & Rosenberg 2006; Camacho, Roberts, Lazarus, Kaplan, & Cohen 1991; Strawbridge, Deleger, Roberts, & Kaplan 2002; van Gool et al. 2003; Stephens 1988; Jacka et al. 2011; Dugan, Bromberger, Segawa, Avery, & Sternfeld 2015). Some studies, however, did not find evidence for a longitudinal association (Cooper-Patrick, Ford, Mead, Chang, & Klag 1997; Birkeland, Torsheim, & Wold 2009; Strohle et al. 2007; Kritz-Silverstein, Barrett-Connor, & Corbeau 2001; Weyerer 1992), found the association in subgroups only (Farmer et al. 1988; Edman, Lynch, & Yates 2014), or found the association to be limited to symptom counts but not extending to clinical diagnosed depression (Stavrakakis, de Jonge, Ormel, & Oldehinkel 2012; Stavrakakis et al. 2013). The strong point of these longitudinal studies is that they rule out reverse causality as the *only* explanation for the prospective association; if an individual that started exercising and shows increased well-being later on, that increase in well-being clearly cannot have influenced the past decision to start exercising.

We need to stress, however, that an association between exercise at baseline and mental health at follow-up does not mean that reverse causality cannot be present as well. Mental health at baseline may in parallel be influencing exercise at follow-up. Various studies have confirmed the existence of such bidirectional prospective associations between exercise behavior and mental health (Azevedo Da et al. 2012; Jerstad et al. 2010; Stavrakakis et al. 2012; Lindwall, Larsman, & Hagger 2011; Pinto Pereira, Geoffroy, & Power 2014; Ku, Fox, & Chen 2012). In the field of depression, these studies conclude that these bidirectional associations provide simultaneous support for the "protection hypothesis" that regular exercise can decrease depressive symptoms through its biological and social actions and for the "inhibition hypothesis" that the lack of energy, anhedonia, and social withdrawal seen in depression is negatively influencing exercise behavior (Stavrakakis et al. 2012; Pinto Pereira et al. 2014). This conclusion is appealing but premature. The cited observational longitudinal studies do not truly resolve causality. The adage "correlation is not causation" applies in full to *longitudinal* correlations. They too cannot rule out the effects of underlying factors on both exercise and well-being, which persist across time. Lower psychological well-being at follow-up in individuals who were non-exercisers at baseline may, for instance, reflect a reluctance of individuals high in the stable personality trait neuroticism to take up exercise behavior in the first place.

To test causality in population-based samples, a different approach is needed than simply relying on prospective associations. Below we show that studying these associations in genetically informative samples provides a solid step forwards.

Causality in Genetically Informative Designs

Genetically informative designs are designs that include pairs of relatives for which the extent to which they are genetically related is known *and* varies for different types of relatives. The most well-known genetically informative study is the twin study, but other studies are also available and include adoption studies or extensions of twin or adoption studies with other types of relatives, such as parents, children, or siblings of twin pairs or adopted siblings. In this chapter, we will focus on twin studies. Twin studies are traditionally being employed to infer whether the familial resemblance observed for a specific behavior, disease, or trait ("the phenotype") represents environmental influences that are shared within a family ("nurture") or genetic influences ("nature"). This is the so-called classical twin study, and it is univariate (only one phenotype is studied at a time). The univariate twin study can also be extended to a bivariate study, enabling the investigation of why two phenotypes are associated. In this section, we will first discuss the general principle of the univariate twin study, and then discuss in detail the different approaches that can be taken in bivariate twin studies to study causality.

In univariate twin studies, “nature” and “nurture” can be separated by comparing the phenotypic resemblance between monozygotic (MZ) and dizygotic (DZ) twins. When twins are reared together in the same home they share part of their environment and this sharing of the family environment is the same for MZ and DZ twins. The important difference between the two types of twins is that MZ twins share (close to) all of their genotypes, whereas DZ twins share on average only half of the genotypes segregating in that family (Falconer & Mackay 1996). This distinction is the basis of the classical twin study. If the phenotypic resemblance within MZ pairs is larger than in DZ pairs, this suggests that ***additive genetic factors*** (A, Figure 3.2) influence the phenotype. If the phenotypic resemblance is as large in DZ twins as it is in MZ twins, this indicates ***common/shared environmental factors*** (C) as the cause of the familial resemblance (Van Dongen, Slagboom, Draisma, Martin, & Boomsma 2012). The extent to which MZ and DZ twins do not resemble each other is ascribed to the ***unique (or non-shared) environmental factors*** (E). These include all unique experiences like differential jobs, accidents or other life events, and in childhood, differential treatment by the parents, and non-shared friends and peers. Shared environmental factors are defined as those environmental factors that are shared within a family, such as parenting style, family functioning, neighbourhood, or socio-economic status. Additive genetic factors represent the sum of all linear effects of the genetic loci that influence the phenotype. The heritability of a phenotype is defined as the relative proportion of the total variance explained by genetic factors. The univariate twin model is sometimes also called the *ACE model*, referring to the three different types of factors that are typically modelled as affecting the phenotype

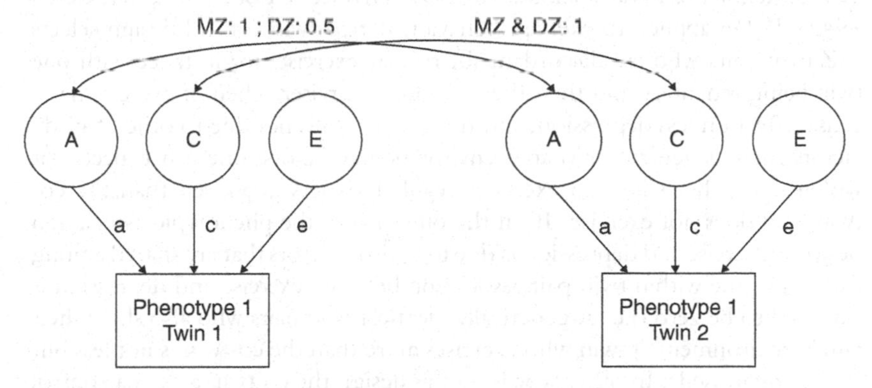

Figure 3.2 The univariate twin model (ACE model)

MZ=Monozygotic twin pairs; DZ=Dizygotic twin pairs; A=Additive genetic factors; C=Shared environmental factors; E=Unique environmental factors; a=effect of additive genetic factors on the phenotype; c=effect of shared environmental factors on the phenotype; e=effect of unique environmental factors on the phenotype; heritability=$a^2/(a^2+c^2+e^2)$.

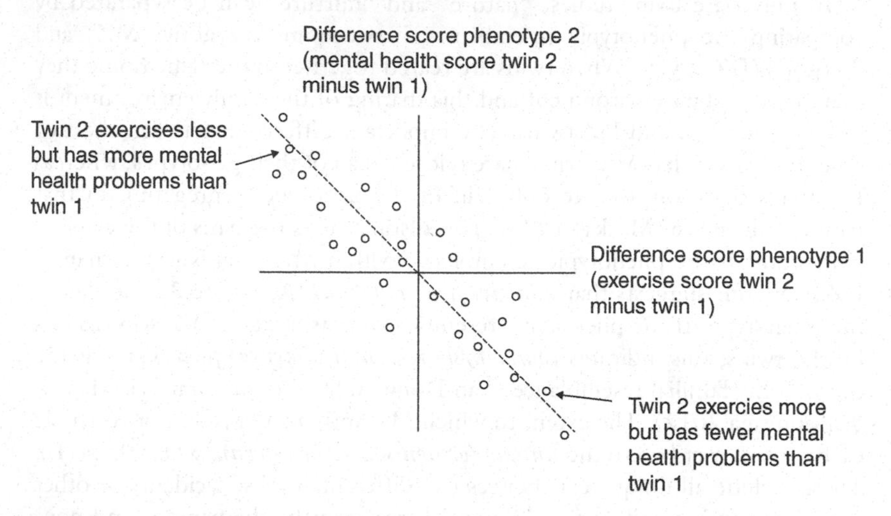

Figure 3.3 Illustration of the MZ intra-pair difference approach to test causality in the association

of interest (Neale & Cardon 1992). The univariate ACE model is depicted in Figure 3.2.

Bivariate twin studies have been employed to study different explanations for why two phenotypes are correlated. Different approaches can be taken (some are illustrated in Figures 3.3 and 3.4). The oldest, most widely used, and perhaps also most simple and intuitive approach is the *co-twin control design* (Kendler & KarkowskiShuman 1997; Hrubec & Robinette 1984; Gesell 1942). When applied to exercise behavior, the co-twin control design selects MZ twin pairs who are discordant for regular exercise, for instance with one twin being sedentary and the other a regular exerciser. Then, if exercise has a causal effect on less depression, and the association is not due to other "third" factors such as genetic or shared environmental factors, one can expect that within pairs, the twin who exercises regularly is less depressed than the co-twin who does not exercise. If on the other hand, the phenotypic association between exercise and depression is due to "third" factors that are shared among MZ twins, the within twin pair association between exercise and depression is expected to be zero (i.e. in genetically identical twin pairs who also share their family environment, a twin who exercises more than the co-twin is not less but equally depressed). In other words, in this design the co-twin acts as a control of the other twin and the design is highly suitable to *falsify the causal hypothesis.* If a within MZ twin pair association is found, this gives credence to the causal hypothesis, but does not formally prove it, because "third" factors that are not shared between twins but influence both phenotypes could also explain why the twin who exercises more is less depressed compared to the co-twin.

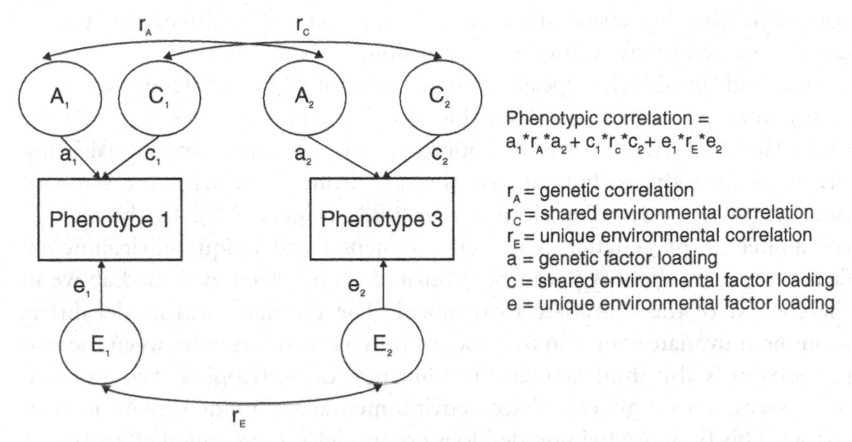
r_A
r_C
A_1
C_1
A_2
C_2
a_1
c_1
a_2
c_2
Phenotype 1
Phenotype 3
e_1
e_2
E_1
E_2
r_E
Phenotypic correlation =
$a_1{*}r_A{*}a_2 + c_1{*}r_c{*}c_2 + e_1{*}r_E{*}e_2$
r_A = genetic correlation
r_C = shared environmental correlation
r_E = unique environmental correlation
a = genetic factor loading
c = shared environmental factor loading
e = unique environmental factor loading
A casual effect of phenotype 1 on phenotype 2 predicts a significant overlap (i.e., correlations) in all factors (here A.c and E) influencing phenotype 1 and phenitype 2 in the bivariale genetic model.
The model can be applied for cross-sectional date (both phenotypes measured at the same time point), or for longitudinal data (phenotype 1 is asscssed at baseline, and phenotype 2 at follow-up).

Figure 3.4 Illustration of the bivariate approach to test causality in the association between two phenotypes in twin studies

Note: For simplification the model is drawn for a single twin of the pair only.

The co-twin control design can also be applied to two phenotypes that are both continuous, in which case no discordant twins are selected, but rather intra-pair differences for both phenotypes are computed for all twin pairs with available data. Causality should induce a correlation between these intra-pair differences, which can be tested for significance, a type of co-twin control design that we call the MZ intra-pair differences design (De Moor, Boomsma, Stubbe, Willemsen, & de Geus 2008) (Figure 3.3). The co-twin control design is commonly applied to cross-sectional data, but has also been used for longitudinal data (Bartels, De Moor, van der Aa, Boomsma, & de Geus 2012). A longitudinal correlation between two phenotypes is compared with a cross-sectional association, as we discussed above, often taken as stronger evidence that an association between two phenotypes is causal. The reason is that underlying factors that fluctuate over time cannot account for the longitudinal association. However, a longitudinal association can still be confounded by "third" factors that influence both phenotypes *and* are stable over time. The longitudinal MZ intra-pair differences design partly overcomes this problem by controlling for shared environmental and genetic factors that could impact both phenotypes at each of the two time points studied. Thus, if for example intra-pair differences in the change in the amount of exercise across time points are not correlated with intra-pair differences in the change in the amount of depression across time points, in spite of the fact that in the full set of twins exercise at the first time point is correlated with depression at the second time point, this falsifies the hypothesis that more exercise causes less depression.

Although intuitively appealing, the discordant MZ twin design uses only a very small part of the data available in twin-family registries (i.e. only the discordant MZ twins). Another approach that similarly aims at falsifying the causal hypothesis, but can use the data from all twins, is the standard bivariate twin model (De Moor et al. 2008) (Figure 3.4). In this model, the influences of genetic, shared environmental and unique environmental factors on each phenotype can be estimated, as has been explained above in the context of the univariate twin model. The bivariate twin model differs from the univariate model in that the phenotypic association between the two phenotypes is also modelled, and is a function of overlapping ("common to both phenotypes") genetic, shared environmental and unique environmental factors. This bivariate twin model does not model a direct causal effect of one phenotype on the other, but is still informative on whether the phenotypic association could be causal. The line of reasoning is as follows. If the observed phenotypic association in reality reflects a causal effect, and the bivariate twin model is fitted to the data, all factors that influence the causative phenotype (for example A and E that influence exercise in adulthood), will through the causal chain carry over to the effect phenotype (for example depression) and will lead to the finding of both genetic and unique environmental factors that are common to both phenotypes. In other words, and provided that there is enough statistical power to detect the effects, significant genetic and environmental correlations will be found (r_e and r_g in Figure 3.4). Again, this finding will only be consistent with the causal hypothesis but not prove it. If either the genetic or environmental correlation is not significant, the causal hypothesis is falsified (e.g. it is unlikely that the observed phenotypic association reflects a causal effect).

Exercise and Anxiety/Depression

The different approaches as described in the previous section have been applied to thoroughly test for causality in the association between exercise behavior and symptoms of anxiety and depression in a large sample of 5952 Dutch twins, their siblings and their parents aged 18–50 years old (De Moor et al. 2008). In keeping with the many previous cross-sectional studies demonstrating that increased exercise behavior is associated with decreased anxious and depressive symptoms, the correlations between exercise behavior and symptoms of anxiety and depression were significant but small, ranging from −0.06 to −0.14. However, the MZ intra-pair differences method showed that intra-pair differences in exercise behavior were not significantly correlated to intra-pair differences in symptoms of anxiety and depression. The bivariate twin models further showed that genetic correlations were significant and ranged from –0.16 to –0.24. Unique environmental correlations, ranging from –0.07 to 0.05, were not significantly different from zero. This pattern of findings strongly suggests that the association between exercise behavior and symptoms of anxiety and depression is not simply explained by a causal effect.

The longitudinal correlations between exercise behavior at baseline and symptoms of anxiety and depression at follow-up ranged between –0.07 to –0.14 and were all significant. As estimated in bivariate twin models, with exercise at baseline and symptoms of anxiety or depression at follow-up, the genetic correlations ranged from –0.21 to –0.40 and were all significantly different from zero. This was not the case for the unique environmental correlations, ranging from –0.12 to 0.22, which were not significantly different from zero. Taken together, the results from this study point to common (i.e. overlapping) genetic factors that explain the cross-sectional and longitudinal correlations between exercise and symptoms of anxiety and depression, rather than a causal effect of exercise on anxiety/depression being the driving mechanism behind these correlations.

Similar findings were obtained in a study on the association between regular exercise behavior and internalizing problem behavior (a composite measure of anxious and depressed symptoms, withdrawn behavior and somatic complaints) in a sample of 6317 Dutch adolescent twins and 1180 of their non-twin siblings (Bartels et al. 2012). The cross-sectional phenotypic correlation between exercise and internalizing problems was –0.13, and did not differ between girls and boys. The MZ intra-pair differences model was applied to cross-sectional and longitudinal data. This showed that the twin who exercises more at baseline did not have less internalizing problems at baseline compared to the co-twin who exercises less. Longitudinally, the twins who increased frequency and intensity of exercise over time did not show lower levels of internalizing problems over time than their co-twins with decreasing levels of exercise over time.

Bivariate genetic modelling again corroborated the MZ intra-pair differences design. The sources of individual differences in exercise behavior and internalizing problems were different in boys versus girls, with for boys genetic and unique environmental factors influencing both phenotypes, while for girls also shared environmental factors played a role. In boys, it was found that the association between exercise behavior and internalizing problems seems to be mainly due to genetic factors that are common (i.e. overlapping) for exercise behavior and internalizing problems. For girls, genetic and shared environmental factors both account for the association. The unique environmental correlation was not significantly different from zero, thus the causal hypothesis was not supported in either boys or girls. Similar results were found for the longitudinal phenotypic correlation between exercise and internalizing problems two years later (–0.15), which was best explained by genetic influences common to both traits.

Exercise and Well-Being

Several studies have focused on positive mental health outcomes, such as subjective well-being, rather than on poor mental health. In the study of Stubbe et al. (2006), the relationship between exercise participation (yes/no) with life satisfaction (satisfied/not satisfied) and happiness (happy/not happy) was examined in around 8000 subjects between 18 and 65 years old using the co-twin control method. It was found that adults who participate in exercise are happier than and more satisfied with their lives than adults who do not participate in exercise. However, in MZ twins discordant for exercise participation, the twins who exercised were not happier or more satisfied with their lives than the co-twins who did not exercise.

In the study conducted in adolescents discussed in the previous section (Bartels et al. 2012), the association between exercise behavior and subjective well-being was also investigated by applying the same methods as for internalizing problems. The findings were highly comparable. The cross-sectional phenotypic correlation between exercise and subjective well-being was 0.12, and the longitudinal phenotypic correlation was 0.16. These associations disappeared when controlling for genetic and shared environmental factors that MZ twins share (MZ intra-pair differences design). Furthermore, in the bivariate genetic models, the associations were best explained by genetic and/or shared environmental factors that are common to exercise and subjective well-being, but unique environmental factors are not shared among exercise and subjective well-being, again falsifying the causal hypothesis.

Taken together the studies that investigated the associations between exercise and different indices of mental health in adults and adolescence (lack of anxiety, depression and internalizing problems, and high subjective well-being) make it clear that these associations do not simply reflect a causal effect of exercise on mental health. Rather, "third" factors influenced by genetic factors are of importance in explaining these associations. We hypothesize that personality constitutes the underlying factor that itself causally affects both exercise behavior and mental health. Because the stability in personality traits is mostly caused by genetic factors, these genetic factors may explain the genetic correlation between exercise behavior and mental health. In the next section we investigate our hypothesis in a novel model applied to longitudinal twin data. We expect to find that personality traits (neuroticism and extraversion) causally influence exercise behavior.

Causality in the Association Between Personality and Exercise

Although a major step forward to more robustly testing causality among phenotypes in non-experimental designs, the above reviewed bivariate twin designs are limited in that they do not allow for the possibility that combinations of mechanisms explain the observed phenotypic associations. For example, it could be that the association between personality and exercise is partly causal, and partly explained by genetic factors that the two phenotypes have in common. Longitudinal genetically informative designs with large sample sizes are needed to test for these more complex explanations of phenotypic associations. We developed a longitudinal bivariate twin model to examine causal effects among two phenotypes in both directions, in addition to modelling genetic and environmental factors that both phenotypes can have in common. The model is described in more detail below and applied to the association between the personality traits neuroticism and extraversion on the one hand and exercise behavior on the other hand. The model was fitted to data from adolescent and adult twins who are voluntarily registered with the Netherlands Twin Register and who take part in a longitudinal study on health, lifestyle, and personality conducted in twins and their family members (Boomsma et al. 2002).

Longitudinal data on exercise, neuroticism, and extraversion from twins were obtained from surveys collected in 1991, 1993, 1995, 1997, 2000, 2002, 2004, and 2009. These surveys included a number of questions on exercise that were used for the current analyses. First it was asked whether the participant exercised regularly ("yes" or "no"). If yes, further information on type of exercise, frequency (times per week), and duration (minutes per time) was gathered. Each type of exercise was coded with a metabolic energy equivalent (MET) score. A MET score of 1 corresponds to the rate of energy expended when at rest (1 kcal/kg.h). Subsequently, the MET score was multiplied by the hours per week spent in this type of exercise to calculate the number of METhours spent weekly in this type of exercise. If participants reported more than one type of exercise, METhours were summed up to obtain the total METhours per week. The 6-month test-retest reliability of METhours was 0.82 (De Moor et al. 2008). Since the distribution of METhours was highly skewed (many twins do not regularly exercise, and a relatively small number of twins exercise very frequently and intensively), a square-root transformation was applied to the data to render the data more normally distributed. Next, sex differences in METhours were taken out by performing a linear regression of METhours on sex and saving the unstandardized residuals, which were the input for all reported analyses on METhours in this chapter.

The surveys collected in 1991, 1993, 1997, 2000, and 2002, contained the Amsterdamse Biografische Vragenlijst (ABV) (Wilde 1970), a Dutch personality questionnaire that is similar to the Eysenck Personality Questionnaire (Eysenck & Eysenck 1964). There are 21 items in the ABV that measure extraversion and 30 items that measure neuroticism; a summed score was computed for each personality trait for all participants for all time points. The Cronbach's alphas for neuroticism and extraversion in this sample have been previously reported and were respectively 0.89 and 0.84 (De Moor et al. 2006). Again, sex differences in neuroticism and extraversion were regressed out and unstandardized residuals were saved and input for all reported analyses in this chapter.

Since there was large age variation across twin pairs in each survey, the data were restructured such that age rather than survey represented each time point. Given that the survey data collections were always at least 2 years apart, and to avoid excessive missingness in the age-aligned longitudinal dataset, the following seven age bins of 2 years were created: 17–18 years; 19–20 years;

21–22 years; 23–24 years; 25–26 years; 27–28 years; and 29–30 years. This led to a final total sample size of 10,105 twins.

The statistical analyses were carried out in three steps. In the first step, descriptives were calculated: means and standard deviations for the untransformed METhours, neuroticism and extraversion scores across age, and phenotypic and twin correlations for the transformed METhours, neuroticism and extraversion scores across age. In a second step, univariate longitudinal twin models were fitted to the data on METhours, neuroticism and extraversion separately (i.e. in three different models), using the structural equation modeling software package OpenMx as implemented in R (Boker et al. 2011). The aim of this step was to model the stability and change of genetic and environmental influences on each phenotype, while not yet taking into account the bivariate associations of METhours with neuroticism or extraversion. In the third step, two bivariate longitudinal twin models were fitted to the data: the first model modelled the association between METhours and neuroticism (see Figure 3.5), while the second model modelled the association between

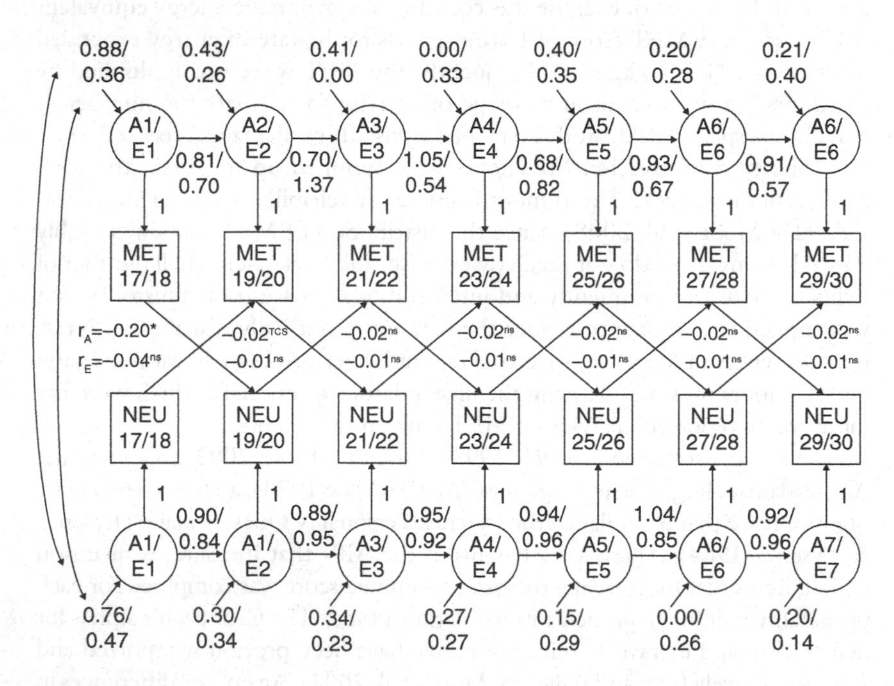

Figure 3.5 The estimated bivariate longitudinal causal model for weekly METhours and neuroticism

MET=Weekly METhours; NEU=Neuroticism; A=Additive genetic factors; E=Unique environmental factors; r_A=Additive genetic correlation; r_E=Unique environmental correlation; * $p<0.001$; ns=non significant ($p>0.05$); Note: for reasons of clarity A and E factors are drawn as one latent variable, with the parameter estimates given twice for each A and E factors. In addition, the measurement error path loadings for both variables (0.55 for METhours and 0.49 for neuroticism) at each time point are not drawn in the figure.

METhours and extraversion (see Figure 3.6). The genetic and environmental influences on METhours, neuroticism and extraversion were modelled according to the outcomes of step two. In addition, the cross-sectional and longitudinal associations between METhours and the personality trait (neuroticism or extraversion) were modelled as possibly arising from three sources. First, they could be the result of genetic factors that the two phenotypes have in common (*the hypothesis of genetic pleiotropy*). Second, the observed phenotypic associations could be explained by causal influences of the personality trait on METhours (*the dominant causal hypothesis*). Third, they could occur because of a causal effect of METhours on the personality trait (*the hypothesis of reverse causality*). Each hypothesis can be tested by constraining the relevant parameter(s) to zero and comparing the fit of this constrained model with the fit of the unconstrained model by using a likelihood-ratio test. The model comes with a number of assumptions. First, it is assumed that the phenotypic correlation between METhours and the personality trait at the first time point,

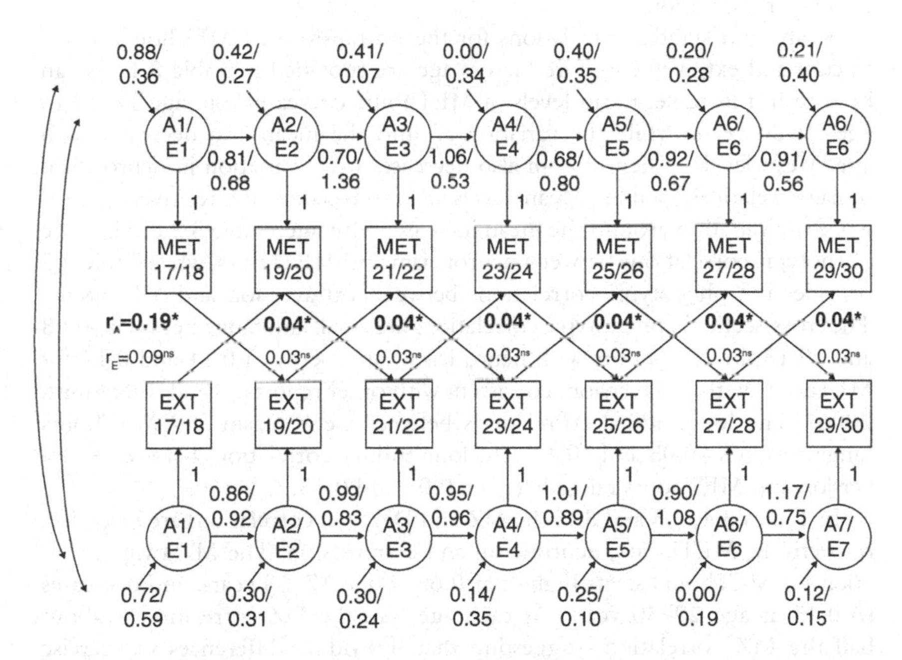

Figure 3.6 The estimated bivariate longitudinal causal model for weekly METhours and extraversion

MET=Weekly METhours; EXT=Extraversion; A=Additive genetic factors; E=Unique environmental factors; r_A=Additive genetic correlation; r_E=Unique environmental correlation; * $p<0.001$; ns=non significant ($p>0.05$); Note: for reasons of clarity A and E factors are drawn as one latent variable, with the parameter estimates given twice for each A and E factors. In addition, the measurement error path loadings for both variables (0.55 for METhours and 0.48 for extraversion) at each time point are not drawn in the figure.

Table 3.1 Means (and standard deviations) of weekly METhours, neuroticism and extraversion at each age bin

	17–18 years	*19–20 years*	*21–22 years*	*23–24 years*	*25–26 years*	*27–28 years*	*29–30 years*
METhours	16.44 (22.88)	13.97 (21.28)	13.57 (22.42)	12.85 (19.01)	12.21 (18.97)	11.23 (17.14)	10.81 (17.79)
Neuroticism	54.38 (24.15)	52.70 (24.47)	50.19 (24.14)	49.67 (24.96)	47.65 (24.88)	46.63 (24.93)	45.29 (23.95)
Extraversion	61.18 (15.78)	61.87 (15.69)	62.83 (15.89)	61.89 (16.61)	62.17 (16.27)	61.70 (16.44)	61.46 (16.66)

17–18 years, fully arises from common genetic factors. Second, it is assumed that the magnitude of the causal effects of the personality trait on METhours, and of METhours on the personality trait does not vary as a function of age. These assumptions might not completely hold but they are necessary in order to identify the model and avoid problems with convergence of the model and parameter estimation.

Means and standard deviations for the untransformed METhours, neuroticism and extraversion scores across age are provided in Table 3.1. As can be seen in this table, mean levels of METhours decrease from age 17–18 to age 29–30 years, while the variation around the mean also decreases with age. Mean levels of neuroticism also decrease, while variation in neuroticism remains relatively stable. Mean levels of extraversion are relatively stable, while the variation around the mean increases with age. Table 3.2 displays the phenotypic correlations between neuroticism and METhours, while Table 3.3 provides the phenotypic correlations between extraversion and METhours. The cross-sectional phenotypic correlations are small and range between –0.08 and –0.13 for METhours with neuroticism, and between 0.11 and 0.17 for METhours with extraversion, consistent with other reports (Rhodes a& Smith 2006). The longitudinal correlations between neuroticism and METhours range between –0.05 and –0.13. The longitudinal correlations between extraversion and METhours range between 0.05 and 0.15.

In Tables 3.4, 3.5 and 3.6, the MZ and DZ twin correlations are given for, respectively, METhours, neuroticism and extraversion. The MZ twin correlation for METhours starts highest at 0.65 at age 17–18 years, and decreases to 0.45 at age 29–30 years. At each age bin, the DZ correlation is about half the MZ correlation, suggesting that individual differences in exercise behavior during young adulthood are influenced by a combination of additive genetic factors and unique environmental factors. For neuroticism, the MZ correlations at each age bin range between 0.43 and 0.55 without a clear trend across age, while the DZ correlations are typically about half of the MZ correlations and range between 0.20 and 0.29. This suggests that individual differences in neuroticism during young adulthood are also influenced by a combination of additive genetic factors and unique environmental factors. For extraversion, the MZ correlations at each age bin are similar to those for

Table 3.2 Cross-sectional (diagonal) and longitudinal (off-diagonal) phenotypic correlations between weekly METhours and neuroticism

		METhours at:						
		17–18 years	*19–20 years*	*21–22 years*	*23–24 years*	*25–26 years*	*27–28 years*	*29–30 years*
Neuroticism at:	17–18 years	−0.11	−0.09	−0.09	−0.10	−0.05	−0.05	−0.05
	19–20 years	−0.13	−0.12	−0.11	−0.09	−0.06	−0.06	−0.06
	21–22 years	−0.05	−0.07	−0.08	−0.09	−0.08	−0.08	−0.08
	23–24 years	−0.15	−0.11	−0.07	−0.13	−0.09	−0.09	−0.09
	25–26 years	−0.10	−0.12	−0.07	−0.08	−0.12	−0.12	−0.12
	27–28 years	−0.10	−0.12	−0.07	−0.08	−0.13	−0.13	−0.13
	29–30 years	−0.10	−0.12	−0.07	−0.08	−0.11	−0.11	−0.11

Table 3.3 Cross-sectional (diagonal) and longitudinal (off-diagonal) phenotypic correlations between weekly METhours and extraversion

		METhours at:						
		17–18 years	*19–20 years*	*21–22 years*	*23–24 years*	*25–26 years*	*27–28 years*	*29–30 years*
Extraversion at:	17–18 years	0.11	0.13	0.07	0.05	0.09	0.09	0.09
	19–20 years	0.11	0.17	0.11	0.10	0.09	0.09	0.09
	21–22 years	0.07	0.13	0.14	0.11	0.14	0.14	0.14
	23–24 years	0.10	0.09	0.10	0.11	0.06	0.06	0.06
	25–26 years	0.09	0.11	0.12	0.09	0.13	0.13	0.13
	27–28 years	0.09	0.11	0.12	0.09	0.11	0.11	0.11
	29–30 years	0.09	0.11	0.12	0.09	0.15	0.15	0.15

neuroticism (with a range between 0.41 and 0.53), but the DZ correlations are typically less than half of the MZ correlations and range between 0.07 and 0.19. This suggests that besides additive genetic influences and unique environmental influences, there are also non-additive genetic influences at work that explain individual differences in extraversion. The twin correlations for these three phenotypes are consistent with what has been reported in the literature before (Vink et al. 2011; Vukasovic & Bratko 2015; Polderman et al. 2015; van den Berg et al. 2014).

Table 3.4 Longitudinal twin correlations for weekly METhours

	17–18 years	*19–20 years*	*21–22 years*	*23–24 years*	*25–26 years*	*27–28 years*	*29–30 years*
17–18 years	**0.65** / *0.30*	*0.28*	*0.23*	*0.32*	*0.09*	*0.10*	*0.10*
19–20 years	**0.52**	**0.61** / *0.32*	*0.22*	*0.26*	*0.23*	*0.22*	*0.10*
21–22 years	**0.36**	**0.41**	**0.49** / *0.21*	*0.20*	*0.12*	*0.21*	*0.22*
23–24 years	**0.45**	**0.44**	**0.52**	**0.55** / *0.27*	*0.12*	*0.24*	*0.14*
25–26 years	**0.29**	**0.48**	**0.35**	**0.40**	**0.46** / *0.15*	*0.17*	*0.27*
27–28 years	**0.30**	**0.26**	**0.45**	**0.22**	**0.45**	**0.46** / *0.25*	*0.14*
29–30 years	**0.22**	**0.35**	**0.22**	**0.30**	**0.38**	**0.37**	**0.45** / *0.09*

MZ twin correlations are given on diagonal and lower off-diagonal **in bold**, and DZ twin correlations are given on diagonal and upper off-diagonal *in italics*.

Table 3.5 Longitudinal twin correlations for neuroticism

	17–18 years	*19–20 years*	*21–22 years*	*23–24 years*	*25–26 years*	*27–28 years*	*29–30 years*
17–18 years	**0.55** / *0.29*	*0.25*	*0.27*	*0.18*	*0.16*	*0.26*	*0.17*
19–20 years	**0.48**	**0.52** / *0.28*	*0.27*	*0.24*	*0.16*	*0.25*	*0.15*
21–22 years	**0.44**	**0.43**	**0.49** / *0.29*	*0.21*	*0.26*	*0.20*	*0.25*
23–24 years	**0.39**	**0.46**	**0.44**	**0.53** / *0.20*	*0.26*	*0.12*	*0.22*
25–26 years	**0.36**	**0.43**	**0.46**	**0.43**	**0.43** / *0.22*	*0.21*	*0.27*
27–28 years	**0.41**	**0.42**	**0.51**	**0.49**	**0.47**	**0.46** / *0.25*	*0.16*
29–30 years	**0.32**	**0.34**	**0.41**	**0.43**	**0.48**	**0.49**	**0.51** / *0.20*

MZ twin correlations are given on diagonal and lower off-diagonal **in bold**, and DZ twin correlations are given on diagonal and upper off-diagonal *in italics*.

Univariate longitudinal twin models were fitted to the data on METhours, neuroticism and extraversion separately, in order to model the stability and change of genetic and environmental influences on each phenotype. For METhours, a Simplex model with additive genetic and unique environmental influences provided the best fit to the data (χ^2(df=57)=30.71, p=1.00). For

Table 3.6 Longitudinal twin correlations for extraversion

	17–18 years	*19–20 years*	*21–22 years*	*23–24 years*	*25–26 years*	*27–28 years*	*29–30 years*
17–18 years	**0.53** / *0.07*	*0.05*	*0.08*	*0.07*	*0.06*	*0.06*	*0.09*
19–20 years	**0.43**	**0.48** / *0.12*	*0.12*	*0.11*	*0.15*	*0.11*	*0.11*
21–22 years	**0.45**	**0.47**	**0.52** / *0.16*	*0.19*	*0.11*	*0.14*	*0.14*
23–24 years	**0.44**	**0.42**	**0.48**	**0.48** / *0.18*	*0.16*	*0.16*	*0.14*
25–26 years	**0.42**	**0.40**	**0.51**	**0.48**	**0.53** / *0.18*	*0.12*	*0.14*
27–28 years	**0.41**	**0.38**	**0.46**	**0.45**	**0.51**	**0.41** / *0.15*	*0.21*
29–30 years	**0.40**	**0.43**	**0.52**	**0.50**	**0.54**	**0.50**	**0.59** / *0.19*

MZ twin correlations are given on diagonal and lower off-diagonal **in bold**, and DZ twin correlations are given on diagonal and upper off-diagonal *in italics*.

neuroticism, again a Simplex model with additive genetic and unique environmental influences fitted the data best ($\chi^2(df=57)=43.15$, $p=0.91$). Similarly, for extraversion a Simplex model with additive genetic and unique environmental influences fitted the data best ($\chi^2(df=57)=36.24$, $p=0.99$).

Neuroticism and Exercise

A bivariate longitudinal causal model was fitted to the data on neuroticism and exercise. With regard to the genetic and environmental influences on neuroticism and METhours, an AE Simplex structure was used to model the stability and change of both phenotypes in this model. The phenotypic associations between neuroticism and METhours could arise from four sources in this model: additive genetic factors shared by both phenotypes, unique environmental factors shared by both phenotypes, a causal effect of neuroticism on METhours, and a causal effect of METhours on neuroticism. The estimated model is depicted in Figure 3.5. The genetic correlation was estimated at –0.20 and could not be omitted from the model ($\chi^2(df=1)=24.02$, $p<0.01$). The environmental correlation was estimated at –0.03 and could be omitted ($\chi^2(df=1)=0.17$, $p=0.68$). The causal effect of neuroticism on METhours was estimated at –0.02 but not significant ($\chi^2(df=1)=2.06$, $p=0.15$). The reverse causal effect of METhours on neuroticism was –0.01 and also not significant ($\chi^2(df=1)=0.64$, $p=0.42$). In summary, these results suggest that genetic pleiotropy, rather than causal effects of neuroticism on METhours or vice versa, explains the small phenotypic correlations between METhours and neuroticism. This is consistent with earlier findings (De Moor et al. 2008).

Extraversion and Exercise

The results of the bivariate longitudinal causal model fitted to data on extraversion and METhours are displayed in Figure 3.6. The non-additive component for either trait was not significant, possibly reflecting low power to detect this component (Posthuma & Boomsma 2000). The genetic correlation was estimated at 0.19 and could not be omitted from the model (χ^2(df=1)=19.34, p<0.01). The environmental correlation was estimated at 0.10 and could be omitted (χ^2(df=1)=1.36, p=0.24). The causal effect of extraversion on METhours was 0.04 and this effect was significant (χ^2(df=1)=8.76, p<0.01). The reverse causal effect of METhours on extraversion was 0.03 and not significant (χ^2(df=1)=2.96, p=0.09). Together, these results suggest that a combination of genetic pleiotropy and a causal effect of extraversion on METhours explains the small phenotypic correlations between METhours and extraversion.

Summary of Findings and Conclusions

Based on the literature reviewed and the new work described above we conclude that, in the population at large, exercise participation is associated with higher levels of life satisfaction and happiness and lower levels of anxiety and depression. This association does not, however, reflect a causal effect of regular exercise on mental health. Instead, the associations are best explained by pleiotropic genetic factors that have an influence on regular exercise behavior while at the same time, but independently from their effect on exercise, have an influence on various aspects of mental well-being.

In a population of (young) adults we investigated to what extent these pleiotropic factors could work through an effect on the personality traits neuroticism and extraversion, under the assumption that neuroticism and extraversion exert a causal effect on regular exercise behavior. This latter assumption was supported for extraversion only. Neuroticism was associated with exercise behavior, but this association was not causal. Pleiotropic genetic factors independently affecting neuroticism and exercise behavior were the source of their association. For extraversion we could not reject the causal hypothesis. In other words, the personality trait extraversion partly explains why some people participate in regular exercise and others don't. This causal effect is, however, not the only source of the extraversion-exercise association. Here too we found pleiotropic genetic factors that independently influence extraversion and regular exercise behavior.

Relevance to Exercise Intervention Programmes

With these conclusions we can re-address the conundrum stated at the beginning of this chapter: If regular exercise provides a consistent reduction in anxious and depressive symptoms and makes everybody "feel better," why does not everybody engage in it? The problem seems to be in the question itself. As do many current population-based intervention campaigns, the question tacitly assumes that the associations found at a population level mainly reflect causal effects of exercise. In this chapter we have challenged this assumption. In the population at large, regular leisure time exercise seems associated with better mental health largely through pleiotropic genetic effects. This opens up the possibility that beneficial effects of exercise are more easily unlocked by some genetic profiles than by others. Favorable genetic profiles may for instance cause a larger sensitivity to the rewarding or a smaller sensitivity to the punishing effects of a broad classes of activities, including exercise.

This would also explain why the favorable genetic profile for exercise is associated with a higher probability of being low in neuroticism and high in extraversion. For exercisers, who also tend to be more extraverted, exercising may be associated with a strong "feel good" experience and constitute an excellent short-term coping strategy that helps to unwind more rapidly from daily pressures experienced in the school, job, or home environment, up to a point where stopping exercise could lead to loss of well-being and self-esteem. For persistent sedentary individuals, who also tend to score higher on neuroticism, the aversive effects of exercise, at least in the forms that they tried so far, may greatly overwhelm the rewarding effects, causing them to drop out.

Acknowledgment of the differential genetic sensitivity to the psychological effects of exercise is of great importance to help create better exercise programs to engage more individuals in some form of regular exercise. Some individuals may require a specific exercise program (with respect to social setting, intensity of exercise, absence or presence of competitive elements, type of exercise) to create a situation in which the rewarding effects of exercise can predominate. This may ensure that these individuals continue to be engaged in regular exercise while maximizing their psychological benefits in terms of increased feelings of well-being and decreased levels of anxiety and depression.

Limitations and Directions for Future Research

In this chapter we have limited our review and our analyses almost completely to regular voluntary exercise behavior in leisure time. We deliberately chose to confine ourselves to this behavioral phenotype, fully knowing that it comprises only a modest part of the total volume of energy expenditure through physical activity for most individuals with the possible exception of vigorous exercisers (Levine 2005; Westerterp 2003). This comes at the price that, from a health perspective, total physical activity may be the more relevant phenotype although the verdict on that is really still out. Moderate-to-vigorous activities have been shown to have the largest protective effect on mortality (Samitz, Egger, & Zwahlen 2011) and leisure time exercise activities are – for most individuals – the major source of bouts that have a sufficient intensity and duration to induce these effects. The advantages of focusing on this subcategory of physical activity are twofold. First, in order to reliably study the causes of individual differences of exercise behavior in genetically informative designs, large amounts of population-based data are needed which is currently only feasible using self-report in surveys. Total physical activity behavior, let alone energy expenditure, cannot be reliably measured by self-report (Prince et al. 2008; Adamo, Prince, Tricco, Connor-Gorber, & Tremblay 2009). Objective methods, for instance using accelerometers, are needed to reliably assess total physical activity and energy expenditure, although the latter may require additional and more expensive technology to account for individual differences in basal metabolic rate and thermogenesis. As exercise activities that are performed during leisure time are deliberately initiated and often clearly defined in time, self-reports on this salient behavior are much more accurate. Indeed, excellent test-retest reliability has been established in our own data (De Moor et al. 2008; Stubbe et al. 2006).

However, ours may have been a too simplistic perspective on a far more complex reality. Recent evidence suggested, for instance, that in college women, where self-report of physical activity may be extra susceptible to a social desirability response bias, extraversion was related to self-reported physical activity but not accelerometer-derived physical activity. In contrast, neuroticism was related to accelerometer-derived physical activity but not self-reported physical activity (Wilson et al. 2015). In our sample, using self-report, we find a causal effect of extraversion on regular exercise behavior but we did not find a causal effect of neuroticism. This raises the question of what the pattern of findings would have been if we also had used an accelerometer-derived measure of regular exercise behavior. No twin data are currently available to our knowledge that could address this question. With the increased availability of user-friendly accelerometer technology, large-scale genetic epidemiological research on the association between mental health and total physical activity is now becoming feasible.

A second advantage of focusing on sports and exercise is that it almost always represents a voluntary behavior, in contrast to many forms of physical activity like bicycling, walking, gardening, household, or work-related physical activity which are an unknown mix of obligatory and voluntary choices (e.g. the choice to bike and not to take the car or bus may have a financial or practical cause, but also be reflective of an innate activity-drive). As we sought to link personality to voluntary behavioral choices, regular exercise seemed a more meaningful choice than total physical activity. The choice for a divide-and-conquer approach with a focus on physical activity in only one of its SLOTH domains (sleep, leisure-time, occupation, transportation, home-based) was further reinforced by evidence that the determinants of physical activity can be specific to each domain (Sallis et al. 2006).

Another limitation of this chapter is that its reasoning was based strongly on findings using the twin design. This design is based on a number of critical assumptions including random mating, i.e. mating that is not related to the phenotype studied, equal environment for DZ and MZ twins, and generalizability of results based on twins to singletons. Each of these assumptions seems to hold rather well for exercise behavior (Eriksson, Rasmussen, & Tynelius

2006; Huppertz 2015; De Moor et al. 2011). Moreover, although part of the twin-based estimated heritability is still missing, the overwhelming message from the recent Cambrian explosion of genome-wide association (GWA) studies is that twin studies correctly estimated the contribution of (additive) genetic variation to many biological and behavioral phenotypes (Visscher & Montgomery 2009; Visscher et al. 2006; Visscher, Brown, McCarthy, & Yang 2012), and their cross-phenotype correlations (Lee et al. 2013; Bulik-Sullivan et al. 2015).

A more serious practical limitation of the twin design is that it requires access to data from a longitudinal twin registry, which is not feasible for all researchers in this field. Currently, only a few research groups have the resources to investigate causality in this specific genetically informative design. However, a different genetically informative design is becoming available through the online repositories with the results of hundreds of GWA studies (www.ebi.ac.uk/gwas/). The availability of these data on a large number of individuals makes it possible to not only test causality while controlling for latent genetic confounds, but to directly assess and hence control for the effects of specific genes. The *Mendelian randomization* technique has gained popularity as a means to causality testing that is in principle similar to the approach that we have used (Lawlor, Harbord, Sterne, Timpson, & Davey-Smith 2008; Davey-Smith & Hemani 2014). Instead of correlating latent genetic and environmental factors that are thought to influence two phenotypes, it is based on actual measured genetic variants. More specifically, a genetic variant that influences an exposure variable (such as exercise behavior) should also, through the causal chain, predict an outcome variable (such as mental health). The big advantage of the Mendelian randomization technique is that it is based on measured genetic variants and can be applied to *any* large population-based sample, whereas our methods for causality testing rely on latent (unmeasured) genetic and environmental factors that need large *twin* samples. The method however requires the identification of solid associations of genetic markers with the exposure variables, and this is not yet the case for most of the phenotypes concerned in this chapter. For neuroticism and extraversion, large-scale GWA analyses have been conducted, which has led to some promising results for neuroticism but no robust associations of genetic variants for extraversion (van den Berg et al. 2016; De Moor et al. 2015). For exercise behavior, robust associations with genetic markers remain to be identified (De Moor et al. 2009). It will hopefully not take long, however, for the large worldwide consortia to tackle physical activity phenotypes with a GWA approach.

This is a perspective to look forward to, as it will yield the required instrumental variables to test the many acclaimed beneficial effects of physical activity. It will also provide unique opportunities to understand the genetic pleiotropic associations between personality, exercise, and mental health at a more mechanistic level.

References

Adamo, K. B., S. A. Prince, A. C. Tricco, S. Connor-Gorber, & M. Tremblay. 2009. "A comparison of indirect versus direct measures for assessing physical activity in the pediatric population: a systematic review." *International Journal of Pediatric Obesity 4*: 2–27.

Azevedo Da Silva, M., A. Singh-Manoux, E. J. Brunner, S. Kaffashian, M. J. Shipley, M. Kivimaki, & H. Nabi. 2012. "Bidirectional association between physical activity and symptoms of anxiety and depression: The Whitehall II study." *European Journal of Epidemiology 27*: 537–46.

Babyak, M., J. A. Blumenthal, S. Herman, P. Khatri, M. Doraiswamy, K. Moore,... Krishnan, K.R. 2000. "Exercise treatment for major depression: Maintenance of therapeutic benefit at 10 months." *Psychosomatic Medicine 62*: 633–8.

Barbour, K. A., & J. A. Blumenthal. 2002. "Exercise training and depression in older adults." *Neurobiology of Aging 26*: S118–23.

Barnett, T. A., L. Gauvin, C. L. Craig, & P. T. Katzmarzyk. 2008. "Distinct trajectories of leisure time physical activity and predictors of trajectory class membership: A 22 year cohort study." *International Journal of Behavior Nutrition and Physical Activity 5*: 57.

Bartels, M., M. H. M. De Moor, N. van der Aa, D. I. Boomsma, & E. J. C. de Geus. 2012. "Regular exercise, subjective wellbeing, and internalizing problems in adolescence: Causality or genetic pleiotropy?" *Frontiers in Genetics 3*: 4.

Birkeland, M. S., T. Torsheim, & B. Wold. 2009. "A longitudinal study of the relationship between leisure-time physical activity and depressed mood among adolescents." *Psychology of Sport and Exercise 10*: 25–34.

Blumenthal, J. A., M. A. Babyak, P. M. Doraiswamy, L. Watkins, B. Hoffman, K. A. Barbour,...A. Sherwood. 2007. "Exercise and pharmacotherapy in the treatment of major depressive disorder." *Psychosomatic Medicine 69*: 587–96.

Boker, S., M. Neale, H. Maes, M. Wilde, M. Spiegel, T. Brick,... J. Fox. 2011. "OpenMx: An Open Source Extended Structural Equation Modeling Framework." *Psychometrika 76*: 306–17.

Boomsma, D. I., J. M. Vink, T. C. van Beijsterveldt, E. J. C. de Geus, A. L. Beem, E. J. Mulder,... G. C. van Baal. 2002. "Netherlands Twin Register: A focus on longitudinal research." *Twin Research and Human Genetics 5*: 401–6.

Brosse, A. L., E. S. Sheets, H. S. Lett, & J. A. Blumenthal. 2002. "Exercise and the treatment of clinical depression in adults – Recent findings and future directions." *Sports Medicine 32*: 741–60.

Brown, W. J., J. H. Ford, N. W. Burton, A. L. Marshall, & A. J. Dobson. 2005. "Prospective study of physical activity and depressive symptoms in middle-aged women." *American Journal of Preventive Medicine, 29*: 265–72.

Bulik-Sullivan, B., H. K. Finucane, V. Anttila, A. Gusev, F. R. Day, P. R. Loh,...B. M. Neale. 2015. "An atlas of genetic correlations across human diseases and traits." *Nature Genetics 47* (11): 1236–41.

Camacho, T. C., R. E. Roberts, N. B. Lazarus, G. A. Kaplan, & R. D. Cohen. 1991. "Physical activity and depression: Evidence from the Alameda County Study." *American Journal of Epidemiology 134*: 220–31.

Chalder, M., N. J. Wiles, J. Campbell, S. P. Hollinghurst, A. Searle, A. Haase,...Lewis, G. 2012. "A pragmatic randomised controlled trial to evaluate the cost-effectiveness of a physical activity intervention as a treatment for depression: The treating depression with physical activity (TREAD) trial." *Health Technology Assessment 16*: 1–164.

Cloninger, C. R., D. M. Svrakic, & T. R. Przybeck. 2006. "Can personality assessment predict future depression? A twelve-month follow-up of 631 subjects." *Journal of Affective Disorders 92*: 35–44.

Cooney, G. M., K. Dwan, C. A. Greig, D. A. Lawlor, J. Rimer, F. R. Waugh,...G. E. Mead. 2013. "Exercise for depression." *Cochrane Database Systematic Reviews 9*: CD004366.

Cooper-Patrick, L., D. E. Ford, L. A. Mead, P. P. Chang, & M. J. Klag. 1997. "Exercise and Depression in Midlife: A Prospective Study." *American Journal of Public Health 87*: 670–3.

Cuijpers, P., & F. Smit. 2002. "Excess mortality in depression: A meta-analysis of community studies." *Journal of Affective Disorders 72*: 227–36.

Cuijpers, P., F. Smit, J. Oostenbrink, R. de Graaf, M. ten Have, & A. Beekman. 2007. "Economic costs of minor depression: a population-based study." *Acta Psychiatrica Scandinavica 115*: 229–36.

Cuijpers, P., F. Smit, B. W. Penninx, R. de Graaf, M. ten Have, & A. T. Beekman. 2010. "Economic costs of neuroticism: A population-based study." *Archives of General Psychiatry 67*: 1086–93.

Daley, A. 2008. "Exercise and depression: A review of reviews." *Journal of ClinicalPsychology in Medical Settings 15*: 140–7.

Davey-Smith, G., & G. Hemani. 2014. "Mendelian randomization: Genetic anchors for causal inference in epidemiological studies." *Human Molecular Genetics 23*: R89–R98.

de Geus, E. J. C., M. Bartels, J. Kaprio, J. T. Lightfoot, & M. Thomis. 2014. "Genetics of regular exercise and sedentary behaviors." *Twin Research in Human Genetics 17*: 262–71.

de Geus, E. J. C., L. J. van Doornen, & J. F. Orlebeke. 1993. "Regular exercise and aerobic fitness in relation to psychological make-up and physiological stress reactivity." *Psychosomatic Medicine 55*: 347–63.

de Geus, E. J. C., & M. H. M. De Moor. 2011. "Genes, exercise, and psychological factors." In C. Bouchard & E. P. Hoffman (Eds.) *Genetic and Molecular Aspects of Sport Performance* (pp. 294–305). Chichester: Wiley and Sons.

de Geus, E. J. C., & M. H. M. De Moor. 2008. "A genetic perspective on the association between exercise and mental health." *Mental Health and Physical Activity 1*: 53–61.

De Moor, M. H. M., D. I. Boomsma, J. H. Stubbe, G. Willemsen, & E. J. C. de Geus. 2008. "Testing causality in the association between regular exercise and symptoms of anxiety and depression." *Archives of General Psychiatry 65*: 897–905.

De Moor, M. H. M., S. M. van den Berg, K. J. Verweij, R. F. Krueger, M. Luciano, A. Arias Vasquez,...D. I. Boomsma. 2015. "Meta-analysis of genome-wide association studies for neuroticism, and the polygenic association with major depressive disorder." *JAMA Psychiatry 72*: 642–50.

De Moor, M. H. M., A. L. Beem, J. H. Stubbe, D. I. Boomsma, & E. J. C. de Geus. 2006. "Regular exercise, anxiety, depression and personality: A population-based study." *Preventive Medicine 42*: 273–9.

De Moor, M. H. M., & E. J. C. de Geus. 2012. "Genetic influences on exercise behavior." In J. M. Rippe (Ed.), *Lifestyle Medicine* (pp. 1367–78). Boca Raton, FL: Taylor and Francis.

De Moor, M. H. M., Y-J. Liu, D. I. Boomsma, J. Li, J. J. Hamilton, J-J. Hottenga,... H-W. Deng. 2009. "Genome-wide association study of exercise behavior in Dutch and American adults." *Medicine and Science in Sports and Exercise 41* (10): 1887–95.

De Moor, M. H. M., G. Willemsen, I., Rebollo-Mesa, J. H. Stubbe, E. J. C. de Geus, & D. I. Boomsma. 2011. "Exercise participation in adolescents and their parents: Evidence for genetic and generation specific environmental effects." *Behavior Genetics 41*: 211–22.

Dugan, S. A., J. T. Bromberger, E. Segawa, E. Avery, & B. Sternfeld. 2015. "Association between physical activity and depressive symptoms: Midlife women in SWAN." *Medicine and Science in Sports and Exercise 47*: 335–42.

Dunn, A. L., M. H. Trivedi, J. B. Kampert, C. G. Clark, & H. O. Chambliss. 2005. "Exercise treatment for depression – Efficacy and dose response." *American Journal of Preventive Medicine 28*: 1–8.

Edman, J. L., W. C. Lynch, & A. Yates. 2014. "The impact of exercise performance dissatisfaction and physical exercise on symptoms of depression among college students: A gender comparison." *Journal of Psychology 148*: 23–35.

Ekkekakis, P. 2008. "The genetic tidal wave finally reached our shores: Will it be the catalyst for a critical overhaul of the way we think and do science?" *Mental Health and Physical Activity 1*: 47–52.

Eriksson, M., F. Rasmussen, & P. Tynelius. 2006. "Genetic factors in physical activity and the equal environment assumption – the Swedish young male twins study." *Behavior Genetics 36*: 238–47.

Eysenck, H. J. 1967. *The biological basis of personality*. Springfield, IL: Thomas.

Eysenck, H. J. 1990. "Biological dimensions of personality." In L. A. Pervin (Eds.) *Handbook of personality: Theory and research*, (pp. 100–2). New York: Guilford Press.

Eysenck, H. J., & M. W. Eysenck. 1985. *Personality and individual differences*. New York: Plenum Press.

Eysenck, H. J., & S. B. G. Eysenck. 1964. *Eysenck personality inventory*. San Diego, CA: Educational Testing Services.

Eysenck, H. J., D. K. B. Nias, & D. N. Cox. 1982. "Sport and personality." In S. Rachman & T. Wilson (Eds.), *Advances in behavior research and therapy* (pp. 1–56). Oxford: Pergamon Press.

Falconer, D. S., & T. F. C. Mackay. 1996. *Introduction to quantitative genetics*. Essex: Pearson Education Limited.

Farmer, M. E., B. Z. Locke, E. K. Moscicki, A. L. Dannenberg, D. B. Larson, & L. S. Radloff. 1988. "Physical activity and depressive symptoms: The NHANES I epidemiologic follow-up study." *American Journal of Epidemiology 128*: 1340–51.

Gesell, A. 1942. "The method of co-twin control." *Science 95*: 446–8.

Gillespie, N. A., D. E. Evans, M. M. Wright, & N. G. Martin. 2004. "Genetic simplex modeling of Eysenck's dimensions of personality in a sample of young Australian twins." *Twin Research 7*: 637–48.

Gray, J. A. 1970. "Psychophysiological basis of introversion-extraversion." *Behaviour Research and Therapy 8*: 249–66.

Gray, J. A. 1985. "Emotional behaviour and the limbic system." *Advances in Psychosomatic Medicine 13*: 1–25.

Gray, J. A., M. R. Liebowitz, & M. G. Gelder. 1987. "Discussions arising from: Cloninger, CR. A unified biosocial theory of personality and its role in the development of anxiety-states." *Psychiatric Developments 5*: 377–92.

Gray, J. A., & N. McNaughton. 1983. "Comparison between the behavioural effects of septal and hippocampal lesions: A review." *Neuroscience Biobehavioral Review 7*: 119–88.

Gray, J. A., & N. McNaughton. 2000. *The Neuropsychology of Anxiety: An enquiry into the functions of the Septo-hippocampal System.* Oxford: Oxford University Press.

Hallal, P. C., L. B. Andersen, F. C. Bull, R. Guthold, W. Haskell, & U. Ekelund. 2012. "Global physical activity levels: Surveillance progress, pitfalls, and prospects." *Lancet 380*: 247–57.

Heath, G. W., D. C. Parra, O. L. Sarmiento, L. B. Andersen, N. Owen, S. Goenka,... Lancet Physical Activity Series Working Group. 2012. "Evidence-based intervention in physical activity: Lessons from around the world." *Lancet 380*: 272–81.

Hettema, J. M., M. C. Neale, J. M. Myers, C. A. Prescott, & K. S. Kendler. 2006. "A population-based twin study of the relationship between neuroticism and internalizing disorders." *American Journal of Psychiatry 163*: 857–64.

Hoyt, A. L., R. E. Rhodes, H. A. Hausenblas, & P. R. Giacobbi. 2009. "Integrating five-factor model facet-level traits with the theory of planned behavior and exercise." *Psychology of Sport and Exercise 10*: 565–72.

Hrubec, Z., & C. D. Robinette. 1984. "The study of human twins in medical research." *The New England Journal of Medicine 310*: 435–41.

Huppertz, C. 2015. *How voluntary exercise behavior runs in families: Twin studies and beyond.* Ph.D. thesis, VU Amsterdam.

Jacka, F. N., J. A. Pasco, L. J. Williams, E. R. Leslie, S. Dodd, G. C. Nicholson,...M. Berk. 2011. "Lower levels of physical activity in childhood associated with adult depression." *Journal of Science and Medicine in Sport 14*: 222–6.

Jerstad, S. J., K. N. Boutelle, K. K. Ness, & E. Stice. 2010. "Prospective reciprocal relations between physical activity and depression in female adolescents." *Journal of Consulting in Clinical Psychology 78*: 268–72.

Johnson, W., M. McGue, & R. F. Krueger. 2005. "Personality stability in late adulthood: A behavioral genetic analysis." *Journal of Personality 73*: 523–51.

Kendler, K. S., & L. KarkowskiShuman. 1997. "Stressful life events and genetic liability to major depression: Genetic control of exposure to the environment." *Psychological Medicine 27*: 539–47.

Kendler, K. S., M. C. Neale, R. C. Kessler, A. C. Heath, & L. J. Eaves. 1993. "A longitudinal twin study of personality and major depression in women." *Archives of General Psychiatry 50*: 853–62.

King, A. C., C. B. Taylor, W. L. Haskell, & R. F. Debusk. 1989. "Influence of regular aerobic exercise on psychological health – A randomized, controlled trial of healthy middle-aged adults." *Health Psychology 8*: 305–24.

Kritz-Silverstein, D., E. Barrett-Connor, & C. Corbeau. 2001. "Cross-sectional and prospective study of exercise and depressed mood in the elderly – The Rancho Bernardo Study." *American Journal of Epidemiology 153*: 596–603.

Krogh, J., B. Saltin, C. Gluud, & M. Nordentoft. 2009. "The DEMO trial: A randomized, parallel-group, observer-blinded clinical trial of strength versus aerobic versus relaxation training for patients with mild to moderate depression." *Journal of Clinical Psychiatry 70*: 790–800.

Krogh, J., P. Videbech, C. Thomsen, C. Gluud, & M. Nordentoft. 2012. "DEMO-II trial. Aerobic exercise versus stretching exercise in patients with major depression-a randomised clinical trial." *PLoS One 7*: e48316.

Ku, P. W., K. R. Fox, L. J. Chen, & P. Chou. 2012. "Physical activity and depressive symptoms in older adults: 11-year follow-up." *American Journal of Preventive Medicine 42*: 355–62.

Lawlor, D. A., R. M. Harbord, J. A. Sterne, N. Timpson, & G. Davey-Smith. 2008. "Mendelian randomization: Using genes as instruments for making causal inferences in epidemiology." *Statistics Medicine 27*: 1133–63.

Lawlor, D. A., & S. W. Hopker. 2001. "The effectiveness of exercise as an intervention in the management of depression: Systematic review and meta-regression analysis of randomised controlled trials." *British Medical Journal 322*: 763–7.

Lee, S. H., S. Ripke, B. M. Neale, S. V. Faraone, S. M. Purcell, R. H. Perlis,... N. R. Wray. 2013. "Genetic relationship between five psychiatric disorders estimated from genome-wide SNPs." *Nature Genetics 45*: 984–94.

Levine, J. A. 2005. "Measurement of energy expenditure." *Public Health Nutrition 8*: 1123–32.

Lindwall, M., P. Larsman, & M. S. Hagger. 2011. "The reciprocal relationship between physical activity and depression in older European adults: A prospective cross-lagged panel design using SHARE data." *Health Psychology 30*: 453–62.

Loehlin, J. C., & N. G. Martin. 2001. "Age changes in personality traits and their heritabilities during the adult years: Evidence from Australian Twin Registry samples." *Personality and Individual Differences 30*: 1147–60.

McKercher, C. M., M. D. Schmidt, K. A. Sanderson, G. C. Patton, T. Dwyer, & A. J. Venn. 2009. "Physical activity and depression in young adults." *American Journal of Preventive Medicine 36*: 161–4.

Martinez-Gonzalez, M. A., J. J. Varo, J. L. Santos, J. De Irala, M. Gibney, J. Kearney, & J. A. Martínez. 2001. "Prevalence of physical activity during leisure time in the European Union." *Medicine and Science in Sports and Exercise 33*: 1142–6.

Middeldorp, C. M., D. C. Cath, A. L. Beem, G. Willemsen, & D. I. Boomsma. 2008. "Life events, anxious depression and personality: A prospective and genetic study." *Psychological Medicine 38* (11):1557–65.

Motl, R. W., A. S. Birnbaum, M. Y. Kubik, & R. K. Dishman. 2004. "Naturally occurring changes in physical activity are inversely related to depressive symptoms during early adolescence." *Psychosomatic Medicine 66*: 336–42.

Neale, M. C., & L. R. Cardon. 1992. *Methodology for genetic studies of twins and families.* Dordrecht: Kluwer Academic Publishers.

Patten, S. B., J. V. Williams, D. H. Lavorato, & M. Eliasziw. 2009. "A longitudinal community study of major depression and physical activity." *General Hospital Psychiatry 31*: 571–5.

Pedersen, N. L., & C. A. Reynolds. 1998. "Stability and change in adult personality: Genetic and environmental components." *European Journal of Personality 12*: 365–86.

Picavet, H. S., G. C. Wendel-vos, H. L. Vreeken, A. J. Schuit, & W. M. Verschuren. 2011. "How stable are physical activity habits among adults? The Doetinchem Cohort Study." *Medicine and Science in Sports and Exercise 43*: 74–9.

Pinto Pereira, S. M., M. C. Geoffroy, & C. Power. 2014. "Depressive symptoms and physical activity during 3 decades in adult life: Bidirectional associations in a prospective cohort study." *JAMA Psychiatry 71*: 1373–80.

Pinto Pereira, S. M., L. Li, & M. C. Power. 2015. "Early life factors and adult leisure time physical inactivity stability and change." *Medicine and Science in Sports and Exercise 47*: 1841–8.

Pirkola, S., S. Saarni, J. Suvisaari, M. Elovainio, T. Partonen, A. M. Aalto,...J. Lönnqvist. 2009. "General health and quality-of-life measures in active, recent, and comorbid mental disorders: A population-based health 2000 study." *Comprehensive Psychiatry 50*: 108–14.

Polderman, T. J. C., B. Benyamin, C. A. de Leeuw, P. F. Sullivan, A. van Bochoven, P. M. Visscher, & D. Posthuma. 2015. "Meta-analysis of the heritability of human traits based on fifty years of twin studies." *Nature Genetics 47* (7): 702–9.

Posthuma, D., & D. I. Boomsma. 2000. "A note on the statistical power in extended twin designs." *Behavior Genetics 30*: 147–58.

Prince, S. A., K. B. Adamo, M. E. Hamel, J. Hardt, G. S. Connor, & M. Tremblay. 2008. "A comparison of direct versus self-report measures for assessing physical activity in adults: A systematic review." *International Journal of Behavior Nutrition and Physical Activity 5*: 56.

Rethorst, C. D., B. M. Wipfli, & D. M. Landers. 2009. "The antidepressive effects of exercise: A meta-analysis of randomized trials." *Sports Medicine 39*: 491–511.

Rhodes, R. E., & N. E. I. Smith. 2006. "Personality correlates of physical activity: A review and meta-analysis." *British Journal of Sports Medicine 40*: 958–65.

Rowland, T. W. 1998. "The biological basis of physical activity." *Medicine and Science in Sports and Exercise 30*: 392–9.

Sagatun, A., A. J. Sogaard, E. Bjertness, R. Selmer, & S. Heyerdahl. 2007. "The association between weekly hours of physical activity and mental health: A three-year follow-up study of 15-16-year-old students in the city of Oslo, Norway." *BMC Public Health 7*: 155.

Sallis, J. F., R. B. Cervero, W. Ascher, K. A. Henderson, M. K. Kraft, & J. Kerr. 2006. "An ecological approach to creating active living communities." *Annual Reviews Public Health 27*: 297–322.

Samitz, G., M. Egger, & M. Zwahlen. 2011. "Domains of physical activity and all-cause mortality: Systematic review and dose-response meta-analysis of cohort studies." *International Journal of Epidemiology 40*: 1382–1400.

Schneider, M. L., & D. J. Graham. 2009. "Personality, physical fitness, and affective response to exercise among adolescents." *Medicine and Science in Sports and Exercise 41*: 947–55.

Stavrakakis, N., P. de Jonge, J. Ormel, & A. J. Oldehinkel. 2012. "Bidirectional prospective associations between physical activity and depressive symptoms. The TRAILS Study." *Journal of Adolescent Health 50*: 503–8.

Stavrakakis, N., A. M. Roest, E. Verhulst, J. Ormel, P. de Jonge, & A. J. Oldehinkel. 2013. "Physical activity and onset of depression in adolescents: A prospective study in the general population cohort TRAILS." *Journal of Psychiatric Research 47*: 1304–8.

Stephens, T. 1988. "Physical activity and mental health in the United States and Canada – evidence from 4 population surveys." *Preventive Medicine 17*: 35–47.

Stephens, T., C. L. Craig, & B. F. Ferris. 1986. "Adult physical activity in Canada: Findings from the Canada Fitness Survey I." *Canadian Journal of Public Health 77*: 285–90.

Steptoe, A., S. Edwards, J. Moses, & A. Mathews. 1989. "The effects of exercise training on mood and perceived coping ability in anxious adults from the general population." *Journal of Psychosomatic Research 33*: 537–47.

Strawbridge, W. J., S. Deleger, R. E. Roberts, & G. A. Kaplan. 2002. "Physical activity reduces the risk of subsequent depression for older adults." *American Journal of Epidemiology 156*: 328–34.

Strohle, A., M. Hofler, H. Pfister, A. G. Muller, J. Hoyer, H. U. Wittchen, & R. Lieb. 2007. "Physical activity and prevalence and incidence of mental disorders in adolescents and young adults." *Psychological Medicine 37*: 1657–66.

Stubbe, J. H., D. I. Boomsma, J. M. Vink, B. K. Cornes, N. G. Martin, A. Skytthe,... E. J. C. de Geus. 2006. "Genetic influences on exercise participation: A comparative study in adult twin samples from seven countries." *PLoS One 1:* e22.

Stubbe, J. H., M. H. M. De Moor, D. Boomsma, & E. J. C. de Geus. 2006. "The association between exercise participation and wellbeing: A co-twin study." *Preventive Medicine 44* (2): 148–52.

Troiano, R. P., D. Berrigan, K. W. Dodd, L. C. Masse, T. Tilert, & M. McDowell. 2008. "Physical activity in the United States measured by accelerometer." *Medicine and Science in Sports and Exercise 40*: 181–8.

Ustun, T. B., J. L. Ayuso-Mateos, S. Chatterji, C. Mathers, & C. J. Murray. 2004. "Global burden of depressive disorders in the year 2000." *British Journal of Psychiatry 184*: 386–92.

van den Berg, S. M., M. H. M. De Moor, M. McGue, E. Pettersson, A. Terracciano, K. J. Verweij,...D. I. Boomsma. 2014. "Harmonization of Neuroticism and Extraversion phenotypes across inventories and cohorts in the Genetics of Personality Consortium: An application of Item Response Theory." *Behavior Genetics 44*: 295–313.

van den Berg, S. M., M. H. M. De Moor, K. J. Verweij, R. F. Krueger, M. Luciano, A. Arias Vasquez,...D. I. Boomsma. 2016. "Meta-analysis of genome-wide association studies for extraversion: Findings from the Genetics of Personality Consortium." *Behavior Genetics 46* (2): 170–82.

van der Aa, N., E. J. C. de Geus, T. C. van Beijsterveldt, D. I. Boomsma, & M. Bartels. 2010. "Genetic influences on individual differences in exercise behavior during adolescence." *Twin Research and Human Genetics 16* (6): 1015–25.

van Dongen, J., P. E. Slagboom, H. H. Draisma, N. G. Martin, & D. I. Boomsma. 2012. "The continuing value of twin studies in the omics era." *Nature Reviews Genetics 13*: 640–53.

van Gool, C. H., G. I. J. M. Kempen, B. W. J. H. Penninx, D. J. H. Deeg, A. T. F. Beekman, & J. T. M. van Eijk. 2003. "Relationship between changes in depressive symptoms and unhealthy lifestyles in late middle aged and older persons: Results from the Longitudinal Aging Study Amsterdam." *Age and Ageing 32*: 81–7.

Vink, J. M., D. I. Boomsma, S. E. Medland, M. H. M. De Moor, J. H. Stubbe, B. K. Cornes,... E. J. C. de Geus. 2011. "Variance components models for physical activity with age as modifier: A comparative twin study in seven countries." *Twin Research in Human Genetics 14*: 25–34.

Visscher, P. M., M. A. Brown, M. I. McCarthy, & J. Yang. 2012. "Five years of GWAS discovery." *American Journal of Human Genetics 90*: 7–24.

Visscher, P. M., S. E. Medland, M. A. Ferreira, K. I. Morley, G. Zhu, B. Cornes,... N. G. Martin. 2006. "Assumption-free estimation of heritability from genome-wide identity-by-descent sharing between full siblings." *PLoS Genetics 2*: e41.

Visscher, P. M., & G. W. Montgomery. 2009. "Genome-wide association studies and human disease: From trickle to flood." *JAMA 302*: 2028–9.

Vukasovic, T., & D. Bratko. 2015. "Heritability of personality: A meta-analysis of behavior genetic studies." *Psychological Bulletin 141*: 769–85.

Weiss, A., T. C. Bates, & M. Luciano. 2008. "Happiness is a personal(ity) thing – The genetics of personality and wellbeing in a representative sample." *Psychological Science 19*: 205–10.

Westerterp, K. R. 2003. "Impacts of vigorous and non-vigorous activity on daily energy expenditure." *Proceedings of the Nutrition Society 62*: 645–50.

Weyerer, S. 1992. "Physical inactivity and depression in the community: Evidence from the Upper Bavarian Field Study." *International Journal of Sports Medicine 13*: 492–6.

Whiteford, H. A., A. J. Ferrari, L. Degenhardt, V. Feigin, & T. Vos. 2015. "The global burden of mental, neurological and substance use disorders: An analysis from the Global Burden of Disease Study 2010." *PLoS One 10*: e0116820.

Wichers, M., Z. Kasanova, J. Bakker, E. Thiery, C. Derom, N. Jacobs, & J. Van Os. 2015. "From affective experience to motivated action: Tracking reward-seeking and punishment-avoidant behaviour in real-life." *PLoS One 10*: e0129722.

Wilde, G. J. S. 1970. *Neurotische labiliteit gemeten volgens de vragenlijstmethode (The questionnaire method as a means of measuring neurotic instability)*. Amsterdam: van Rossen.

Wilkinson, A. V., K. P. Gabriel, J. Wang, M. L. Bondy, Q. Dong, X. Wu,...M. R. Spitz. 2013. "Sensation-seeking genes and physical activity in youth." *Genes Brain Behavior 12*: 181–8.

Wilson, K. E., B. M. Das, E. M. Evans, & R. K. Dishman. 2015. "Personality correlates of physical activity in college women." *Medicine and Science in Sports and Exercise 47*: 1691–7.

Wilson, K. E., & R. K. Dishman. 2015. "Personality and physical activity: A systematic review and meta-analysis." *Personality and Individual Differences 72*: 230–42.

Wipfli, B. M., C. D. Rethorst, & D. M. Landers. 2008. "The anxiolytic effects of exercise: A meta-analysis of randomized trials and dose-response analysis." *Journal Sport and Exercise Psychology 30*: 392–410.

Wise, L. A., L. L. Adams-Campbell, J. R. Palmer, & L. Rosenberg. 2006. "Leisure time physical activity in relation to depressive symptoms in the Black Women's Health Study." *Annals of Behavioral Medicine 32*: 68–76.

Wittchen, H. U., F. Jacobi, J. Rehm, A. Gustavsson, M. Svensson, B. Jonsson,...H. C. Steinhausen. 2011. "The size and burden of mental disorders and other disorders of the brain in Europe 2010." *European Neuropsychopharmacology 21*: 655–79.

4 Treating Depression with Exercise
An Immune Perspective

Aderbal Silva Aguiar, Jr. and Alexandra Latini

Introduction: Depression and Inflammation

The World Health Organization (Kolappa 2013) has stated that there is no health without mental health. This was endorsed by the Pan American Health Organization, the EU Council of Ministers, the World Federation of Mental Health, and the UK Royal College of Psychiatrists (Insel 2011). Mental disorders make a substantial independent contribution to the burden of disease worldwide (Bramer 1988). The WHO's Global Burden of Disease report highlighted the role of mental disorders for disability-adjusted life-year, which is the sum of years lived with disability and years of life lost because of disabilities. Mental illness accounts for a quarter of all disability-adjusted life-years and one-third of those attributed to non-communicable diseases, such as cardiovascular disease, neurodegeneration, and diabetes (Bramer 1988; Insel 2011). Neuropsychiatric disorders contribute to the greatest global burden of disease, more than cancer or cardiovascular diseases (Bramer 1988; Insel 2011).

Depression is the most common serious mental illness. Depression is a state of low mood and activity aversion that affects thoughts, behavior, feelings, and sense of well-being. Major depressive disorder (MDD) is a common neurological disorder, widely distributed in the population, and usually associated with substantial symptom severity and life impairment (Kessler et al. 2003). The prevalence of MDD for lifetime was 16.2% and for 12-month was 6.6%, according to WHO's Composite International Diagnostic Interview (CIDI) (Kessler et al. 2003). A depressed individual feels sad, anxious, empty, hopeless, helpless, worthless, guilty, irritable, ashamed, or restless. They may lose interest in activities that were pleasurable, experience overeating or loss of appetite, have problems concentrating, remembering details or making decisions, and may contemplate, attempt, or commit suicide. Insomnia, excessive sleeping, fatigue, aches, pains, digestive problems, or reduced energy are also related (Kessler et al. 2003).

In the last 50 years, occupational physical activity decreased to 120 kcal. day^{-1} range, and sedentarism has emerged as an additional risk factor, along with physical inactivity, for mental disorders (Gonzalez-Gross & Melendez 2013). Sedentarism and physical inactivity are both modifiable risk factors for depression (Warburton, Nicol, & Bredin 2006, Babyak, Roberts, Lazarus, Kaplan, & Cohen 2000, Camacho et al. 1991). Like depression (Babyak et al. 2000), a sedentary lifestyle is becoming a prominent risk factor for a variety of non-communicable diseases: diabetes mellitus, cancer (colon and breast), obesity, and hypertension (Warburton et al. 2006). Conversely, regular exercise is associated with improved mental and psychological health (Warburton et al. 2006; Camacho et al. 1991).

There is a strong association between chronic non-communicable diseases, including cardiovascular, diabetes, cancer, and respiratory illness – the ones with the highest worldwide rate of mortality – and emotional disorders. There is much evidence showing that the immune system is the molecular connection between these conditions. Inflammation is a protective and coordinated response to cellular stress that results from the exquisite communication among different types of immune cells. Acute inflammation is an early and almost immediate tissue response, namely to injury. It is nonspecific, of short duration, and occurs before the immune response is established. At this stage, the main objective of the immune response is to remove the cellular stress: injury, injurious agents, foreign bodies or resolve hypersensitivity reactions. On the other hand, chronic inflammation is not a part of the natural healing process, and eventually chronic inflammation will cause organ damage, including mental illness, since the body is not prepared to cope with persistent unfocused immune activity.

Cytokines are immunomodulatory molecules, typically produced by immune cells (Svensson, Lexell, & Deierborg 2015) that coordinate the inflammatory response. The immune response can be either neurotoxic or neuroprotective, depending on which signals the immune cells receive. The functions of different cytokines are very complex and depend on the context and concentration of cytokines in relation to one another. However, cytokines such as interleukin-6 (IL-6), tumor necrosis factor-α (TNF-α), interferon-γ (IFN-γ), and interleukin-1β (IL-1β) are considered to have profound pro-inflammatory functions. On the contrary, cytokines such as interleukin-4 (IL-4) and interleukin-10 (IL-10) are regarded as anti-inflammatory (Svensson et al. 2015). The seminal observation that the activation of the immune system by an infectious agent (i.e. malaria inoculation) can affect psychiatric function (Julius Wagner-Jauregg, Nobel Prize, 1927), was essential to understand that cytokines signal the brain and can serve as mediators between the immune and central nervous system. Thus, it is conceived that the inflammatory status in the periphery corresponds to the level of inflammation in the brain.

Studies showing that (1) increased pro-inflammatory cytokine plasma concentrations correlated with the severity of illness in MDD-affected patients, who also had hypothalamic–pituitary–adrenal (HPA) axis hyperactivity (Dantzer, O'Connor, Freund, Johnson, & Kelley 2008; Raison, Capuron, & Miller 2006; Ridker, Cushman, Stampfer, Tracy, & Hennekens 1997), and (2) increased brain inflammatory factors, including TNF-α, IL-6, C-reactive protein (CRP), and IL-1β induces depression-like and sickness behaviors in animals (Garcia-Bueno, Caso, & Leza 2008; Anisman 2009), were the basis for proposing the cytokine-induced model of depression. It is worthwhile to highlight that most of the studies that generated the concept that systemic inflammation is a pathological mechanism for the development of mental health, including MDD (Dantzer et al. 2008; Raison et al. 2006), have been conducted in subjects with sedentary lifestyle.

Treating Depression with Exercise: The Role of IL-6

Exercise is a promising intervention for the prevention and treatment of various diseases characterized by chronic inflammation, including neurological disorders. Aerobic exercise is widely used to reduce inflammation in the periphery and in the brain (Petersen & Pedersen 2005; Svensson et al. 2015). Many experimental studies have shown that exercise decreased proliferation of microglia (Vukovic, Colditz, Blackmore, Ruitenberg, & Bartlet 2012; Kohman, Bhattacharya, Wojcik, & Rhodes 2013; Svensson et al. 2015), decreased hippocampal expression of immune-related genes (Parachikova, Nichol, & Cotman 2008; Martin et al. 2013), reduced nuclear NF-κB activation (Parachikova et al. 2008), and reduced the expression of inflammatory cytokines, such as TNF-α (Gomes da Silva et al. 2013), IFN-g (Svensson et al. 2015) and IL-1β (Erion et al. 2014) (see Figure 4.1).

Exercise is a readily available therapeutic option, effective as a first-line treatment in mild to moderate depression. Additionally, besides preventing depression, physical activity has extra beneficial effects on common comorbidities, like cardiovascular diseases and diabetes. Strong evidence shows that the success of exercise against depression is mainly attributed to the neurobiological induced-mechanisms that promote or enhance monoamine metabolism (Aguiar et al. 2014; Eyre & Baune 2012), neurotrophic factors formation and neurogenesis (Figueiredo et al. 2010; Aguiar, Speck, Prediger, Kapczinski, & Pinho 2008), antioxidant status, appropriate HPA axis function and reduced neuroinflammation (Sigwalt et al. 2011; Eyre & Baune 2012).

Inflammatory diseases are commonly associated with depressed mood; however, the opposite has also been recently demonstrated. Neuroinflammation as well as microglial activation has been shown to occur in patients affected by neurological disorders. The translocator protein density, a marker of neuroinflammation, was found to be increased in the prefrontal cortex,

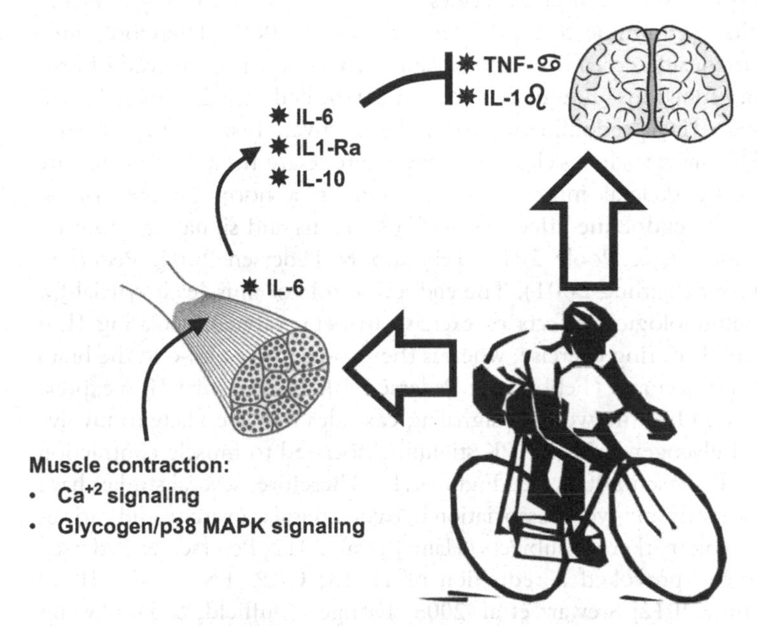

Figure 4.1 Muscle increase of interleukin 6 (IL-6) levels as an exercise signal. Increased IL-6 levels increases the expression of the anti-inflammatory interleukin 10 (IL-10) and the interleukin-1 receptor antagonist (IL-1Ra). These molecules reduce the brain levels of inflammatory molecules interleukin 1-beta (IL-1β) and tumor necrosis factor-alfa (TNF-α). These exercise-induced immune changes are associated with improved overall health and well-being, including antidepressant effects

anterior cingulate cortex, and insula in patients with major depressive episodes secondary to MDD (Setiawan et al. 2015).

The impact of the immune system activation on mental health has been extensively studied in animal models. The classical experimental system involves the challenge with Gram-negative bacteria lipopolysaccharide (LPS). LPS-treated mice display increased IL-6 and IL-1β levels in the prefrontal cortex and hippocampus, associated with depressive-like behavior, lack of motivation, and decreased locomotion. Six weeks of voluntary exercise in running wheels attenuates the levels of the neuroinflammatory mediators and also the sickness behavior (Littlefield, Setti, Priester, & Kohman 2015).

During exercise, the cascade in the cytokine response differs from the "classical" response to infections. It has been reported that muscle and blood IL-6 increase up to 100-fold during exercise, which is followed by increased levels of anti-inflammatory and cytokine inhibitors such as soluble interleukin-1 receptor antagonist (IL-1RA) and IL-10 levels (Svensson et al. 2015; Steensberg, Fischer, Keller, Moller, & Pedersen 2003) (see Figure 4.1). Increased anti-inflammatory IL-10 levels can cause suppression of the pro-inflammatory cytokines IL-1α, IL-1β, TNF-α and also chemokines, as well as a downregulation of adhesion molecules, such as ICAM-1, thereby reducing the infiltration of immune cells (Maynard & Weaver 2008). Therefore, muscular IL-6 is considered to have anti-inflammatory properties instead of pro-inflammatory functions (Brandt & Pedersen 2010; Pedersen 2011).

In the brain, IL-6 predominantly comes from activated astrocytes (Svensson et al. 2015). The cytokines released during exercise are thought to originate from exercising skeletal muscles, which work in a hormone-like fashion exerting specific endocrine effects on various organs and signaling pathways (Hamer, Endrighi, & Poole 2012; Febbraio & Pedersen 2002; Pedersen, Steensberg, & Schjerling 2001). The endocrine role of muscles is crucial for the neuroimmunological effects of exercise in depression. Circulating IL-6 increased rapidly during exercise, whereas the production of IL-6 in the brain increases at a slower rate (Febbraio & Pedersen 2002). Muscular IL-6 expression is regulated by a network of signaling cascades that are likely to involve the Ca^{+2} and glycogen/p38 MAPK stimuli, associated to muscle contraction (Febbraio & Pedersen 2002) (see Figure 4.1). Therefore, several studies have shown consistently an inverse association between regular exercise and various inflammatory biomarkers in subjects (Hamer et al. 2012; Petersen & Pedersen 2005). Exercise provoked a reduction of IL-18, CRP, TNF-α and IL-1β (Eyre & Baune 2012; Stewart et al. 2005; Donges, Duffield, & Drinkwater 2010), and a marked increase in anti-inflammatory mediators, such as IL-10 (Kadoglou et al. 2007).

Exercise-induced cellular neuroimmune factors also include increased CD11b and CD66b enhanced gene expression from peripheral blood mononuclear cells (i.e. IL-5, IL-8, IL-2), increased regulatory T cells (Tregs) and increased $CD14^+$, $CD16^+$ monocytes (Timmerman, Flynn, Coen, Markofski, & Pence 2008; Coen, Flynn, Markofski, Pence, & Hannemann 2010; Wang et al. 2012; Stewart et al. 2005). Exercise has also shown to increase, in parallel to IL-10, the levels of chemokine (C-X-C motif) ligand 1 (CXCL1), CXCL12, and systemic macrophage-released MAPK phosphatise-1 (MKP-1) in rodent brain tissues (Parachikova et al. 2008). CXCL1 is considered to be a neuroprotective biomolecule, and the up-regulation of CXCL12 is known to exert several enhancing effects on: (1) glutamate release from astrocytes, hence regulating neuronal excitability, (2) signal propagation within glial networks, and (3) synaptic transmission; while MKP-1 negatively regulates pro-inflammatory macrophages activation. Finally, exercise also increases hippocampal T cells population due to C-C chemokine receptor type 2 (CCR2), a microglial chemoattractant factor (Parachikova et al. 2008). T cells are responsible for neuroregeneration and modulation of microglia. These effects are associated with other benefits such as a decrease in pro-inflammatory visceral white fat mass (Esposito et al. 2003; Petersen et al. 2005; Donges et al. 2010). Thus, it could be concluded that the better mental well-being experienced by individuals who regularly exercise might be partly explained by IL-6-induced anti-inflammatory mechanisms.

The therapeutic effects of exercise on depression have been shown in animal studies as well as in humans. A protective effect of sustained and regular exercise has been demonstrated in various experimental models of depression (Duman 2005; Aguiar et al. 2014; Cunha et al. 2013; Greenwood, Strong, Dorey, & Fleshner 2007). In addition, randomized and crossover clinical trials demonstrate the efficacy of aerobic or resistance training exercise (2–4 months) as a treatment for depression in both young (Nabkasorn et al. 2006) and older (Blumenthal et al. 1999) individuals. The association between depressive symptoms and low-grade inflammation was partly attenuated by physical activity in 3609 older adults from The English Longitudinal Study of Ageing (Hamer, Molloy, de Oliveira, & Demakakos 2009). In addition to the therapeutic effect, evidence from human studies shows that exercise can also provide some protection against the development of depression (Strawbridge, Deleger, Roberts, & Kaplan 2002).

Treating Depression with Exercise: The Hypothalamic–Pituitary–Adrenal Axis

HPA axis dysregulation and increased blood cortisol levels are implicated in mental stress (Appelhans, Pagoto, Peters, & Spring 2010; Holsen et al. 2013), and complex, bidirectional relationships operate between the HPA axis and the immune system (Savastano et al. 1994). Cytokines influence the neuroendocrine system, in particular the HPA axis. IL-6 and TNF-α share HPA-activating activity, although they are less potent and effective than IL-1β, or IL-2 and IFN, which have no influence on HPA (Dunn 2000). On the other hand, glucocorticoid hormones provoke a depression of the axis function (Savastano et al. 1994). Thus, cytokine-HPA interactions can control the maintenance of homeostasis and the development of diseases (Silverman, Pearce, Biron, & Miller 2005).

Mental stress reduces the sensitivity of the immune system to dexamethasone (synthetic glucocorticoid) inhibition, and cortisol cannot suppress the production of inflammatory cytokines (Rohleder, Schommer, Hellhammer, Engel, & Kirschbaum 2001). Thus, depressed patients present HPA dysfunction, in parallel to high cortisol secretion, and impairment in responsiveness to glucocorticoids (Pariante & Lightman 2008). Major depression also increases the size and activity of the pituitary and adrenal glands (Rubin & Phillips 1993). Successful treatment with antidepressants is associated with a normalization of HPA axis activity and restoration of glucocorticoid receptor function (Anacker et al. 2011; Lopresti, Hood, & Drummond 2013).

In addition to antidepressants, regular exercise can induce changes to the HPA axis, as well as physical and mental stress. Acute exercise-stimulated adrenocorticotropic hormone, cortisol, and lactate responses were attenuated in trained individuals and the effect was proportional to the exercise effort (Luger et al. 1987; Budde et al. 2010). In addition, physical training increases tissue sensitivity to glucocorticoids, a mechanism that prevents excessive muscle and brain inflammatory reactions (Duclos, Gouarne, & Bonnemaison 2003; Wittert, Livesey, Espiner, & Donald 1996; Sigwalt et al. 2011). Therefore, based on much evidence (Luger et al. 1987; DeRijk et al. 1997; Adlard & Cotman 2004; Mastorakos, Pavlatou, Diamanti-Kandarakis, & Chrousos 2005), the normalization of the HPA system may be considered the final step necessary for stable remission of depression (Anacker et al. 2011; Lopresti, Hood, & Drummond 2013).

Summary

Immunomodulation is a biological mechanism responsible for the anti-inflammatory effects of regular exercise. At the same time, robust evidence shows that exercised subjects are more resistant to mental stress and development of depression, and new studies will probably better explain the molecular basis of these findings. On the other hand, lifestyle changes and healthy habits, including exercise, are strongly associated with physical and mental health. Still, exercise can also bring health risks, including sudden death. Thus, an exercise-based therapeutic intervention should be always supervised by a specialist.

References

Adlard, P. A., & C. W. Cotman. 2004. "Voluntary exercise protects against stress-induced decreases in brain-derived neurotrophic factor protein expression." *Neuroscience 124* (4): 985–92. doi:10.1016/j.neuroscience.2003.12.039

Aguiar, A. S., Jr., A. E. Speck, R. D. Prediger, F. Kapczinski, & R. A. Pinho. 2008. "Downhill training upregulates mice hippocampal and striatal brain-derived neurotrophic factor levels." *J Neural Transm 115* (9): 1251–5. doi:10.1007/s00702-008-0071-2

Aguiar, A. S., Jr., E. Stragier, D. da Luz Scheffer, A. P. Remor, P. A. Oliveira, R. D. Prediger,...L. Lanfumey. 2014. "Effects of exercise on mitochondrial function, neuroplasticity and anxio-depressive behavior of mice." *Neuroscience 271*: 56–63. doi:10.1016/j.neuroscience.2014.04.027

Anacker, C., P. A. Zunszain, A. Cattaneo, L. A. Carvalho, M. J. Garabedian, S. Thuret,...C. M. Pariante. 2011. "Antidepressants increase human hippocampal neurogenesis by activating the glucocorticoid receptor." *Mol Psychiatry 16* (7): 738–50. doi:10.1038/mp.2011.26

Anisman, H. 2009. "Cascading effects of stressors and inflammatory immune system activation: implications for major depressive disorder." *J Psychiatry Neurosci 34* (1): 4–20.

Appelhans, B. M., S. L. Pagoto, E. N. Peters, & B. J. Spring. 2010. "HPA axis response to stress predicts short-term snack intake in obese women." *Appetite 54* (1): 217–20. doi:10.1016/j.appet.2009.11.005

Babyak, M., J. A. Blumenthal, S. Herman, P. Khatri, M. Doraiswamy, K. Moore,...K. R. Krishnan. 2000. "Exercise treatment for major depression: Maintenance of therapeutic benefit at 10 months." *Psychosom Med 62* (5): 633–8.

Blumenthal, J. A., M. A. Babyak, K. A. Moore, W. E. Craighead, S. Herman, P. Khatri,...K. R. Krishnan. 1999. "Effects of exercise training on older patients with major depression." *Arch Intern Med 159* (19): 2349–56.

Bramer, G. R. 1988. "International statistical classification of diseases and related health problems. Tenth revision." *World Health Stat Q 41* (1): 32–6.

Brandt, C., & B. K. Pedersen. 2010. "The role of exercise-induced myokines in muscle homeostasis and the defense against chronic diseases." *J Biomed Biotechnol.* doi:10.1155/2010/520258

Budde, H., C. Voelcker-Rehage, S. Pietrassyk-Kendziorra, S. Machado, P. Ribeiro, & A. M. Arafat. 2010. "Steroid hormones in the saliva of adolescents after different exercise intensities and their influence on working memory in a school setting." *Psychoneuroendocrinology 35* (3): 382–91. doi:10.1016/j.psyneuen.2009.07.015

Camacho, T. C., R. E. Roberts, N. B. Lazarus, G. A. Kaplan, & R. D. Cohen. 1991. "Physical activity and depression: Evidence from the Alameda County Study." *Am J Epidemiol 134* (2): 220–31.

Coen, P. M., M. G. Flynn, M. M. Markofski, B. D. Pence, & R. E. Hannemann. 2010. "Adding exercise to rosuvastatin treatment: Influence on C-reactive protein, monocyte toll-like receptor 4 expression, and inflammatory monocyte (CD14+CD16+) population." *Metabolism 59* (12): 1775–83. doi:10.1016/j.metabol.2010.05.002

Cunha, M. P., A. Oliveira, F. L. Pazini, D. G. Machado, L. E. Bettio, J. Budni,... A. L. Rodrigues. 2013. "The antidepressant-like effect of physical activity on a voluntary running wheel." *Med Sci Sports Exerc 45* (5): 851–9. doi:10.1249/MSS.0b013e31827b23e6

Dantzer, R., J. C. O'Connor, G. G. Freund, R. W. Johnson, & K. W. Kelley. 2008. "From inflammation to sickness and depression: When the immune system subjugates the brain." *Nature Reviews Neuroscience 9* (1): 46–57. doi:10.1038/nrn2297

DeRijk, R., D. Michelson, B. Karp, J. Petrides, E. Galliven, P. Deuster,...E. M. Sternberg. 1997. "Exercise and circadian rhythm-induced variations in plasma cortisol differentially regulate interleukin-1 beta (IL-1 beta), IL-6, and tumor necrosis factor-alpha (TNF alpha) production in humans: High sensitivity of TNF alpha and resistance of IL-6." *J Clin Endocrinol Metab 82* (7): 2182–91. doi:10.1210/jcem.82.7.4041

Donges, C. E., R. Duffield, & E. J. Drinkwater. 2010. "Effects of resistance or aerobic exercise training on interleukin 6, C-reactive protein, and body composition." *Med Sci Sports Exerc 42* (2): 304–13. doi:10.1249/MSS.0b013e3181b117ca

Duclos, M., C. Gouarne, & D. Bonnemaison. 2003. "Acute and chronic effects of exercise on tissue sensitivity to glucocorticoids." *J Appl Physiol (1985) 94* (3): 869–75. doi:10.1152/japplphysiol.00108.2002

Duman, R. S. 2005. "Neurotrophic factors and regulation of mood: Role of exercise, diet and metabolism." *Neurobiol Aging 26* Suppl 1: 88–93. doi:10.1016/j.neurobiolaging.2005.08.018

Dunn, A. J. 2000. "Cytokine activation of the HPA axis." *Ann N Y Acad Sci 917*: 608–17.

Erion, J. R., M. Wosiski-Kuhn, A. Dey, S. Hao, C. L. Davis, N. K. Pollock, & A. M. Stranahan. 2014. "Obesity elicits interleukin 1-mediated deficits in hippocampal synaptic plasticity." *J Neurosci 34* (7): 2618–31. doi:10.1523/JNEUROSCI.4200-13.2014

Esposito, K., A. Pontillo, C. Di Palo, G. Giugliano, M. Masella, R. Marfella, & D. Giugliano. 2003. "Effect of weight loss and lifestyle changes on vascular inflammatory markers in obese women: A randomized trial." *JAMA 289* (14): 1799–804. doi:10.1001/jama.289.14.1799

Eyre, H., & B. T. Baune. 2012. "Neuroimmunological effects of physical exercise in depression." *Brain Behav Immun 26* (2): 251–66. doi:10.1016/j.bbi.2011.09.015

Febbraio, M. A., & B. K. Pedersen. 2002. "Muscle-derived interleukin-6: Mechanisms for activation and possible biological roles." *FASEB J 16* (11): 1335–47. doi:10.1096/fj.01-0876rev

Figueiredo, C. P., F. A. Pamplona, T. L. Mazzuco, A. S. Aguiar, Jr., R. Walz, & R. D. Prediger. 2010. "Role of the glucose-dependent insulinotropic polypeptide and its receptor in the central nervous system: Therapeutic potential in neurological diseases." *Behav Pharmacol 21* (5–6): 394–408. doi:10.1097/FBP.0b013e32833c8544

Garcia-Bueno, B., J. R. Caso, & J. C. Leza. 2008. "Stress as a neuroinflammatory condition in brain: Damaging and protective mechanisms." *Neurosci Biobehav Rev 32* (6): 1136–51. doi:10.1016/j.neubiorev.2008.04.001

Gomes da Silva, S., P. S. Simoes, R. A. Mortara, F. A. Scorza, E. A. Cavalheiro, M. da Graca Naffah-Mazzacoratti, & R. M. Arida. 2013. "Exercise-induced hippocampal anti-inflammatory response in aged rats." *J Neuroinflammation 10*: 61. doi:10.1186/1742-2094-10-61

Gonzalez-Gross, M., & A. Melendez. 2013. "Sedentarism, active lifestyle and sport: Impact on health and obesity prevention." *Nutr Hosp 28* Suppl 5: 89–98. doi:10.3305/nh.2013.28.sup5.6923

Greenwood, B. N., P. V. Strong, A. A. Dorey, & M. Fleshner. 2007. "Therapeutic effects of exercise: Wheel running reverses stress-induced interference with shuttle box escape." *Behav Neurosci 121* (5): 992–1000. doi:10.1037/0735-7044.121.5.992

Hamer, M., R. Endrighi, & L. Poole. 2012. "Physical activity, stress reduction, and mood: Insight into immunological mechanisms." *Methods Mol Biol 934*: 89–102. doi:10.1007/978-1-62703-071-7_5

Hamer, M., G. J. Molloy, C. de Oliveira, & P. Demakakos. 2009. "Persistent depressive symptomatology and inflammation: To what extent do health behaviours and weight control mediate this relationship?" *Brain Behav Immun 23* (4): 413–8. doi:10.1016/j.bbi.2009.01.005

Holsen, L. M., K. Lancaster, A. Klibanski, S. Whitfield-Gabrieli, S. Cherkerzian, S. Buka, & J. M. Goldstein. 2013. "HPA-axis hormone modulation of stress response circuitry activity in women with remitted major depression." *Neuroscience 250*: 733–42. doi:10.1016/j.neuroscience.2013.07.042

Insel, T. 2011. "No health without mental health." *Journal of Psychopharmacology 25* (8): A78–A78.

Kadoglou, N. P., F. Iliadis, N. Angelopoulou, D. Perrea, G. Ampatzidis, C. D. Liapis, & M. Alevizos. 2007. "The anti-inflammatory effects of exercise training in patients with type 2 diabetes mellitus." *Eur J Cardiovasc Prev Rehabil 14* (6): 837–43. doi:10.1097/HJR.0b013e3282efaf50

Kessler, R. C., P. Berglund, O. Demler, R. Jin, D. Koretz, K. R. Merikangas,...P. S. Wang. 2003. "The epidemiology of major depressive disorder: Results from the National Comorbidity Survey Replication (NCS-R)." *JAMA 289* (23): 3095–105. doi:10.1001/jama.289.23.3095

Kohman, R. A., T. K. Bhattacharya, E. Wojcik, & J. S. Rhodes. 2013. "Exercise reduces activation of microglia isolated from hippocampus and brain of aged mice." *J Neuroinflammation 10*: 114. doi:10.1186/1742-2094-10-114

Kolappa, K., D. C. Henderson, & S. P. Kishore. 2013. "No physical health without mental health: Lessons unlearned?" *Bull World Health Organ 91*: 3–3A. doi:10.2471/BLT.12.115063

Littlefield, A. M., S. E. Setti, C. Priester, & R. A. Kohman. 2015. "Voluntary exercise attenuates LPS-induced reductions in neurogenesis and increases microglia expression of a proneurogenic phenotype in aged mice." J Neuroinflammation *12*: 138. doi:10.1186/s12974-015-0362-0

Lopresti, A. L., S. D. Hood, & P. D. Drummond. 2013. "A review of lifestyle factors that contribute to important pathways associated with major depression: Diet, sleep and exercise." *J Affect Disord 148* (1): 12–27. doi:10.1016/j.jad.2013.01.014

Luger, A., P. A. Deuster, S. B. Kyle, W. T. Gallucci, L. C. Montgomery, P. W. Gold,... G. P. Chrousos. 1987. "Acute hypothalamic-pituitary-adrenal responses to the stress of treadmill exercise. Physiologic adaptations to physical training." *N Engl J Med 316* (21): 1309–15. doi:10.1056/NEJM198705213162105

Martin, S. A., B. D. Pence, R. M. Greene, S. J. Johnson, R. Dantzer, K. W. Kelley, & J. A. Woods. 2013. "Effects of voluntary wheel running on LPS-induced sickness behavior in aged mice." *Brain Behav Immun 29*: 113–23. doi:10.1016/j.bbi.2012.12.014

Mastorakos, G., M. Pavlatou, E. Diamanti-Kandarakis, & G. P. Chrousos. 2005. "Exercise and the stress system." *Hormones (Athens) 4* (2): 73–89.

Maynard, C. L., & C. T. Weaver. 2008. "Diversity in the contribution of interleukin-10 to T-cell-mediated immune regulation." *Immunol Rev 226*: 219–33. doi:10.1111/j.1600-065X.2008.00711.x

Nabkasorn, C., N. Miyai, A. Sootmongkol, S. Junprasert, H. Yamamoto, M. Arita, & K. Miyashita. 2006. "Effects of physical exercise on depression, neuroendocrine stress hormones and physiological fitness in adolescent females with depressive symptoms." *Eur J Public Health 16* (2): 179–84. doi:10.1093/eurpub/cki159

Parachikova, A., K. E. Nichol, & C. W. Cotman. 2008. "Short-term exercise in aged Tg2576 mice alters neuroinflammation and improves cognition." *Neurobiol Dis 30* (1): 121–9. doi:10.1016/j.nbd.2007.12.008

Pariante, C. M., & S. L. Lightman. 2008. "The HPA axis in major depression: Classical theories and new developments." *Trends Neurosci 31* (9): 464–8. doi:10.1016/j.tins.2008.06.006

Pedersen, B. K. 2011. "Exercise-induced myokines and their role in chronic diseases." *Brain Behav Immun 25* (5): 811–6. doi:10.1016/j.bbi.2011.02.010

Pedersen, B. K., A. Steensberg, & P. Schjerling. 2001. "Muscle-derived interleukin-6: Possible biological effects." *J Physiol 536* (Pt 2): 329–37.

Petersen, A. M., & B. K. Pedersen. 2005. "The anti-inflammatory effect of exercise." *J Appl Physiol (1985) 98* (4): 1154–62. doi:10.1152/japplphysiol.00164.2004

Petersen, E. W., A. L. Carey, M. Sacchetti, G. R. Steinberg, S. L. Macaulay, M. A. Febbraio, & B. K. Pedersen. 2005. "Acute IL-6 treatment increases fatty acid turnover in elderly humans in vivo and in tissue culture in vitro." *Am J Physiol Endocrinol Metab 288* (1): E155–62. doi:10.1152/ajpendo.00257.2004

Raison, C. L., L. Capuron, & A. H. Miller. 2006. "Cytokines sing the blues: Inflammation and the pathogenesis of depression." *Trends Immunol 27* (1): 24–31. doi:10.1016/j.it.2005.11.006

Ridker, P. M., M. Cushman, M. J. Stampfer, R. P. Tracy, & C. H. Hennekens. 1997. "Inflammation, aspirin, and the risk of cardiovascular disease in apparently healthy men." *N Engl J Med 336* (14): 973–9. doi:10.1056/NEJM199704033361401

Rohleder, N., N. C. Schommer, D. H. Hellhammer, R. Engel, & C. Kirschbaum. 2001. "Sex differences in glucocorticoid sensitivity of proinflammatory cytokine production after psychosocial stress." *Psychosom Med 63* (6): 966–72.

Rubin, R. T., & J. J. Phillips. 1993. "Adrenal gland enlargement in major depression." *Arch Gen Psychiatry 50* (10): 833–5.

Savastano, S., A. P. Tommaselli, R. Valentino, M. T. Scarpitta, G. D'Amore, A. Luciano,... G. Lombardi. 1994. "Hypothalamic-pituitary-adrenal axis and immune system." *Acta Neurol (Napoli) 16* (4): 206–13.

Setiawan, E., A. A. Wilson, R. Mizrahi, P. M. Rusjan, L. Miler, G. Rajkowska,...J. H. Meyer. 2015. "Role of translocator protein density, a marker of neuroinflammation, in the brain during major depressive episodes." *JAMA Psychiatry 72* (3): 268–75.

Sigwalt, A. R., H. Budde, I. Helmich, V. Glaser, K. Ghisoni, S. Lanza,...A. Latini. 2011. "Molecular aspects involved in swimming exercise training reducing anhedonia in a rat model of depression." *Neuroscience 192*: 661–74. doi:10.1016/j.neuroscience.2011.05.075

Silverman, M. N., B. D. Pearce, C. A. Biron, & A. H. Miller. 2005. "Immune modulation of the hypothalamic-pituitary-adrenal (HPA) axis during viral infection." *Viral Immunol 18* (1): 41–78. doi:10.1089/vim.2005.18.41

Steensberg, A., C. P. Fischer, C. Keller, K. Moller, & B. K. Pedersen. 2003. "IL-6 enhances plasma IL-1ra, IL-10, and cortisol in humans." *Am J Physiol Endocrinol Metab 285* (2): E433–7. doi:10.1152/ajpendo.00074.2003

Stewart, L. K., M. G. Flynn, W. W. Campbell, B. A. Craig, J. P. Robinson, B. K. McFarlin,...E. Talbert. 2005. "Influence of exercise training and age on CD14+ cell-surface expression of toll-like receptor 2 and 4." *Brain Behav Immun 19* (5): 389–97. doi:10.1016/j.bbi.2005.04.003

Strawbridge, W. J., S. Deleger, R. E. Roberts, & G. A. Kaplan. 2002. "Physical activity reduces the risk of subsequent depression for older adults." *Am J Epidemiol 156* (4): 328–34.

Svensson, M., J. Lexell, & T. Deierborg. 2015. "Effects of physical exercise on neuroinflammation, neuroplasticity, neurodegeneration, and behavior: What we can learn from animal models in clinical settings." *Neurorehabil Neural Repair 29* (6): 577–89. doi:10.1177/1545968314562108

Timmerman, K. L., M. G. Flynn, P. M. Coen, M. M. Markofski, & B. D. Pence. 2008. "Exercise training-induced lowering of inflammatory (CD14+CD16+) monocytes: A role in the anti-inflammatory influence of exercise?" *J Leukoc Biol 84* (5): 1271–8. doi:10.1189/jlb.0408244

Vukovic, J., M. J. Colditz, D. G. Blackmore, M. J. Ruitenberg, & P. F. Bartlett. 2012. "Microglia modulate hippocampal neural precursor activity in response to exercise and aging." *J Neurosci 32* (19): 6435–43. doi:10.1523/JNEUROSCI.5925-11.2012

Wang, J., H. Song, X. Tang, Y. Yang, V. J. Vieira, Y. Niu, & Y. Ma. 2012. "Effect of exercise training intensity on murine T-regulatory cells and vaccination response." *Scand J Med Sci Sports 22* (5): 643–52. doi:10.1111/j.1600-0838.2010.01288.x

Warburton, D. E., C. W. Nicol, & S. S. Bredin. 2006. "Health benefits of physical activity: the evidence." *CMAJ 174* (6): 801–9. doi:10.1503/cmaj.051351

Wittert, G. A., J. H. Livesey, E. A. Espiner, & R. A. Donald. 1996. "Adaptation of the hypothalamopituitary adrenal axis to chronic exercise stress in humans." *Med Sci Sports Exerc 28* (8): 1015–9.

Section 2

Age-Related Effects of Exercise on Mental Health

5 The Exercise Effect on Mental Health in Children and Adolescents

Mirko Wegner and Henning Budde

Introduction

Current estimates state that children and adolescents growing up in the twenty-first century will experience a lower life expectancy than previous generations (Olshansky et al. 2005). Lower life expectancy, reduced mental and physical health may partly be attributed to increasing physical inactivity in industrial countries like the United States (Hillman, Erickson, & Kramer 2008; Secretary of Health and Human Services and the Secretary of Education 2007). Mental disorders like depression and anxiety, for example, are on the rise in a young age group (Viner & Booy 2005). The economic cost for this societal inactivity is tremendous (Colditz 1999; Pratt, Macera, & Wang 2000), see also the epidemiological chapter by Kohn (2018) in this edition.

Moreover, this inactivity does not only decrease health but may also interfere with children's and adolescents' cognitive development (Vaynman & Gomez-Pinilla 2006). These expected trends and anticipated costs of inactivity are found, although policy makers may integrate physical activity and exercise more effectively in different public health sectors (e.g. in schools, programs in professional companies, or health programs financed by insurance companies). Previous research has strongly suggested that physical activity (PA) and exercise benefit different areas of mental health (Hughes 1984; Taylor, Sallis, & Needle 1985) including depression, anxiety, cognitive functioning, and psychological well-being in adults (Gauvin & Spence 1996; Hillman et al. 2008; Wegner, Helmich, Machado, Arias-Carrión, & Budde 2014) and in children and adolescents (Lagerberg 2005; Donaldson & Ronan 2006; Biddle & Asare 2011; Sibley & Etnier 2003). The present chapter aims at illustrating how physical activity and exercise contribute different areas of mental health in children and adolescents, including their general well-being, reductions in depression and anxiety, and benefits to cognitive functioning. We will first define the terms physical activity, exercise, and well-being. Following this we will present evidence primarily from high-quality studies (e.g. randomized controlled trials) on the effects of PA and exercise on psychological well-being, anxiety, depression, and cognitive functioning. In the concluding part of this chapter we will present existing evidence for neurobiological explanations of the benefits of exercise for mental health that might play a role in children and adolescents (see Figure 5.1).

Physical activity (PA) refers to body movement that leads to energy expenditure and is initiated by skeletal muscles (Caspersen, Powell, & Christenson 1985; Budde et al. 2016). *Exercise* has been previously defined as a disturbance of homeostasis through muscle activity resulting in movement and increased energy expenditure (Scheuer & Tipton 1977). However, the critical difference between both terms refers to the planned and structured nature of exercise (Caspersen et al. 1985). Additionally, it has to be distinguished between acute and chronic exercise (Budde et al. 2016). While acute exercise is the physiological response associated with the immediate effects of a single bout of exercise, the term chronic exercise refers to the repeated performance of acute exercise and is often referred to as training (Scheuer & Tipton 1977). *Physical fitness* in turn is the result of exercise or a planned, structured, and repetitive training process and can typically address health components like cardiorespiratory endurance, strength, muscular and skeletal flexibility, or body composition (Howley 2001; Blair, Kohl, & Powell 1987).

It is agreed that *mental health* is not just the absence of mental disorders. Mental health means mental functioning and has a physiological base. It is interconnected with physical and social functioning as well as with health outcomes (World Health Organization 2001). This definition of mental health also includes concepts like subjective well-being, self-efficacy, autonomy, competence, intergenerational dependence, and being able to utilize one's intellectual (e.g. cognitive functioning) and emotional potential (e.g. absence

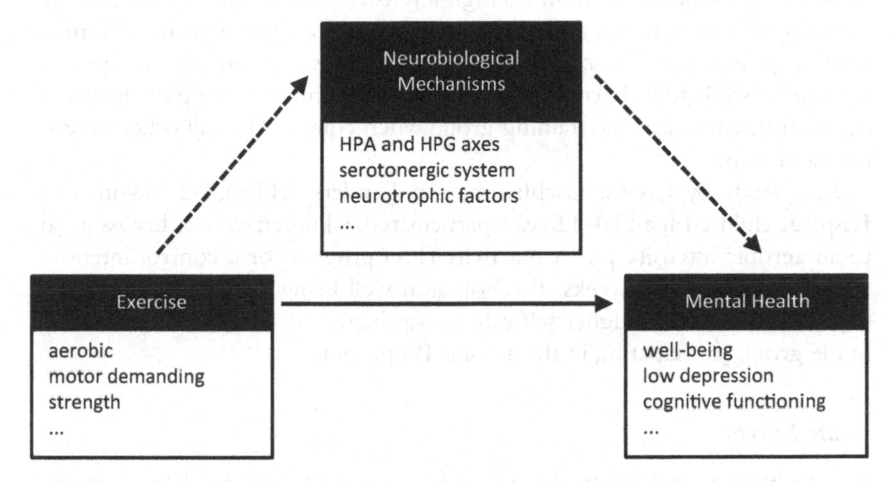

Figure 5.1 Exercise and mental health link and assumed neurobiological mechanisms

of depression and anxiety), and ensures social functioning (World Health Organization 2001, 2004) and life satisfaction (Gauvin & Spence 1996). The present chapter focuses on four concepts included in mental health – psychological well-being; the absence of anxiety; the absence of depression; and cognitive functioning as a sign of mental health.

The Exercise Effect on Psychological Well-Being in Children and Adolescents

Subjective or psychological *well-being* is typically referred to as individuals' positive evaluation of their lives regarding life satisfaction or current life accomplishments and the emotional experience (Diener & Fujita 1995; Diener 1984). Thus, a person who evaluates his or her current life as satisfying and who frequently experiences positive emotions like joy and happiness in the absence of negative emotions like sadness and anger may be described as high in subjective well-being. It has been previously suggested that moderately to highly intense aerobic exercise of 20–30 minutes embedded in programs of 10–12 weeks is most effective (Gauvin & Spence 1996). In the following we will present study results of chronic and acute exercise effects on well-being.

Chronic Effects

In a study with 147 adolescents, Norris, Carroll, and Cochrane (1992) reported a cross-sectional positive correlation between self-reported levels of physical activity and different measures of well-being. They further investigated this relationship by assigning the participants to four different groups for a

10-week intervention program – a high-intensity aerobic training, a moderate intensity aerobic training, a flexibility training, and a control group. All three training groups met for approximately 30 minutes of exercise, two times a week. This study found beneficial effects on well-being only for participants in the high-intensity aerobic training group when compared to all other experimental groups.

In a study by Crews, Lochbaum, and Landers (2004), 62 low-income Hispanic children aged 10–11 years participated. Children were either assigned to an aerobic intensity physical activity (PA) program or a control intensity PA program lasting 6 weeks. Psychological well-being as expressed in lower depression scores and higher self-esteem was increased after the program only in the group participating in the aerobic PA program.

Acute Effects

In a study with 16–19 year-old low-active male and female smokers, Everson, Daley, and Ussher (2006) investigated the effect of exercise on well-being on the day after quitting to smoke. Participants were either assigned to 10 minutes of moderate intensity cycle ergometry or a placebo control condition including very light intensity cycle ergometry. Participants' well-being was measured using the Mood and Physical Symptoms Scale (MPSS; West & Hajek 2004). Well-being (as expressed in psychological distress) scores decreased after participating in moderate exercise when compared to the control condition. Contrary to expectations, adolescents' desire to smoke was not altered through the intervention.

Similarly, in a study by Lofrano-Prado and colleagues (2012) eight physically inactive obese adolescents aged 15 years performed (1) a high-intensity exercise at 10% above their individual ventilatory threshold for 30 minutes, (2) a low-intensity exercise at 10% below the ventilatory threshold, and were (3) seated in a control condition for the same amount of time. Participants' well-being was measured using the Profile of Mood States (POMS; McNair, Lorr, & Droppleman 1992). Both exercise conditions led to decreased well-being (as expressed in vigor scores) in obese participants in this study.

The results above suggest negative acute effects of exercise on well-being in rather inactive adolescent groups. Petruzzello, Jones, and Tate (1997) in their study were interested in different effects of exercise depending on participants' physical activity status. Regularly active vs. non-active participants performed a 24-minute moderately intense bicycle exercise. Well-being measures were taken before, every six minutes during, and after the intervention period. The authors could show that exercise effects on participants' well-being depended on how physically active adolescents were in life. Regularly active individuals gained in well-being (positive affect compared to negative affect) through the exercise intervention. However, non-active participants showed a sharp drop in positive compared to negative affect, suggesting negative acute effects on their well-being.

Overall, the research base of high-quality studies on the link between exercise interventions and different well-being measures in children and adolescents is weak; especially studies with children younger than 14 years are missing. Studies investigating exercise programs of several weeks suggest positive chronic effects on well-being. From studies on the acute effect it can be derived that exercise may affect well-being negatively when participants are not regularly participating in sports.

The Exercise Effect on Anxiety in Children and Adolescents

The chronic and acute anxiolytic effects of exercise are better researched in children and adolescents. *Anxiety* is a warning sign to the individual that his or her resources are limited in face of the appraisal of a certain situation. This unpleasant emotional state includes experiential (e.g. worry cognitions and emotions), physiological, and behavioral components (Liebert & Morris 1967; Lazarus & Folkman 1984). According to Spielberger (1972) anxiety can be perceived as an emotional state as well as a personality disposition (trait). If people are high in *trait anxiety* they are predisposed to perceive many different situations as threatening and respond with experiencing anxiety states (Spielberger 1983). If people are high in *state anxiety* they will respond with higher levels of apprehension, tension, negative emotions, and organismic arousal to threatening situations (Spielberger 1983). Anxiety is also categorized as a mental disorder (American Psychiatric Association 2013) and it includes diagnoses like panic disorder, agoraphobia, generalized anxiety disorder, post-traumatic stress disorder (PTSD), social phobia, acute stress disorder, obsessive compulsive disorder, and disorders due to medical conditions or substance use. More details on the definition of anxiety can be found in the chapter by Mumm, Bischoff, and Ströhle (2018) in this book.

A previous review of existing meta-analyses on the effect of exercise and physical activity on anxiety disorders identified a small average effect size of 0.34 in adults (Wegner et al. 2014). The only meta-analysis focusing on the anxiety-reducing effect of exercise in a younger population (Larun, Nordheim, Ekeland, Hagen, & Heian 2006) found a slightly stronger effect (0.48) but also included young adults up to an age of 20 years. The following published randomized controlled studies addressed the chronic and acute anxiolytic effects of exercise in children and adolescents.

Chronic Effects

A RCT conducted by Brown, Welsh, Labbe, Vitulli, and Kulkarni (1992) studied the effect of a 9-week exercise program with physical education classes three times a week in adolescents diagnosed with dysthemia and conduct disorders (average age: M = 15.6 years). Only the girls in the study decreased their anxiety scores in the Profile of Mood State (POMS) inventory (McNair et al. 1992) while the control group's anxiety scores remained the same or slightly increased.

Juvenile delinquents (*N* = 60) in a community counseling program participated in a RCT study examining the effect of exercise on anxiety (Hilyer et al. 1982). The students between the ages of 15 and 18 were part of an experimental exercise group with a physical fitness program of 90 minutes, three times a week, or a control group. The program lasted 20 weeks. Students in the experimental group displayed lower state as well as trait anxiety scores in the State-Trait Anxiety Inventory for children (Spielberger 1973).

In a school-based study in Chile conducted by Bonhauser and colleagues (2005), 198 students of low socioeconomic status aged 15 years were randomly selected for an intervention group or a control group. In the intervention group, students participated in 90 minutes of exercise three times a week, including stretching, dynamic large muscle movements, and skill training. The control group participated in regular school physical education classes of 90 minutes, one day per week. The pre to post test delay was 10 weeks. Changes in anxiety levels were measured using the Hospital Anxiety Depression Scale (HADS) (Bjelland, Dahl, Haug, & Neckelmann 2002). The decrease in the anxiety scores was significantly stronger in the intervention group when compared to the control group.

In a study examining the anxiolytic effect of resistance training (Lau, Yu, Lee, & Sung 2004) 36 obese Chinese teenagers aged 10–17 years participated. Students were either assigned to the resistance-training group or a control group. Within the resistance-training program they practiced at a level of 70–85% of 1RM (repetition maximum). Both groups additionally attended a nutrition program. Anxiety scores were measured before and after the 6-week exercise program using the HADS. Although the intervention group showed a decrease in anxiety scores, this change was not significantly different from the change in the control group.

Acute Effects

Few studies have focused on acute effects of exercise on state anxiety levels. Recently, Wegner and colleagues (2014) performed an experimental study with 14-year-old high school students who were randomly assigned to three experimental groups. In the exercise group they ran for 15 minutes at a medium intensity level of 65–75% of their individual maximum heart rate (HR_{max}). In the psychosocial stress condition they worked on an intelligence test for the same time period under the assumption that their intelligence quotient will be made public in front of the class immediately after the test. And in the control group they participated in a regular teacher-centered class session. State anxiety was measured with an adopted form of the Competitive State Anxiety Inventory (Cox, Martens, & Russell 2003), using questions referring to the stress situation students experienced within the experiment. Although the students in the exercise group showed no increases in anxiety levels while participants in the control group showed increases from pre to post test, this result was not significant, speaking against an anxiolytic effect of acute moderately intense exercise in this experiment.

In a study by Bahrke and Smith (1985), 65 children aged 9–12 years participated in one of three experimental groups. In the exercise group they walked and ran for 15 minutes. In the resting control group they were sitting quietly reading a book and in the activity control group they cut out paper shapes for the same amount of time. State anxiety levels were measured before, immediately after, and in a 10-minute follow-up using Spielberger's scale for children (Spielberger 1973). Non-significant decreases in anxiety levels were found in all three groups 10 minutes after the intervention. Girls and boys responded the same way.

Lofrano-Prado and colleagues (2012) had eight obese adolescents aged 15 years undergo three experimental trials in randomized order. In the first condition they performed high-intensity exercise at 10% above their individual ventilatory threshold for 30 minutes, which corresponded to approximately 75% of the individual VO_{2max} in this study. The second condition included low-intensity exercise at 10% below the ventilatory threshold (approx. 55% of VO_{2max}). And in the control condition they were seated for the same amount of time. Experimenters measured state anxiety using the STAI (Spielberger, Gorsuch, & Luchene 1976). In this group of obese adolescents both exercise trials significantly increased anxiety levels when compared to the control condition.

Overall, findings from high-quality studies examining the anxiolytic effect of exercise are inconsistent. Studies prefer adolescents aged at least 14 years or older. Children younger than 14 years are hardly found in these studies. Here, research may address the exercise–anxiety link in children more strongly in the future. The chronic effects of exercise programs on anxiety seem to be small. Acute effects are hardly found in experimental studies.

The Exercise Effect on Depression in Children and Adolescents

In the Diagnostic and Statistical Manual of Mental Disorders (DSM-5; American Psychiatric Association 2013) *depression* is characterized as a disorder with different emotional, motivational, somatic, motor-behavioral, and interactive symptoms. A certain amount of symptoms needs to be present for at least two weeks in order to be identified as a depressive episode. Among these symptoms are depressed mood, helplessness, self-devaluation, loss of interest or pleasure, increased tiredness, abnormalities in appetite and sleep, psychomotor retardation, mimic stiffness, withdrawal from social interaction as well as sexual inactivity. For a more detailed definition of depression and depressive disorders please refer to Mutrie, Richards, Lawrie, and Mead's chapter (Mutrie, Richards, Lawrie, & Mead 2018) in this edition.

Budde and colleagues recently reviewed seven meta-analyses of 31 studies on the effect of physical activity and exercise on depression disorders in children and adolescents (Budde et al. forthcoming). They found a medium average effect size of 0.38. This effect is slightly lower than the effect found in adults in a different review of meta-analyses performed for adults (Wegner et al. 2014).

Here the average effect of exercise on depression was 0.56. Meta-analyses and reviews of the exercise effect on depression in children and adolescents that included less controlled studies (e.g. no randomized controlled trials, RCT) consistently reported lower effect sizes (Brown, Pearson, Braithwaite, Brown, & Biddle 2013; Cairns, Yap, Pilkington, & Jorm 2014; Craft & Landers 1998) than meta-analyses including only RCT studies. The following RCT studies are examples for the research findings in children and adolescents and exclusively focus on the chronic effect of exercise on depression.

Petty, Davis, Tkacz, Young-Hyman, and Waller (2009) performed a randomized controlled trial with 207 children aged 7–11 years for an average of 13 weeks. Children participated either in 20 or 40 minutes of aerobic exercise per day or were part of a control group. Children participating in 40 minutes of aerobic exercise per day (highest dose) showed the strongest decrease in the Reynolds Child Depression Scale (Reynolds 1989) when compared to children in the low dose group (20 minutes per day) and children in the control group.

Amnesi (2005) investigated how depression scores developed within a 12-week period in 90 children aged 9–12 years participating either in a physical activity after-school program or a no-exercise program control condition. Participants were asked to enter their depression scores every week. Results suggest that children in the exercise program group showed significant reductions in depression scores, which were not found in the control group. The effect sizes were small, deviating from what can be expected from adult samples.

MacMahon and Gross (1988) performed a RCT study with 69 adolescent delinquent males who either participated in an aerobic exercise program or in a comparison condition with limited exertion. In the exercise condition they performed 40 minutes of aerobic exercise three times a week for three months. Participants in the aerobic exercise group showed significantly decreased scores in the Beck Depression Inventory (BDI) (Beck, Steer, Ball, & Ranieri 1996) after the intervention and when compared to the control group.

Hilyer and colleagues (1982) assigned 60 juvenile delinquents aged 15 to 18 years to an experimental exercise group and a control group within a community counseling program. In the experimental group, students attended a physical fitness program of 90 minutes per day, three days a week, for 20 weeks. Results show that participants of the experimental group benefitted significantly regarding their depression scores in the BDI when compared to the control group.

Brown et al. (1992) examined a group (N = 27) of psychiatrically institutionalized adolescents (diagnoses: dysthymia, conduct disorder) aged M = 15.6 years on average. Participants were assigned either to the exercise or a control group. The exercise program lasted 9 weeks with physical education classes three times a week and BDI scores were taken prior as well as post intervention and in a 4-week follow-up. The authors report that only the girls benefitted from the exercise intervention regarding their depression scores in the BDI consistently decreasing in level from pre to post, and to the follow-up measure.

Despite these positive findings on the effect of exercise on depression, some studies also reported null effects. In a study by Mendelson and associates (2010), 97 children aged 9–11 years participated either in an intervention group in which they performed yoga-based physical activity (e.g. fluid movement, bending, stretching) or in a control group. In this RCT no significant difference between the intervention and the control group (ES = 0.13) could be found for children's depressive symptoms as assessed with the child version of the Short Mood and Feelings Questionnaire (SMFQ-C) (Angold et al. 1995). One reason why there was no effect found is that the intensity of the exercise program was too low.

Obese children and adolescents ($N = 81$) aged between 11 and 16 years were the focus in a RCT conducted by Daley, Copeland, Wright, Roalfe, and Wales (2006). Participants were assigned to one of three experimental groups (exercise therapy, equal-contact exercise placebo intervention, usual care) for a 6-week program. Although participants' self-perceptions changed significantly over the course of the exercise program, the authors could not find significant effects on participants' scores in the Children's Depression Inventory (CDI) (Kovacs 2005) after 8, 14, and 28 weeks although the experimental exercise group consistently scored lower in the CDI on a descriptive level.

Another study on obese teenagers was conducted by Lau and colleagues (2004) in China. The students ($N = 36$) participated either in a 6-week resistance exercise training set at a level of 70–85% of 1RM (repetition maximum) or in a control group. Depression was measured using the Hospital Anxiety and Depression Scale (Bjelland et al. 2002). Participants in the exercise group did not show a significant decrease in their depression scores compared to the control groups over the program period.

Compared to psychological well-being and anxiety, the research base on the effect of exercise on depression in children and adolescents is stronger. However, studies on preadolescent children are also rare for this particular aspect of well-being. Moreover, the results of qualitatively high studies on the chronic effect for children and adolescents suggest a small effect being lower compared to adult samples.

The Exercise Effect on Cognitive Functions in Children and Adolescents

The term cognition includes processes of perception, attention, thinking/problem solving, memory, and language and is typically referred to as how the mind works (Pinker 1999). Cognitive control processes, also called executive functions, include different cognitive functions such as self-control, selective attention, cognitive inhibition, working memory, and cognitive flexibility (Diamond 2013; Miyake et al. 2000). Executive functions are usually subsumed into the three categories of self-control, working memory, and cognitive flexibility. *Self-control* involves resisting temptations and avoiding impulsive acting. The *working memory* supports keeping information in mind and allows working with this information mentally (e.g. to solve a problem). And *cognitive flexibility* refers to the ability to change perspectives on how to solve a problem, and the flexibility to adjust to changing priorities, rules, or demands (Diamond 2013). Executive functions further contribute to the higher-order cognitive processes of planning, problem-solving, and reasoning and are linked to mental health (Collins and Koechlin 2012; Diamond 2013). Individuals suffering from mental disorders (e.g. attention deficit hyperactivity, conduct disorder, depression) show decreased executive functioning (Diamond 2005; Fairchild, van Goozen, Stollery, Aitken, & Savage 2009; Taylor Tavares et al. 2007).

Three previous meta-analyses illustrated the strength of the link between exercise or physical activity and cognitive functioning in children and adolescents showing small to medium effects (0.28 – 0.52) (Sibley & Etnier 2003; Fedewa & Ahn 2011; Verburgh, Königs, Scherder, & Oosterlaan 2014). Sibley and Etnier (2003) did not find differences between types of exercise on cognition. Fedewa and Ahn (2011) found the strongest effect on cognition for aerobic exercise programs. However, there was also some evidence that programs involving perceptual motor aspects may also positively affect cognitive performance. This point was addressed in several recent studies arguing for the importance of cognitively involving (Crova et al. 2014; Pesce 2012) or coordinative exercise (Budde, Voelcker-Rehage, Pietrassyk-Kendziorra, Ribeiro, & Tidow 2008; Koutsandréou, Wegner, & Budde et al. 2016) for improving cognitive performance. Moreover, Fedewa and Ahn (2011) found that physical activity performed three times per week showed the strongest effects on cognitive performance in children and adolescents. Sibley and Etnier (2003) pointed to an age effect with 11–13-year-old children benefitting most from exercise interventions followed by very young children aged 4–7 years. However, Verburgh and colleagues (2014) in their meta-analysis found no age effect of exercise on executive functions, thus the effect did not differ between studies examining children, adolescents, and young adults. Among the three executive functions they found the strongest effect for self-control/inhibition while the effect on working memory was only marginally significant and that on cognitive flexibility was not significant. Sibley and Etnier (2003) found the strongest effects of exercise on perceptual skills and measures of intelligence. Furthermore, Verburgh et al. (2014) did not find an overall meta-analytic effect for chronic exercise interventions on cognitive function. For acute exercise interventions, however, they found a moderate effect on cognitive functioning. Sibley and Etnier (2003), however, could not find differences regarding the cognition effect between chronic and acute exercise interventions. The following paragraph includes a few examples of recent randomized controlled studies published on the chronic and acute links between exercise and cognition not published earlier than 1990.

Chronic Effects

Zervas, Danis, and Klissouras (1991) had $N = 26$ boys (9 pairs of monozygotic twins and 8 normal) aged between 11–14 years participate in an exercise program for 25 weeks. The experiment consisted of two control groups and one experimental group. In the experimental group, one twin of each of the nine pairs performed 90 minutes of interval or continuous running aligned in intensity to their individual anaerobic thresholds three times a week. The other single twins and eight normal boys were part of the control group performing their regular physical education lessons 2–3 times per week. All participants performed the Cognitrone Test (Schuhfried 1984) to assess self-control/inhibition prior and post the intervention. The authors report that the trained single twins showed an increased number of correct responses in the task, implying improved cognitive performance after the 25-week intervention period when compared to the control groups.

In a study with 92 obese children at the age of 9–10 years, Davis and colleagues (2007) tested the effects of a low-dose and a high-dose exercise program compared to a control situation. The programs lasted 15 weeks. In the low-dose program children participated in 20 minutes of exercise 5 days per week. In the high-dose program children performed 40 minutes of exercise on 5 days a week. The Cognitive Assessment System (CAS; Naglieri & Das 1997) was used to assess children's executive functions. The authors found significant improvement in the aspect of planning in the CAS only in the high-dose group compared to the control group, suggesting that higher doses of exercise are needed to benefit executive functioning in obese children. In a different study by the same research group they confirmed these results, additionally showing that exercise also benefitted math performance in school (Davis et al. 2011).

In a 9-month intervention period, 43 children aged 7–9 years were either randomly assigned to an afterschool physical activity program or to a waitlist control group (Kamijo et al. 2011). Children's working memory was measured using a modified Sternberg Task (Sternberg 1966) before and after the intervention period. Compared to the control group, children in the intervention group significantly improved their working memory performance over the course of the intervention program.

Recently, a study with 71 children aged 9–10 years was conducted to examine the chronic effects of exercise on working memory (Koutsandréou et al. 2016). Children were randomly assigned to (1) a cardiovascular exercise group, (2) to a motor exercise group, or (3) to a control group. The program lasted 10 weeks. Every week, children participated in this afterschool program for three sessions of 45 minutes each. In the cardiovascular exercise group the children performed aerobic exercise at an intensity of 60–70% of their individual HR_{max}, in the motor exercise group they worked at 55–65% of HR_{max} performing motor-demanding tasks like juggling or balancing. The control group did assisted homework at this time. Children's working memory was measured using the letter digit span task (Gold, Carpenter, Randolph, Goldberg, & Weinberger 1997). Results revealed that children's working memory in both exercise groups increased from pre to post test while it was not different in the control group. However, only participants in the motor exercise group showed increased working memory scores compared to the control group after the intervention period, indicating that motor demands may additionally benefit cognitive development.

In a study with 70 lean and obese 9–10-year-old children, Crova and colleagues (2014) compared the effects of cognitively demanding and regular physical education programs. They measured children's inhibition and working memory skills prior and post intervention. The intervention period lasted 6 months. In this study only the overweight but not the lean children benefitted from the cognitively enriched physical education program regarding their inhibition but not their working memory scores. The regular physical education classes did not foster executive functions in lean or obese children.

Acute Effects

Caterino and Polak (1999) report a study with 177 participants aged 7–10 years who were either assigned to a classroom activity (control) or a physical activity, including 15 minutes of stretching and aerobic walking. All participants performed the Woodcock-Johnson Test of Concentration (Woodcock & Johnson 1989) to measure cognitive self-control. The authors report that only the oldest children (age 9–10) benefitted significantly from the physical activity intervention regarding their performance in the concentration task.

In a study testing the effects of acute exercise intensity on working memory performance (Letter Digit Span; Gold et al. 1997), Budde et al. (2010) tested 60 adolescents aged 15–16 years and randomly assigned them to two experimental and one control group. In the low-intensity experimental group participants performed 12 minutes of aerobic exercise at 50–65% of their individual maximum heart rate (HR_{max}). In the high-intensity exercise group, participants worked at 70–85% of their individual HR_{max}. The authors found slightly positive effects of moderately intense exercise of 50–65% individual HR_{max} on participants' working memory performance. This effect was especially pronounced in participants scoring low in the pre test of working memory.

In a different study, Budde et al. (2008) had 115 participants aged 13–16 years randomly assigned to an experimental and a control group. In the experimental condition, participants performed 10 minutes of coordinative/motor-demanding tasks. The control group participated in a regular physical education lesson with no motor focus but with the same intensity for the cardiovascular system (HR of 120 bpm). All participants performed a d2 test of attention (Brickenkamp 2002) prior and post the experimental condition. The authors found a positive effect on cognitive performance in the attention task in favor of the coordinative exercise group.

Finally, Niemann and colleagues (2013) investigated $N = 42$ primary school children aged 9–10 years participating in 12 minutes of intensive physical activity at a heart rate of 180–190 bpm or a control group watching a non-arousing movie. Children performed a d2 test of attention prior and post the intervention. Results show no differences between the experimental groups. All participants gained in their cognitive performance. However, the authors state that participants reporting high levels of physical activity in their everyday life show stronger increases from pre to post test in their cognitive performance scores, indicating that chronic physical activity might affect the exercise effect on cognition.

Overall, findings regarding the effects of exercise on different cognitive functions suggest a small, positive relationship (Trudeau & Shephard 2010) with working memory and inhibitory functions (self-control) benefitting most in this age group. More recent findings suggest that cognitively enriched or motor-demanding exercises may additionally benefit cognitive aspects like inhibition and working memory in children and adolescents.

Physiological Mechanisms Underlying the Positive Effects of Exercise in Children and Adolescents

Exercise may positively affect well-being through different neurobiological mechanisms, including effects on the functioning of the hypothalamic-pituitary-adrenal (HPA) axis, through effects on testosterone production (HPG axis) and other sex hormones, through development in brain regions like the limbic system, the hippocampus, and the amygdala, and through BDNF growth factors, serotonin transporter (SERT), and other neurotransmitters (5-HIAA). Reviews regarding these neurobiological mechanisms are given by Mutrie and colleagues (2018) as well as McMorris (2018) in this edition and can be found elsewhere (Helmich et al. 2010; Wegner et al. 2014).

Hypothalamic-Pituitary-Adrenal (HPA) Axis

Mood disorders (e.g. depression, anxiety) have been previously linked to elevated HPA axis activity in adults and to stressful life events occurring early in life (Pariante & Lightman 2008; Pruessner, Hellhammer, & Kirschbaum 1999; Rubin, Poland, Lesser, Winston, & Blod 1987; Gotlib, Joormann, Minor, & Hallmayer 2008). Although less research exists examining the link between HPA axis activity and well-being in children and adolescents, the existing findings point to similar processes in the young age group when compared to older adults (Kudielka, Buske-Kirschbaum, Hellhammer, & Kirschbaum 2004). Depressed children and adolescents, for example, show higher baseline cortisol levels and stronger HPA axis reactivity to psychological stress (Lopez-Duran, Kovacs, & George 2009), although there are also findings suggesting that younger dysphoric children (e.g. 10 years and younger) may respond with hypo-reactivity of the HPA axis while older dysphoric adolescents (e.g. 14 years and older) show a hyper-reactivity to stress (Hankin, Badanes, Abela, & Watamura 2010; Stroud et al. 2009). These different reactivity patterns may be due to maturity changes in the HPA axis throughout puberty (Gunnar, Wewerka, Frenn, Long, & Griggs 2009). Different factors were named that contribute to higher HPA axis activity, including altered regulation of adrenocorticotropin (ACTH) and cortisol secretory activity (Holsboer 2000; Parker, Schatzberg, & Lyons 2003), increased levels of corticotropin-releasing factor (CRF) in the brain and an increased number of neurons secreting CRF in the limbic system (Raadsheer, Hoogendijk, Stam, Tilders, & Swaab 1994; Nemeroff et al. 1984). Previous studies linked physical activity to HPA axis activity and reactivity in different young age groups (Martikainen et al. 2013) with participants who are more physically active showing higher levels of well-being and less depression. Children that are not very physically active, however, show higher reactivity to stress. Exercise supports glucocorticoid secretion (Budde et al. 2010a; Budde et al. 2010b) and changes tissue sensitivity to glucocorticoids (Duclos, Gouarne, & Bonnemaison 2003). Cortisol responses after acute exercise in physically active individuals are dampened and consumed more quickly than in the physically inactive (Rudolph & McAuley 1998; Mathur, Toriola, & Dada 1986). Exercise also increases plasma concentrations in the atrial natriuretic peptide (ANP) (Mandroukas et al. 1995), which has been shown to inhibit the HPA axis (Kellner, Wiedemann, & Holsboer 1992) and reduce, for example, anxiety (Ströhle, Kellner, Holsboer, & Wiedemann 2001).

Sex Hormones

In addition to the activity of the HPA axis, the hypothalamic-pituitary-gonadal (HPG) axis expressed in the activity of testosterone (Wegner, Niemann, & Budde 2015; Wegner, Windisch, & Budde 2012) and other sex hormones like progesterone and estrogen have been discussed to be involved in depression and mood disorders in adolescents (Angold, Costello, Erkanli, & Worthman 1999; Angold, Costello, & Worthman 1998). For cognitive functioning moderate levels of testosterone have been suggested to be beneficial (Hampson 1995; Wolf & Kirschbaum 2002). In adolescents, for example, it was shown that stress-induced changes in testosterone may be linked to manual dexterity when performing a fine motor task (Wegner, Koedijker, & Budde, 2014) as well as to working memory (Budde et al. 2010b). The effects have been attributed to teststerone binding to androgen receptors in the cytoplasm. These androgen receptors are often found in brain areas responsible for memory and learning such as the hippocampus and the prefrontal cortex (Janowsky 2006). Receptors connecting with testosterone result in increased biosynthesis of specific proteins needed for synaptogenesis (Frye, Edinger, Seliga, & Wawrzycki 2004). Additionally, Granger and colleagues (2003) found that low levels of testosterone or a steeper decline in testosterone throughout the day was associated with mood disorders and depression levels in 13–14 year-old adolescents. By contrast, Steiner, Dunn, and Born (2003) suggest that increased levels of sex hormones like testosterone within puberty could be associated with higher depression levels. However, there are also studies that do not find these associations in adolescents (Buchanan, Eccles, & Becker 1992; Brooks-Gunn & Warren 1989).

Brain Structure

The human brain is subject to significant development of different regions from the age of 4 to 21 (Gogtay et al. 2004; Weir, Zakama, & Rao 2012). Different studies have associated mood disorder diseases like depression with reduced volumes in different brain regions like the white matter in the frontal lobe (Steingard et al. 2002) and the limbic system (e.g. Rosso et al. 2005; McKinnon, Yucel, Nazarov, & MacQueen 2009). Compared to healthy individuals children and adolescents suffering from depression show a reduced volume of the amygdala (Rosso et al. 2005; Hamilton, Siemer, & Gotlib 2008). Additionally, the hippocampus has been shown to be reduced in volume in children who experienced an ongoing depression for at least two years (McKinnon et al. 2009), which can be explained in parts with a higher concentration of the catabolic cortisol. However, MacMillan and colleagues (2003) found that increased volumes of the amygdala in relation to hippocampus volume is rather linked to anxiety disorders than to severity or duration of depression. Regarding the effects of exercise on well-being, it was found that children and adolescents with higher fitness levels showed better performances in memory tasks compared to less fit children, which could be linked to an increased volume of the hippocampus (Chaddock, Pontifex, Hillman, & Kramer 2011). Davis and colleagues (2011) additionally claim that exercise increases activity in the prefrontal cortex and may therefore benefit cognitive performance. Using functional magnetic resonance imaging (fMRI), chronic exercise programs were further shown to positively affect children's brain activation in the surrounding areas like the anterior cingulate and superior frontal gyrus that also support executive functioning (Krafft et al. 2014).

Serotonergic System

The functioning of the serotonin system (5-Hydroxytryptamin; 5HT) including its main metabolite 5-hydroxyindoleacetic acid (5HIAA), decreased plasma tryptophan, and low tryptophan-amino acid ratio have been previously identified as one of the major causes of mood disorders and depression (Meltzer 1989; Coccaro, Siever, & Klar 1989; Quintana 1992; Träskman, Åsberg, Bertilsson, & Sjüstrand 1981). Although much of this research has been conducted with adults, it was also shown that adolescents' mood is associated with peripheral serotonin levels (Crowell et al. 2008). Moreover, treatment of depressed adolescents with selective serotonin reuptake inhibitors (SSRI) has been shown to be very effective (Ryan 2005). Dahlström et al.

(2000) pointed to a difference in serotonin activation in depressed children compared to adults. Children's serotonin transporter availability to the presynaptic neuron, which adds to the serotonergic effect, is increased in the midbrain and hypothalamus. It has been shown in adults that physical activity and exercise increase 5HT and 5HIAA neurotransmitter activity (Young 2007; Post and Goodwin 1974). Meta-analytic evidence supports the view that physical activity potentially lowers depressive symptoms also in children and adolescents (Brown et al. 2013). To date, it is assumed that serotonin only expresses its health benefits through its effects on neurotrophic growth factors, which explains why acute serotonergic effects on mental health cannot be observed (Groves 2007).

Brain-Derived Neurotrophic Growth Factor (BDNF)

From studies with animals we know that physical activity may increase BDNF levels in the hippocampus (Cotman & Berchtold 2002; Erickson, Miller, & Roecklein 2012), the cortex and the cerebellum (Neeper, Gomez-Pinilla, Choi, & Cotman et al. 1996; Gomez-Pinilla, Ying, Opazo, & Edgerton 2001), which benefit health and the survival of nerve cells in these areas. In fact, physical activity in this sense acts much like antidepressants enhancing the expression of neurotrophic growth factors, for example, in the hippocampus (Duman & Monteggia 2006). In adults, reduced levels of BDNF could be associated with higher likelihood of developing depressive symptoms (Neves-Pereira et al. 2002). It was also shown that not every individual benefits the same from physical activity and that a certain genotype is more prone for exercise-induced benefits to health (Mata, Thompson, & Gotlib 2010). This effect is pronounced in girls – girls that genetically show higher risks for depression actually benefit more from physical activity.

Still little is known about the exercise–mental health relationship regarding physiological mechanisms in the young age group. Much of the research suggesting positive effects stems from animal research or studies with adults. Although bidirectional links between exercise and neurobiological activity, exercise and mental health, as well as neurobiological activity and mental health have been repeatedly reported, seldom do studies identify mediating neurobiological mechanisms for the exercise–mental health link. Future studies are strongly needed that more closely investigate these mechanisms in children and adolescents.

References

American Psychiatric Association. 2013. *Diagnostic and Statistical Manual of Mental Disorders: DSM-V*. 5th ed. Arlington: American Psychiatric Association.

Amnesi, J. J. 2005. "Correlations of depression and total mood disturbance with physical activity and self-concept in preadolescents enrolled in an after-school exercise program." *Psychological Reports 96*: 891–8.

Angold, A., E. J. Costello, A. Erkanli, & C. M. Worthman. 1999. "Pubertal changes in hormone levels and depression in girls." *Psychological Medicine 29* (5): 1043–53.

Angold, A., E. J. Costello, S. C. Messer, A. Pickles, F. Winder, & D. Silver. 1995. "Development of a short questionnaire for use in epidemiological studies of depression in children and adolescents." *International Journal of Methods in Psychiatric Research 5*: 237–49.

Angold, A., E. J. Costello, & C. M. Worthman. 1998. "Puberty and depression: The roles of age, pubertal status and pubertal timing." *Psychological Medicine 28* (1): 51–61.

Bahrke, M. S., & R. G. Smith. 1985. "Alterations in anxiety of children after exercise and rest." *American Corrective Therapy Journal 39* (4): 90–4.

Beck, A. T., R. A. Steer, R. Ball, & W. F. Ranieri. 1996. "Comparison of Beck Depression Inventories-IA and-II in Psychiatric Outpatients." *Journal of Personality Assessment 67* (3): 588–97.

Biddle, S. J. H., & M. Asare. 2011. "Physical activity and mental health in children and adolescents: A review of reviews." *British Journal of Sports Medicine 45*: 886–95.

Bjelland, I., A. A. Dahl, T. T. Haug, & D. Neckelmann. 2002. "The validity of the Hospital Anxiety and Depression Scale. An updated literature review." *Journal of Psychosomatic Research 52*: 69–77.

Blair, S. N., H. W. Kohl, & K. E. Powell. 1987. "Physical activity, physical fitness, exercise, and the public's health." In *The cutting edge in physical education and exercise science research*, edited by M. J. Safrit, & H. M. Eckert. Champaign, IL: Human Kinetics.

Bonhauser, M., G. Fernandez, K. Püschel, F. Yañez, J. Montero, B. Thompson, & G. Coronado. 2005. "Improving physical fitness and emotional well-being in adolescents of low socioeconomic status in Chile: Results of a school-based controlled trial." *Health Promotion International 20* (2): 113–22.

Brickenkamp, R. 2002. *d2-Aufmerksamkeits-Belastungs-Test: Manual [The d2 test of attention: Manual]*. Göttingen: Hogrefe.

Brooks-Gunn, J., & M. P. Warren. 1989. "Biological and social contributions to negative affect in young adolescent girls." *Child Development 60* (1): 40–55.

Brown, H. E., N. Pearson, R. E. Braithwaite, W. J. Brown, & S. J. H. Biddle. 2013. "Physical activity interventions and depression in children and adolescents." *Sports Medicine 43* (3): 1–12.

Brown, S. W., M. C. Welsh, E. E. Labbe, W. F. Vitulli, & P. Kulkarni. 1992. "Aerobic exercise in the psychological treatment of adolescents." *Perceptual and Motor Skills 74*: 555–60.

Buchanan, C. M., J. S. Eccles, & J. B. Becker. 1992. "Are adolescents the victims of raging hormones? Evidence for activational effects of hormones on moods and behavior at adolescence." *Psychological Bulletin 111* (1): 62–107.

Budde, H., S. Pietrassyk-Kendziorra, S. Bohm, & C. Voelcker-Rehage. 2010a. "Hormonal responses to physical and cognitive stress in a school setting." *Neuroscience Letters 474* (3): 131–4.

Budde, H., R. Schwarz, B. Velasques, P. Ribeiro, M. Holzweg, S. Machado,...M. Wegner. 2016. "The need for differentiating between exercise, physical activity, and training." *Autoimmunity Reviews 15*: 110–11.

Budde, H., A. Kaulitzky, S. Amatriain Fernández, S. Machado, A. Emeljanovas, M. Wegner, & E. Murillo-Rodriguez. Forthcoming. "Effects of physical exercise on depression disorders in children and adolescents: A systematic review of meta-analyses and neurobiological mechanisms."

Budde, H., C. Voelcker-Rehage, S. Pietrassyk-Kendziorra, S. Machado, P. Ribeiro, & A. M. Arafat. 2010b. "Steroid hormones in the saliva of adolescents after different exercise intensities and their influence on working memory in a school setting." *Psychoneuroendocrinology 35* (3): 382–91.

Budde, H., C. Voelcker-Rehage, S. Pietrassyk-Kendziorra, P. Ribeiro, & G. Tidow. 2008. "Acute coordinative exercise improves attentional performance in adolescents." *Neuroscience Letters 441* (2): 219–23.

Cairns, K., M. B. H. Yap, P. D. Pilkington, & A. F. Jorm. 2014. "Risk and protective factors for depression that adolescents can modify: A systematic review and meta-analysis of longitudinal studies." *Journal of Affective Disorders 169*: 61–75.

Caspersen, C. J., K. E. Powell, & G. M. Christenson. 1985. "Physical activity, exercise, and physical fitness: Definitions and distinctions for health-related research." *Public Health Reports 100* (2): 131.

Caterino, M. C., & E. D. Polak. 1999. "Effects of two types of activity on the performance of second-, third-, and fourth-grade students on a test of concentration." *Perceptual and Motor Skills 89* (1): 245–8.

Chaddock, L., M. B. Pontifex, C. H. Hillman, & A. F. Kramer. 2011. "A review of the relation of aerobic fitness and physical activity to brain structure and function in children." *Journal of the International Neuropsychological Society 17*: 1–11.

Coccaro, E. F., L. J. Siever, & H. M. Klar. 1989. "Serotonergic studies in patients with affective and personality disorders. Correlates with suicidal and impulsive aggressive behavior." *Archives of General Psychiatry 47* (7): 587–99.

Colditz, G. A. 1999. "Economic costs of obesity and inactivity." *Medicine and Science in Sports and Exercise 31*: 663–7.

Collins, A., & E. Koechlin. 2012. "Reasoning, learning, and creativity: Frontal lobe function and human decision-making." *PLoS Biology 10*: e10011293.

Cotman, C. W., & N. C. Berchtold. 2002. "Exercise: A behavioral intervention to enhance brain health and plasticity." *Trends Neurosci 25* (6): 295–301.

Cox, R. H., M. P. Martens, & W. D. Russell. 2003. "Measuring anxiety in athletics: The revised Competitive State Anxiety Inventory-2." *Journal of Sport and Exercise Psychology 25* (4): 519–33.

Craft, L. L., & D. M. Landers. 1998. "The effect of exercise on clinical depression and depression resulting from mental illness: A meta-analysis." *Journal of Sport and Exercise Psychology 20* (4): 339–57.

Crews, D. J., M. R. Lochbaum, & D. M. Landers. 2004. "Aerobic physical activity effects on psychological well-being in low-income hispanic children." *Perceptual and Motor Skills 98*: 319–24.

Crova, C., I. Struzzolino, R. Marchetti, I. Masci, G. Vannozzi, R. Forte, & C. Pesce. 2014. "Cognitively challenging physical activity benefits executive function in overweight children." *Journal of Sports Sciences 32* (3): 201–11.

Crowell, S. E., T. P. Beauchaine, E. McCauley, C. J. Smith, C. A. Vasilev, & A. L. Stevens. 2008. "Parent–child interactions, peripheral serotonin, and self-inflicted injury in adolescents." *Journal of Consulting and Clinical Psychology 76* (1): 15–21.

Dahlström, M., A. Ahonen, H. Ebeling, P. Torniainen, J. Heikkilä, & I. Moilanen. 2000. "Elevated hypothalamic/midbrain serotonin (monoamine) transporter availability in depressive drug-naive children and adolescents." *Molecular Psychiatry 5* (5): 514–22.

Daley, A. J., R. J. Copeland, N. P. Wright, A. Roalfe, & J. K. Wales. 2006. "Exercise therapy as a treatment for psychopathologic conditions in obese and morbidly obese adolescents: A randomized, controlled trial." *Pediatrics 118* (5): 2126–34. doi:10.1542/peds.2006-1285

Davis, C. L., P. D. Tomporowski, C. A. Boyle, J. L. Waller, P. H. Miller, J. A. Naglieri, & M. Gregoski. 2007. "Effects of aerobic exercise on overweight children's cognitive functioning: A randomized controlled trial." *Research Quarterly for Exercise and Sport 78* (5): 510–19.

Davis, C. L., P. D. Tomporowski, J. E. McDowell, B. P. Austin, P. H. Miller, N. E. Yanasak,…J. A. Naglieri. 2011. "Exercise improves executive function and achievement and alters brain activation in overweight children: A randomized, controlled trial." *Health Psychology 30* (1): 91–8.

Diamond, A. 2005. "Attention-deficit disorder (attention-deficit/hyperactivity disorder without hyperactivity): A neurobiologically and behaviorally distinct disorder from attention-deficit/hyperactivity disorder (with hyperactivity)." *Developmental Psychopathology 17*: 807–25.

Diamond, A. 2013. "Executive functions." *Annual Review of Psychology 64*: 135–68.

Diener, E. 1984. "Subjective well-being." *Psychological Bulletin 95*: 542–75.

Diener, E., & F. Fujita. 1995. "Resources, personal strivings, and subjective well-being: A nomothetic and idiographic approach." *Journal of Personality and Social Psychology 68*: 926–35.

Donaldson, S. J., & K. R. Ronan. 2006. "The effects of sports participation on young adolescents' emotional well-being." *Adolescence 41* (162): 369–89.

Duclos, M., C. Gouarne, & D. Bonnemaison. 2003. "Acute and chronic effects of exercise on tissue sensitivity to glucocorticoids." *Journal of Applied Physiology 94* (3): 869–75.

Duman, R. S., & L. M. Monteggia. 2006. "A neurotrophic model for stress-related mood disorders." *Biological Psychiatry 59* (12): 1116–27.

Erickson, K. I., D. L. Miller, & K. A. Roecklein. 2012. "The aging hippocampus interactions between exercise, depression, and BDNF." *The Neuroscientist 18* (1): 82–97.

Everson, E. S., A. J. Daley, & M. Ussher. 2006. "Does exercise have an acute effect on desire to smoke, mood and withdrawal symptoms in abstaining adolescent smokers?" *Addictive Behaviors 31*: 1547–58.

Fairchild, G., S. H. van Goozen, S. J. Stollery, M. R. Aitken, & J. Savage. 2009. "Decision making and executive function in male adolescents with early-onset or adolescence-onset conduct disorder and control subjects." *Biological Psychiatry 66*: 162–8.

Fedewa, A. L., & S. Ahn. 2011. "The effects of physical activity and physical fitness on children's achievement and cognitive outcomes: A meta-analysis." *Research Quarterly for Exercise and Sport 82* (3): 521–35.

Frye, C. A., K. L. Edinger, A. M. Seliga, & J. M. Wawrzycki. 2004. "5α-reduced androgens may have actions in the hippocampus to enhance cognitive performance of male rats." *Psychoneuroendocrinology 29*: 1019–27.

Gauvin, L., & J. C. Spence. 1996. "Physical activity and psychological well-being: Knowledge base, current issues, and caveats." *Nutrition Reviews 54* (4): S53–S67.

Gogtay, N., J. N. Giedd, L. Lusk, K. M. Hayashi, D. Greenstein, A. C. Vaituzis,…P. M. Thompson. 2004. "Dynamic mapping of human cortical development during childhood through early adulthood." *PNAS 101* (21): 8174–9.

Gold, J. M., C. Carpenter, C. Randolph, T. E. Goldberg, & D. R. Weinberger. 1997. "Auditory working memory and Wisconsin Card Sorting test performance in schizophrenia." *Archives of General Psychiatry 54* (2): 159–65.

Gomez-Pinilla, F., Z. Ying, P. Opazo, & V. R. Edgerton. 2001. "Differential regulation by exercise of BDNF and NT-3 in rat spinal cord and skeletal muscle." *European Journal of Neuroscience 13* (6): 1078–84.

Gotlib, I. H., J. Joormann, K. L. Minor, & J. Hallmayer. 2008. "HPA axis reactivity: A mechanism underlying the associations among 5-HTTLPR, stress, and depression." *Biological Psychiatry 63*: 847–51.

Granger, D. A., E. A. Shirtcliff, C. Zahn-Waxler, B. Usher, B. Klimes-Dougan, & P. Hastings. 2003. "Salivary testosterone diurnal variation and psychopathology in adolescent males and females: Individual differences and developmental effects." *Development and Psychopathology 15* (2): 431–49.

Groves, J. O. 2007. "Is it time to reassess the BDNF hypothesis of depression?" *Molecular Psychiatry 12*: 1079–88.

Gunnar, M. R., S. Wewerka, K. Frenn, J. D. Long, & C. Griggs. 2009. "Developmental changes in hypothalamus-pituitary-adrenal activity over the transition to adolescence: Normative changes and associations with puberty." *Development and Psychopathology 21* (1): 69–85.

Hamilton, J. P., M. Siemer, & I. H. Gotlib. 2008. "Amygdala volume in major depressive disorder: A meta-analysis of magnetic resonance imaging studies." *Molecular Psychiatry 13* (11): 993–1000.

Hampson, E. 1995. "Spatial cognition in humans: Possible modulation by androgens and estrogens." *Journal of Psychiatry and Neuroscience 20* (5): 397–404.

Hankin, B. L., L. S. Badanes, J. R. Z. Abela, & S. E. Watamura. 2010. "Hypothalamic–pituitary–adrenal axis dysregulation in dysphoric children and adolescents: Cortisol reactivity to psychosocial stress from preschool through middle adolescence." *Biological Psychiatry 68*: 484–90.

Helmich, I., A. S. Latini, A. Sigwalt, M. G. Carta, S. Machado, B. Velasques,...H. Budde. 2010. "Neurobiological alterations induced by exercise and their impact on depressive disorders." *Clinical Practice and Epidemiology in Mental Health 6*: 115–25.

Hillman, C. H., K. I. Erickson, & A. F. Kramer. 2008. "Be smart, exercise your heart: Exercise effects on brain and cognition." *Nature Reviews Neuroscience 9* (1): 58–65.

Hilyer, J. C., D. G. Wilson, C. Dillon, L. Caro, C. D. Jenkins, W. A. Spencer,...W. Booker. 1982. "Physical fitness training and counseling as treatment for youthful offenders." *Journal of Counseling Psychology 29* (3): 292–303.

Holsboer, F. 2000. "The corticosteroid receptor hypothesis of depression." *Neuropsychopharmacology 23* (5): 477–501.

Howley, E. T. 2001. "Type of activity: Resistance, aerobic and leisure versus occupational physical activity." *Medicine and Science in Sports and Exercise 33* (6): S364–S369.

Hughes, J. R. 1984. "Psychological effects of habitual aerobic exercise: A critical review." *Preventive Medicine 13*: 66–78.

Janowsky, J. S. 2006. "Thinking with your gonads: Testosterone and cognition." *Trends in Cognitive Sciences 10* (2): 77–82.

Kamijo, K., M. B. Pontifex, K. C. O'Leary, M. R. Scudder, C. T. Wu, D. M. Castelli, & C. H. Hillman. 2011. "The effects of an afterschool physical activity program on working memory in preadolescent children." *Developmental Science 14* (5): 1046–58.

Kellner, M., K. Wiedemann, & F. Holsboer. 1992. "Atrial natriuretic factor inhibits the CRH-stimulated secretion of ACTH and cortisol in man." *Life Sciences 50* (24): 1835–42.

Kohn, R. 2018. "Epidemiology of common mental disorders." In H. Budde & M. Wegner (Eds.), *Exercise and mental health: Neurobiological mechanisms of the exercise effect on depression, anxiety, and well-being*. New York: Taylor and Francis.

Koutsandréou, F., M. Wegner, C. Neumann, & H. Budde. In press. "Effects of motor versus cardiovascular exercise training on children's working memory." *Medicine & Science in Sports & Exercise 48* (6): 1144–52.

Kovacs, M. 2005. *The Children's Depression Inventory (CDI)*. North Tonowanda, NY: MHS.

Krafft, C. E., N. F. Schwarz, L. Chi, A. L. Weinberger, D. J. Schaeffer, J. E. Pierce,... J. E. McDowell. 2014. "An 8-month randomized controlled exercise trial alters brain activation during cognitive tasks in overweight children." *Obesity 22* (1): 232–42.

Kudielka, B. M., A. Buske-Kirschbaum, D. H. Hellhammer, & C. Kirschbaum. 2004. "HPA axis responses to laboratory psychosocial stress in healthy elderly adults, younger adults, and children: Impact of age and gender." *Psychoneuroendocrinology 29* (1): 83–98.

Lagerberg, D. 2005. "Physical activity and mental health in schoolchildren: A complicated relationship." *Acta Paediatrica 94*: 1699–701.

Larun, L., L. V. Nordheim, E. Ekeland, K. B. Hagen, & F. Heian. 2006. "Exercise in prevention and treatment of anxiety and depression among children and young people." *Cochrane Database of Systematic Reviews* (3): CD004691.

Lau, P. W. C., C. W. Yu, A. M. Lee, & R. Y. T. Sung. 2004. "The physiological and psychological effects of resistance training on Chinese obese adolescents." *Journal of Exercise Science and Fitness 2* (2): 115–20.

Lazarus, R. S., & S. Folkman. 1984. *Stress, appraisal, and coping*. New York: Springer.

Liebert, E. M., & L. W. Morris. 1967. "Cognitive and emotional components of test anxiety: A distinction and some intitial data." *Psychological Reports 20*: 975–8.

Lofrano-Prado, M. C., J. O. Hill, H. J. Silva, C. R. Freitas, S. Lopes-de-Souza, T. A. Lins, & W. L. do Prado. 2012. "Acute effects of aerobic exercise on mood and hunger feelings in male obese adolescents: A crossover study." *International Journal of Behavioral Nutrition and Physical Activity 9* (38): 1–6.

Lopez-Duran, N. L., M. Kovacs, & C. J. George. 2009. "Hypothalamic–pituitary–adrenal axis dysregulation in depressed children and adolescents: A meta-analysis." *Psychoneuroendocrinology 34* (9): 1272–83.

McKinnon, M. C., K. Yucel, A. Nazarov, & G. M. MacQueen. 2009. "A meta-analysis examining clinical predictors of hippocampal volume in patients with major depressive disorder." *Journal of Psychiatry and Neuroscience 34* (1): 41–54.

MacMahon, J., & R. T. Gross. 1988. "Physical and psychological effects of aerobic exercise in delinquent adolescent males." *American Journal of Diseases of Children 142* (12): 1361–6.

MacMillan, S., P. R. Szeszko, G. J. Moore, R. Madden, E. Lorch, J. Ivey,...D. R. Rosenberg. 2003. "Increased amygdala: Hippocampal volume ratios associated with severity of anxiety in pediatric major depression." *Journal of Child and Adolescent Psychopharmacology 13* (1): 65–73.

McNair, D., M. Lorr, & L. Droppleman. 1992. *Profile of Mood States (POMS) manual (rev.)*. San Diego, CA: Educational and Industrial Testing Service.

Mandroukas, K., A. Zakas, N. Aggelopoulou, K. Christoulas, G. Abatzides, & M. Karamouzis. 1995. "Atrial natriuretic factor responses to submaximal and maximal exercise." *British Journal of Sports Medicine 29* (4): 248–251.

Martikainen, S., A.-K. Pesonen, J. Lahti, K. Heinonen, K. Feldt, R. Pyhälä, T. Tammelin, E. Kajantie, J. G. Eriksson, T. E. Strandberg, & K. Räikkönen. 2013. "Higher levels of physical activity are associated with lower hypothalamic-pituitary-adrenocortical axis reactivity to psychosocial stress in children." *The Journal of Clinical Endocrinology and Metabolism 98* (4): E619–E627.

Mata, J., R. J. Thompson, & I. H. Gotlib. 2010. "BDNF genotype moderates the relation between physical activity and depressive symptoms." *29* (2): 130–3.

Mathur, D. N., A. L. Toriola, & O. A. Dada. 1986. "Serum cortisol and testosterone levels in conditioned male distance runners and nonathletes after maximal exercise." *Journal of Sports Medicine and Physical Fitness 26* (3): 245–50.

Meltzer, H. 1989. "Serotonergic dysfunction in depression." *British Journal of Psychiatry* Suppl. *8*: 25–31.

Mendelson, T., M. T. Greenberg, J. K. Dariotis, L. F. Gould, B. L. Rhoades, & P. J. Leaf. 2010. "Feasibility and preliminary outcomes of a school-based mindfulness intervention for urban youth." *Journal of Abnormal Child Psychology 38* (7): 985–94.

Miyake, A., N. P. Friedman, M. J. Emerson, A. H. Witzki, A. Howerter, & T. D. Wager. 2000. "The unity and diversity of executive functions and their contributions to complex "frontal lobe" tasks: A latent variable analysis." *Cognitive Psychology 41* (1): 49–100.

Mumm, J. L. M., S. Bischoff, & A. Ströhle. 2018. "Exercise and anxiety disorders." In H. Budde & M. Wegner (Eds.), *Exercise and mental health: Neurobiological mechanisms of the exercise effect on depression, anxiety, and well-being*. New York: Taylor and Francis.

Mutrie, N., K. Richards, S. Lawrie, & G. Mead. 2018. "Can physical activity prevent or treat clinical depression?" In H. Budde & M. Wegner (Eds.), *Exercise and mental health: Neurobiological mechanisms of the exercise effect on depression, anxiety, and well-being*. New York: Taylor and Francis.

Naglieri, J. A., & J. P. Das. 1997. *Cognitive assessement system: Interpretive handbook*. Itasca, IL: Riverside Publishing.

Neeper, S. A., F. Gomez-Pinilla, J. Choi, & C. W. Cotman. 1996. "Physical activity increases mRNA for brain-derived neurotrophic factor and nerve growth factor in rat brain." *Brain Research 726* (1–2): 49–56.

Nemeroff, C. B., E. Widerlov, G. Bissette, H. Walleus, I. Karlsson, K. Eklund,...W. Vale. 1984. "Elevated concentrations of CSF corticotropin-releasing factor-like immunoreactivity in depressed patients." *Science 226* (4680): 1342–4.

Neves-Pereira, M., E. Mundo, P. Muglia, N. King, F. Macciardi, & J. L. Kennedy. 2002. "The brain-derived neurotrophic factor gene confers susceptibility to bipolar disorder: Evidence from a family-based association study." *American Journal of Human Genetics 71* (3): 651–5.

Niemann, C., M. Wegner, C. Voelcker-Rehage, A. M. Arafat, & H. Budde. 2013. "Influence of physical activity and acute exercise on cognitive performance and saliva testosterone in preadolescent school children." *Mental Health and Physical Activity 6* (3): 197–204.

Norris, R., D. Carroll, & R. Cochrane. 1992. "The effects of physical activity and exercise training on psychological stress and well-being in an adolescent population." *Journal of Psychosomatic Research 36* (1): 55–65.

Olshansky, S. J., D. J. Passaro, R. C. Hershow, J. Layden, B. A. Carnes, J. Brody,...D. S. Ludwig. 2005. "A potential decline in life expectancy of the United States in the 21st century." *New England Journal of Medicine 352*: 1138–45.

Pariante, C., & S. L. Lightman. 2008. "The HPA axis in major depression: Classical theories and new developments." *Trends in Neurosciences 31* (9): 464–8.

Parker, K. J., A. F. Schatzberg, & D. M. Lyons. 2003. "Neuroendocrine aspects of hypercortisolism in major depression." *Hormones and Behavior 43* (1): 60–6.

Pesce, C. 2012. "Shifting the focus from quantitative to qualitative exercise characteristics in exercise and cognition research." *Journal of Sport and Exercise Psychology 34*: 766–86.

Petruzzello, S. J., A. C. Jones, & A. K. Tate. 1997. "Affective responses to acute exercise: A test of opponent-process theory." *Journal of Sports Medicine and Physical Fitness 37* (3): 205–12.

Petty, K. H., C. L. Davis, J. Tkacz, D. Young-Hyman, & J. L. Waller. 2009. "Exercise effects on depressive symptoms and self-worth in overweight children: A randomized controlled trial." *Journal of Pediatric Psychology 34* (9): 929–39.

Pinker, S. 1999. "How the mind works." *Annals of the New York Academy of Sciences 882* (1): 119–27.

Post, R. M., & F. K. Goodwin. 1974. "Simulated behavior states: An approach to specificity in psychobiological research." *Biological Psychiatry 7* (3): 237–54.

Pratt, M., M. A. Macera, & G. Wang. 2000. "Higher direct medical costs associated with physical inactivity." *The Physician and Sportsmedicine 28*: 63–79.

Pruessner, J. C., D. H. Hellhammer, & C. Kirschbaum. 1999. "Burnout, perceived Stress, and cortisol responses to awakening." *Psychosomatic Medicine 61* (2): 197–204.

Quintana, J. 1992. "Platelet serotonin and plasma tryptophan decreases in endogenous depression." *Journal of Affective Disorders 24* (2): 55–62.

Raadsheer, F. C., W. J. Hoogendijk, F. C. Stam, F. J. H. Tilders, & D. F. Swaab. 1994. "Increased numbers of corticotropin-releasing hormone expressing neurons in the hypothalamic paraventricular nucleus of depressed patients." *Neuroendocrinology 60* (4): 436–44.

Reynolds, W. M. 1989. *Reynolds Child Depression Scale*. Odessa, TX: Psychological Assessment Resources.

Rosso, I. M., C. M. Cintron, R. J. Steingard, P. F. Renshaw, A. D. Young, & D. A. Yurgelun-Todd. 2005. "Amygdala and hippocampus volumes in pediatric major depression." *Biological Psychiatry 57* (1): 21–6.

Rubin, R. T., R. E. Poland, I. M. Lesser, R. A. Winston, & A. N. Blodgett. 1987. "Neuroendocrine aspects of primary endogenous depression. Cortisol secretory dynamics in patients and matched controls." *Archives of General Psychiatry 44* (4): 328–36.

Rudolph, D. L., & E. McAuley. 1998. "Cortisol and affective responses to exercise." *Journal of Sports Sciences 16* (2): 121–8.

Ryan, N. D. 2005. "Treatment of depression in children and adolescents." *Lancet 366* (9489): 933–40.

Scheuer, J., & C. M. Tipton. 1977. "Cardiovascular adapations to physical training." *Annual Review of Physiology 39*: 221–51.

Schuhfried, G. 1984. *Vienna Test System: Version 10.85/B – A-2340*. Mödling, Austria: Schuhfried.

Secretary of Health and Human Services and the Secretary of Education. 2007. "Promoting better health for young people through physical activity and sports." *Centers for Disease Control and Prevention*. www.cdc.gov/healthyyouth/physicalactivity/promoting_health.

Sibley, B. A., & J. L. Etnier. 2003. "The relationship between physical activity and cognition in children: A meta analysis." *Pediatric Exercise Science 15*: 243–56.

Spielberger, C. D. (Ed.). 1972. *Anxiety: Current trends in theory and research.* New York: Academic Press.

Spielberger, C. D. 1973. *State-Trait Anxiety Inventory for Children preliminary manual.* Palo Alto, CA: Consulting Psychologists Press.

Spielberger, C. D. 1983. *Manual for the State-Trait Anxiety Inventory: STAI (Form Y).* Palo Alto, CA: Consulting Psychologists Press.

Spielberger, C. D., R. L. Gorsuch, & R. E. Luchene. 1976. *Manual for the State-Trait Anxiety Inventory.* Palo Alto, CA: Consulting Psychologist Press.

Steiner, M., E. Dunn, & L. Born. 2003. "Hormones and mood: From menarche to menopause and beyond." *Journal of Affective Disorders 74* (1): 67–83.

Steingard, R. J., P. F. Renshaw, J. Hennen, M. Lenox, C. B. Cintron, A. D. Young,...D. A. Yurgelun-Todd. 2002. "Smaller frontal lobe white matter volumes in depressed adolescents." *Biological Psychiatry 52* (5): 413–17.

Sternberg, S. 1966. "High-speed scanning in human memory." *Science 153*: 652–4.

Ströhle, A., M. Kellner, F. Holsboer, & K. Wiedemann. 2001. "Anxiolytic activity of atrial natriuretic peptide in patients with panic disorder." *American Journal of Psychiatry 158* (9): 1514–16.

Stroud, L. R., E. Foster, G. D. Papandonatos, K. Handwerger, D. A. Granger, K. T. Kivlighan, & R. Niaura. 2009. "Stress response and the adolescent transition: Performance versus peer rejection stressors." *Development and Psychopathology 21* (1): 47–68.

Taylor, C. B., J. F. Sallis, & R. Needle. 1985. "The relation of physical activity and exercise to mental health." *Public Health Reports 100* (2): 195–202.

Taylor Tavares, J. V., L. Clark, D. M. Cannon, K. Erickson, W. C. Drevets, & B. J. Sahakian. 2007. "Distinct profiles of neurocognitive function in unmedicated unipolar depression and bipolar II depression." *Biological Psychiatry 62*: 917–24.

Träskman, L., M. Åsberg, L. Bertilsson, & L. Sjüstrand. 1981. "Monoamine metabolites in csf and suicidal behavior." *Archives of General Psychiatry 38* (6): 631–6.

Trudeau, F., & R. J. Shephard. 2010. "Relationships of physical activity to brain health and the academic performance of schoolchildren." *American Journal of Lifestyle Medicine 4*: 138–50.

Vaynman, S., & F. Gomez-Pinilla. 2006. "Revenge of the 'sit': How lifestyle impacts neuronal and cognitive health through molecular systems that interface energy metabolism with neuronal plasticity." *Journal of Neuroscience Research 84*: 699–715.

Verburgh, L., M. Königs, E. J. A. Scherder, & J. Oosterlaan. 2014. "Physical exercise and executive functions in preadolescent children, adolescents and young adults: A meta-analysis." *British Journal of Sports Medicine 48* (12): 973–9.

Viner, R., & R. Booy. 2005. "Emidemiology of health and illness." *British Medical Journal 330*: 411–14.

Wegner, M., I. Helmich, S. Machado, O. Arias-Carrión, & H. Budde. 2014. "Effects of exercise on anxiety and depression disorders: Review of meta-analyses and neurobiological mechanisms." *CNS and Neurological Disorders – Drug Targets 13* (6): 1002–14.

Wegner, M., J. M. Koedijker, & H. Budde. 2014. "The effect of acute exercise and psychosocial stress on fine motor skills and testosterone concentration in the saliva of high school students." *PLoS ONE 9* (3): e92953.

Wegner, M., C. Niemann, & H. Budde. 2015. "Physiological and psychological associations of testosterone in sports and exercise with due regard to adolescents." In L. Sher & T. Rice (Eds.), *Neurobiology of men's mental health* (pp. 67–82). New York: Nova Publishers.

Wegner, M., C. Windisch, & H. Budde. 2012. "Psychophysische Auswirkungen von akuter körperlicher Belastung im Kontext Schule: Ein Überblick [The psychological effects of acute physical stress in the school context: An overview]." *Zeitschrift für Sportpsychologie 19* (1): 37–47.

Weir, J. M., A. Zakama, & U. Rao. 2012. "Developmental risk I: Depression and the developing brain." *Child and Adolescent Psychiatric Clinics of North America 21* (2): 237–59.

West, R., & P. Hajek. 2004. "Evaluation of the mood and physical symptoms scale (MPSS) to assess cigarette withdrawal." *Psychopharmacology 177*: 195–9.

Wolf, O. T., & C. Kirschbaum. 2002. "Endogenous estradiol and testosterone levels are associated with cognitive performance in older women and men." *Hormones and Behavior 41* (3): 259–66.

Woodcock, R. N., & M. B. Johnson. 1989. *Revised tests of cognitive ability, test 10. Standard and supplemental batteries.* Allen, TX: DLM Teaching Resources.

World Health Organization. 2001. *The World Health Report 2001: Mental health: New understanding, new hope.* Geneva: World Health Organization.

World Health Organization. 2004. "Prevalence, severity, and unmet need for treatment of mental disorders in the World Health Organization: World Mental Health Survey." *Journal of the American Medical Association 291*: 2581–90.

Young, S. N. 2007. "How to increase serotonin in the human brain without drugs." *Journal of Psychiatry and Neuroscience 32* (6): 394–9.

Zervas, Y., A. Danis, & V. Klissouras. 1991. "Influence of physical exertion on mental performance with reference to training." *Perceptual and Motor Skills 72*: 1215–21.

6 The Exercise Effect on Mental Health in Older Adults

Inna Bragina, Claudia Niemann, and Claudia Voelcker-Rehage

Introduction

Preventing cognitive impairments and maintaining psychological well-being are essential parts of healthy and satisfied aging. Promising evidence exists that physical activity (PA) and exercise can thereby make crucial contributions. PA is an important lifestyle factor, which has not only physical benefits, but is also associated with the preservation of mental health across the entire life span (Kramer & Erickson 2007; Netz, Wu, Becker, & Tenenbaum 2005). For instance, studies examining the effect of PA and exercise on the prevention or postponement of dementia/MCI reveal promising results (see also Chapter 18 in this book). Furthermore it seems that PA or exercise can prevent depression and anxiety disorders, or at least reduce the symptoms (see also Chapter 13 and Chapter 18 in this book). Additionally, the exposure to PA leads to an increased self-perception and self-evaluation, which is important for a successful and satisfying life (McAuley et al. 2005).

Examinations on the PA–mental health relationship investigate either the association between an individual's *overall physical activity level* with, or the effect of a (long-term) *targeted and structured exercise intervention* on cognitive impairment, depression, anxiety disorders, or self-perception. However, the explaining mechanisms underlying the effects of regular physical activity and exercise on the different facets of mental health are not completely understood up to now (see also Chapter 2 in this book).

Cognition

Healthy Aging of Cognitive Functions and the Brain

In mean, the aging process is characterized by a decline of important cognitive functions and changes in brain structure and function. Those cognitive functions that belong to the so-called "fluid intelligence" (for example, the speed and accuracy of perceptual processes, working memory, and inhibitory control functions) seem to start to decline very early in life (from 25–30 years of age) (Hedden & Gabrieli 2004; Hommel, Li, & Li 2004). Similarly, episodic and long-term memory performance show an age-related decline from middle adulthood on (Park et al. 2002). In contrast, cognitive functions that belong to the "crystallized intelligence" (for example, knowledge and wisdom based abilities, strategies of processing and learning, and learned skills like reading, writing, or occupational skills), remain stable or even increase until old age and are often able to compensate for the decline in the abilities of fluid intelligence during activities of daily living. Nevertheless, particularly under laboratory settings, differences in performance levels between young and old subjects increase with task difficulty and/or complexity.

On the brain level, grey matter and white matter volumes decrease with age. Declining grey matter volume has been related to a decrease in the quality and quantity of connections between neurons (Peters 2002). Particularly, dendritic branches and spines show age-related decline. Further, a reduction of blood capillaries and glial cells contributes to brain volume decline. Strongest brain volume decline is observed in the caudate nucleus (as part of the basal ganglia), the cerebellum, the hippocampus, and the PFC, and these changes are well correlated to deficits in executive control and memory processes. Spatial memory seems to be particularly sensitive to a loss of axodendritic synapses in the dentate gyrus. On the contrary, there is much less decline in the limbic system and occipital (visual) cortex (Park & Reuter-Lorenz 2009; Raz & Rodrigue 2006). Also, the microstructure of the white matter changes with age. Density and integrity of axons as well as their myelinization get impaired and are often regarded as basis for slowed and less efficient processing of cognitive, motor, and sensory information.

In addition to structural brain data, functional neuroimaging data reveal "over- and underactivation" as compared to younger adults in the aging cortex as well as changed activity patterns. Underactivation of the prefrontal cortex (PFC) is often observed in older adults with difficulties in working memory and executive control (Hedden & Gabireli 2004) and may be conceived as equivalent of reduced integrity of cortical areas and neuronal circuitries. On the other hand, overactivation has been shown in those brain regions representing executive functions, motor control, and episodic, autobiographical, and working memory (Reuter-Lorenz & Lustig 2005; Seidler et al. 2010). It has been suggested that increased activity in the PFC of older adults compensates for processing deficits in the sensory or other domains of functioning. Further task-specific age effects include a decreased lateralization in activation of the PFC, which might be either the consequence of compensation processes to enable normal cognitive functioning by recruiting contralateral resources, or a decreased specialization of brain processes reflecting difficulties in recruiting specialized neuronal processes (Li et al. 2004). Various findings suggest, however, that changed processing strategies in well-performing older adults may lead to youth-like activation patterns. Based on these findings, behaviorally derived theories suggest that along with aging a failure in top-down (self-initiated) control and regulation of activation of task-specific brain regions like the occipital and mediotemporal cortices occurs (Reuter-Lorenz & Lustig 2005).

Cognitive and Neuronal Plasticity

The described age-related changes in brain and cognition show remarkable individual differences (inter-individual variability) but also differences within a person between different functions and structures (intra-individual variability). Aging trajectories may be delayed or reveal changes in slope in both a positive and a negative direction and reveal the plasticity of the aging process. Plasticity denotes an individual's potential for modifications in his or her developmental trajectory throughout the lifespan. Cognitive plasticity refers to the potential for the modifiability of the trajectory of cognitive development within one individual (Baltes, Lindenberger, & Staudinger 1998). Brain plasticity denotes the fact that the brain shows structural and/or functional changes when people are faced with new or altering demands (Lövdén, Bäckman, Lindenberger, Schaefer, & Schmiedek 2010; Staudinger 2012) or as a reaction to a loss of neural resources as a consequence of lesions, diseases, or even aging. The latter type of neural plasticity is strongly tight to compensational mechanisms in the aging brain. The degree of plasticity depends on the available individual (physiological, psychological) or contextual (social, cultural) developmental resources available (cf. Staudinger, Marsiske, & Baltes 1995). The high inter- and intra-individual variability of cognitive impairment during aging indicates that besides genetic predisposition individual lifestyle is a crucial factor. PA or exercise is an important and successful possibility to stimulate cognitive and neuronal plasticity.

Dementia and Mild Cognitive Impairment

In the western world approximately 6–8% of the older population suffers from a mild form and another 6–8% from more severe forms of dementia (Förstl & Lang 2011). Due to prolonged life expectancy, the incidence and prevalence of Alzheimer's disease and other forms of dementia are expected to quadruple over the next 50 years (Hebert, Beckett, Scherr, & Evans 2001). Dementia is associated with major declines in cognitive functioning through neuronal loss, especially in the hippocampus, substantia innominata, locus coeruleus, and in the temporo-parietal and frontal cortex (International Statistical Classification of Diseases and Related Health Problems-10 (ICD-10). In accordance to the ICD-10, the prerequisites for a diagnosis of dementia are decreases in memory and intellectual capacity affecting the person's everyday life. The most common forms of dementia occurring in older age are the neurodegenerative forms Alzheimer dementia (AD) and Lewy-body dementia (aggregation of specific proteins within the brain (β-amyloid and microtubule-associated protein tau (MAPT) in AD and α-synuclein in Lewy body dementia) and vascular dementia (difficulties with the supply of blood to the brain).

Mild cognitive impairments (MCI) are regarded and diagnosed as autonomous psychological diseases. MCIs have to be distinguished from "normal" age-related decline and early stages of dementia. It is a transitional state between the cognitive changes of normal cognitive aging and dementia. Persons with MCIs suffer from cognitive decline, memory impairment, and problems with concentration. Additionally, decline of those cognitive functions, which belong to the "fluid intelligence" can occur. In contrast to dementia, MCI does not lead to impairments in everyday life (Petersen et al. 2001). MCIs are classified into two types. MCI that primarily affects memory is known as amnestic MCI. Persons with amnestic MCI may start to forget important information that they would previously have recalled easily. MCI that affects cognitive skills other than memory is known as nonamnestic MCI. Persons with MCI have a probability of developing a dementia within 5 years of about 50% (e.g. Boyle, Wilson, Aggarwal, Tang, & Bennett 2006; Zaudig 2005, 2011).

A successful medical treatment to prevent the development or progression of MCI and dementia is so far not available (Ahlskog, Geda, Graff-Radford, & Petersen 2011). There is, however, evidence that PA and exercise may have a positive influence on the development of dementia and MCI even if considering the fact that some types cannot be inhibited but at least attenuated (see also Chapter 18 in this book). PA seems to be especially promising for older adults with a genetic risk for dementia (Smith et al. 2011).

Physical Activity to Prevent or Postpone MCI and Dementia

An increasing number of studies have suggested that lifestyle factors like a socially integrated network, cognitive leisure activity, and regular PA can prevent or postpone dementia and MCI (Fratiglioni, Paillard-Borg, & Winblad 2004). The cognitive reserve theory suggests that these factors may have an influence on the cognitive reserves, which allows to cope better with the pathologic changes in dementia (Scarmeas et al. 2003; Stern 2002). Among the lifestyle factors PA seems to have the greatest protective effect against MCI and dementia (for review cf. Bherer, Erickson, & Teresa Liu-Ambrose 2013; Hertzog, Kramer, Wilson, & Lindenberger 2008; Kramer, Bherer, Colcombe, Dong, & Greenough 2004). In this vein, research on the individual genetic predisposition underscores the meaning of PA. Generally, persons who possess one or more apolipoprotein E-ε4 (APOE-ε4) alleles are at an increased risk to develop AD (Kim, Basak, & Holtzman 2009). However, APOE-ε4 carriers who are physically active exhibit a reduced risk of MCI (Geda et al. 2010) and AD (Laurin, Verreault, Lindsay, MacPherson, & Rockwood 2001) in comparison to sedentary carriers and even non-carriers (Etnier et al. 2007).

Also a number of longitudinal cohort studies have demonstrated that midlife PA or exercise may contribute to the maintenance of cognitive functioning and delay or reduce the risk of late-life dementia (Andel et al. 2008; Geda et al. 2010; Chang et al. 2010). In this vein, a recent population-based cohort study of over 1.1 million Swedish male conscripts revealed that poor cardiovascular fitness was associated with a seven-fold increased risk for early-onset mild cognitive impairment and dementia (Nyberg et al. 2014). A meta-analysis of 15 prospective studies (n = 33,816) documented a significantly reduced risk of dementia associated with midlife PA (Sofi et al. 2011). Twelve years later 3,210 persons exhibited impairments in cognitive functioning. Physically active persons had a 38% reduced risk of cognitive decline as compared to inactive individuals. Even a moderate level of activity resulted in a 35% reduced risk. Other systematic reviews and meta-analyses of prospective cohort studies revealed also positive but less convincing results with respect to dementia (Blondell, Hammersley-Mather, & Veerman 2014: -18% reduced risk; Hamer & Chida 2009: -28% reduced risk). PA performed later in life has beneficial effects as well. Boyle, Buchman, Wilson, Leurgans, and Bennett (2010) observed 761 healthy older persons over a period of 12 years. They demonstrated that persons with a better physical condition are at lower risk of cognitive decline or MCI. Sumic, Michael, Carlson, Howieson, and Kaye (2007) supported these results revealing that women (85+ years) who reported exercising for more than 4 h per week showed a by 88% reduced risk to develop cognitive impairments as compared to those less active women. In a cross-sectional study, Burns and colleagues (2008) used an objective measure to assess cardiovascular fitness and were able to show that cardiorespiratory fitness was negatively related to brain atrophy in the earliest clinical stages of AD and may moderate AD-related brain volume decline. Despite these positive results it should be taken into account that the association of PA with cognitive preservation does not allow to draw causal conclusions and could be also explained by reverse causality. That is, persons with very early, preclinical neurodegenerative disease might not be motivated to engage in regular PA.

Nevertheless, reverse causality does not explain improvements in cognitive functions in short-term randomized controlled trials (RCT). For instance, Baker et al. (2010) could prove this positive effect after a 6-month high-intensity aerobic exercise (75–85% of maximum heart rate) in comparison to a stretching control program with 33 older patients suffering from amnestic MCI aged 55–85 years. Their results demonstrated beneficial effects of aerobic exercise, especially on speed of processing and executive functioning. However, in some tests they observed gender differences in cognitive improvement despite comparable improvement in cardiorespiratory fitness. But the effect of PA or exercise on cognitive functioning does not seem to be so clear. Whereas Heyn, Abreu, and Ottenbacher (2004) prove by meta-analysis that exercise interventions lead to increases in fitness, physical functions and cognition in persons suffering from dementia, in a more recent systematic review of 22 RCTs Öhman, Savikko, Strandberg, and Pitkälä (2014) revealed more conflicting results. On the one hand, exercise in subjects with MCI can lead to positive cognitive outcomes, mainly on global cognition, executive functions, and attention. On the other hand, results of exercise intervention studies among persons with dementia showed no effect of exercise on cognition. It has to be taken into consideration that some of the studies had methodological problems in defining dementia/MCI diagnosis, in blinding the group assignment, in recruiting adequate sample sizes (very small samples) and in reporting dropouts, compliance, or complications. Furthermore, it is noteworthy that physical frailty (including walking speed, muscle strength, body composition, and fatigue) seems to have an influence on the risk of developing cognitive decline and MCI.

Types of Exercise and Doses

Most of the studies dealing with the positive effects of PA and exercise on MCI/dementia investigated the effect of aerobic exercise (e.g. Baker et al. 2010; Eggermont, Swaab, Hol, & Scherder 2009; Bherer et al. 2013; Ahlskog et al. 2011 for a review). Other studies examined the effects of other types of exercise on MCI/dementia. For instance, Lam et al. (2011) assessed the cognitive impact of a Tai-Chi intervention in comparison to stretching and toning. Their results demonstrated that both groups improved in global cognitive function, delayed recall, and subjective cognitive complaints. However, improvements in Clinical Dementia Rating scores were observed only in the Tai-Chi group. Similarly, in their systematic review and meta-analysis, Wayne and colleagues (2014) concluded that Tai Chi training was effective in improving global cognitive functions in cognitively intact but also cognitively impaired older adults. Nagamatsu et al. (2013) compared the effect of aerobic and strength exercise with a sedentary control group in older adults with possible MCI. The results showed that both experimental groups improved in spatial memory performance in comparison to the control group. However, only the aerobic training group exhibited a significant correlation between spatial memory performance and overall physical capacity after the intervention. The meta-analysis by Heyn, Johnsons, and Kramer (2008) confirmed the results by Nagamatsu and colleagues (2013) revealing no differences in effect sizes of RCTs between strength or aerobic training or a combination of both in cognitively impaired individuals.

Current literature provides evidence that different types of exercise, like cardiovascular exercise, Tai Chi, and strength training, might benefit cognitive functions in older adults and reduce the risk for developing dementia. However, future studies are needed to further specify which type of intervention – aerobic, strength training, coordinative elements (such as in Tai Chi interventions) – is most beneficial to improve cognition in MCI/dementia patients.

Only a few studies examined the dose-response relationship of exercise in persons with MCI/dementia and there is no consensus in literature. For instance, Larson et al. (2006) found a dose-response effect for exercise frequency if they considered the physical functioning of the demented persons. Patients with low physical functioning, exercising three or more times per week, exhibited a decreased risk for dementia in comparison with those who exercised fewer than three times. Sofi and colleagues (2011) found similar estimates of effectiveness for both high and low-to-moderate intensity of exercise whereas Laurin et al. (2001) revealed that with increased intensity the preventive effect of PA increases. Further, Scarmeas and colleagues (2003) support the assumption that higher levels of PA are more effective.

Mechanisms of the Exercise Effect on MCI and Dementia

Many common putative biological mechanisms explaining the positive effect of exercise on MCI/dementia are based on the underlying assumption that PA may facilitate neuroplasticity, which is fundamental for learning, memory, and general cognition (Ahlskog et al. 2011). The upregulation of neurotrophic factors (e.g. BDNF, IGF-1) by cardiovascular exercise are in turn a possible underlying mechanism of neuroplastic processes through PA. Further mechanisms of the positive exercise effect in MCI and dementia are the regulating effects on brain β-amyloid, MAPT, and brain vascularization.

NEUROPLASTICITY

Various animal studies identified several mechanisms whereby PA and exercise may facilitate neuroplasticity. Exercise seems to increase the expression of synaptic plasticity genes, gene products (synapsin, synaptophysin), and various neuroplasticity-related transcription factors (Berchtold, Castello, & Cotman 2010; Shen, Tong, Balazs, & Cotman 2001). Furthermore, hippocampal dendritic length and dendritic spine complexity can be enhanced through exercise (Eadie, Redila, & Christie 2005; Redila & Christie 2006). Additional research provides evidence that exercise induces neurogenesis within the hippocampal dentate gyrus (van Praag, Kempermann, & Gage 1999; Eadie et al. 2005; Redila & Christie 2006; Fabel et al. 2003) even in older ages (van Praag, Shubert, Zhao, & Gage 2005; Kronenberg et al. 2006).

NEUROTROPHIC FACTORS

It has been demonstrated that BDNF is able to modulate brain plasticity, including increased neurotic outgrowth and synaptic function in vitro. Additionally, it seems to promote in vitro survival of a vast array of neurons affected by neurodegenerative diseases, including AD (Murer, Yan, & Raisman-Vozari 2001). Patients with AD have reduced circulating BDNF levels (e.g. Laske et al. 2007). Furthermore, AD patients whose state is declining rapidly exhibit significantly lower serum BDNF levels than those whose condition is declining slowly (Laske et al. 2011). Acute and chronic exposure to PA was shown to result in increased peripheral levels of BDNF (for reviews see Coelho et al. 2012; Huang, Larsen, Ried-Larsen, Møller, & Andersen 2014).

Insufficiency of IGF-1 (insulin-like growth factor) seems to be another risk factor for AD (Vega et al. 2006). Castellano and White (2008) demonstrated in a small cross-sectional study that patients with AD have significantly lower circulating IGF-1 levels than the healthy control group. IGF-1 is one of the most important hormones for growth and development in humans (Sonntag, Ramsey, & Carter 2005). Studies revealed that IGF-1 enhances synaptic plasticity and neuronal survival and increase concentrations of BDNF (Cotman & Berchtold 2002; Vaynman, Ying, Yin, & Gomez-Pinilla 2006). Additionally Thornton, Ingram, and Sonntag (2000) indicated that IGF-1 replacement enhanced learning and memory in rats. Long-term resistance training seems to be able to increase serum IGF-1 concentrations (Adamo & Farrar 2006; Hameed et al. 2004).

BRAIN β-AMYLOID AND MICROTUBULE-ASSOCIATED PROTEIN TAU (MAPT)

AD as the most common form of dementia is associated with the accumulation of neuritic plaques (primary component β-amyloid) and neurofibrillary tangles which are aggregates of the hyperphosphorylated microtubule-associated protein tau (MAPT). Liang et al. (2010) revealed that cognitively normal older adults who are physically active have a lower density of β-amyloid. This indicates a lower risk for neuritic plaques. However, Baker et al. (2010) could not confirm this dependence. Further, in an animal study, Belarbi et al. (2011) were able to prevent the phosphorylation of the tau protein in the hippocampus and delay memory impairment through 9 months of aerobic exercise.

VASCULAR CONTRIBUTIONS

The development of AD and vascular dementia is associated with glucose intolerance and diabetes mellitus, hypertension, hyperlipidemia, obesity, and different types of cardiovascular deficits (Knopman & Roberts 2010). Regular PA is well known to prevent or at least attenuate each of these risk factors (Smith 2001; Pitsavos, Panagiotakos, Weinem, & Stefanadis 2006; Kokkinos, Sheriff, & Kheirbek 2011). For example, the maintenance of adequate cerebral blood flow is essential to maintain a constant supply of oxygen and nutrients to the metabolically active brain. It is among others influenced by the partial pressure of arterial carbon dioxide (PaCO2, termed cerebrovascular reactivity) and by the cardiac output (Murrell et al. 2013). A reduced cerebral blood flow was associated with AD and dementia (de la Torre 2012; Murrell et al. 2013).

There is conflicting evidence on the effect of PA on cerebral blood flow and cerebrovascular reactivity. Ainslie et al. (2008) were able to demonstrate that a greater aerobic capacity achieved through regular PA leads to higher cerebral blood flow in healthy older adults. Furthermore, after a 3-month cardiovascular training, cerebral blood volume in the hippocampus of middle-aged adults was also enhanced, pointing to a better vascularization of the tissue, and was in line with better memory performance of the participants (Pereira et al. 2007). However Ivey, Ryan, Hafer-Macko, and Macko (2011) could not confirm the findings with regard to cerebral blood flow in stroke patients after a 6-month aerobic exercise intervention. But they were able to reveal increases in cerebrovascular reactivity. Additionally, animal studies showed that vascular endothelial growth factor (VEGF) expression is enhanced after low- and moderate-intensity exercise (Lou, Liu, Chang, & Chen 2008).

Psychological Well-Being

Maintaining psychological well-being in the face of the changes and losses of later life (e.g. cognitive and physical aging) is a crucial part of healthy ageing (Baltes & Baltes 1990). In this vein, Gale, Cooper, Deary, and Sayer (2014) demonstrated that psychological well-being at the age of 60+ may be a protective factor against becoming frail over a 4-year follow-up period.

The literature provides a variety of interchangeable terms for psychological well-being. For instance, psychological health, mental health (Netz et al. 2005), and mental well-being (Black et al. 2015). Although there is no general applicable definition of psychological health, it is commonly accepted that it is a multifaceted construct including hedonic (pleasure) and eudaimonic (control, autonomy, and self-realization) dimensions (Gale et al. 2014; Netz et al. 2005; Ware 2003).

However, so far research on the positive effects of PA and exercise on psychological well-being has not focused on the multifaceted construct, but on selected components like depression, anxiety disorders, and self-perception (including self-efficacy and self-esteem). In this part of the chapter we will summarize the effects on these domains and give an overview of proposed effect mechanism.

Depression and Anxiety

Interaction of Dementia and Depression

Depression is one of the most frequently diagnosed comorbid psychiatric disorders of Alzheimer disease and other types of dementia (Starkstein, Mizrahi, & Power 2008). Up to 87% of older adults with AD are estimated to be depressed (Carpenter, Ruckdeschel, Ruckdeschel, & Van Haitsma 2003). Dementia in combination with depression is associated with greater disability of daily living, a worse quality of life, a faster cognitive decline, and higher mortality and institutionalization rates as compared to non-depressed individuals with dementia (Lyketsos et al. 1997; Kales, Chen, Blow, Welsh, & Mellow 2005). For example, Starkstein, Jorge, Mizrahi, and Robinson (2014) examined the clinical correlates of depression in 670 patients with AD and revealed that depressed patients had more severe social dysfunction and greater impairment in activities of daily living than AD patients without depression. It is noteworthy that even mild levels of depression are significantly associated with more functional impairments in AD.

It has been shown that exercise is effective in reducing depressive symptoms in cognitively intact depressed persons (cf. below). However, less is known about the effects of exercise on depression in older adults with AD. Williams and Tappen (2008) compared the effect of a 16-week comprehensive exercise treatment (including strength, balance, flexibility, and aerobic exercise, n=17) with a supervised walking-group (n=17) and an attention-control group (n=12) in depressed patients with AD. Depression was reduced in all three groups with some evidence of superior benefit from both exercise groups. The findings of this study were limited because of the small sample size. Further research with a larger sample is necessary to confirm these results and to draw conclusions about relative benefits of different types of exercise.

Depression

Depression is one of the most prevalent mental disorders in old age (Luppa et al. 2012). In most epidemiological investigations, the prevalence of major depressive disorder in community samples of adults aged 65+ ranges from 1% to 5% (e.g. Hasin, Goodwin, Stinson, & Grant 2005). Especially patients with (auto) inflammatory diseases (e.g. rheumatoid arthritis) have an increased incidence of depressive disorders (Naarding et al. 2005).

Depressive disorders are associated with an increased risk of morbidity and suicide as well as decreased physical, cognitive, and social functioning (Blazer 2003). The prevalence and the underlying mechanisms of late-life depression seem not to differ from depression in younger age but the phenomenology may be partly different. In older adults, depression seems to be especially associated with cognitive decline, somatic symptoms, and loss of interest. In younger adults, emotional conspicuity seems to be more prevalent (Fiske, Wetherell, & Gatz 2009; Hegeman, Kok, van der Mast, & Giltay 2012).

Currently, antidepressant medication is the accepted treatment of choice for depression, to some extent in combination with psychological therapy (Robertson, Robertson, Jepson, & Maxwell 2012; Hollon, Thase, & Markowitz 2002). However, the success of the described treatments is limited and questioned for mild depression (Moncrieff & Kirsch 2005). Mura, Moro, Patten, & Carta (2014) emphasize that a high percentage of patients do not respond to a first-line antidepressant drug and that augmentation strategies increase the risk of side effects. Additionally, the effectiveness of antidepressant medication seems to be less effective in older patients (> age 75; Roose et al.

2015). Therefore, there is a need for an additional therapy, which is effective and has minimal side effects (Mura et al. 2014).

Exercise seems to be such a promising treatment method (Brenes et al. 2007; Blumenthal et al. 2007) (see also Chapter 18 in this book). Blumenthal and colleagues (1999) conducted one of the most well-known investigations on this topic. They compared the antidepressant effect of aerobic exercise and medical treatment in a sample of 156 older patients. They were able to demonstrate that the efficacy of exercise was generally comparable with the effect of antidepressant medication (Blumenthal et al. 1999). The initial response was faster with antidepressant medication; however, after 16 weeks of treatment exercise was equally effective (Blumenthal et al. 1999). Based on this result, several studies used exercise as treatment for depressive disorders in older adults (e.g. Brenes et al. 2007; Kerse et al. 2010) and confirmed the efficiency of exercise interventions on the reduction of depressive symptoms in old adults (for a review and meta-analysis on 18 RTCs in older adults (cf. Park, Han, & Kang 2014). Similarly, Sjösten and Kivelä (2006) confirmed in a systematic review of RCTs that for older subjects exercise interventions might be efficient in the reduction of depressive symptoms.

PA also seems to have preventive effects for the development of depressive symptoms. In older adults it has been shown that depression is associated with low PA levels (Hassmén, Koivula, & Uutela 2000; Farmer et al. 1988). Strawbridge, Deleger, Roberts, & Kaplan (2002) investigated the relationship between PA and the prevalence and incidence of depression over 5 years in a sample of 1.947 community-dwelling adults aged 50–94 years. Even with adjustments for age, sex, ethnicity, financial strain, chronic conditions, disability, body mass index, alcohol consumption, smoking, and social relations, they showed a negative correlation between the level of PA and the occurrence of depressive symptoms. In a recent review Mammen and Faulkner (2013) emphasized that the maintenance of PA over the life span may serve as a valuable mental health promotion strategy in reducing the risk of developing depression.

Which Types and Doses of Exercises Lead to Reductions in Depression?

Rethorst, Wipfli, and Landers (2009) analyzed the moderating variables of exercise programs in terms of exercise type. They emphasize that exercise programs with combined aerobic and resistance exercise resulted in greater antidepressant effects than aerobic or resistance training alone. However, they included healthy as well as clinically depressed subjects and had no focus on older adults.

In regard to intervention parameters, a meta-analysis conducted by Park et al. (2014) revealed that significant antidepressant benefits of exercise interventions in older adults were apparent after 3 months. Another study compared the effectiveness of four different doses of aerobic exercise (3 or 5 days per week with 7.0 kcal/kg/week or 17.5 kcal/kg/week). The results revealed that only performing PA at least 5 times per week with high dose achieved the response rates of a medication intervention (Dunn, Trivedi, Kampert, Clark, & Chambliss 2005). However, further research on the optimal duration, the type of exercise, the frequency, and the intensity of the exercise programs for the specific target group of older adults is necessary.

Anxiety Disorders

Anxiety is an umbrella term covering several different forms of abnormal and pathological fear. Anxiety is typified by several symptoms depending on the degree of severity including emotional (worry, self-doubt, apprehension), behavioral (nervousness, trembling, tics) and physiological (hyperventilation, increases in heart rate, blood pressure, muscle tension, perspiration, stress hormone levels) conspicuities (ICD-10; Knapen & Vancampfort 2013). Anxiety disorders can vary from mild discomfort to panic disorder. Anxiety which only occurs in response to specific circumstances in which a person perceives a lack of control or uncertainty has been defined as state anxiety, whereas a person's predisposition to become anxious across many situations has been termed trait anxiety (Taylor 2000). While most studies including older adults focus on the effects of physical activity on trait anxiety levels, less research has been performed for the effects of physical activity on state anxiety in older adults.

Bryant, Jackson, and Ames (2008) reported that the prevalence of anxiety disorders in community-dwelling adults aged 55+ range from 1.2% to 15% and is almost twice as high in clinical samples (geriatric inpatients, nursing home residents). The prevention and treatment of anxiety disorders is of high importance because permanent anxiety may lead to the inability to perform activities of daily living (Trollor, Anderson, Sachdev, Brodaty, and Andrews 2007) and to fear of social interaction or even suicide (Taylor 2000). The common treatment for anxiety is similar to that for depression (Wipfli, Rethorst, & Landers 2008) consisting of pharmacological and psychotherapeutic interventions, also for older adults (Gonçalves & Byrne 2012). Population-based evidence indicates that regular exercise also has an anxiolytic effect and may be used at least as an additional therapy (Herring, Lindheimer, & O'Connor 2013). This has been supported by a number of meta-analyses which investigated the effect of exercise on trait anxiety across all age groups (see also Mumm et al. in this book). These studies report that exercise leads to decreases in anxiety symptoms and that its effectiveness is nearly as great as that of the common treatments (psychotherapy and pharmacotherapy; Kugler, Seelbach, & Krüskemper 1994; Long & Van Stavel 1995; Petruzzello, Landers, Hatfield, Kubitz, & Salazar 1991; Wipfli et al. 2008). Early meta-analyses indicated that middle-aged adults benefitted the most from the anxiolytic effect of exercise (Schlicht 1994; Wipfli et al. 2008). However, more recent analysis did not identify any age-related differences and confirm that the anxiolytic effect of exercise is also valid for older individuals as well (Coventry & Hind 2007; Netz et al. 2005; Wegner et al. 2014). However there are also contradictory results in regard of the exercise effect on anxiety. For instance a recent systematic review conducted by Snowden and others (2015) did not found sufficient evidence for the positive exercise effect on emotional health (and especially anxiety).

Types of Exercise and Doses

Early studies indicated the anxiolytic effect (on trait and state anxiety) of exercise only for aerobic exercise and not for resistance training (e.g. Petruzzello et al. 1991). In contrast, Herring and colleagues (2013) revealed also significant reductions in trait anxiety after 6 weeks of resistance training (effect size = 0.52; and effect size = 0.54 6 after 6 weeks of aerobic exercise) in 37 adults with generalized anxiety disorder. The anxiolytic effect of nonaerobic training was supported by an investigation with older adults, who tend to be anxious (aged 65 to 75 years). Results revealed that participants who performed a high resistance training over a period of 24 weeks exhibited a reduction in anxiety symptoms and improved mood in comparison to a control group (Cassilhas, Antunes, Tufik, & de Mello 2010). Considering the different effects of acute and chronic resistance training, Bibeau and colleagues (2010) showed that acute vigorous resistance training resulted in a temporary increase in state anxiety immediately after the training and a return to baseline level after 20–60 minutes. Chronic resistance training, on the other hand, was associated with decreased state and trait anxiety (Asmundson et al. 2013). Also yoga and tai chi seem to have anxiolytic effects (Asmundson et al. 2013). These types of exercise may be especially interesting for older adults, who may be unable to participate in more vigorous or high-impact exercise (e.g. jogging). Only a few studies investigated the dose-response relation of exercise and anxiety reduction, but none of them was conducted with older persons. As for MCI/dementia, the power of the anxiolytic effects seems to depend on the duration of the exercise intervention (Petruzzello et al. 1991). Further, for meaningful reductions in trait anxiety the duration of the interventions seem to need to be performed in bouts of at least 10 weeks and not less than 21 minutes of exercise (Petruzzello et al. 1991). With regard to the frequency, exercise programs with a frequency of 3 or 4 times per week resulted in significantly higher effect sizes than programs with 1–2 or 5 exercise sessions per week (Wipfli et al. 2008). Further, the degree of the acute anxiolytic benefit seems to be linked to the fitness level of a person. During exercises with higher intensities only fit participants showed increases in exhilaration (Steptoe & Bolton 1988).

Mechanisms of the Exercise Effect on Depression and Anxiety

Depression and anxiety in late life are frequently comorbid with other physical and psychological disorders like arthritis, heart disease, diabetes, and cancer (Blazer 2000; Roy-Byrne et al. 2008). Many bodily symptoms of these disorders could be alleviated through PA or exercise as well. In turn

PA may have a positive influence on depression and anxiety symptoms. Thus, PA influences depression indirectly by the alleviation of comorbid symptoms and directly by altering the underlying neurobiological processes (Moylan et al. 2013).

The literature presents various possible neurobiological mechanisms of PA and exercise on the reduction in depressive and anxiolytic symptoms (Medina, Jacquart, & Smits 2015; Wegner et al. 2014) (see also Chapter 2 in this book). Nevertheless, there is no consensus in literature up to now. Considerable evidence exists that PA leads to changes in neurotransmitter and cortisol levels, influences the inflammatory system and results in an upregulation of neurotrophic factors.

NEUROTRANSMITTER

Depressed persons commonly exhibit imbalances in the production and transmission of neurotransmitters such as serotonin, dopamine, noradrenaline, and glutamate (Maletic et al. 2007). The most extensive research has been conducted on the dysfunction of the serotonin system (Kari, Davidson, Kohl, & Kochhar 1978; Lopresti, Hood, & Drummond 2013). Depression is associated with deficits in the serotonin availability and abnormalities of serotonin receptors (Carr & Lucki 2011). The role of serotonin in anxiety disorders is not fully elucidated up to now. Indeed, as mentioned before, the common successful treatment of anxiety is similar to that of depression (Wegner et al. 2014), aiming to increase the reuptake of serotonin by increasing its availability in the synaptic cleft (Blier & de Montigny 1994). Animal studies revealed that chronic exercise could increase neurotransmitter levels like serotonin as well (e.g. Kiuchi, Lee, & Mikami 2012).

HPA AXIS ACTIVITY

Chronic psychological stress and aging per se are associated with a hyperactivity of the HPA (hypothalamic-pituitary-adrenocortical) axis and increased levels of glucocorticoids (cortisol) (Born, Ditschuneit, Schreiber, Dodt, & Fehm 1995; Seeman, Singer, Wilkinson, & McEwen 2001). Prolonged high activation of the HPA axis is a health risk for the organism because high levels of glucocorticoids seem to induce brain region specific damage (e.g. Fuss et al. 2010).

Approximately 50–60% of persons with depressive disorders exhibit changes in the HPA system (Ströhle 2003) and an increased cortisol release (Deuschle et al. 1997). In comparison with depressive patients, persons with anxiety disorders have lower cortisol secretion (Curtis, Cameron, & Nesse 1982), but they still exhibit higher cortisol levels than healthy persons (Schreiber, Lauer, Krumrey, Holsboer, & Krieg 1996).

On the one hand, in healthy young adults acute moderate PA leads to an increased cortisol release (e.g. Kirschbaum & Hellhammer 1994; Brownlee,

Moore, & Hackney 2005; Daly, Seegers, Rubin, Dobridge, & Hackney 2005; Hackney 2006). On the other hand improved physical fitness is associated with a lower reactivity of the HPA axis to an acute stressor (Rimmele et al. 2009) and lower baseline cortisol levels (measured in young depressive women: Nabkasorn et al. 2006). It is proposed that exercise might reduce symptoms of anxiety and depression through a reduction in the reactivity of the HPA system and attenuated glucocorticoid responses to psychological stress (Rimmele et al. 2009; Wegner et al. 2014).

INFLAMMATION AND OXIDATIVE AND NITROGEN STRESS

Persons suffering from a depressive disorder seem to have a dysregulation of the inflammatory system (Hiles, Baker, de Malmanche, & Attia 2012). Dowlati et al. (2010) confirmed elevated levels of pro-inflammatory cytokines in depressed persons. The most reliable elevations were observed in Interleukin-6 (IL-6). Along with the elevated levels of pro-inflammatory cytokines among depressed individuals, several studies show lower than average levels of anti-inflammatory cytokines such as Interleukin-10 (IL-10) (e.g. Mesquita et al. 2008). Furthermore, older patients with auto inflammatory (e.g. rheumatoid arthritis) diseases have an increased risk for an incident depression (Naarding et al. 2005)

In contrast to depressive disorders the role of inflammation in anxiety disorder pathogenesis has received less attention so far, although research is growing (Moylan et al. 2013; Hou & Baldwin 2012). Current literature on this topic is heterogeneous, including various comorbidities and use of different inflammatory markers. Overall, anxiety disorders also seem to be associated with the described imbalance in pro- and anti-inflammatory cytokines resulting in an elevated inflammatory state of the central nervous system (Moylan et al. 2013). This inflammatory state might contribute to increased oxidative and nitrogen stress. Reactive oxygen and nitrogen species (ROS and RNS) are produced during normal physiologic processes. However, when their levels exceed a critical level of antioxidant capacity of cells ROS and RNS are associated with the damage of cells (Moylan et al. 2013; Moylan et al. 2014).

To sum up, anxiety and depressive disorders are both associated with an increased level of oxidative and nitrogen stress markers (Hovatta, Juhila, & Donner 2010; Leonard & Maes 2012). This state of enhanced stress leads to brain-region specific, stress-related damage (Hovatta et al. 2010). In addition, chronic low-grade inflammation is associated with a reduced neurotransmitter (e.g. serotonin) biosynthesis (Sperner-Unterweger, Kohl, & Fuchs 2014).

Silverman and Deuster (2014) described an anti-inflammatory effect of regular physical exercise through promoting an anti-inflammatory environment (see also Eyre, Papps, & Baune 2013). Acute high-intensity exercise temporally leads to an exponential increase in the production of the pro-inflammatory cytokine IL-6 in skeletal muscle. This, in turn, leads to an activation of the synthesis of anti-inflammatory cytokines (IL-10) and inhibits the release of other pro-inflammatory cytokines such as TNF-alpha (Silverman & Deuster 2014). Chronic exercise seems to trigger the same anti-inflammatory processes (Kohut et al. 2006; Silverman & Deuster 2014) like acute exercise does. Furthermore, first animal studies revealed that regular moderate exercise is protective against oxidative stress damage through enhanced antioxidant enzyme activity, e.g. MDA (marker of lipid peroxidation) (Salim et al. 2010).

NEUROTROPHIC FACTORS

The "neurotrophin hypothesis of depression" is based on observations that decreases in hippocampal BDNF levels are correlated with stress-induced depressive behaviors and that antidepressant treatment enhances the expression of BDNF (Duman & Monteggia 2006). Additional studies showed that depressed persons have lower peripheral BDNF concentrations in comparison to non-depressed individuals (Karege et al. 2002; Molendijk et al. 2014). The role of BDNF in anxiety disorders is ambiguous. A model described by Salim et al. (2011) assumes that sub-chronic oxidative stress, as observed in both depression and anxiety, could lead to a degradation of BDNF. Chan, Wu, Chang, Hsu, and Chan (2010) proposed an antioxidant role of BDNF, meaning that as an immediate protective mechanism, BDNF levels increase during acute oxidative stress. When, however, the oxidative stress becomes chronic, this protective mechanism is lacking due to the reduction of BDNF. Additionally, BDNF is also considered to be essential for the maintenance of the synaptic plasticity (Rothman & Mattson 2013, see section "Neurotrophic factors" in the Dementia section). On this account, Salim et al. (2011) concluded that a decreased BDNF level favors a condition with high oxidative stress and reduced plasticity which are both associated with a potentially high anxiety state.

Acute and chronic exposure to PA is able to increase the synthesis, release, and expression of BDNF (Huang et al. 2014). A meta-analysis by Szuhany and colleagues (2015) revealed that regularly performed PA showed greater increases of peripheral BDNF compared to baseline level than a single session of exercise. PA of moderate intensity might induce enhancement of peripheral BDNF levels even in older adults (Coelho et al. 2013). However, exercise-induced expression of BDNF is age-dependent and less pronounced in older than in younger individuals (Adlard, Perreau, & Cotman 2005; for a review see Mora, Segovia, & del Arco 2007).

Self-Efficacy and Self-Esteem

PA and global quality of life in older adults seems to be indirectly associated by way of self-efficacy and physical self-esteem (White, Wójcicki, & McAuley 2009). Self-esteem is widely accepted as a key indicator of emotional stability and adjustment to life demands. Self-esteem is regarded as one of the strongest predictors of subjective well-being and is therefore an important element of mental well-being and quality of life (Fox 2000). It is unclear whether self-esteem shows normative changes with increasing age. However, if a decline occurs in old age, a high socioeconomic status and physical health seem to be protective factors (Orth, Trzesniewski, & Robins 2010). PA is an important component of positive self-evaluation (McAuley et al. 2005). Elavsky et al. (2005) emphasized that self-esteem of older adults has consistently been shown to be influenced by PA.

Self-efficacy beliefs regulate human functioning through cognitive, motivational, affective, and decisional processes. They affect whether individuals think in self-enhancing or self-debilitating ways, how well they motivate themselves and persevere in the face of difficulties; the quality of their emotional life and vulnerability to stress and depression (Benight & Bandura 2004). Research suggests that self-efficacy can act as a determinant and a consequence of PA participation. On the one hand, self-efficacy seems to be a significant predictor of exercise adherence and compliance (McAuley & Blissmer 2000). For instance, Sallis et al. (1986) were able to demonstrate that changes in self-efficacy over time were linked to changes in PA behavior. On the other hand, exercise could be a situation in which persons become aware of their own efficacy (McAuley & Blissmer 2000). McAuley, Courneya, and Lettunich (1991) demonstrated that acute and chronic exercise result in significant increases in self-efficacy. McAuley et al. (2006) analyzed the interaction between PA, self-efficacy, physical/mental health status, and global well-being. Their results indicated that older women who were more active showed greater self-efficacy levels as well as better physical and mental health. In turn, health status was positively related to satisfaction with life. Additionally, Kaplan, Ries, Prewitt, and Eakin (1994) concluded that self-efficacy is a significant predictor of long-term survival for patients with chronic obstructive pulmonary disease (COPD). Seeman and Chen (2002) supported this assumption by demonstrating that changes in levels of functioning in older adults with chronic health conditions (e.g. diabetes, hypertension high blood pressure, heart disease, cancer, broken bones) were predicted not only by health status or disease state but also by PA and self-efficacy. These findings are very promising concerning life-prolonging measures for those patients.

Sonstroem and Morgan (1989) developed a generally applicable model, which describes the relationship between exercise/PA and self-esteem (EXSEM; they used the terms *exercise* and *physical activity* interchangeably, see Figure 6.1). This model consists of four key constructs which could be influenced by PA: physical self-efficacy, physical competence (perception and evaluation of the body and its capacity), physical acceptance (body acceptance and satisfaction), and global self-esteem. Changes in PA are only indirectly associated with global self-esteem through the effect on self-efficacy (Edmunds & Clow 2014).

McAuley et al. (2005) have tested the EXSEM over a 4-year period in a sample of 174 older adults and revised the EXSEM so that both PA and

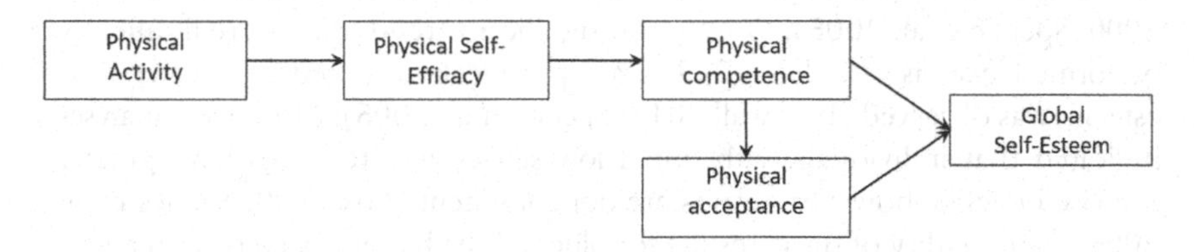

Figure 6.1 Model of four key constructs which could be influenced by PA
Source: Adapted from Sonstroem and Morgan (1989). Original EXSEM

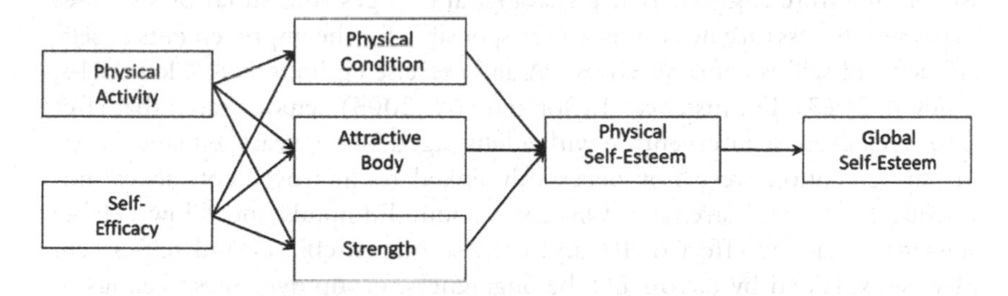

Figure 6.2 PA's indirect influence on self-esteem
Source: Adapted from McAuley et al. (2006). Revised EXSEM

self-efficacy directly influence the domains of self-esteem (and its subdomain measures "physical condition," "attractive body," "strength"), rather than PA indirectly influences self-esteem through its effect on self-efficacy (see Figure 6.2).

The EXSEM and its revision has created a basis for examining the underlying pathways by which PA behavior may lead to improvements in positive self-perception.

Types of Exercise and Doses

To the best of our knowledge no review or meta-analysis investigated specific exercise effects or dose-response relations with a special focus on older adults as there is a lack of RTCs examining the exercise effect on self-efficacy and self-esteem in older adults. A meta-analysis conducted by Fox (2000) included all age groups and several types of exercise (aerobic, resistance, flexibility, martial arts, expressive dance) and demonstrated that aerobic and resistance training tend to be the most beneficial in improving physical self-esteem (Fox 2000). However, a more recent meta-analysis examining adults older than 18 years revealed no differences in the efficacy of various types of exercise (aerobic, strength, flexibility, martial arts, mix) on self-esteem (Spence, McGannon, & Poon 2005). Up to now, no significant dose-response relationship for intensity, duration, or length of the exercise program has been detected (Fox 2000; Spence et al. 2005). Only a non-significant trend, that more frequently performed exercise (at least 2–3 times per week) resulted in greater self-esteem, was observed (Lindwall 2013; Spence et al. 2005). Moreover, analyses indicated that individuals with initial low self-esteem tends to have greater exercise benefits; however, results are not consistent (Fox 2000; Spence et al. 2005). The validity of these results for older adults has not been proven and is only speculative and further research is needed.

Mechanisms of the Exercise Effect on Self-Efficacy and Self-Esteem

Recent literature suggests that physiological changes (measured by increases in physical fitness) are not necessarily responsible for the improvements in self-efficacy and self-esteem caused by PA and exercise (Edmunds & Clow 2014; Lindwall 2013). For instance, Taylor and Fox (2005) demonstrated that after a 10-week exercise intervention with adults (aged 40–70 years) enhancements in self-perception were not necessarily linked to improvements in cardio-vascular fitness and strength. On this account Edmunds and Clow (2014) suggested that the effect of PA and exercise on self-efficacy and self-esteem may be explained by factors like belongingness, group dynamics, feelings of self-control, and competence.

Conclusion and Future Directions

Although exercise is a promising "easy to conduct treatment," exercise research comes with some limitations. That is, exercise and PA studies should be reflected critically. First, exercise research – like most research with (older) adults – could exhibit quite high non-response rates. Second, exercise interventions depend upon the participants' compliance. When participants exhibit high non-compliance rates the validity of the results is restricted. Third, it may be assumed that persons suffering from anxiety disorders, depression or cognitive impairments may have a diminished motivation to be physically active. Thus research in this area might represent a positive selection. Forth, an important restriction of studies with older adults is the sample selectivity. In the case of aging populations, the generalizability of results can be impaired in the ways that the average level of functional competence is overestimated as individuals with lower levels of functioning are less likely to participate in a study than individuals with higher levels of functioning.

Nevertheless, exercise and PA seem to be promising means to maintain older adults' mental health. Thereby the effects of exercise are multi-factorial and interrelated. Many research questions, however, still remain unanswered. For instance, further research is necessary to clarify in more detail which type of exercise (aerobic, strength, coordinative training) is most beneficial for the respective domains of mental health. Also more clarity about the dose-response effects is required. Moreover, future studies should target special populations of older adults and older adults of different age groups – older adults after retirement age and (very) old adults (e.g. 80+ years) – as all groups might have special needs. Further, one important question is how to bring research results into the community; that is, how interventions and community approaches should be designed to reach/motivate persons at need and how interventions should be provided to be easily used by the older adults/patients, caregivers or physicians. For example, one might think of IT-based instruments to outreach older adults in the communities, also the oldest old and older adults in rural areas. A general integration of exercise interventions for older adults in the health system might be another approach.

References

Adamo, M. L., & R. P. Farrar. 2006. "Resistance training, and IGF involvement in the maintenance of muscle mass during the aging process." *Ageing Research Reviews 5*: 310–31.

Adlard, P. A., V. M. Perreau, & C. W. Cotman. 2005. "The exercise-induced expression of BDNF within the hippocampus varies across life-span." *Neurobiology of Aging 26*: 511–20.

Ahlskog, J. E., Y. E. Geda, N. R. Graff-Radford, & R. C. Petersen. 2011. "Physical exercise as a preventive or disease-modifying treatment of dementia and brain aging." *Mayo Clinic Proceedings 86*: 876–84.

Ainslie, P. N., J. D. Cotter, K. P. George, S. Lucas, C. Murrell, R. Shave,…G. Atkinson. 2008. "Elevation in cerebral blood flow velocity with aerobic fitness throughout healthy human ageing." *The Journal of Physiology 586*: 4005–10.

Andel, R., M. Crowe, N. L. Pedersen, L. Fratiglioni, B. Johansson, & M. Gatz. 2008. "Physical exercise at midlife and risk of dementia three decades later: A population-based study of Swedish twins." *The Journals of Gerontology Series A: Biological Sciences and Medical Sciences 63*: 62–6.

Asmundson, G. J. G., M. G. Fetzner, L. B. DeBoer, M. B. Powers, M. W. Otto, & J. A. J. Smits. 2013. "Let's get physical: A contemporary review of the anxiolytic effects of exercise for anxiety and its disorders." *Depression and Anxiety 30*: 362–73.

Baker, L. D., L. L. Frank, K. Foster-Schubert, P. S. Green, C. W. Wilkinson, A. McTiernan,…S. Craft. 2010. "Effects of aerobic exercise on mild cognitive impairment: A controlled trial." *Archives of Neurology 67*: 71–9.

Baltes, P. B., & M. M. Baltes. 1990. "Psychological perspectives on successful aging: The model of selective optimization with compensation." *Successful Aging: Perspectives From the Behavioral Sciences 1*: 1–34.

Baltes, P. B., U. Lindenberger, & U. M. Staudinger. 1998. "Life span theory in developmental psychology." In W. Damon & R. M. Lerner (Eds.), *Handbook of Child Psychology, Sixth Edition* (pp. 569–664). New York: Wiley.

Belarbi, K., S. Burnouf, F.-J. Fernandez-Gomez, C. Laurent, S. Lestavel, M. Figeac,…L. Buée. 2011. "Beneficial effects of exercise in a transgenic mouse model of Alzheimer's disease-like Tau pathology." *Neurobiology of Disease 43*: 486–94.

Benight, C. C., & A. Bandura. 2004. "Social cognitive theory of post-traumatic recovery: The role of perceived self-efficacy." *Behaviour Research and Therapy 42*: 1129–48.

Berchtold, N. C., N. Castello, & C. W. Cotman. 2010. "Exercise and time-dependent benefits to learning and memory." *Neuroscience 167*: 588–97.

Bherer, Louis, K. I. Erickson, & T. Liu-Ambrose. 2013. "A review of the effects of physical activity and exercise on cognitive and brain functions in older adults." *Journal of Aging Research.*

Bibeau, W. S., J. B. Moore, N. G. Mitchell, T. Vargas-Tonsing, & J. B. Bartholomew. 2010. "Effects of acute resistance training of different intensities and rest periods on anxiety and affect." *The Journal of Strength and Conditioning Research 24*: 2184–91.

Black, S. V., R. Cooper, K. R. Martin, S. Brage, D.. Kuh, & M. Stafford. 2015. "Physical activity and mental well-being in a cohort aged 60–64 years." *American Journal of Preventive Medicine.*

Blazer, D. G. 2000. "Psychiatry and the oldest old." *American Journal of Psychiatry.*

Blazer, D. G. 2003. "Depression in late life: review and commentary." *Journals of Gerontology Series A 58*: 249–65.

Blier, P., & C. de Montigny. 1994. "Current advances and trends in the treatment of depression." *Trends in Pharmacological Sciences 15*: 220–26.

Blondell, S. J., R. Hammersley-Mather, & J. L. Veerman. 2014. "Does physical activity prevent cognitive decline and dementia? A systematic review and meta-analysis of longitudinal studies." *BMC Public Health 14*: 510.

Blumenthal, J. A., M. A. Babyak, P. M. Doraiswamy, L. Watkins, B. M. Hoffman, K. A. Barbour,...A. Sherwood. 2007. "Exercise and pharmacotherapy in the treatment of major depressive disorder." *Psychosomatic Medicine 69*: 587.

Blumenthal, J. A., M. A. Babyak, K. A. Moore, W. E. Craighead, S. Herman, P. Khatri,...K. R. Krishnan. 1999. "Effects of exercise training on older patients with major depression." *Archives Of Internal Medicine 159*: 2349–56.

Born, J., I. Ditschuneit, M. Schreiber, C. Dodt, & H. L. Fehm. 1995. "Effects of age and gender on pituitary-adrenocortical responsiveness in humans." *European Journal of Endocrinology 132*: 705–11.

Boyle, P. A., A. S. Buchman, R.S. Wilson, S. E. Leurgans, & D. A. Bennett. 2010. "Physical frailty is associated with incident mild cognitive impairment in community-based older persons." *Journal of the American Geriatrics Society 58*: 248–55.

Boyle, P. A., R. S. Wilson, N. T. Aggarwal, Y. Tang, & D. A. Bennett. 2006. "Mild cognitive impairment Risk of Alzheimer disease and rate of cognitive decline." *Neurology 67*: 441–5.

Brenes, G. A., J. D. Williamson, S. P. Messier, W. J. Rejeski, M. Pahor, E. Ip, & B. W. J. H. Penninx. 2007. "Treatment of minor depression in older adults: A pilot study comparing sertraline and exercise." *Aging and Mental Health 11*: 61–8.

Brownlee, K. K., A. W. Moore, & A. C. Hackney. 2005. "Relationship between circulating cortisol and testosterone: Influence of physical exercise." *Journal of Sports Science and Medicine 4*: 76.

Bryant, C., H. Jackson, & D. Ames. 2008. "The prevalence of anxiety in older adults: Methodological issues and a review of the literature." *Journal of Affective Disorders 109*: 233–50.

Burns, J. M., B. B. Cronk, H. S. Anderson, J. E. Donnelly, G. P. Thomas, A. Harsha,...R. H. Swerdlow. 2008. "Cardiorespiratory fitness and brain atrophy in early Alzheimer disease." *Neurology 71*: 210–16.

Carpenter, B., K. Ruckdeschel, H. Ruckdeschel, & K.van Haitsma. 2003. "REM psychotherapy: A manualized approach for long-term care residents with depression and dementia." *Clinical Gerontologist 25*: 25–49.

Carr, G. V., & I. Lucki. 2011. "The role of serotonin receptor subtypes in treating depression: A review of animal studies." *Psychopharmacology 213*: 265–87.

Cassilhas, R. C., H. K. M. Antunes, S. Tufik, & M. T. de Mello. 2010. "Mood, anxiety, and serum IGF-1 in elderly men given 24 weeks of high resistance exercise 1, 2." *Perceptual and Motor Skills 110*: 265–76.

Castellano, V., & L. J. White. 2008. "Serum brain-derived neurotrophic factor response to aerobic exercise in multiple sclerosis." *Journal of the Neurological Sciences 269*: 85–91.

Chan, S. H., C-W. J. Wu, Alice Y.W. Chang, K-S. Hsu, & J. Y.H. Chan. 2010. "Transcriptional upregulation of brain-derived neurotrophic factor in rostral ventrolateral medulla by angiotensin II significance in superoxide homeostasis and neural regulation of arterial pressure." *Circulation Research 107*: 1127–39.

Chang, M., P. V. Jonsson, J. Snaedal, S. Bjornsson, J. S. Saczynski, T. Aspelund, & L. J. Launer. 2010. "The effect of midlife physical activity on cognitive function among older adults: AGES—Reykjavik Study." *The Journals of Gerontology Series A: Biological Sciences and Medical Sciences 65*: 1369–74.

Coelho, F. M., D. S. Pereira, L. P. Lustosa, J. P. Da Silva, J. M. D. Dias, R. C. D. Dias, & L. S.M. Pereira. 2012. "Physical therapy intervention (PTI) increases plasma brain-derived neurotrophic factor (BDNF) levels in non-frail and pre-frail elderly women." *Archives of Gerontology and Geriatrics 54*: 415–20.

Cotman, C. W., & N. C. Berchtold. 2002. "Exercise: A behavioral intervention to enhance brain health and plasticity." *Trends in Neurosciences 25*: 295–301.

Coventry, P.A., & D. Hind. 2007. "Comprehensive pulmonary rehabilitation for anxiety and depression in adults with chronic obstructive pulmonary disease: Systematic review and meta-analysis." *Journal of Psychosomatic Research 63*: 551–65.

Curtis, G. C., O. G. Cameron, & R. M. Nesse. 1982. "The dexamethasone suppression test in panic disorder and agoraphobia." *The American Journal of Psychiatry.*

Daly, W. J., C. A. Seegers, D. A. Rubin, J.D. Dobridge, & A. C. Hackney. 2005. "Relationship between stress hormones and testosterone with prolonged endurance exercise." *European Journal of Applied Physiology 93*: 375–80.

de la Torre, J. C. 2012. "Cardiovascular risk factors promote brain hypoperfusion leading to cognitive decline and dementia." *Cardiovascular Psychiatry and Neurology.*

Deuschle, M., U. Schweiger, B. Weber, U. Gotthardt, A. Körner, J. Schmider,...I. Heuser. 1997. "Diurnal activity and pulsatility of the hypothalamus-pituitary-adrenal system in male depressed patients and healthy controls." Journal of Clinical Endocrinology and Metabolism *82* (1): 234–8.

Dowlati, Y., N. Herrmann, W. Swardfager, H. Liu, L.Sham, E. K. Reim, & K. L. Lanctôt. 2010. "A meta-analysis of cytokines in major depression." *Biological Psychiatry 67*: 446–57.

Duman, R. S., & L. M. Monteggia. 2006. "A neurotrophic model for stress-related mood disorders." *Biological Psychiatry 59*: 1116–27.

Dunn, A.. L., M. H. Trivedi, J. B. Kampert, C. G. Clark, & H. O. Chambliss. 2005. "Exercise treatment for depression: Efficacy and dose response." *American Journal of Preventive Medicine 28*: 1–8.

Eadie, B. D., V. A. Redila, & B. R. Christie. 2005. "Voluntary exercise alters the cytoarchitecture of the adult dentate gyrus by increasing cellular proliferation, dendritic complexity, and spine density." *Journal of Comparative Neurology 486*: 39–47.

Edmunds, S., & A. Clow. 2014. *Physical Activity and Mental Health in Long Term Conditions.* Champaign, IL: Human Kinetics.

Eggermont, L. H. P., D. F. Swaab, E. M. Hol, & E. J. A. Scherder. 2009. "Walking the line: A randomised trial on the effects of a short-term walking programme on cognition in dementia." *Journal of Neurology, Neurosurgery and Psychiatry 80*: 802–4.

Elavsky, S., E. McAuley, R. W. Motl, J. F. Konopack, D. X. Marquez, L. Hu,...E. Diener. 2005. "Physical activity enhances long-term quality of life in older adults: Efficacy, esteem, and affective influences." *Annals of Behavioral Medicine 30*: 138–45.

Etnier, J. L., R. J. Caselli, E. M. Reiman, G. E. Alexander, B. A. Sibley, D. Tessier, & E. C. McLemore. 2007. "Cognitive performance in older women relative to ApoE-epsilon4 genotype and aerobic fitness." *Medicine and Science in Sports and Exercise 39*: 199–207.

Eyre, H. A., E. Papps, & B. T. Baune. 2013. "Treating depression and depression-like behavior with physical activity: An immune perspective." *Frontiers in Psychiatry 4.*

Fabel, K., K. Fabel, B. Tam, D. Kaufer, A. Baiker, N. Simmons,...T. D. Palmer. 2003. "VEGF is necessary for exercise-induced adult hippocampal neurogenesis." *European Journal of Neuroscience 18*: 2803–12.

Farmer, M. E., B. Z. Locke, E. K. Moscicki, A. L. Dannenberg, D. B. Larson, & L. S. Radloff. 1988. "Physical activity and depressive symptoms: The NHANES I Epidemiologic Follow-up Study." *American Journal of Epidemiology 128*: 1340–51.

Fiske, A., J. L. Wetherell, & M. Gatz. 2009. "Depression in older adults." *Annual Review of Clinical Psychology 5*: 363.

Förstl, H., & C. Lang. 2011. *Demenzen in Theorie und Praxis.* Berlin: Springer-Verlag.

Fox, K. R. 2000. "The effects of exercise in self-perceptions and self-esteem." In S. J. H. Biddle, K. R. Fox, & S. H. Boutcher (Eds.), *Physical Activity and Psychological Well-Being* (pp. 88–117). London and New York: Routledge.

Fratiglioni, L., S. Paillard-Borg, & B. Winblad. 2004. "An active and socially integrated lifestyle in late life might protect against dementia." *The Lancet Neurology 3*: 343–53.

Fuss, J., N. M. B. Abdallah, M. A. Vogt, C.. Touma, P. G. Pacifici, R.. Palme,...P. Gass. 2010. "Voluntary exercise induces anxiety-like behavior in adult C57BL/6J mice correlating with hippocampal neurogenesis." *Hippocampus 20*: 364–76.

Gale, C. R., C. Cooper, I. J. Deary, & A.. A. Sayer. 2014. "Psychological well-being and incident frailty in men and women: The English Longitudinal Study of Ageing." *Psychological Medicine 44*: 697–706.

Geda, Y. E., R. O. Roberts, D. S. Knopman, T. J. Christianson, V.S. Pankratz, R. J. Ivnik,...W.A. Rocca. 2010. "Physical exercise, aging, and mild cognitive impairment: A population-based study." *Archives of Neurology 67*: 80–6.

Gonçalves, D. C., & G. J. Byrne. 2012. "Interventions for generalized anxiety disorder in older adults: Systematic review and meta-analysis." *Journal of Anxiety Disorders 26*: 1–11.

Hackney, A. C. 2006. "Exercise as a stressor to the human neuroendocrine system." *Medicina (Kaunas) 42*: 788–97.

Hameed, K. M, K. H. W. Lange, J. L. Andersen, P. Schjerling, M. Kjaer, S.. D.R. Harridge, & G. Goldspink. 2004. "The effect of recombinant human growth hormone and resistance training on IGF-I mRNA expression in the muscles of elderly men." *The Journal of Physiology 555*: 231–40.

Hamer, M., & Y. Chida. 2009. "Physical activity and risk of neurodegenerative disease: A systematic review of prospective evidence." *Psychological Medicine 39*: 3–11.

Hasin, D. S., R. D. Goodwin, F. S. Stinson, & B. F. Grant. 2005. "Epidemiology of major depressive disorder: Results from the National Epidemiologic Survey on Alcoholism and Related Conditions." *Archives of General Psychiatry 62*: 1097–1106.

Hassmén, P., N. Koivula, & A. Uutela. 2000. "Physical exercise and psychological well-being: A population study in Finland." *Preventive Medicine 30*: 17–25.

Hebert, L. E., L. A. Beckett, P. A. Scherr, & D. A. Evans. 2001. "Annual incidence of Alzheimer disease in the United States projected to the years 2000 through 2050." *Alzheimer Disease and Associated Disorders 15*: 169–73.

Hedden, T., & J. D. Gabrieli. 2004. "Insights into the ageing mind: A view from cognitive neuroscience." *Nature Reviews Neuroscience 5*: 87–96.

Hegeman, J. M., R. M. Kok, R. C. Van der Mast, & E. J. Giltay. 2012. "Phenomenology of depression in older compared with younger adults: Meta-analysis." *The British Journal of Psychiatry 200*: 275–81.

Herring, M. P., J. B. Lindheimer, & P. J. O'Connor. 2013. "The effects of exercise training on anxiety." *American Journal of Lifestyle Medicine*. doi:10.1177/1559827613508542

Hertzog, C., A. F. Kramer, R. S. Wilson, & U. Lindenberger. 2008. "Enrichment effects on adult cognitive development can the functional capacity of older adults be preserved and enhanced?" *Psychological Science in the Public Interest 9*: 1–65.

Heyn, P., B. C. Abreu, & K. J. Ottenbacher. 2004. "The effects of exercise training on elderly persons with cognitive impairment and dementia: A meta-analysis." *Archives of Physical Medicine and Rehabilitation 85*: 1694–1704.

Heyn, P., K. E. Johnsons, & A. F. Kramer. 2008. "Endurance and strength training outcomes on cognitively impaired and cognitively intact older adults: a meta-analysis." *The Journal of Nutrition Health and Aging 12*: 401–9.

Hiles, S. A., A. L. Baker, T. de Malmanche, & J. Attia. 2012. "A meta-analysis of differences in IL-6 and IL-10 between people with and without depression: Exploring the causes of heterogeneity." *Brain, Behavior, and Immunity 26*: 1180–8.

Hollon, S. D., M. E. Thase, & J. C. Markowitz. 2002. "Treatment and prevention of depression." *Psychological Science in the Public Interest 3*: 39–77.

Hommel, B., K. Z. H. Li, & S.C. Li. 2004. "Visual search across the life span." *European Journal of Development Psychology 40*: 545–58.

Hou, R., & D. S. Baldwin. 2012. "A neuroimmunological perspective on anxiety disorders." *Human Psychopharmacology: Clinical and Experimental 27*: 6–14.

Hovatta, I., J. Juhila, & J. Donner. 2010. "Oxidative stress in anxiety and comorbid disorders." *Neuroscience Research 68*: 261–75.

Huang, T., K. T. Larsen, M. Ried-Larsen, N. C. Møller, & L. B. Andersen. 2014. "The effects of physical activity and exercise on brain-derived neurotrophic factor in healthy humans: A review." *Scandinavian Journal of Medicine and Science in Sports 24*: 1–10.

Ivey, F. M., A. S. Ryan, C. E. Hafer-Macko, & R. F. Macko. 2011. "Improved cerebral vasomotor reactivity after exercise training in hemiparetic stroke survivors." *Stroke 42*: 1994–2000.

Kales, H. C., P. Chen, F. C. Blow, De. E. Welsh, & A. M. Mellow. 2005. "Rates of clinical depression diagnosis, functional impairment, and nursing home placement in coexisting dementia and depression." *American Journal of Geriatric Psychiatry 13*: 441–9.

Kaplan, R. M., A. L. Ries, L. M. Prewitt, & E. Eakin. 1994. "Self-efficacy expectations predict survival for patients with chronic obstructive pulmonary disease." *Health Psychology 13*: 366.

Karege, F., G. Perret, G. Bondolfi, M. Schwald, G. Bertschy, & J-M. Aubry. 2002. "Decreased serum brain-derived neurotrophic factor levels in major depressed patients." *Psychiatry Research 109*: 1438.

Kari, H. P., P. P. Davidson, H. H. Kohl, & M. M. Kochhar. 1978. "Effects of ketamine on brain monoamine levels in rats." *Research Communications in Chemical Pathology and Pharmacology 20*: 475–88.

Kerse, K., K. J. Hayman, S. A. Moyes, K. Peri, E. Robinson, A. Dowell,…B. Arroll. 2010. "Home-based activity program for older people with depressive symptoms: DeLLITE – a randomized controlled trial." *Annals of Family Medicine 8*: 214–23.

Kim, J., J. M. Basak, & D. M. Holtzman. 2009. "The role of apolipoprotein E in Alzheimer's disease." *Neuron 63*: 287–303.

Kirschbaum, C., & D. H. Hellhammer. 1994. "Salivary cortisol in psychoneuroendocrine research: recent developments and applications." *Psychoneuroendocrinology 19*: 313–33.

Kiuchi, T., H. Lee, & T. Mikami. 2012. "Regular exercise cures depression-like behavior via VEGF-Flk-1 signaling in chronically stressed mice." *Neuroscience 207*: 208–17.

Knapen, J., & D. Vancampfort. 2013. "Evidence for exercise therapy in the treatment of depression and anxiety." *Depression and Anxiety 17*: 75–87.

Knopman, D. S., & R. Roberts. 2010. "Vascular risk factors: Imaging and neuropathologic correlates." *Journal of Alzheimer's Disease 20*: 699.

Kohut, M. L., D. A. McCann, D. W. Russell, D.N. Konopka, J. E. Cunnick, W. D. Franke, & E. M. Vanderah. 2006. "Aerobic exercise, but not flexibility/resistance exercise, reduces serum IL-18, CRP, and IL-6 independent of β-blockers, BMI, and psychosocial factors in older adults." *Brain, Behavior, and Immunity 20*: 201–209.

Kokkinos, P., H. Sheriff, & R. Kheirbek. 2011. "Physical inactivity and mortality risk." *Cardiology Research and Practice.*

Kramer, A. F., L. Bherer, S. J. Colcombe, W. Dong, & W. T. Greenough. 2004. "Environmental influences on cognitive and brain plasticity during aging." *Journals of Gerontology Series A: Biological Sciences and Medical Sciences 59*: 940–57.

Kramer, A. F., & K. I. Erickson. 2007. "Capitalizing on cortical plasticity: Influence of physical activity on cognition and brain function." *Trends in Cognitive Sciences 11*: 342–8.

Kronenberg, G., A. Bick-Sander, E. Bunk, C. Wolf, D. Ehninger, & G. Kempermann. 2006. "Physical exercise prevents age-related decline in precursor cell activity in the mouse dentate gyrus." *Neurobiology of Aging 27*: 1505–13.

Kugler, J., H. Seelbach, & G. M. Krüskemper. 1994. "Effects of rehabilitation exercise programmes on anxiety and depression in coronary patients: A meta-analysis." *British Journal of Clinical Psychology 33*: 401–10.

Lam, L. C., R. Chau, B. M. Wong, A. W. Fung, V. W. Lui, C. C. Tam,…W. M. Chan. 2011. "Interim follow-up of a randomized controlled trial comparing Chinese style mind body (Tai Chi) and stretching exercises on cognitive function in subjects at risk of progressive cognitive decline." *International Journal of Geriatric Psychiatry 26*: 733–40.

Larson, E. B., L. Wang, J. D. Bowen, W. C. McCormick, L. Teri, P. Crane, & W. Kukull. 2006. "Exercise is associated with reduced risk for incident dementia among persons 65 years of age and older." *Annals of Internal Medicine 144*: 73–81.

Laske, C., K. Stellos, N. Hoffmann, E. Stransky, G. Straten, G. W. Eschweiler, & Thomas Leyhe. 2011. "Higher BDNF serum levels predict slower cognitive decline in Alzheimer's disease patients." *International Journal of Neuropsychopharmacology 14*: 399–404.

Laske, C., E. Stransky, T. Leyhe, G. W. Eschweiler, W. Maetzler, A. Wittorf,…Schott. 2007. "BDNF serum and CSF concentrations in Alzheimer's disease, normal pressure hydrocephalus and healthy controls." *Journal of Psychiatric Research 41*: 387–94.

Laurin, D., R. Verreault, J. Lindsay, K. MacPherson, & K. Rockwood. 2001. "Physical activity and risk of cognitive impairment and dementia in elderly persons." *Archives of Neurology 58*: 498–504.

Leonard, B., & M. Maes. 2012. "Mechanistic explanations how cell-mediated immune activation, inflammation and oxidative and nitrosative stress pathways and their sequels and concomitants play a role in the pathophysiology of unipolar depression." *Neuroscience and Biobehavioral Reviews 36*: 764–85.

Li, S-C., U. Lindenberger, B. Hommel, G. Aschersleben, W. Prinz, & P. B. Baltes. 2004. "Transformations in the couplings among intellectual abilities and constituent cognitive processes across the life span." *Psychological Science 15*: 155–63.

Liang, K. Y., M. A. Mintun, A. M. Fagan, A. M. Goate, J. M. Bugg, D. M. Holtzman,...D. Head. 2010. "Exercise and Alzheimer's disease biomarkers in cognitively normal older adults." *Annals of Neurology 68*: 311–18.

Lindwall, M. 2013. "Exercise, self-esteem and self-perceptions." In D. Tod & D. Lavallee (Eds.), *Psychology of strength and conditioning* (pp. 82–108). London: Routledge.

Long, B. C., & R. V. Stavel. 1995. "Effects of exercise training on anxiety: A meta-analysis." *Journal of Applied Sport Psychology 7*: 167–89.

Lopresti, A. L., S. D. Hood, & P. D. Drummond. 2013. "A review of lifestyle factors that contribute to important pathways associated with major depression: Diet, sleep and exercise." *Journal of Affective Disorders 148*: 12–27.

Lou, S., J. Liu, H. Chang, & P. Chen. 2008. "Hippocampal neurogenesis and gene expression depend on exercise intensity in juvenile rats." *Brain Research 1210*: 48–55.

Lövdén, M., L. Bäckman, U. Lindenberger, S. Schaefer, & F. Schmiedek. 2010. "A theoretical framework for the study of adult cognitive plasticity." *Psychological Bulletin 136*: 659.

Luppa, M., C. Sikorski, T. Luck, L. Ehreke, A. Konnopka, B. Wiese, & S. G. Riedel-Heller. 2012. "Age-and gender-specific prevalence of depression in latest-life – systematic review and meta-analysis." *Journal of Affective Disorders 136*: 212–21.

Lyketsos, C. G., C. Steele, L. Baker, E. Galik, S. Kopunek, M. Steinberg, & A. Warren. 1997. "Major and minor depression in Alzheimer's disease: Prevalence and impact." *Journal of Neuropsychiatry and Clinical Neurosciences 9*: 556–61.

McAuley, E., & B. Blissmer. 2000. "Self-efficacy determinants and consequences of physical activity." *Exercise and Sport Sciences Reviews 28*: 85–8.

McAuley, E., K. S. Courneya, & J. Lettunich. 1991. "Effects of acute and long-term exercise on self-efficacy responses in sedentary, middle-aged males and females." *The Gerontologist 31*: 534–42.

McAuley, E., S. Elavsky, R. W. Motl, J. F. Konopack, L. Hu, & D. X. Marquez. 2005. "Physical activity, self-efficacy, and self-esteem: Longitudinal relationships in older adults." *Journals of Gerontology Series B: Psychological Sciences and Social Sciences 60*: 268–75.

McAuley, E., J. F. Konopack, R. W. Motl, K. S. Morris, S. E. Doerksen, & K. R. Rosengren. 2006. "Physical activity and quality of life in older adults: Influence of health status and self-efficacy." *Annals of Behavioral Medicine 31*: 99–103.

Maletic, V., M. Robinson, T. Oakes, S. Iyengar, S. G. Ball, & J. A. Russell. 2007. "Neurobiology of depression: An integrated view of key findings." *International Journal of Clinical Practice 61*: 2030–40.

Mammen, G., & G. Faulkner. 2013. "Physical activity and the prevention of depression: A systematic review of prospective studies." *American Journal of Preventive Medicine 45*: 649–57.

Medina, J. L., J. Jacquart, & J. A. Smits. 2015. "Optimizing the exercise prescription for depression: The search for biomarkers of response." *Current Opinion in Psychology 4*: 43–7.

Mesquita, A. R., M. Correia-Neves, S. Roque, A. G. Castro, P. Vieira, J. Pedrosa,...N. Sousa. 2008. "IL-10 modulates depressive-like behavior." *Journal of Psychiatric Research 43*: 89–97.

Molendijk, M. L., P. Spinhoven, M. Polak, B. A. A. Bus, B. W. J. H. Penninx, & B. M. Elzinga. 2014. "Serum BDNF concentrations as peripheral manifestations of depression: Evidence from a systematic review and meta-analyses on 179 associations." *Molecular Psychiatry 19*: 791–800.

Moncrieff, J., & I. Kirsch. 2005. "Efficacy of antidepressants in adults." British Medical Journal *331* (7509): 155.

Mora, F., G. Segovia, & A. del Arco. 2007. "Aging, plasticity and environmental enrichment: Structural changes and neurotransmitter dynamics in several areas of the brain." *Brain Research Reviews 55*: 78–88.

Moylan, S., M. Berk, O. M. Dean, Y. Samuni, L. J. Williams, A. O'Neil,... M. Maes. 2014. "Oxidative and nitrosative stress in depression: Why so much stress?" *Neuroscience and Biobehavioral Reviews 45*: 46–62.

Moylan, S., H. A. Eyre, M. Maes, B. T. Baune, F. N. Jacka, & M. Berk. 2013. "Exercising the worry away: How inflammation, oxidative and nitrogen stress mediates the beneficial effect of physical activity on anxiety disorder symptoms and behaviours." *Neuroscience and Biobehavioral Reviews 37*: 573–84.

Mura, G., M. F. Moro, S. B. Patten, & M. G. Carta. 2014. "Exercise as an add-on strategy for the treatment of major depressive disorder: A systematic review." *CNS Spectrums 19*: 496–508.

Murer, M. G., Q. Yan, & R. Raisman-Vozari. 2001. "Brain-derived neurotrophic factor in the control human brain, and in Alzheimer's disease and Parkinson's disease." *Progress in Neurobiology 63*: 71–124.

Murrell, C. J., J. D. Cotter, K. N. Thomas, S. J. Lucas, M. J. Williams, & P. N. Ainslie. 2013. "Cerebral blood flow and cerebrovascular reactivity at rest and during sub-maximal exercise: effect of age and 12-week exercise training." *Age 35*: 905–20.

Naarding, P., R. A. Schoevers, J. G. Janzing, C. Jonker, P.J. Koudstaal, & A. T. Beekman. 2005. "A study on symptom profiles of late-life depression: The influence of vascular, degenerative and inflammatory risk-indicators." *Journal of Affective Disorders 88*: 155–62.

Nabkasorn, C., N. Miyai, A. Sootmongkol, S. Junprasert, Hi. Yamamoto, M. Arita, & K. Miyashita 2006. "Effects of physical exercise on depression, neuroendocrine stress hormones and physiological fitness in adolescent females with depressive symptoms." *European Journal of Public Health 16*: 179–84.

Nagamatsu, L. S., A. Chan, J. C. Davis, B. L. Beattie, P. Graf, M. W. Voss,...T. Liu-Ambrose. 2013. "Physical activity improves verbal and spatial memory in older adults with probable mild cognitive impairment: A 6-month randomized controlled trial." *Journal of Aging Research*. doi:10.1155/2013/861893

Netz, Y., M-J. Wu, B. J. Becker, & G. Tenenbaum. 2005. "Physical activity and psychological well-being in advanced age: A meta-analysis of intervention studies." *Psychology and Aging 20*: 272.

Nyberg, J., M. A. Åberg, L. Schiöler, M. Nilsson, A. Wallin, K. Torén, & K. G. Kuhn. 2014. "Cardiovascular and cognitive fitness at age 18 and risk of early-onset dementia." *Brain*.

Öhman, H., N. Savikko, T. E. Strandberg, & K. H. Pitkälä. 2014. "Effect of physical exercise on cognitive performance in older adults with mild cognitive impairment or dementia: a systematic review." *Dementia and Geriatric Cognitive Disorders, 38*: 347–65.

Orth, U., K. H. Trzesniewski, & R. W. Robins. 2010. "Self-esteem development from young adulthood to old age: A cohort-sequential longitudinal study." *Journal of Personality and Social Psychology 98*: 645.

Park, D. C., G. Lautenschlager, T. Hedden, N. S. Davidson, A. D. Smith, & P. K. Smith. 2002. "Models of visuospatial and verbal memory across the adult life span." *Psychology and Aging 17*: 299–320.

Park, D. C., & P. A. Reuter-Lorenz. 2009. "The adaptive brain: Aging and neurocognitive scaffolding." *Annual Review of Psychology 60*: 173–96.

Park, S-H., K. S. Han, & C-B. Kang. 2014. "Effects of exercise programs on depressive symptoms, quality of life, and self-esteem in older people: A systematic review of randomized controlled trials." *Applied Nursing Research 27*: 219–26.

Pereira, A. C., D. E. Huddleston, A. M. Brickman, A. A. Sosunov, R. Hen, G. M. McKhann,...S. A. Small. 2007. "An in vivo correlate of exercise-induced neurogenesis in the adult dentate gyrus." *Proceedings of the National Academy of Sciences 104*: 5638–43.

Peters, A. 2002. "Structural changes in the normally aging cerebral cortex of primates." *Progress in Brain Research 136*: 455–65.

Petersen, R. C., R. Doody, A. Kurz, R. C. Mohs, J. C. Morris, P. V. Rabins,...B. Winblad. 2001. "Current concepts in mild cognitive impairment." *Archives of Neurology 58*: 1985–92.

Petruzzello, S. J., D. M. Landers, B. D. Hatfield, K. A. Kubitz, & W. Salazar. 1991. "A meta-analysis on the anxiety-reducing effects of acute and chronic exercise." *Sports Medicine 11*: 143–82.

Pitsavos, C., D. Panagiotakos, M. Weinem, & C. Stefanadis. 2006. "Diet, exercise and the metabolic syndrome." *Review of Diabetic Studies 3*: 118.

Raz, N., & K. M. Rodrigue. 2006. "Differential aging of the brain: Patterns, cognitive correlates and modifiers." *Neuroscience and Biobehavioral Reviews 30*: 730–48.

Redila, V. A., & B. R. Chrisrie. 2006. "Exercise-induced changes in dendritic structure and complexity in the adult hippocampal dentate gyrus." *Neuroscience 137*: 1299–1307.

Rethorst, C. D., B. M. Wipfli, & D. M. Landers. 2009. "The antidepressive effects of exercise." *Sports Medicine 39*: 491–511.

Reuter-Lorenz, P. A., & C. Lustig. 2005. "Brain aging: Reorganizing discoveries about the aging mind." *Current Opinion in Neurobiology 15*: 245–51.

Rimmele, U., R. Seiler, B. Marti, P. H. Wirtz, U. Ehlert, & M. Heinrichs. 2009. "The level of physical activity affects adrenal and cardiovascular reactivity to psychosocial stress." *Psychoneuroendocrinology 34*: 190–8.

Robertson, R., A. Robertson, R. Jepson, & M. Maxwell. 2012. "Walking for depression or depressive symptoms: A systematic review and meta-analysis." *Mental Health and Physical Activity 5*: 66–75.

Roose, S. P., H. A. Sackeim, K. R. R. Krishnan, B. G. Pollock, G. Alexopoulos, H. Lavretsky,...Old-Old Depression Study Group. 2015. "Antidepressant pharmacotherapy in the treatment of depression in the very old: A randomized, placebo-controlled trial." *American Journal of Psychiatry 161* (11): 2050–9.

Rothman, S. M., & M.P. Mattson. 2013. "Activity-dependent, stress-responsive BDNF signaling and the quest for optimal brain health and resilience throughout the lifespan." *Neuroscience 239*: 228–40.

Roy-Byrne, P. P., K. W. Davidson, R. C. Kessler, G. J. G. Asmundson, R. D. Goodwin, L. Kubzansky,…M. B. Stein. 2008. "Anxiety disorders and comorbid medical illness." *General Hospital Psychiatry 30*: 208–25.

Salim, S., M. Asghar, M. Taneja, I.. Hovatta, G. Chugh, C. Vollert, & A. Vu. 2011. "Potential contribution of oxidative stress and inflammation to anxiety and hypertension." *Brain Research 1404*: 63–71.

Salim, S., N. Sarraj, M. Taneja, K. Saha, M. V. Tejada-Simon, & G. Chugh. 2010. "Moderate treadmill exercise prevents oxidative stress-induced anxiety-like behavior in rats." *Behavioural Brain Research 208*: 545–52.

Sallis, J. F., W. L. Haskell, S. P. Fortmann, K. M. Vranizan, C. Barr Taylor, & D. S. Solomon. 1986. "Predictors of adoption and maintenance of physical activity in a community sample." *Preventive Medicine 15*: 331–41.

Scarmeas, N., E. Zarahn, K. E. Anderson, C. G. Habeck, J. Hilton, J. Flynn, … Y. Stern. 2003. "Association of life activities with cerebral blood flow in Alzheimer disease: Implications for the cognitive reserve hypothesis." *Archives of Neurology 60*: 359–65.

Schlicht, W. 1994. "Does physical exercise reduce anxious emotions? A meta-analysis." *Anxiety, Stress and Coping 6*: 275–88.

Schreiber, W., C. J. Lauer, K. Krumrey, F. Holsboer, & J-C. Krieg. 1996. "Dysregulation of the hypothalamic-pituitary-adrenocortical system in panic disorder." *Neuropsychopharmacology 15*: 7–15.

Seeman, T., & X. Chen. 2002. "Risk and protective factors for physical functioning in older adults with and without chronic conditions macarthur studies of successful aging." *Journals of Gerontology Series B: Psychological Sciences and Social Sciences 57*: 135–44.

Seeman, T., B. Singer, C. W. Wilkinson, & B. McEwen. 2001. "Gender differences in age-related changes in HPA axis reactivity." *Psychoneuroendocrinology 26*: 225–40.

Seidler, R. D., J. A. Bernard, T. B. Burutolu, B. W. Fling, M. T. Gordon, J. T. Gwin,… D. B. Lipps. 2010. "Motor control and aging: Links to age-related brain structural, functional, and biochemical effects." *Neuroscience and Biobehavioral Reviews 34*: 721–33.

Shen, H., L. Tong, R. Balazs, & C. W. Cotman. 2001. "Physical activity elicits sustained activation of the cyclic AMP response element-binding protein and mitogen-activated protein kinase in the rat hippocampus." *Neuroscience 107*: 219–29.

Silverman, M. N., & P. A. Deuster. 2014. "Biological mechanisms underlying the role of physical fitness in health and resilience." *Journal of the Royal Society Interface.*

Sjösten, N., & S-L. Kivelä. 2006. "The effects of physical exercise on depressive symptoms among the aged: A systematic review." *International Journal of Geriatric Psychiatry 21*: 410–18.

Smith, J. C., K. A. Nielson, J. L. Woodard, M. Seidenberg, S. Durgerian, P. Antuono,… S. M. Rao. 2011. "Interactive effects of physical activity and APOE-ε4 on BOLD semantic memory activation in healthy elders." *Neuroimage 54*: 635–44.

Smith, J. K. 2001. "Exercise and atherogenesis." *Exercise and Sport Sciences Reviews 29*: 49–53.

Snowden, M. B., L. E. Steinman, W. L. Carlson, K. N. Mochan, A. F. Abraido-Lanza, L. L. Bryant,…E. J. Lenze. 2015. "Effect of physical activity, social support, and skills training on late-life emotional health: A systematic literature review and implications for public health research." Frontiers in Public Health *3*: 213.

Sofi, F., D. Valecchi, D. Bacci, R. Abbate, G. F. Gensini, A. Casini, & C. Macchi. 2011. "Physical activity and risk of cognitive decline: A meta-analysis of prospective studies." *Journal of Internal Medicine 269*: 107–17.

Sonntag, W. E., M. Ramsey, & C. S. Carter. 2005. "Growth hormone and insulin-like growth factor-1 (IGF-1) and their influence on cognitive aging." *Ageing Research Reviews 4*: 195–212.

Sonstroem, R. J., & W. P. Morgan. 1989. "Exercise and self-esteem: Rationale and model." *Medicine and Science in Sports and Exercise.*

Spence, J. C., K. R. McGannon, & P. Poon. 2005. "The effect of exercise on global self-esteem: A quantitative review. Exercise physiology." *Journal of Sport and Exercise Psychology 27*: 311–34.

Sperner-Unterweger, B., C. Kohl, & D. Fuchs. 2014. "Immune changes and neurotransmitters: Possible interactions in depression?" *Progress in Neuro-Psychopharmacology and Biological Psychiatry 48*: 268–76.

Starkstein, S. E., R. Jorge, R. Mizrahi, & R. G. Robinson. 2014. "The construct of minor and major depression in Alzheimer's disease." *American Journal of Psychiatry.*

Starkstein, S. E., R. Mizrahi, & B. D. Power. 2008. "Depression in Alzheimer's disease: Phenomenology, clinical correlates and treatment." *International Review of Psychiatry 20*: 382–88.

Staudinger, U. M. 2012. "Möglichkeiten und Grenzen menschlicher Entwicklungen über die Lebensspanne." In J. Hacker & M. Hecker (Eds.), *Was ist Leben? Nova Acta Leopoldina 116* (pp. 255–66). Stuttgart: Wissenschaftliche Verlagsgesellschaft.

Staudinger, U. M., M. Marsiske, & P. B. Baltes. 1995. "Resilience and reserve capacity in later adulthood: Potentials and limits of development across the life span." *Developmental Psychopathology 2*: 801–47.

Stern, Y. 2002. "What is cognitive reserve? Theory and research application of the reserve concept." Journal of the International Neuropsychological Society *8* (3): 448–60.

Strawbridge, W. J., S. Deleger, R. E. Roberts, & G. A. Kaplan. 2002. "Physical activity reduces the risk of subsequent depression for older adults." *American Journal of Epidemiology 156*: 328–34.

Ströhle, A. 2003. "Die Neuroendokrinologie von Stress und die Pathophysiologie und Therapie von Depression und Angst." *Der Nervenarzt 74*: 279–92.

Sumic, A., Y. L. Michael, N. E. Carlson, D. B. Howieson, & J. A. Kaye. 2007. "Physical activity and the risk of dementia in oldest old." *Journal of Aging and Health 19*: 242–59.

Szuhany, K. L., M. Bugatti, M. W. Otto. 2015. "A meta-analytic review of the effects of exercise on brain-derived neurotrophic factor." *Journal of psychiatric research 60*: 56–64.

Taylor, A. H. 2000. "Physical activity, anxiety and stress." In S. J. H. Biddle, K. R. Fox & S. H. Boutcher (Eds.), *Physical Activity and Psychological Well-Being* (pp. 10–45). London and New York: Routledge.

Taylor, A. H., & K. R. Fox. 2005. "Effectiveness of a primary care exercise referral intervention for changing physical self-perceptions over 9 months." *Health Psychology 24*: 11.

Thornton, P. L., R. L. Ingram, & W. E. Sonntag. 2000. "Chronic [D-Ala^ 2]-growth hormone-releasing hormone administration attenuates age-related deficits in spatial memory." *Journals of Gerontology-Biological Sciences and Medical Sciences 55*: 106.

Trollor, J. N., T. M. Anderson, P. S. Sachdev, H. Brodaty, & G. Andrews. 2007. "Prevalence of mental disorders in the elderly: The Australian National Mental Health and Well-Being Survey." *American Journal of Geriatric Psychiatry 15*: 455–66.

van Praag, H., G. Kempermann, & F. H. Gage. 1999. "Running increases cell proliferation and neurogenesis in the adult mouse dentate gyrus." *Nature Neuroscience 2*: 266–70.

van Praag, H., T. Shubert, C. Zhao, & F. H. Gage. 2005. "Exercise enhances learning and hippocampal neurogenesis in aged mice." *Journal of Neuroscience 25*: 8680–5.

Vaynman, S. S., Z. Ying, D. Yin, & F. Gomez-Pinilla. 2006. "Exercise differentially regulates synaptic proteins associated to the function of BDNF." *Brain Research 1070*: 124–30.

Vega, S. R., H. K. Strüder, B. Vera Wahrmann, A. Schmidt, W. Bloch, & W. Hollmann. 2006. "Acute BDNF and cortisol response to low intensity exercise and following ramp incremental exercise to exhaustion in humans." *Brain Research 1121*: 59–65.

Ware, J. E. 2003. "Conceptualization and measurement of health-related quality of life: comments on an evolving field." *Archives of Physical Medicine and Rehabilitation 84*: 43–51.

Wayne, P. M., J. N. Walsh, R. E. Taylor-Piliae, R.E. Wells, K.n V. Papp, N. J. Donovan, & G. Y. Yeh. 2014. "Effect of Tai Chi on cognitive performance in older adults: Systematic review and meta-analysis." *Journal of the American Geriatrics Society 62*: 25–39.

Wegner, M., I. Helmich, Se. Machado, A. E. Nardi, O. Arias-Carrión, & H. Budde. 2014. "Effects of exercise on anxiety and depression disorders: Review of meta-analyses and neurobiological mechanisms." *CNS and Neurological Disorders-Drug Targets 13*: 1002–14.

White, S. M., T.R. Wójcicki, & E. McAuley. 2009. "Physical activity and quality of life in community dwelling older adults." *Health and Quality of Life Outcomes 7.*

Williams, C. L., & R. M. Tappen. 2008. "Exercise training for depressed older adults with Alzheimer's disease." *Aging and Mental Health 12*: 72–80.

Wipfli, B. M., C. D. Rethorst, & D. M. Landers. 2008. "The anxiolytic effects of exercise: A meta-analysis of randomized trials and dose-response analysis." *Journal of Sport and Exercise Psychology 30*: 392–410.

Zaudig, M. 2005. "The prodromes and early detection of Alzheimer's disease." In M. Maj, J. J. Lopes-Ibor, N. Sartorius, M. Sato, & A. Okasha (Eds.), *Early detection and management of mental disorders* (pp. 277–94). West Sussex: J. Wiley.

Zaudig, M. 2011. "Leichte kognitive Beeinträchtigung im Alter." In Hans Förstl (Ed.), *Demenzen in Theorie und Praxis* (pp. 25–46). Berlin, Heidelberg: Springer-Verlag.

Section 3

Exercise Effects in Cognition and Motor Learning

7 Physical Exercise and Cognitive Enhancement

David Moreau

Introduction

Perhaps counterintuitively, one of the most effective ways to look after one's brain is physical exercise. This idea is not yet fully engrained in common views of cognition – despite current scientific efforts emphasizing the interplay between brain and body, the Cartesian notion of a duality between the two is long lived. However, evidence showing the importance of a healthy body for a healthy mind is rapidly accumulating, and the influence of physical fitness and exercise in this regard is well documented. Combining behavioral, neural, and neurobiological findings, this line of research is complemented by theoretical frameworks in cognitive science, together providing convincing support for the relationship between cognitive and motor systems. In this chapter, we review the evidence for the effect of physical exercise on cognition and the brain, discuss current trends of work exploring the potential of physical exercise – and particularly of complex motor activities – to enhance cognition and to remediate learning difficulties, and finally suggest promising directions for future research.

Cognitive Benefits of Physical Exercise

Numerous benefits of physical exercise have been documented over the last few decades, including a wide range of physical and mental health improvements.

For example, in observational studies, physical fitness appears to be associated with reduced occurrences of potentially fatal conditions, such as cardiovascular diseases, cancers, strokes, and diabetes (Blair 1995). Even more aligned with the perspective of this chapter, reduced risks in several cognitive or neurological conditions have also been reported. Neurological conditions such as autism, attention deficit/hyperactive disorder (ADHD), schizophrenia, dementia, and Alzheimer's disease all appear to be worsened by poor physical fitness, and to benefit from exercise interventions (Penedo & Dahn 2005). Importantly, this effect is not contingent on preexisting deficits, and can be generalized to non-clinical populations – physical exercise seems to be related to better cognitive function regardless of cognitive impairment. As such, individuals with better indices of physical fitness tend to perform above average on executive function tasks and spatial reasoning problems (Colcombe & Kramer 2003; Hillman, Erickson, & Kramer 2008). Although a few studies have suggested that the association between physical activity and performance is weaker when considering tasks that do not primarily tap executive function (Kramer, Hahn, & Gopher 1999), complementary findings support the idea of a strong relationship between physical exercise and a wide range of cognitive tasks (see for a review Moreau & Conway 2013).

Building upon this line of research, additional work has intended to elucidate the mechanisms underlying the association between physical fitness and cognitive health. In particular, and consistent with typical concerns when considering findings from observational studies, an important question is whether the link is causal or driven by confounding factors. Several studies have attempted to shed light on this issue. For example, exercise interventions have led to attention improvements in children with developmental coordination disorders (Tsai, Wang, & Tseng 2012) and in overweight children (Davis et al. 2007). A substantial body of research has also explored the relationship between physical activity and academic performance. Students who follow physical activity guidelines tend to have better grades than those who do not (Coe, Pivarnik, Womack, Reeves, & Malina 2006), while reading and mathematics competencies tend to correlate with measures of physical fitness (Castelli, Hillman, Buck, & Erwin 2007). These are not isolated cases – the literature concerning the relationship between physical exercise and academic achievement is well documented, with clear and robust findings (Keeley & Fox 2009).

Therefore, and perhaps contrary to popular belief, decreases in the amount of physical activity at school *negatively* impacts academic performance (see for a review Trudeau & Shephard 2008). Common responses to poor school achievement such as replacing physical education classes with so-called core academic components (i.e. English, mathematics) are thus profoundly counterproductive and worsen the problem they are intended to fix (Goh, Hannon, Webster, Podlog, & Newton 2016; Mahar et al. 2006; Trudeau & Shephard 2008). This is an important point of this chapter, since the work linking physical exercise and cognition is perhaps not appreciated as much as it deserves outside of exercise science and kinesiology, especially considering the vast amount of work that has been dedicated to explaining its mechanisms and ramifications.

At the other end of the lifespan continuum, physical exercise has also been found to be fundamental to healthy aging, leading to better associative learning (Fabre, Chamari, Mucci, Massé-Biron, & Préfaut 2002), better quality of life and general cognition (Cancela Carral & Ayán Pérez 2007), and enhanced mood and emotional stability (Blumenthal et al. 1991), although it should be noted that these findings have been questioned by a recent meta-analysis (Young, Angevaren, Rusted, & Tabet 2015). To be fair, we should point out that exercise interventions is an umbrella term and that these regimens come in many different flavors, in terms of type of training, duration, frequency, sample population, outcome measures, and several other factors. Therefore, pooling such disparate interventions together using meta-analytic techniques is only informative up to a point – eventually, what needs to be determined is not so much the absolute effectiveness of exercise interventions but the potential mediators of sizeable effects (Moreau 2014). What works and for whom, and under which circumstances, remains to be informed by future research.

When examined in isolation, the literature on young and middle-aged adults is slightly less consistent, for several reasons. Arguably, such individuals might be at their natural cognitive peak, thus making any large improvement difficult (Salthouse & Davis 2006). Given the lack of power most intervention studies suffer from by design, effects often need to be quite substantial to be detected. Adults typically experience environments that are typically complex and stimulating – these provide naturally enriched conditions, leaving little room for cognitive gains. From a research standpoint, there might also be smaller incentives to target non-clinical adult populations, as opposed to children, the elderly, or clinical populations, who all can greatly benefit from exercise interventions.

Structural and Functional Changes in the Brain

Importantly, cognitive improvements are typically accompanied by changes in brain structure and function. Due to the aforementioned difficulties of detecting such changes in adult populations, this line of research is mostly based on evidence in children and the elderly. For example, several studies indicate that children show greater integrity in two main neural networks after a physical exercise intervention: the uncinate fasciculus (Schaeffer et al. 2014), a white matter tract connecting parts of the limbic system and the prefrontal cortex; and in the superior longitudinal fasciculus (Krafft et al. 2014), a long bundle of axons linking frontal and occipital lobes, and part of parietal and temporal lobes. This line of research is based on Diffusion Tensor Imaging (DTI), an MRI technique that allows probing the brain's white matter tracts through detailed mapping of water molecules diffusion. Cross-sectional studies support the hypothesis of an effect of physical exercise on brain structure, via evidence for differences in white matter integrity between higher and lower fit children (Chaddock-Heyman et al. 2014).

In older adults, studies have shown that exercise interventions can counteract age-related brain atrophy (Colcombe et al. 2003), and that this effect might mediate the association between aerobic fitness and executive function (Weinstein et al. 2012). Higher physical fitness has also been associated with larger cortical areas, especially frontal regions (Weinstein et al. 2012), and larger hippocampus (Erickson et al. 2009; Makizako et al. 2015). Training interventions corroborate these findings, showing volume increases in frontal (Colcombe et al. 2006; Ruscheweyh et al. 2011) and hippocampal (Erickson et al. 2011; Ruscheweyh et al. 2011) areas after undergoing physical exercise training.

More subtle functional changes have also been documented. Until recently, most of this evidence in children was based on correlational findings, reporting an association between aerobic fitness or physical exercise and brain function mostly based on fMRI and EEG techniques (Chaddock et al. 2012; Moore, Drollette, Scudder, Bharij, & Hillman 2014). However, this line of research has also been strengthened by experiments (RCTs) highlighting the benefits of a physical exercise intervention on brain function (Chaddock-Heyman et al. 2013; Davis et al. 2011; Hillman et al. 2014; Kamijo et al. 2011). In parallel, cross-sectional evidence for the relationship between physical exercise and brain function is available for older populations (Berchicci, Lucci, Perri, Spinelli, & Di Russo 2014; Prakash et al. 2011; Smith et al. 2011), and these are also supported by more direct experimental findings (Chapman et al. 2013; Colcombe et al. 2004; Smith, Nielson, Woodard, Seidenberg, & Rao 2013). Specifically, physical exercise is associated with increased connectivity between prefrontal, cingulate, and hippocampal areas, which results in better performance on several cognitive tasks (Burdette et al. 2010; Voss et al. 2010).

Besides promising evidence at the behavioral and neural levels, the association between physical exercise and cognition is also consistent with findings at the neurobiological level, which help explain the mechanisms responsible for the observed correlations or improvements. Seeking corroborating evidence at the neurophysiological level is akin to zooming in further, from overt behaviors and neural correlates at the system level to their underlying mechanisms. This line of work is discussed in the next section.

Neurobiological Mechanisms of Exercise-Induced Improvements

Most of the neurobiological correlates of exercise-induced cognitive enhancement are now well understood (for a detailed account of the changes associated with exercise, see McMorris & Corbett 2016). Physical exercise leads to an increase in cerebral vascularization (Black, Isaacs, Anderson, Alcantara, & Greenough 1990), proteins and neurotransmitters (Mora, Segovia, & del Arco

2007), heightened insult resistance (Stummer, Weber, Tranmer, Baethmann, & Kempski 1994), enhanced neurogenesis (van Praag et al. 2002), synaptic metabolism (Vaynman, Ying, Yin, & Gomez-Pinilla 2006), angiogenesis (Black et al. 1990), neuronal survival (Vaynman et al. 2006), and enhanced overall brain volume (Colcombe et al. 2006). More specifically, brain-derived neurotrophic factor (BDNF) appears to play an active role in mediating the effect of physical exercise on cognition. Animal studies have found increases in hippocampal BDNF post-exercise (Neeper, Gómez-Pinilla, Choi, & Cotman 1995), particularly significant considering the central role of the hippocampus in learning and memory and its deterioration in many degenerative diseases including Alzheimer's. These effects have been found to last at least several weeks (Berchtold, Kesslak, Pike, Adlard, & Cotman 2001), therefore potentially playing a critical role in exercise-induced neural plasticity. Besides its effect on hippocampal areas, BDNF also increases post-exercise in the spinal cord (Gómez-Pinilla, Ying, Opazo, Roy, & Edgerton 2001), the cerebellum and several cortical regions (Neeper, Gómez-Pinilla, Choi, & Cotman 1996), possibly through increased levels of Insulin-like Growth Factor-1 (IGF-1), a growth factor involved in neuronal development (Arsenijevic & Weiss 1998). Based on this evidence, it has been suggested that IGF-1 might be a key determinant in the effect of physical exercise on BDNF levels (Cotman & Berchtold 2002), and thus on cognition more generally.

Increased BDNF levels in cortical areas are consistent with the documented effects of exercise on serum BDNF (sBDNF). Several studies have found that exercise induces an augmentation in sBDNF levels (Griffin et al. 2011; Schmolesky, Webb, & Hansen 2013), typically within a few hours, with the magnitude of the increase dependent on exercise intensity (Ferris, Williams, & Shen 2007). Thus, these findings provide additional support for the idea that aerobic exercise might not be the most optimal way to target cognitive improvement in healthy individuals, and that more intense forms of exercise could induce larger gains.

Overall, the remarkable consequence of these potent underlying mechanisms is that physical exercise provides a powerful way to stimulate general health. In addition to favoring positive changes, exercise also allows controlling harmful factors, such as stress. Corticosteroids, or stress hormones, have a damaging effect on BDNF concentration, and can therefore, if sustained, lead to neuronal degradation and dendritic atrophy (Gould, Woolley, Frankfurt, & McEwen 1990). Exercise prevents this debilitating effect by blocking downregulations of BDNF, particularly in the hippocampus (Russo-Neustadt, Ha, Ramirez, & Kesslak 2001). Further evidence has highlighted the impact of exercise on levels of monoamine neurotransmitters (e.g. dopamine, epinephrine, norepinephrine) and tryptamine neurotransmitters (e.g. serotonin, melatonin), which could be responsible for some of the benefits typically observed after physical training (Winter et al. 2007). This line of research therefore provides compelling evidence for the role of neurobiological factors in mediating the behavioral changes typically observed post-exercise.

The Embodied Cognition Framework

The coherent view emerging from the integration of behavioral, neural, and neurobiological findings also resonates well with motor theories currently getting traction in cognitive science (e.g. Jeannerod 2001). In particular, the embodied cognition framework (e.g. Barsalou 2008; Gallese & Sinigaglia 2011; Glenberg 2010) argues that the motor system is involved in most of our actions, and that motor processes influence cognitive function drastically.[1] Although legitimate criticisms have been raised against several findings and their embodied interpretation (Wilson 2002) or about the way the embodied framework is typically defined (Wilson & Golonka 2013), the framework remains informative to further understand the link between physical exercise, motor activities, and cognition.

Moreover, the embodied cognition framework is corroborated by numerous compelling findings, which indirectly support the link between exercise and cognition, and by extension the potential of motor activities to enhance cognition. For example, prior work has shown that different levels of motor expertise are associated with differences in performance on motor processing (Güldenpenning, Koester, Kunde, Weigelt, & Schack 2011). These findings also extend beyond motor tasks – spatial ability (Moreau 2012) and working memory capacity tasks (Moreau 2013), good indicators of cognitive abilities known to be central to human cognition (Carroll 1993) and to have tremendous ramifications on numerous everyday tasks (Hegarty & Waller 2005; Kane et al. 2004), are also influenced positively by motor expertise.

In addition, other components of cognition, such as the ability to solve complex problems, are improved when gestures are consistent with the motor actions naturally associated with a particular task, suggesting that gestures ground mental representation in action (Beilock & Goldin-Meadow 2010). Further research indicates that such associations are also found in linguistic domains, with motor actions influencing language comprehension (Holt & Beilock 2006), a finding corroborated by neural changes in language comprehension depending on motor experience (Beilock, Lyons, Mattarella-Micke, Nusbaum, & Small 2008).

Beyond these correlational associations, experimental evidence has also demonstrated that the link is causal – motor training leads to enhanced performance on spatial ability and working memory capacity tasks (Moreau, Clerc, Mansy-Dannay, & Guerrien 2012; Moreau, Morrison, & Conway 2015), and these behavioral changes are mediated by cortical changes (Adkins, Boychuk, Remple, & Kleim 2006). This line of research paves the way for interesting applications intended to target cognitive gains via behavioral interventions, which we discuss in the next section.

Beyond Mere Physical Exercise: Complex Motor Activities

The effect of physical exercise on cognition is well documented and largely replicated – alternative forms of cognitive enhancement based on behavioral interventions can hardly compare with the extensive literature linking physical exercise and cognition (see for a review Moreau & Conway 2013). However, that physical exercise appears to be one of the most effective ways to trigger substantial and durable changes does not mean that it should be exclusive – adequate combinations can provide interesting benefits over interventions solely based on physical exercise.

Approaches seeking to combine cognitive training activities have flourished in the past few years. For example, research has investigated the potential of combining physical exercise with meditation (Astin et al. 2003), with cognitive training (Shatil 2013), or with transcranial Direct Current Stimulations (tDCS; Ditye, Jacobson, Walsh, & Lavidor 2012; Madhavan & Shah 2012; Martin et al. 2013; Moreau, Wang, Tseng, & Juan 2015). These approaches all have merit of their own, as they represent a step toward more advanced frameworks of cognitive enhancement. Yet they typically combine two successful approaches separately – interventions focus on the different types of regimen either successively or in an interleaved fashion. This approach might be suboptimal considering the opportunity costs associated with cognitive training and techniques of enhancement (Moreau & Conway 2014). In addition, the potential side effects of these interventions can possibly be concerning, whether it is the psychiatric risks of mindfulness meditation (Lazarus 1976; Shapiro 1992) or the inherent blur over techniques that stimulate the brain with direct current (Davis 2014).

In a recent study, we tested the idea of integrating cognitive demands within a physical exercise regimen as an optimal solution to cognitive enhancement (Moreau, Morrison, & Conway 2015). The rationale was that aerobic workouts, typically favored in cognitive training interventions based on physical exercise (Hillman et al. 2008), might benefit from additional cognitive challenges. Because aerobic exercise is merely demanding cognitively, most of the cognitive gains typically observed post-training are due to neurophysiological mechanisms, either direct (e.g. BDNF, IGF-1) or in response to changes in a mediating factor, such as sleep or stress. These types of gains are potent, yet in our view they could be further maximized by the addition of specific components directly tapping cognitive abilities. Following this idea, we presented participants with perceptive, motor, and cognitive problems, to solve in a movement-based framework, while continuously sustaining moderate physical activity. Specifically, perceptive problems included situations where participants had to rely on different sensory inputs to make decisions, often including limited visual inputs. In everyday life, most of our motor actions are typically vision-centered; therefore, we hypothesized that depriving individuals of this source of information had the potential to improve processing quality of alternate inputs (see also Landry, Shiller, & Champoux 2013), along with favoring complex computations to circumvent atypical demands. Motor problems were intended to promote unusual motor coordination in three-dimensional space, for example with transitions from different levels of motion (e.g. standing, kneeling, lying). In a way, the underlying idea is somewhat similar to that used in Constraint-Induced (CI) therapy, either in stroke recovery or in motor impairment (Taub, Uswatte, & Mark 2014). Ad hoc motor problems prompted unfamiliar situations that had to be dealt with in a timely manner, based on little prior experience. Finally, cognitive problems included memorizing sequences of movements for subsequent recall and execution, or the presentation of situations where different strategies had to be considered and evaluated in order for participants to respond successfully. Throughout the program, difficulty was adjusted to participants' performance to ensure sustained motivation and continuously challenging material.

Consistent with our initial hypothesis, we found that an integrated approach combining physical and cognitive demands within a single activity was the most effective to induce cognitive and physiological gains. Specifically, we looked at transfer to working memory capacity and spatial ability, cognitive constructs with important ramifications to numerous activities (Moreau 2015). For both of these constructs, the integrated approach based on complex motor activities was superior to either cognitive or physical exercise components alone. In addition to cognitive changes, complex motor training also induced improvements in biomarkers associated with general health, namely blood pressure and resting heart rate, similar to those following the aerobic exercise regimen.

This training intervention lies within a broader line of research, with encouraging early findings. For example, a study assessing the potential for games combining exercise and cognitive challenges has shown promises in school settings, and provide additional evidence for the relevance of an integrated approach (Tomporowski, Lambourne, & Okumura 2011). In addition, several studies have demonstrated that a combination of executive function and physical demands (e.g. martial arts) might be particularly beneficial, especially in children (see for a brief review Diamond & Lee 2011).

Clearly, combining interventions also has pitfalls – researchers need to identify the effect of components in isolation, and a combined regimen can obscure their respective effects or require large sample sizes to compensate for the additional cells in a factorial design. Good experiments allow a comparison of the integrated approach with all the components in isolation – for example, a combination of physical and cognitive training compared with either physical or cognitive training alone, as described previously (Moreau et al. 2015). Obviously, this approach is costly in terms of sample size to achieve adequate power, and brings up problems of its own, such as matching duration, frequency, and motivational factors between regimens. Yet in our view it represents a necessary line of research if this field is to impact society meaningfully.

Taking Advantage of Developmental Plasticity: Applications in the Classroom

As we alluded to in the previous section, adopting an integrated approach also has ramifications in the classroom. In a period of intense neural plasticity, the influence of environmental factors on children is potentially greater than at any other age across the lifespan. This brings tremendous possibilities to teachers, educators, and the educational system in general, but also important responsibilities: the type and strength of learning children experienced will have large effects and repercussions on their lives.

Situations that favor multi-sensory integration of motor and cognitive demands in the classroom should therefore be encouraged. For example, research at the intersection of spatial and embodied cognition shows that the addition of passive or active motor features (e.g. action observation, gestures) helps reinforce learning and ensure it is integrated meaningfully with existing knowledge (Broaders, Cook, Mitchell, & Goldin-Meadow 2007; Cook, Mitchell, & Goldin-Meadow 2008). In addition, structured plays combining cognitive challenges and physical motion are also essential to optimal cognitive development, and school environments are especially suitable to this type of learning. This can come from blends across subjects – for example, physical exercise with mathematics, physics or biology, to understand *and* experience the concepts that are being taught. Facilitating these kinds of translational approaches also provide additional motivational components. More than at any other age, children are interested in novel and diverse items, a feature largely exploited in the preferential looking paradigm on which most psychology research in infants is based (Golinkoff, Hirsh-Pasek, Cauley, & Gordon 1987). Thus, an approach emphasizing diversity in learning content has the potential to remain more appealing to children in the long run.

Finally, through the process of knowledge generalization, transfer emerges given sufficient encounters with particular content (Christiansen & Curtin 1999). This means that multiple explanations of the same ideas are often needed before an idea or a concept is well understood and can be applied more generally. In the context of school learning, it follows that the process of extracting rules can be favored by concurrent or subsequent presentation of the same principle in different situations, either across subjects (gravitational forces in physics and mathematics, heart rate in biology and PE), or within a particular subject (e.g. concept introduced through geometry and further strengthened with algebra). The approach proposed here can help diversify learning content in this regard, a facet that increases chances of proposing content adapted to each individual, based on individual abilities and preferences. The implementation of original pedagogical situations requires creativity, willingness, and effort, but can have tremendous impact on children's experience in the classroom.

Other novel approaches are promising and could contribute further to bringing physical exercise to the classroom (e.g. High Intensity Training, HIT; see Moreau 2015). Most importantly, it is worth reiterating that common reductions in physical exercise classes in favor of "core" subjects are profoundly ill-conceived, and have shown disastrous consequences when implemented (Strong et al. 2005). On the contrary, more time should be dedicated to activities that challenge children cognitively and physically, and that encourage the development of complex motor coordination.

Concluding Remarks: Toward Personalized Regimens

The line of research discussed in this chapter comes at a time of sizeable excitement and promises, but also of greater awareness concerning the subtleties of the interaction between physical exercise, brain, and cognition. As the field of cognitive training matures, researchers are transitioning from dichotomous beliefs about the absolute effectiveness of training regimens toward a finer understanding of the inherent mechanisms at play in such complex dynamics. In particular, many would now argue that the possibility to impact cognitive abilities through training is not the core of the debate, which rather is currently focused on the effectiveness of respective components for given populations (e.g. Moreau 2015). This idea implies that cognitive abilities thrive in environments that are individualized, and thus underline the need to deliver content adapted to specific deficits and imbalances (Moreau 2014).

In addition, if the neurobiological and neural correlates of the association between physical exercise and cognitive improvement are well understood, the meaning of cognitive enhancement in general is not. What are the underlying mechanisms leading to improvement in cognitive abilities? Can gains last? Why do some individuals fail to improve at all? Are there potential tradeoffs? These are just a few of the numerous questions this trend of work is bound to explore and try to answer. Limitations of the current picture have been raised elsewhere (e.g. Hills & Hertwig 2011), and can be circumvented only with the establishment of a theoretical framework of cognitive enhancement. Until then, cognitive training studies will remain a heterogeneous collection of work with no clear common mechanism and connection.

Eventually, one important goal for research at the crossroad of cognitive neuroscience and exercise science is to inform policies more directly, so that individuals can have access to the tools and knowledge they need to make more informed choices when it comes to cognitive health and to maximizing their cognitive potential. Advances in this regard, although less dramatic than a novel drug or a groundbreaking treatment in the eyes of the public, are nonetheless critical, as they can allow dysfunctions or disorders to be alleviated, postponed, or even prevented by early interventions.

Note

1 Caution is required when defining such a broad field, because this oversimplification has often led to a misperception of embodied theories and of their core differences with more traditional models of cognition. One particular point that has done a disservice to the field of embodied cognition is the recurrent tendency to promote embodied theories by opposition to outdated approaches of cognitive processing. Some researchers still present embodied cognition as an alternative to purely amodal and propositional models of cognition, implicitly or explicitly suggested as the current state admitted in cognitive science. This approach is clearly misleading and does not make a compelling case for an embodied alternative to more traditional cognitive theories.

References

Adkins, D. L., J. Boychuk, M. S. Remple, & J. A. Kleim. 2006. "Motor training induces experience-specific patterns of plasticity across motor cortex and spinal cord." *Journal of Applied Physiology 101* (6): 1776–82. doi:10.1152/japplphysiol.00515.2006

Arsenijevic, Y., & S. Weiss. 1998. "Insulin-like growth factor-I is a differentiation factor for postmitotic CNS stem cell-derived neuronal precursors: Distinct actions from those of brain-derived neurotrophic factor." *Journal of Neuroscience : The Official Journal of the Society for Neuroscience 18* (6): 2118–28.

Astin, J. A., B. M. Berman, B. Bausell, W.-L. Lee, M. Hochberg, & K. L. Forys. 2003. "The efficacy of mindfulness meditation plus Qigong movement therapy in the treatment of fibromyalgia: A randomized controlled trial." *J Rheumatol 30* (10): 2257–62.

Barsalou, L. W. 2008. "Grounded cognition." *Annual Review of Psychology 59*: 617–45. doi:10.1146/annurev.psych.59.103006.093639

Beilock, S. L., & S. Goldin-Meadow. 2010. "Gesture changes thought by grounding it in action." *Psychological Science 21* (11): 1605–10. doi:10.1177/0956797610385353

Beilock, S. L., I. M. Lyons, A. Mattarella-Micke, H. C. Nusbaum, & S. L. Small. 2008. "Sports experience changes the neural processing of action language." *Proceedings of the National Academy of Sciences of the United States of America 105* (36): 13269–73. doi:10.1073/pnas.0803424105

Berchicci, M., G. Lucci, R. L. Perri, D. Spinelli, & F. Di Russo. 2014. "Benefits of physical exercise on basic visuo-motor functions across age." *Frontiers in Aging Neuroscience 6*: 48. doi:10.3389/fnagi.2014.00048

Berchtold, N. C., J. P. Kesslak, C. J. Pike, P. A. Adlard, & C. W. Cotman. 2001. "Estrogen and exercise interact to regulate brain-derived neurotrophic factor mRNA and protein expression in the hippocampus." *European Journal of Neuroscience 14* (12): 1992–2002.

Black, J. E., K. R. Isaacs, B. J. Anderson, A. A. Alcantara, & W. T. Greenough. 1990. "Learning causes synaptogenesis, whereas motor activity causes angiogenesis, in cerebellar cortex of adult rats." *Proceedings of the National Academy of Sciences of the United States of America 87* (14): 5568–72.

Blair, S. N. 1995. "Changes in physical fitness and all-cause mortality." *JAMA 273* (14): 1093. doi:10.1001/jama.1995.03520380029031

Blumenthal, J. A., C. F. Emery, D. J. Madden, S. Schniebolk, M. Walsh-Riddle, L. K. George,… R. E. Coleman. 1991. "Long-term effects of exercise on psychological functioning in older men and women." *Journal of Gerontology 46* (6): P352–61.

Broaders, S. C., S. W. Cook, Z. Mitchell, & S. Goldin-Meadow. 2007. "Making children gesture brings out implicit knowledge and leads to learning." *Journal of Experimental Psychology. General 136* (4): 539–50. doi:10.1037/0096-3445.136.4.539

Burdette, J. H., P. J. Laurienti, M A. Espeland, A. Morgan, Q. Telesford, C. D. Vechlekar,… W. J. Rejeski. 2010. "Using network science to evaluate exercise-associated brain changes in older adults." *Frontiers in Aging Neuroscience 2*: 23. doi:10.3389/fnagi.2010.00023

Cancela Carral, J. M., & C. Ayán Pérez. 2007. "Effects of high-intensity combined training on women over 65." *Gerontology 53* (6): 340–6. doi:10.1159/000104098

Carroll, J. 1993. *Human cognitive abilities: A survey of factor-analytical studies.* New York: Cambridge University Press.

Castelli, D. M., C. H. Hillman, S. M. Buck, & H. E. Erwin. 2007. "Physical fitness and academic achievement in third- and fifth-grade students." *Journal of Sport & Exercise Psychology 29* (2): 239–52.

Chaddock, L., K. I. Erickson, R. S. Prakash, M. W. Voss, M. VanPatter, M. B. Pontifex,...A. F. Kramer. 2012. "A functional MRI investigation of the association between childhood aerobic fitness and neurocognitive control." *Biological Psychology 89* (1): 260–8. doi:10.1016/j.biopsycho.2011.10.017

Chaddock-Heyman, L., K. I. Erickson, J. L. Holtrop, M. W. Voss, M. B. Pontifex, L. B. Raine, ...A. F. Kramer. 2014. "Aerobic fitness is associated with greater white matter integrity in children." *Frontiers in Human Neuroscience 8*: 584. doi:10.3389/fnhum.2014.00584

Chaddock-Heyman, L., K. I. Erickson, M. W. Voss, A. M. Knecht, M. B. Pontifex, D. M. Castelli, ...A. F. Kramer. 2013. "The effects of physical activity on functional MRI activation associated with cognitive control in children: A randomized controlled intervention." *Frontiers in Human Neuroscience 7*: 72. doi:10.3389/fnhum.2013.00072

Chapman, S. B., S. Aslan, J. S. Spence, L. F. Defina, M. W. Keebler, N. Didehbani, & H. Lu. 2013. "Shorter term aerobic exercise improves brain, cognition, and cardiovascular fitness in aging." *Frontiers in Aging Neuroscience 5*: 75. doi:10.3389/fnagi.2013.00075

Christiansen, M., S. & Curtin. 1999. "Transfer of learning: Rule acquisition or statistical learning?" *Trends in Cognitive Sciences 3* (8): 289–90.

Coe, D. P., J. M. Pivarnik, C. J. Womack, M. J. Reeves, & R. M. Malina. 2006. "Effect of physical education and activity levels on academic achievement in children." *Medicine and Science in Sports and Exercise 38* (8): 1515–9. doi:10.1249/01.mss.0000227537.13175.1b

Colcombe, S. J., K. I. Erickson, N. Raz, A. G. Webb, N. J. Cohen, E. McAuley, & A. F. Kramer. 2003. "Aerobic fitness reduces brain tissue loss in aging humans." *Journals of Gerontology. Series A, Biological Sciences and Medical Sciences 58* (2): 176–80.

Colcombe, S. J., K. I. Erickson, P. E. Scalf, J. S. Kim, R. Prakash, E. McAuley, ...A. F. Kramer. 2006. "Aerobic exercise training increases brain volume in aging humans." *Journals of Gerontology. Series A, Biological Sciences and Medical Sciences 61* (11): 1166–70.

Colcombe, S. J., & A. F. Kramer. 2003. "Fitness effects on the cognitive function of older adults: A meta-analytic study." *Psychological Science 14* (2): 125–30.

Colcombe, S. J., A. F. Kramer, K. I. Erickson, P. Scalf, E. McAuley, N. J. Cohen,... S. Elavsky. 2004. "Cardiovascular fitness, cortical plasticity, and aging." *Proceedings of the National Academy of Sciences of the United States of America 101* (9): 3316–21. doi:10.1073/pnas.0400266101

Cook, S. W., Z. Mitchell, & S. Goldin-Meadow. 2008. "Gesturing makes learning last." *Cognition 106* (2): 1047–58. doi:10.1016/j.cognition.2007.04.010

Cotman, C. W., & N. C. Berchtold. 2002. "Exercise: A behavioral intervention to enhance brain health and plasticity." *Trends in Neurosciences 25* (6): 295–301.

Davis, C. L., P. D. Tomporowski, C. A. Boyle, J. L. Waller, P. H. Miller, J. A. Naglieri, & M. Gregoski. 2007. "Effects of aerobic exercise on overweight children's cognitive functioning: a randomized controlled trial." *Research Quarterly for Exercise and Sport 78* (5): 510–9. doi:10.1080/02701367.2007.10599450

Davis, C. L., S. Tomporowski, J. E. McDowell, B. P. Austin, P. H. Miller, N. E. Yanasak,...J. A. Naglieri. 2011. "Exercise improves executive function and achievement and alters brain activation in overweight children: A randomized, controlled trial." *Health Psychology : Official Journal of the Division of Health Psychology, American Psychological Association 30* (1): 91–8. doi:10.1037/a0021766

Davis, N. J. 2014. "Transcranial stimulation of the developing brain: A plea for extreme caution." *Frontiers in Human Neuroscience 8*: 600. doi:10.3389/fnhum.2014.00600

Diamond, A., & K. Lee. 2011. "Interventions shown to aid executive function development in children 4 to 12 years old." *Science 333* (6045): 959–64. doi:10.1126/science.1204529

Ditye, T., L. Jacobson, V. Walsh, & M. Lavidor. 2012. "Modulating behavioral inhibition by tDCS combined with cognitive training." *Experimental Brain Research, 219* (3): 363–8. doi:10.1007/s00221-012-3098-4

Erickson, K. I., R. S. Prakash, M. W. Voss, L. Chaddock, L. Hu, K. S. Morris,...A. F. Kramer. 2009. "Aerobic fitness is associated with hippocampal volume in elderly humans." *Hippocampus, 19* (10): 1030–9. doi:10.1002/hipo.20547

Erickson, K. I., M. W. Voss, R. S. Prakash, C. Basak, A. Szabo, L. Chaddock,...A. F. Kramer. 2011. "Exercise training increases size of hippocampus and improves memory." *Proceedings of the National Academy of Sciences of the United States of America, 108* (7): 3017–22. doi:10.1073/pnas.1015950108

Fabre, C., K. Chamari, P. Mucci, J. Massé-Biron, & C. Préfaut. 2002. "Improvement of cognitive function by mental and/or individualized aerobic training in healthy elderly subjects." *International Journal of Sports Medicine 23* (6): 415–21. doi:10.1055/s-2002–33735

Ferris, L. T., J. S. Williams, & C.-L. Shen. 2007. "The effect of acute exercise on serum brain-derived neurotrophic factor levels and cognitive function." *Medicine and Science in Sports and Exercise 39* (4): 728–34. doi:10.1249/mss.0b013e31802f04c7

Gallese, V., & C. Sinigaglia. 2011. "What is so special about embodied simulation?" *Trends in Cognitive Sciences 15* (11): 512–9. doi:10.1016/j.tics.2011.09.003

Glenberg, A. M. 2010. "Embodiment as a unifying perspective for psychology." *Wiley Interdisciplinary Reviews. Cognitive Science 1*(4): 586–96. doi:10.1002/wcs.55

Goh, T. L., J. Hannon, C. Webster, L. Podlog, & M. Newton. 2016. "Effects of a TAKE 10!® classroom-based physical activity intervention on 3rd to 5th grades children's on-task behavior." *Journal of Physical Activity & Health*. doi:10.1123/jpah.2015-0238

Golinkoff, R. M., K. Hirsh-Pasek, K. M. Cauley, & L. Gordon. 1987. "The eyes have it: Lexical and syntactic comprehension in a new paradigm." *Journal of Child Language 14* (1): 23–45.

Gómez-Pinilla, F., Z. Ying, P. Opazo, R. R. Roy, & V. R. Edgerton. 2001. "Differential regulation by exercise of BDNF and NT-3 in rat spinal cord and skeletal muscle." *European Journal of Neuroscience 13* (6): 1078–84.

Gould, E., C. S. Woolley, M. Frankfurt, & V. McEwen. 1990. "Gonadal steroids regulate dendritic spine density in hippocampal pyramidal cells in adulthood." *Journal of Neuroscience : The Official Journal of the Society for Neuroscience 10* (4): 1286–91.

Griffin, É. W., S. Mullally, C. Foley, S. A. Warmington, S. M. O'Mara, & A. M. Kelly. 2011. "Aerobic exercise improves hippocampal function and increases BDNF in the serum of young adult males." *Physiology and Behavior 104* (5): 934–41. doi:10.1016/j.physbeh.2011.06.005

Güldenpenning, I., D. Koester, W. Kunde, M. Weigelt, & T. Schack. 2011. "Motor expertise modulates the unconscious processing of human body postures." *Experimental Brain Research 213* (4): 383–91. doi:10.1007/s00221-011-2788-7

Hegarty, M., & D. A. Waller. 2005. "Individual differences in spatial abilities." In P. Shah & A. Miyake (Eds.), *The Cambridge Handbook of Visuospatial Thinking* (pp. 121–69). New York: Cambridge University Press.

Hillman, C. H., K. I. Erickson, & A. F. Kramer 2008. "Be smart, exercise your heart: Exercise effects on brain and cognition." *Nature Reviews Neuroscience 9* (1): 58–65. doi:10.1038/nrn2298

Hillman, C. H., M. B. Pontifex, D. M. Castelli, N. A. Khan, L. B. Raine, M. R. Scudder,...K. Kamijo. 2014. "Effects of the FITKids randomized controlled trial on executive control and brain function." *Pediatrics 134* (4): e1063–71. doi:10.1542/peds.2013-3219

Hills, T., & R. Hertwig. 2011. "Why aren't we smarter already: Evolutionary trade-offs and cognitive enhancements." *Current Directions in Psychological Science 20* (6): 373–7. doi:10.1177/0963721411418300

Holt, L. E., & S. L. Beilock. 2006. "Expertise and its embodiment: Examining the impact of sensorimotor skill expertise on the representation of action-related text." *Psychonomic Bulletin and Review 13* (4): 694–701.

Jeannerod, M. 2001. "Neural simulation of action: A unifying mechanism for motor cognition." *NeuroImage 14* (1 Pt 2): S103–9. doi:10.1006/nimg.2001.0832

Kamijo, K., M. B. Pontifex, K. C. O'Leary, M. R. Scudder, C.-T. Wu, D. M. Castelli, & C. H. Hillman. 2011. "The effects of an afterschool physical activity program on working memory in preadolescent children." *Developmental Science 14* (5): 1046–58. doi:10.1111/j.1467-7687.2011.01054.x

Kane, M. J., D. Z. Hambrick, S. W. Tuholski, O. Wilhelm, T. W. Payne, & R. Engle. 2004. "The generality of working memory capacity: A latent-variable approach to verbal and visuospatial memory span and reasoning." *Journal of Experimental Psychology. General 133* (2): 189–217. doi:10.1037/0096-3445.133.2.189

Keeley, T. J. H., & K. R. Fox. 2009. "The impact of physical activity and fitness on academic achievement and cognitive performance in children." *International Review of Sport and Exercise Psychology 2* (2): 198–214. doi:10.1080/17509840903233822

Krafft, C. E., N. F. Schwarz, L. Chi, A. L. Weinberger, D. J. Schaeffer, J. E. Pierce,...J. E. McDowell. 2014. "An 8-month randomized controlled exercise trial alters brain activation during cognitive tasks in overweight children." *Obesity 22* (1): 232–42. doi:10.1002/oby.20518

Kramer, A. F., S. Hahn, & D. Gopher. 1999. "Task coordination and aging: Explorations of executive control processes in the task switching paradigm." *Acta Psychologica 101* (2–3): 339–78.

Landry, S. P., D. M. Shiller, & F. Champoux. 2013. "Short-term visual deprivation improves the perception of harmonicity." *Journal of Experimental Psychology: Human Perception and Performance 39* (6): 1503–7.

Lazarus, A. A. 1976. "Psychiatric problems precipitated by transcendental meditation." *Psychological Reports 39* (2): 601–2. doi:10.2466/pr0.1976.39.2.601

McMorris, T., & J. Corbett. 2016. "Neurobiological changes as an explanation of benefits of exercise." In H. Budde & M. Wegner (Eds.), *Exercise and mental health: Neurobiological mechanisms of the exercise effect on depression, anxiety, and well-being*. Oxford: Taylor & Francis.

Madhavan, S., & B. Shah. 2012. "Enhancing motor skill learning with transcranial direct current stimulation – a concise review with applications to stroke." *Frontiers in Psychiatry 3*: 66. doi:10.3389/fpsyt.2012.00066

Mahar, M. T., S. K. Murphy, D. A. Rowe, J. Golden, A. T. Shields, & T. D. Raedeke. 2006. "Effects of a classroom-based program on physical activity and on-task behavior." *Medicine and Science in Sports and Exercise 38* (12): 2086–94. doi:10.1249/01.mss.0000235359.16685.a3

Makizako, H., T. Liu-Ambrose, H. Shimada, T. Doi, H. Park, K. Tsutsumimoto,...T. Suzuki. 2015. "Moderate-intensity physical activity, hippocampal volume, and memory in older adults with mild cognitive impairment." *Journals of Gerontology. Series A, Biological Sciences and Medical Sciences 70* (4): 480–6. doi:10.1093/gerona/glu136

Martin, D. M., R. Liu, A. Alonzo, M. Green, M. J. Player, P. Sachdev, & C. K. Loo. 2013. "Can transcranial direct current stimulation enhance outcomes from cognitive training? A randomized controlled trial in healthy participants." *International Journal of Neuropsychopharmacology / Official Scientific Journal of the Collegium Internationale Neuropsychopharmacologicum (CINP) 16* (9): 1927–36. doi:10.1017/S1461145713000539

Moore, R. D., E. S. Drollette, M. R. Scudder, A. Bharij, & C. H. Hillman. 2014. "The influence of cardiorespiratory fitness on strategic, behavioral, and electrophysiological indices of arithmetic cognition in preadolescent children." *Frontiers in Human Neuroscience 8*: 258. doi:10.3389/fnhum.2014.00258

Mora, F., G. Segovia, & A. del Arco. 2007. "Aging, plasticity and environmental enrichment: Structural changes and neurotransmitter dynamics in several areas of the brain." *Brain Research Reviews 55* (1): 78–88. doi:10.1016/j.brainresrev.2007.03.011

Moreau, D. 2012. "The role of motor processes in three-dimensional mental rotation: Shaping cognitive processing via sensorimotor experience." *Learning and Individual Differences 22* (3): 354–9.

Moreau, D. 2013. "Motor expertise modulates movement processing in working memory." *Acta Psychologica 142* (3): 356–61.

Moreau, D. 2014. "Making sense of discrepancies in working memory training experiments: A Monte Carlo simulation." *Frontiers in Systems Neuroscience 8*: 161. doi:10.3389/fnsys.2014.00161

Moreau, D. 2015. "Brains and brawn: Complex motor activities to maximize cognitive enhancement." *Educational Psychology Review 27* (3): 475–82. doi:10.1007/s10648-015-9323-5

Moreau, D., J. Clerc, A. Mansy-Dannay, & A. Guerrien. 2012. "Enhancing spatial ability through sport practice: Evidence for an effect of motor training on mental rotation performance." *Journal of Individual Differences 33* (2): 83–8. doi:10.1027/1614-0001/a000075

Moreau, D., & A. R. A. Conway. 2013. "Cognitive enhancement: A comparative review of computerized and athletic training programs." *International Review of Sport and Exercise Psychology 6* (1): 155–83. doi:10.1080/1750984X.2012.758763

Moreau, D., & A. R. A. Conway. 2014. "The case for an ecological approach to cognitive training." *Trends in Cognitive Sciences 18* (7): 334–6. doi:10.1016/j.tics.2014.03.009

Moreau, D., A. Morrison, & A. R. A. Conway. 2015. "An ecological approach to cognitive enhancement: Complex motor training." *Acta Psychologica 157*: 44–55. doi:10.1016/j.actpsy.2015.02.007

Moreau, D., A. Wang, P. Tseng, & C.-H. Juan. 2015. "Blending transcranial direct current stimulations and physical exercise to maximize cognitive improvement." *Frontiers in Psychology 6*: 678. doi:10.3389/fpsyg.2015.00678

Neeper, S. A., F. Gómez-Pinilla, J. Choi, & C. Cotman. 1995. "Exercise and brain neurotrophins." *Nature 373* (6510): 109. doi:10.1038/373109a0

Neeper, S. A., F. Gómez-Pinilla, J. Choi, & C. W. Cotman. 1996. "Physical activity increases mRNA for brain-derived neurotrophic factor and nerve growth factor in rat brain." *Brain Research 726* (1–2): 49–56.

Penedo, F. J., & J. R. Dahn. 2005. "Exercise and well-being: a review of mental and physical health benefits associated with physical activity." *Current Opinion in Psychiatry 18* (2): 189–93.

Prakash, R. S., M. W. Voss, K. I. Erickson, J. M. Lewis, L. Chaddock, E. Malkowski, ...A. F. Kramer. 2011. "Cardiorespiratory fitness and attentional control in the aging brain." *Frontiers in Human Neuroscience 4*: 229. doi:10.3389/fnhum.2010.00229

Ruscheweyh, R., C. Willemer, K. Krüger, T. Duning, T. Warnecke, J. Sommer,... Flöel. 2011. "Physical activity and memory functions: An interventional study." *Neurobiology of Aging 32* (7): 1304–19. doi:10.1016/j.neurobiolaging.2009.08.001

Russo-Neustadt, A., T. Ha, R. Ramirez, & J. P. Kesslak. 2001. "Physical activity-antidepressant treatment combination: Impact on brain-derived neurotrophic factor and behavior in an animal model." *Behavioural Brain Research 120* (1): 87–95.

Salthouse, T., & H. Davis. 2006. "Organization of cognitive abilities and neuropsychological variables across the lifespan." *Developmental Review 26* (1): 31–54. doi:10.1016/j.dr.2005.09.001

Schaeffer, D. J., C. E. Krafft, N. F. Schwarz, L. Chi, A. L. Rodrigue, J. Pierce,... McDowell. 2014. "An 8-month exercise intervention alters frontotemporal white matter integrity in overweight children." *Psychophysiology 51* (8): 728–33. doi:10.1111/psyp.12227

Schmolesky, M. T., D. L. Webb, & S. Hansen. 2013. "The effects of aerobic exercise intensity and duration on levels of brain-derived neurotrophic factor in healthy men." *Journal of Sports Science and Medicine 12* (3): 502–11.

Shapiro, D. H. 1992. "Adverse effects of meditation: A preliminary investigation of long-term meditators." *International Journal of Psychosomatics : Official Publication of the International Psychosomatics Institute 39* (1–4): 62–7.

Shatil, E. 2013. "Does combined cognitive training and physical activity training enhance cognitive abilities more than either alone? A four-condition randomized controlled trial among healthy older adults." *Frontiers in Aging Neuroscience 5*: 8. doi:10.3389/fnagi.2013.00008

Smith, J. C., K. A. Nielson, J. L. Woodard, M. Seidenberg, & S. M. Rao. 2013. "Physical activity and brain function in older adults at increased risk for Alzheimer's disease." *Brain Sciences 3* (1): 54–83. doi:10.3390/brainsci3010054

Smith, J. C., K. A. Nielson, J. L. Woodard, M. Seidenberg, M. D. Verber, S. Durgerian,...S. M. Rao. 2011. "Does physical activity influence semantic memory activation in amnestic mild cognitive impairment?" *Psychiatry Research 193* (1): 60–2. doi:10.1016/j.pscychresns.2011.04.001

Strong, W. B., R. M. Malina, C. J. R. Blimkie, S. R. Daniels, R. K. Dishman, B. Gutin,...F. Trudeau. 2005. "Evidence based physical activity for school-age youth." *Journal of Pediatrics 146* (6): 732–7. doi:10.1016/j.jpeds.2005.01.055

Stummer, W., K. Weber, B. Tranmer, A. Baethmann, & O. Kempski. 1994. "Reduced mortality and brain damage after locomotor activity in gerbil forebrain ischemia." *Stroke; a Journal of Cerebral Circulation 25* (9): 1862–9.

Taub, E., G. Uswatte, & V. W. Mark. 2014. "The functional significance of cortical reorganization and the parallel development of CI therapy." *Frontiers in Human Neuroscience 8*: 396. doi:10.3389/fnhum.2014.00396

Tomporowski, P. D., K. Lambourne, & M. S. Okumura. 2011. "Physical activity interventions and children's mental function: An introduction and overview." *Preventive Medicine 52 Suppl 1*: S3–9. doi:10.1016/j.ypmed.2011.01.028

Trudeau, F., & R. J. Shephard. 2008. "Physical education, school physical activity, school sports and academic performance." *International Journal of Behavioral Nutrition and Physical Activity 5*: 10. doi:10.1186/1479-5868-5-10

Tsai, C.-L., C.-H. Wang, & Y.-T. Tseng. 2012. "Effects of exercise intervention on event-related potential and task performance indices of attention networks in children with developmental coordination disorder." *Brain and Cognition 79* (1): 12–22. doi:10.1016/j.bandc.2012.02.004

van Praag, H., Schinder, A. F., Christie, B. R., Toni, N., Palmer, T. D., & Gage, F. H. 2002. "Functional neurogenesis in the adult hippocampus." *Nature, 415*(6875), 1030–4. doi:10.1038/4151030a

Vaynman, S. S., Z. Ying, D. Yin, & F. Gomez-Pinilla. 2006. "Exercise differentially regulates synaptic proteins associated to the function of BDNF." *Brain Research 1070* (1): 124–30. doi:10.1016/j.brainres.2005.11.062

Voss, M. W., R. S. Prakash, K. I. Erickson, C. Basak, L. Chaddock, J. S. Kim,...A. F. Kramer. 2010. "Plasticity of brain networks in a randomized intervention trial of exercise training in older adults." *Frontiers in Aging Neuroscience 2.* doi:10.3389/fnagi.2010.00032

Weinstein, A. M., M. W. Voss, R. S. Prakash, L. Chaddock, A. Szabo, S. White,...K. I. Erickson. 2012. "The association between aerobic fitness and executive function is mediated by prefrontal cortex volume." *Brain, Behavior, and Immunity 26* (5): 811–9. doi:10.1016/j.bbi.2011.11.008

Wilson, A. D., & S. Golonka. 2013. "Embodied cognition is not what you think it is." *Frontiers in Psychology 4*: 58. doi:10.3389/fpsyg.2013.00058

Wilson, M. 2002. "Six views of embodied cognition." *Psychonomic Bulletin and Review 9* (4): 625–36.

Winter, B., C. Breitenstein, F. C. Mooren, K. Voelker, M. Fobker, A. Lechtermann ... S. Knecht. 2007. "High impact running improves learning." *Neurobiology of Learning and Memory 87* (4): 597–609. doi:10.1016/j.nlm.2006.11.003

Young, J., M. Angevaren, J. Rusted, & N. Tabet. 2015. "Aerobic exercise to improve cognitive function in older people without known cognitive impairment." *Cochrane Database of Systematic Reviews 4*, CD005381. doi:10.1002/14651858.CD005381.pub4

8 Exercise-Induced Improvement in Motor Learning

Nico Lehmann and Marco Taubert

Introduction

The optimization of motor learning is of particular importance in many sport-related settings such as disease prevention, rehabilitation after neurological or orthopedic injury, physical education as well as competitive sports. A huge body of literature in movement and training science proposes strategies to optimize motor-skill learning with a strong emphasis on practice distribution (for example massed vs. distributed practice), scheduling (blocked vs. random practice), variation of motor tasks (constant vs. variable practice) as well as movement feedback or attentional focus (Magill 2011; Schmidt & Lee 2014). From a more mechanistic perspective, strategies to enhance motor learning should also take into account the underlying neurobiological mechanisms of skill acquisition, stabilization, and retention in the brain. One strategy to stimulate the brain and to improve the capacity for information processing and learning is physical exercise. Here, we will review physical (endurance) exercise as a new approach to enhance motor skill learning through facilitation of the underlying neurobiological processes.

Mounting evidence demonstrates that physical activity, defined as any bodily movements produced by skeletal muscles that result in energy expenditure

(Caspersen, Powell, & Christenson 1985), affects brain structure and function from the molecular to the systems level. More specifically, exercise also positively affects subsequent neuroplastic changes induced by learning (Hillman, Erickson, & Kramer 2008; Voss, Vivar, Kramer, & van Praag 2013b). Thus, it is expected that brain changes induced by exercise are causally related to exercise-induced improvement in motor learning and the facilitation of specific neuroplastic events.

This chapter attempts to highlight the potential of *endurance exercise*, encompassing physical activities that recruit mainly the oxidative metabolism for energy supply and are sustained for a long period of time (Jones & Poole 2009), to affect neuroplasticity in the motor system in the form of motor skill learning. A definition of the term *motor (skill) learning* typically comprises practice- and/or experience-induced and relatively permanent changes in motor behavior (Schmidt & Lee 2011). The aim of motor skill learning is to acquire new or to stabilize/enhance already practiced *motor skills* (Magill 2011), recognized as voluntary or intentional movements in order to achieve a certain environmental goal.

While the vast majority of studies examining the influence of exercise on cognition focused on non-motor aspects (Smith et al. 2010; Chang, Labban, Gapin, & Etnier 2012; Roig, Nordbrandt, Geertsen, & Nielsen 2013; see also Chapter 7, this volume), its effects on motor skill learning remain poorly understood (Voss et al. 2013b). Thus, this chapter pursues several objectives, amongst them to provide a summary of the current state of research, to highlight important knowledge gaps and to make proposals for future studies to add to the field.

First, we will focus on the neurobiology of exercise and motor learning in the central nervous system (CNS) with the aim to outline potential interactions between exercise- and skill-related brain adaptations. Second, we will discuss the existing behavioral studies examining exercise-induced changes in motor learning. Based on the insights gained in these two sections, we will then present hypothetical mechanisms on a molecular level translating the effects of exercise-induced muscle activity to improved neuroplasticity of the sensorimotor system and infer proposals to optimize exercise interventions. Finally, we want to outline potential future directions of the field.

Throughout this chapter, we follow the notion of neuroplasticity as the neural basis of motor learning in the CNS (Dayan & Cohen 2011). This means that motor learning is based upon changes of neural networks that occur through lasting biochemical modulations in synaptic transmission (Rioult-Pedotti, Friedman, & Donoghue 2000; Butefisch et al. 2000; Donchin, Sawaki, Madupu, Cohen, & Shadmehr 2002; Antonov, Antonova, Kandel, & Hawkins 2003), including the coordinated strengthening (e.g. LTP) and weakening (e.g. LTD) of synaptic connections (Mayford, Siegelbaum, & Kandel 2012; Lee, Im Rhyu, & Pak 2014). In this respect, long-term potentiation (LTP) and long-term depression (LTD) are considered as the cellular analogue of memory. LTP and LTD reflect sustained changes in synaptic efficacy in response to the correlated arrival of action potentials between neurons. Note that the early phase of LTP (E-LTP) mainly in M1 is hypothesized to reflect short-term (motor) memory (Mayford et al. 2012), whereas the late phase of LTP (L-LTP) is considered as the cellular analogue of the neuropsychological concept of long-term (motor) memory (Hillman et al. 2008; Mayford et al. 2012). The central question that runs through the following chapter is how endurance exercise may contribute to the creation of a productive neural environment that supports plastic processes responsible for motor skill learning.

Exercise- and Motor Learning-Induced Adaptations of the Central Nervous System

In this section, we focus on endurance exercise- and motor learning-induced CNS-adaptations associated with LTP or LTP-like plasticity at different levels of brain organization.

Before reviewing the current literature, we will highlight some historical aspects of experience-induced neuroplasticity with the aim to better integrate these new findings into a coherent view on exercise- and motor learning-induced neuroplasticity. Since the earliest days of research in neural plasticity, it was speculated if and how experience and training impact brain structure and function (Bennett, Diamond, Krech, & Rosenzweig 1964). Interestingly, coordinative-demanding physical activities and endurance exercise programs were both incorporated in early animal experiments aiming at identifying neural correlates of experience-dependent brain plasticity as originally hypothesized by Hebb (Rosenzweig & Bennett 1996). Hebb reports that rats he took at home, enabling them to move freely and collect a lot of new experiences, outperformed their cage-reared litters in a problem-solving task (Hebb 1949). In a series of early experiments with rats, Rosenzweig and co-workers demonstrated, for example, an increased weight of the rat brain cortex and increased enzymatic activity after exposure to enriched environments (Bennett et al. 1964). Note that these paradigms did not allow disentangling the relative contribution of endurance exercise and more coordinative-demanding activities to brain plasticity. While Anderson et al. showed that both exercise and motor skill learning, for example, increase the thickness of the motor cortex (Anderson, Eckburg, & Relucio 2002), Black and co-workers first demonstrated that rats exposed to a challenging environment with affordances like a rope ladder or a seesaw for 30 days show distinct alterations in the cerebellum to those who exercised either voluntarily or forced in a running wheel for the same amount of time. Specifically, the acrobatic (motor learning) condition led to an increase in the number of synapses per Purkinje cell, whereas the aerobic exercise conditions led to an increase in capillary density (Black, Isaacs, Anderson, Alcantara, & Greenough 1990). This was the first work suggesting that endurance exercise and more coordinative-demanding activities likely impose specific demands on the brain, that in turn lead to apparently distinct adaptations of the CNS.

Accordingly, the key notion of this section is that endurance exercise and motor skill learning affect the CNS, at least in part, distinctly (Markham & Greenough 2004; Thomas, Dennis, Bandettini, & Johansen-Berg 2012; Voelcker-Rehage & Niemann 2013) and these differences may be exploited to subsequently facilitate motor learning (Adkins, Boychuk, Remple, & Kleim 2006). First, we deal with structural and functional neuroplastic changes on the *systems level* of brain organization using novel techniques to study brain morphometry and large-scale structural and functional connectivity. Such changes are in part based on adaptations at the *cellular level* (Nudo 2008; Zatorre, Fields, & Johansen-Berg 2012), which are in turn promoted, for example, by the action of neuromodulatory transmitters and neurotrophic factors on the *molecular level.* Throughout, we will relate these changes to the proposed mechanisms of motor learning (e.g. LTP). Brain adaptations to endurance exercise are just treated inasmuch as essential for a basic understanding of the assumed mechanisms. For a more comprehensive view, we refer the reader to Chapter 2 (this volume).

Systems Level

Endurance exercise and motor skill learning are accompanied by a bunch of neuroplastic changes at the systems level. We will focus primarily on studies that investigate human participants using transcranial magnetic stimulation (TMS) as well as structural (sMRI) and resting state functional magnetic resonance imaging (rs fMRI).

Transcranial Magnetic Stimulation (TMS)

TMS is a non-invasive technique to focally stimulate superficial cortical brain region across the scalp. Application of a single, suprathreshold TMS pulse over the primary motor cortex (M1) causes peripheral activity of the target muscle that can be recorded via electromyography, typically referred to as motor evoked potential (MEP). The dependent TMS variables used to characterize motor learning-related changes are the size of cortical area from which an MEP could be evoked (movement representation), the lowest stimulus intensity to evoke an MEP (motor threshold), and the size of the MEP at a defined stimulation intensity (1mV MEP). In general, motor learning increases motor map size, decreases motor thresholds, and increase MEP amplitudes (Pascual-Leone, Amedi, Fregni, & Merabet 2005; Adkins et al. 2006). More recently, these indices have also been recorded in response to endurance exercise. Exhaustive exercise lowers the motor threshold indicating increased cortico-spinal excitability (Coco et al. 2010). Also, increased cortical excitability as evidenced by an increased input-output curve slope was observed in very active compared with sedentary subjects (Cirillo, Lavender, Ridding, & Semmler 2009). Moreover, Singh, Neva, and Staines (2016) investigated whether 20 min of moderate continuous exercise (65–70% of age-predicted HR_{max})

could enhance the neural correlates of motor learning. For this purpose, the authors used single-pulse TMS (extensor carpi radialis, ECR) and acquisition performance in a bimanual motor task as dependent variables. Participants were divided into two groups and three experimental conditions. The exercise group performed exercise alone and exercise followed by motor task practice, whereupon these two experimental sessions took place at least one week apart. The training group solely practiced the bimanual motor task. Statistical analysis revealed that the combination of exercise and training increased MEP amplitude to a greater extent than motor task practice alone. Despite the observation of potentially favorable excitability changes when pairing exercise and motor task practice, acquisition performance in the bimanual motor task remained unaffected.

These single-pulse TMS results are complemented by paired pulse TMS studies that allow specific examination of local inhibitory or facilitative processes within the motor cortex (intra-cortical excitability). This technique pairs two TMS pulses with a particular inter-stimulus interval to target inhibitory (5 ms or less) or facilitatory (10–25 ms) intracortical circuits. A decrease in intracortical inhibition, which seems to be dependent on the level of the inhibiting neurotransmitter γ-aminobutyric acid (GABA), is generally assumed to reflect a favorable environment for the induction of neuroplasticity and therefore motor skill learning (Singh & Staines 2015). While the effect of long-term exercise on intracortical networks has, to the best of our knowledge, not been assessed to date in longitudinal studies with younger subjects, the effects of an acute bout of exercise has been examined with within-subjects designs. For example, Yamaguchi et al. reported a decrease in short-interval intracortical inhibition (SICI) of the leg area (tibialis anterior and soleus muscles) after just 7 min of low-intensity active cycling (Yamaguchi, Fujiwara, Liu, & Liu 2012). Similar effects were observed for the upper extremity (first dorsal interosseous muscle) after 30 min of low–moderate or moderate–high intensity cycling (Smith, Goldsworthy, Garside, Wood, & Ridding 2014). Likewise, 20 minutes of continuous biking with moderate intensity decreased SICI and increased intracortical facilitation (ICF) measured in the extensor carpi radialis muscle (Singh, Duncan, Neva, & Staines 2014).

Recent work suggests that reductions in local GABA concentrations in M1 are correlated with motor learning in a serial-reaction time task (Stagg, Bachtiar, & Johansen-Berg 2011) as well as in a sensorimotor adaptation paradigm (Kim, Stephenson, Morris, & Jackson 2014). Moreover, increased GABA inhibition reduces LTP-like plasticity (Butefisch et al. 2000). A rodent study showed that several days of exercise upregulates genes associated with the excitatory glutamatergic system and downregulates genes related to the inhibitory GABA system in the hippocampus (Molteni, Ying, & Gomez-Pinilla 2002). Taken together, these studies provide strong evidence that exercise at low, moderate, or even high intensities prior to motor skill practice reduces intracortical inhibition and that this effect is not limited to the exercised limbs. However, it must be mentioned that an increased intracortical inhibition in the lower extremity (vastus lateralis muscle) was observed during fatiguing cycling exercise (Sidhu, Lauber, Cresswell, & Carroll 2013) indicating that exercise at very high intensities may attenuate learning immediately after termination of exercise.

Finally, we want to present the existing studies that examined the effect of endurance exercise on TMS protocols aiming to experimentally induce changes in synaptic efficacy (LTP- or LTD-like plasticity). This enables researchers to study synaptic plasticity in vivo and to reduce the influence of numerous boundary conditions normally affecting behavior (think, for example, of potentially differing levels of motor abilities between subjects or even the role of strategies). In this regard, four studies are available that examined the effects of exercise on paired-associative stimulation (PAS)-induced changes in MEPs. Basically, this technique induces cortical LTP-like plasticity by first stimulating a peripheral nerve electrically, followed by electromagnetic stimulation of the corresponding motor cortical area several milliseconds later.

In one such study involving 28 subjects (aged 18–38), Cirillo and coworkers aimed to explore the effect of regular physical activity on PAS-evoked neuroplasticity (Cirillo et al. 2009). Participants were divided into two groups dependent on self-reported physical activity level. The sedentary group (n=14) performed physical activity less than 20 min per day on no more than 3 days per week, whereas the active group (n=14) performed moderate-to-vigorous aerobic activity more than 150 min per day on at least 5 days per week. The authors found that the active subjects revealed increased LTP-like plasticity, as measured by the MEP amplitude of the abductor pollicis brevis (APB) muscle (hand muscle). Noticeably, similar effects were also registered in other experiments focusing on the effects of a single bout of exercise. For example, enhanced PAS-induced plasticity in the APB muscle was observed after 20 min of moderate-intensity cycling (Singh, Neva, & Staines 2014). This beneficial effect applies for higher exercise intensities as well, since a potentiation of the immediate response to PAS in the APB muscle has also been noted after 20 min of high-intensity interval cycling (Mang, Snow, Campbell, Ross, & Boyd 2014). However, PAS-induced MEP facilitation was absent in the soleus muscle (lower extremity) of endurance athletes but pronounced in skill athletes (Kumpulainen et al. 2015). Together, the reasons for the diminished plasticity in lower limbs and the enhanced plasticity in upper limbs in endurance athletes remain speculative. One possibility is that the extensive repetitive use of the lower extremity saturated LTP-like plasticity in the lower limb representations in endurance athletes that makes the induction of further LTP difficult (Kumpulainen et al. 2015).

While the previously reported studies examined LTP-like neuroplasticity, one study assessed LTD-like plasticity mechanisms by means of excitability decreasing theta burst stimulation (cTBS) (McDonnell, Snow, Campbell, Ross, & Boyd 2013). The authors tested 25 subjects that went through three experimental conditions: moderate-intensity cycling (15 min at 77% of age-predicted HR_{max}), low-intensity cycling (30 min at 57%), and an inactive control condition. Interestingly, the results showed that low-intensity cycling suppressed MEPs of the first dorsal interosseus muscle, while moderate intensity exercise as well as the control condition failed to do so. While neither exercise condition produced an increase in BDNF levels, a 10% increase in cortisol levels was exclusively registered in the moderate exercising group. Given this, the authors hypothesized that elevated cortisol levels may have interfered with BDNF expression and plasticity induction. However, this finding contrasts the observations of Mang et al. (2014), where a positive effect of acute high-intensity exercise on neuroplasticity was found.

To sum up, studies examining the responses of acute or long-term exercise on several TMS indices mostly suggest a facilitative effect on neuroplasticity. However, the dose-response relationship between exercise, especially regarding intensity, and TMS indices is not unambiguously clear to date (Singh & Staines 2015).

Structural and Functional Magnetic Resonance Imaging

In recent years, a considerable number of studies demonstrated motor learning-induced structural and functional changes in the brain (Zatorre et al. 2012). An excellent method to study changes in brain tissue macrostructure is in vivo magnetic resonance imaging (MRI). MRI allows non-invasive assessment of the shape and size (morphology) of brain regions and to compare these measures between participants and across time. Morphological measures such as grey matter volume or cortical thickness are derived from the segmentation of individual brain scans into distinct tissue types (e.g. grey matter, white matter, and cerebrospinal fluid).

In a cross-sectional study, Schlaffke and colleagues directly compared grey matter volumes between three groups, consisting of 13 male subjects, respectively: long-distance endurance athletes, martial artists, and a non-sport control group not reporting participation in any regular physical activities (Schlaffke et al. 2014). The idea of comparing endurance vs. martial artists is based on their obviously differing metabolic profile (mainly aerobic vs. mainly anaerobic). In comparison to controls, VBM analysis of the whole brain showed higher grey matter volume in the supplementary motor area/dorsal premotor cortex (BA 6) in both athlete groups. Endurance athletes additionally revealed a significantly higher medial temporal lobe volume. Based on prior knowledge, the authors conclude that the higher volumes in these regions in the athlete groups may be related to motor control and motor skill acquisition (c.f. Dayan & Cohen 2011).

However, differences in brain morphology between exercise groups may also be influenced by their genetic predisposition. In a twin study, Rottensteiner et al. assessed how physical activity levels are associated with body composition, glucose homeostasis, and brain morphology (Rottensteiner et al. 2015). Ten young adult male monozygotic twin pairs (age range 32–36) met the inclusion criteria of the study. Based on interviews and questionnaires, several scores for energy expenditure of the participants in commuting and leisure time activities were calculated. For example, the twins had to exhibit a pair-wise difference of 1.5 $MET*h*d^{-1}$ in the last year and of 1 $MET*h*d^{-1}$ in the last 3 years to be included. VBM analysis revealed a higher grey matter volume of the striatum and the prefrontal cortex in the non-dominant hemisphere respectively. While this study provides evidence that long-term physical activity affects brain morphology independent of genetic predisposition, it is not entirely clear whether differences in physical activity or alternative factors across the lifespan of the twins are responsible for brain structure alterations. In this respect, longitudinal studies can provide further insight.

Two such studies with rodents recently demonstrated that endurance exercise affects brain regions well known to be involved in motor function and learning (Dayan & Cohen 2011; Seidler 2010). In the first one, Cahill et al. exposed male mice to four weeks of voluntary exercise and compared alterations in brain structure between this group and inactive controls using high resolution MRI (Cahill et al. 2015). The authors registered exercise-induced grey matter changes in several brain structures, amongst them hippocampus, dentate gyrus, stratum granulosum of the dentate gyrus, cingulate cortex, olivary complex, inferior cerebellar peduncle, and regions of the cerebellum. In a further longitudinal MRI study, Sumiyoshi et al. examined structural grey matter changes in response to a period as short as one week of voluntary wheel-running (Sumiyoshi, Taki, Nonaka, Takeuchi, & Kawashima 2014). Analyses revealed grey matter volume changes in widely distributed regions of the cerebral cortex, including motor, somatosensory, association, and visual areas. Remarkably, these structural changes were kept up for a period of at least one week, as measured by a follow-up scan. Also, a positive correlation between grey matter changes in the motor cortex and the total running distance covered was observed.

Below the cortical sheath, white matter tracts interconnect distant cortical regions to allow information processing in large-scale networks required for complex information processing (Fields 2008). Importantly, the precise timing and speed of neural information transmission, regulated by the structure of white matter tracts (e.g. myelination, c.f. McKenzie et al. 2014), is critical for the induction of LTP in the cortex. Motor learning-related white and grey matter plasticity was observed in humans (Scholz, Klein, Behrens, & Johansen-Berg 2009; Taubert et al. 2010). Interestingly, aerobic fitness in cross-sectional studies as well as endurance exercise interventions have shown to affect white matter tracts (Herting, Colby, Sowell, & Nagel 2014; Chaddock-Heyman et al. 2014; Voss et al. 2013a). A longitudinal intervention study involving 33 patients with schizophrenia and 48 healthy controls (age 18–48 years, 60 males/21 females) randomly assigned the subjects to either a physical exercise or a life-as-usual condition (Svatkova et al. 2015). The intervention lasted 6 months and contained 1 h training sessions conducted twice weekly. The proportion of aerobic (for instance cycling, rowing, treadmill running) to anaerobic exercises (weight-based strengthening exercises) was 2:1. Using diffusion-tensor imaging (DTI), a method that assesses the diffusion properties of water molecules to infer microstructural white matter changes, Svatkova et al. found that 6 months of exercise training alters white matter microstructure specifically in fibre tracts implicated in motor functioning such as the corpus callosum, corticospinal tract and superior longitudinal fascicle. Remarkably, this benefit was comparable for patients and healthy subjects.

Taken together, these studies demonstrate that endurance exercise may lead to structural adaptations in motor function-related brain regions and associated fibre connections. But do these structural alterations translate into an improved functional communication in sensorimotor-related brain networks, too (Will, Dalrymple-Alford, Wolff, & Cassel 2008)? To the best of our knowledge, this question has not been assessed yet with long-term exercise interventions. However, the acute response of the brain's functional connectivity to a single bout of cycling exercise has been examined recently (Rajab et al. 2014). In this study, 15 young healthy adults exercised for 20 min with moderate intensity (70% of age-predicted HR_{max}), while the control group (n=15) rested. Increased connectivity was found within sensorimotor and thalamic-caudate networks immediately after exercise.

While this provides further support for exercise-induced structural and functional brain changes, their relevance for motor learning remains largely speculative to date. This is surprising since the prediction of learning success based on a priori conducted brain scans is a hot topic in contemporary neuroscience (Gabrieli, Ghosh, & Whitfield-Gabrieli 2015) and was successfully applied in the fields of speech and music (Zatorre 2013). For instance, it has not been examined yet whether neural network activity before training is predictive for subsequent motor skill acquisition and learning, as was demonstrated for example in perceptual tasks (Baldassarre et al. 2012). However, an increased functional connectivity of motor circuits may be beneficial for motor control and learning (Dayan & Cohen 2011), especially under conditions of motor diseases like Parkinson's (Wang et al. 2015b). Furthermore, Sampaio-Baptista et al. showed that baseline grey matter volume in medial occipito-parietal areas is associated with the rate of subsequent skill acquisition performance in juggling (Sampaio-Baptista et al. 2014) and manual motor performance in elderly subjects correlated with grey and white matter volume of the cerebellum (Koppelmans, Hirsiger, Mérillat, Jäncke, & Seidler 2015). Nonetheless, longitudinal studies examining exercise-induced brain changes and their relation with subsequent motor learning-induced structural and functional changes were not conducted yet. Beyond that, conclusions about practical significance are hindered, since the MRI-observable changes could be driven by very different cellular changes (Zatorre et al. 2012). To gain more insight about that, the next section will focus on neurobiological adaptations to exercise on a more fine-grained level of observation.

Cellular Level

As already mentioned, changes in synaptic efficacy (LTP/LTD) are considered to reflect the neural basis of motor learning in the CNS (Sanes & Donoghue

2000). LTP is directly linked to morphological alterations on the cellular level (Toni, Buchs, Nikonenko, Bron, & Muller 1999; Harms, Rioult-Pedotti, Carter, & Dunaevsky 2008). These cellular changes rely on *de novo* protein synthesis (Lu, Christian, & Lu 2008) and injecting protein synthesis inhibitors in M1 results in a loss of previously learned skills as well as an impairment in new motor skill acquisition (Kleim et al. 2003). Worsened motor skill learning was correlated with microstructural adaptations like reduced synapse number and size (Kleim et al. 2003). The prevailing and generally accepted view is that motor learning reorganizes neuronal and synaptic connections, whereas endurance exercise mainly influences the supportive vascular components (Churchill et al. 2002). Coordinative-demanding motor activities have shown to elicit structural alterations on dendrites and synapses or neuroglia (extensively reviewed by Markham & Greenough 2004; Thomas et al. 2012; Voelcker-Rehage & Niemann 2013). For example, Xu and co-workers demonstrated a rapid formation of new dendritic spines within only one hour after skilled motor practice in M1. While the overall spine density returns to initial values soon, the newly formed spines are preferentially stabilized through subsequent practice and outlast the end of the training period (Xu et al. 2009). Further studies reported synaptogenesis after a few weeks of motor learning (Black et al. 1990; Kleim et al. 2002) that was specific to the cortical representation of the trained limb and accompanied by an increase of motor map size (Kleim et al. 2002; Kleim et al. 2004). Such changes were not observable in the untrained limb representation and occur as a consequence of skilled motor activity instead of mere repetitive limb use (Kleim, Lussnig, Schwarz, Comery, & Greenough 1996; Plautz, Milliken, & Nudo 2000; Tyč, Boyadjian, & Devanne 2005) or even strength training (Remple, Bruneau, VandenBerg, Goertzen, & Kleim 2001; Jensen, Marstrand, & Nielsen 2005). Given this, novelty of the task to be learnt seems to be a crucial precondition for such changes to take place (Adkins et al. 2006; Nudo 2008).

In contrast to motor learning, Kleim and co-workers showed that 30 days of endurance exercise (wheel running) did not alter the movement representations of the forelimb in comparison to an inactive control group (Kleim, Cooper, & VandenBerg 2002). This finding is in line with earlier findings demonstrating that synaptogenesis does not occur in response to endurance exercise (Black et al. 1990) but results in a greater density of blood vessels in layer V of the forelimb motor cortex (Kleim et al. 2002). In line with this, an increased exercise-induced blood vessel density was reported in other rodent studies using histological methods (Black et al. 1990; Isaacs, Anderson, Alcantara, Black, & Greenough 1992) as well as MRI (Swain et al. 2003). Thus, endurance exercise does likely not lead to adaptations of the cortical circuitry, but the vascular adaptations might contribute to subsequent learning-induced neuroplasticity (Adkins et al. 2006). Since memory formation and consolidation are energy-demanding processes (Tononi & Cirelli 2014), an improved supply of oxygen and other fuels to motor regions might be of relevance.

Molecular Level

A concerted action of key neurochemicals is required for the occurrence of motor learning-related neurophysiological and anatomical adaptations at the micro- and macrostructural levels (Hillman et al. 2008; He, Zhang, Yung, Zhu, & Wang 2013). Biochemical research provides evidence that the levels of many memory-related trophic factors like BDNF, VEGF, and IGF or neuromodulatory transmitters like dopamine, epinephrine or norepinephrine in peripheral blood vessels are typically increased in response to acute endurance exercise (for overview see Rojas Vega, Hollmann, & Strüder 2012; Phillips, Baktir, Srivatsan, & Salehi 2014).

Among the abovementioned neuromodulators and neurotrophic factors, BDNF is likely the best investigated and maybe the most important one. Using BDNF-mutant mice, Korte and colleagues first demonstrated that BDNF has a functional role for LTP expression (Korte et al. 1995). In the same year, it was reported that rats exposed to 7 days of voluntary wheel running exercise showed increased BDNF gene expression in the hippocampus and certain layers of the caudal neocortex (Neeper, Gómez-Pinilla, Choi, & Cotman 1995), providing first evidence that growth factors may be responsible for beneficial effects of exercise on the brain. These observations led to a series of studies examining the effects of exercise on growth factor signalling, brain structure, and cognitive function (see Chapter 2 and Chapter 7, this volume).

Ever since, several studies have demonstrated that BDNF is involved in all steps of memory formation from neuronal excitation to the induction and maintenance of early and late forms of LTP (Gómez-Pinilla & Feng 2012; Bekinschtein, Cammarota, & Medina 2014). Importantly, this also applies for the motor system (reviewed by He et al. 2013). BDNF and its receptor TrKB are important molecular intersections of exercise and motor learning (Klintsova, Dickson, Yoshida, & Greenough 2004). Previous studies indicate a causal link between BDNF function and synaptic plasticity. Subjects carrying the Val66Met polymorphism of the BDNF gene, which is known to affect activity-dependent BDNF secretion, show reduced MEP amplitudes and reduced motor map reorganization in response to motor skill learning (Kleim et al. 2006). Furthermore, the Val66Met polymorphism has shown to negatively affect short-term learning and long-term retention of a motor skill (McHughen et al. 2010). In line with this, it was demonstrated that mice with mutations of BDNF and its receptor TrkB show diminished responses to brain stimulation (Fritsch et al. 2010). Not last, the loss or critically low levels of BDNF are associated with motor system dysfunction, for example with severe neurodegenerative diseases (Teixeira, Barbosa, Diniz, & Kummer 2010; He et al. 2013).

Because the exogenous administration of BDNF is problematic in humans (for discussion see Fumagalli, Racagni, & Riva 2006), natural ways to elevate the levels of BDNF and other neurochemicals are a promising way to counteract motor dysfunction and to enhance motor learning in healthy people. In this respect, intrahippocampal injection of BDNF enhances learning in mice (Alonso et al. 2002) and exercise-induced increases in circulating BDNF levels correlate with behavioral parameters of motor skill learning (Skriver et al. 2014). Knowing that the values of BDNF as well as other trophic factors and neuromodulatory transmitters typically increase through endurance exercise (Knaepen, Goekint, Heyman, & Meeusen 2010; Skriver et al. 2014), exercise represents a promising and natural enhancement strategy for motor learning. In general, enhanced peripheral levels of BDNF obtained immediately after the cessation of exercise return to baseline levels within several minutes (Rojas Vega et al. 2006). However, animal research provides evidence for elevated cortical BDNF levels 5 hours after completion of exercise, with the 5 h values exceeding those obtained immediately after exercise (Takimoto & Hamada 2014). In contrast to this, many human studies examining BDNF levels in the resting state after a long-term exercise intervention report just small increases of circulating BDNF levels (Szuhany, Bugatti, & Otto 2015; Rojas Vega, Hollmann, & Strüder 2012), whereas higher exercise intensities might elicit a more pronounced BDNF increase (Baker et al. 2010). Noticeably, a cross-sectional study examining the link between habitual physical activity and resting BDNF levels report even a negative correlation (Currie, Ramsbottom, Ludlow, Nevill, & Gilder 2009). This discrepancy might be explained by an increased BDNF clearance and uptake in other tissues like the brain (Knaepen et al. 2010; Rojas Vega et al. 2012). Regular exercise may also enhance the BDNF response to an acute bout of exercise (Szuhany et al. 2015). Logically, the question arises how exercise protocols should be designed to affect these biomarkers and in turn potentially evoke beneficial effects on motor learning. We will deal with this question later on in this chapter.

Besides the changes in neurochemicals, exercise influences the energy supply of the brain. For example, recent investigations highlighted that lactate, elevated in response to exercise-induced anaerobiosis in the muscle cells (Robergs, Ghiasvand, & Parker 2004), is increasingly used as energy source for the brain and becomes the preferred fuel if arterial lactate values exceed the lactate values in the brain (Dalsgaard et al. 2004; Kemppainen et al. 2005; Boumezbeur et al. 2010; Dennis et al. 2015). This fact is of particular importance since high lactate levels increased motor cortex excitability (Coco et al. 2010). Moreover, the availability of lactate plays a crucial role in long-term memory formation because the blockade of the expression of monocarbocylate-transporters (MCT), which catalyze the diffusion of lactate, reduces the transfer of lactate to astrocytes and neurons in the brain and in addition impairs long-term memory in rats (Suzuki et al. 2011). Given this, the finding that an acute bout of exercise near or above the lactate threshold results in an elevated expression of MCTs is potentially relevant (Takimoto & Hamada 2014). However, it remains to be determined how regular exercise affects brain energetics and whether this might relate to motor function and memory. Maybe most important, lactate is directly involved in growth-factor signalling in response to exercise.

Behavioral Evidence

As we have already mentioned, the vast majority of studies examined the influence of exercise on cognition while its effects on motor skill learning remain poorly understood. This section reviews studies involving chronic or long-term endurance exercise and studies involving acute exercise to enhance motor learning. Acute protocols comprise endurance exercise activities on a single day that are intense enough to evoke a systemic physiological response. Typically, acute exercise takes place immediately before (think of classical warm-up) or immediately after a skill practice session. The section on long-term exercise includes studies examining the effects of endurance exercise conducted over longer time periods (days, weeks, months) before motor skill learning. While both types of interventions have certain neurobiological mechanisms in common, they represent disparate strategies to affect memory. In general, long-term exercise aspires to enhance the responsiveness of the brain to new environmental stimuli through enhancement of learning-induced neuroplasticity. While this is also true for acute exercise prior to motor skill practice, this type of exercise additionally targets an optimal preparation for high performance in the upcoming training session, for example by increasing arousal. On the contrary, exercising after a practice session selectively impacts motor memory consolidation without affecting other factors like arousal (Snigdha, de Rivera, Milgram, & Cotman 2014). This is especially relevant from a research-methodological perspective, since the effects of acute exercise likely outlast the practice session, thus not just affecting acquisition, but also consolidation (Roig et al. 2013). Therefore, the effects of acute exercise depend on its temporal positioning in relation to motor skill practice (Roig, Skriver, Lundbye-Jensen, Kiens, & Nielsen 2012). Note that it is not possible in all cases to draw conclusions on motor learning defined as relatively permanent changes in motor behavior, since many studies lack delayed retention tests (Schmidt & Lee 2011; Kantak & Winstein 2012).

Acute Exercise

Acute Exercise Before Learning

Generally, warm-up aims to prepare the central nervous, neuromuscular, cardiovascular, and respiratory systems for the upcoming training session and therefore ensures high performance and reduction of injury risk (Shellock & Prentice 1985). If training sessions target motor learning, the conditions for memory encoding should be optimized as well. From a psychological perspective, this may be induced by an optimal level of arousal that is in turn dependent on the nature of the task to be practiced (Schmidt & Lee 2014). Likely, increased arousal is enabled by an exercise-induced elevation of catecholamines like dopamine, epinephrine, or norepinephrine (Winter et al. 2007; Skriver et al. 2014). Additionally, as stated in the previous section, an upregulation of neurotrophic factors like BDNF may benefit subsequent learning-induced synaptic plasticity (Winter et al. 2007; Mang et al. 2014; Skriver et al. 2014). Using magnetic resonance spectroscopy, Maddock, Casazza, Fernandez, and Maddock (2016) showed that the cortical content of neurochemicals like glutamate and GABA increased following strenuous physical activity. Furthermore, endurance exercise has shown to alter cerebral blood flow (Ogoh & Ainslie 2009; Dietrich & Audiffren 2011), enhance cortical excitability (Coco et al. 2010), reduce intracortical inhibition in exercised (Yamaguchi et al. 2012) as well as non-exercised limbs (Singh et al. 2014; Smith et al. 2014) and to improve the conditions for the induction of synaptic plasticity (McDonnell et al. 2013; Singh et al. 2014; Mang et al. 2014).What is the behavioral evidence with reference to the effects of acute endurance exercise prior to motor skill performance and learning?

One study specifically examined the role of acute exercise on motor skill acquisition and long-term motor memory (Roig et al. 2012). In an experimental design with 48 healthy young male subjects split into three groups, interval cycling was conducted either before (PRE) or after learning (POST) a visuomotor tracking task, whereas controls (CON) rested. The dependent variable was the absolute retention performance of a visuomotor tracking task (RMSE) after 1 h, 24 h, and 7 d. While no between-group differences regarding the rate of motor skill acquisition were registered, it was found that both exercise groups showed better retention compared with controls 24 hours and 7 days after practice. In succession, the same working group published an association study correlating the peripheral blood plasma levels of several biomarkers (normalized to baseline levels) with skill acquisition and retention of the tracking task (Skriver et al. 2014). Blood samples were drawn immediately after exercise (PRE condition as introduced above) or rest (CON). Interestingly, in the exercise group the stress-related biomarkers lactate ($r = 0.877$) and norepinephrine ($r = 0.636$) were associated with an improved rate of skill acquisition during practice. Moreover, as could be expected from earlier work, norepinephrine levels (Segal, Cotman, & Cahill 2012) in PRE significantly correlated with skill retention at 7 d after practice ($r = -0.584$), with noticeable trends towards significance for the other measurement points (1 h, 24 h). Likewise, plasma BDNF levels (Winter et al. 2007) in this group were associated with improved skill retention 1h ($r = -0.672$) and 7 d ($r = 0.608$) after practice. An intriguing finding of Skriver's study is the significant correlation of lactate with better skill retention at all measurement points (1 h: $r = -0.658$, 24 h: $r = -0.715$, 7 d: $r = -0.672$). We will discuss the potential role of lactate in (motor) learning in more detail in the next section. In controls, none of the examined biomarkers correlated with neither skill acquisition nor retention with the exception of norepinephrine, which showed, somewhat surprisingly, the opposite pattern as observed in PRE, since higher blood plasma values at each measurement point indicated higher error values at skill retention (1 h: $r = 0.530$, 24 h: $r = 0.535$, 7 d: $r = 0.529$). Significant associations with skill acquisition and retention were not found for dopamine, IGF-1, and VEGF in either group.

Inspired by Roig's study, Mang and co-workers examined the effects of an acute bout of high-intensity exercise on PAS-induced LTP-like plasticity (see previous section) and on learning of an implicit visuomotor tracking task (Mang et al. 2014). In a within-subjects design, 8 female and 8 male participants aged between 19 and 33 passed through several experimental conditions. A motor tracking task had to be acquired under different conditions and memorized approximately 24 h later. The order of exposition to the paired conditions "rest followed by skill practice and 24-h retention test" or "exercise followed by skill practice and 24-h retention test" was balanced across the sample. This balanced variation applied also for the direction of joystick control and the shape of the sequence to be learnt (original or reversed). Additionally, serum BDNF blood samples from the finger were collected immediately before and after exercise. While the spatial task component of the tracking task was not affected by either condition, the temporal elements improved from early to late practice and were preserved 24 h after practice in the exercise condition. Note that such a positive effect of an acute bout of exercise on skill acquisition was not observed by Roig (Roig et al. 2012). However, acquisition and retention of the temporal component revealed no significant change for the rest condition. Given the exercise-induced improvement, especially of the timing-related task component, the authors hypothesized that exercise specifically affected cerebellar function (Mang et al. 2014). Despite the marked 3.4-fold increase in serum BDNF following exercise, significant correlations between normalized BDNF change and any behavioral data were not found. Note that positive effects of acute exercise on skill acquisition were also observed by Statton, Encarnacion, Celnik, and Bastian (2015). Using 30 min of moderate-intensity exercise prior to motor practice of a sequential motor task, Statton et al. (2015) observed improvements in skill acquisition but not skill retention which is in contrast to the above mentioned results of Roig et al. (2012) and may be induced by the different exercise intensities (high vs. moderate intensity).

In another experiment, Snow et al. (2016) used a similar tracking task as applied in the study of Mang et al. (2014) but changed the deployed exercise regimen. Unlike the two studies reported above using high-intensity interval exercise (Roig et al. 2012; Mang et al. 2014), the authors scheduled a continuous exercise protocol with moderate intensity (30 min of continuous cycling at 60% VO_2max). Sixteen healthy adults passed through the experimental conditions of aerobic exercise or seated rest followed by motor task practice in a counter-balanced way. Both conditions were interspersed by a wash-out period of at least 2 weeks. Results revealed no group difference in the timing components of the task for neither acquisition nor consolidation/offline learning indices. However, consolidation and offline learning for the spatial components of the task were improved.

As an intermediate result, the reported studies conducted in laboratory settings observed beneficial effects of an acute bout of high-intensity exercise prior to skill acquisition on motor learning as objectified with delayed retention tests (Roig et al. 2012; Mang et al. 2014). Despite the similar structure and intensity of exercise, a significant association of the behavioral data with BDNF was not consistently reported (Skriver et al. 2014; Mang et al. 2014). Studies using moderate intensity exercise prior to learning suggest an acute positive online-learning effect with respect to immediate improvements in motor skill acquisition (Statton et al. 2015) or to an improved maintenance of performance (spatial accuracy) during practice (Snow et al. 2016). However, neither of these studies registered an exercise-driven effect on offline learning/skill retention (Statton et al. 2015; Snow et al. 2016). Despite an enhanced cortical excitability after exercise and subsequent motor skill training, Singh et al. (2016) registered no measurable effect on skill acquisition when motor practice succeeded an acute bout of moderate intensity exercise.

Given the facts that the mastering of comparably simple skills like tracking does not require high amounts of practice and that it is a part-body movement, questions the ecological validity of such findings, especially with respect to whole-body movements (Wulf & Shea 2002). To gain insight into more complex motor learning processes, a recent meta-analysis focused exclusively on the performance of whole-body, psychomotor tasks following any type of moderate and strenuous acute conditioning exercise (endurance, resistance, balance) (McMorris, Hale, Corbett, Robertson, & Hodgson 2015). The results obtained from 28 studies involving 570 participants revealed a slightly positive effect size for moderate (g=0.15), and a considerably negative effect size for high-intensity exercise (g=-0.86). These results are contrary to the view that moderate, and even more high-intensive, warm-up improves performance.

The reasons why especially resistance and high-intensity endurance exercise might have detrimental effects on performance are not examined systematically to date. Theoretically, this effect could be based on reduced cortical excitability (Takahashi et al. 2011), increased intracortical inhibition (Sidhu et al. 2013), or a diminished potential for neuroplastic change (McDonnell et al. 2013). Noticeably, studies registering a positive effect of high-intensity exercise on motor learning used lower limb exercise to promote skills performed with the upper extremity (Roig et al. 2012; Mang et al. 2014). This suggests that a local peripheral and/or central fatigue mechanism affecting exclusively the pre-strained muscle groups, but not the non-exercised limbs (note that this might just apply for endurance and not for resistance exercise, c.f. Takahashi et al. 2011). In line with this, increased PAS-induced synaptic plasticity after 20 min high-intensity interval cycling was observed in the non-exercised abductor pollicis brevis muscle (Mang et al. 2014). Also remarkably, studies using low to moderate intensity endurance exercise mainly showed facilitative effects on complex motor skill performance like shot putting (Anshel & Novak 1989) or soccer dribbling (McMorris, Gibbs, Palmer, Payne, & Torpey 1994). This suggests a potential facilitative effect of exercise prior to motor skill practice that is not limited to simple motor skills like tracking.

To sum up, evidence for acute exercise-induced improvements in motor skill learning is not uncontroversial and further research examining potential moderating factors like different exercise regimens or the nature of the motor task to be learnt (discrete vs. continuous, simple vs. complex, pre-strained effectors involved in task practice or not, involved brain regions in acquiring and consolidating the task) is required. Based on the existing evidence, a negative effect on motor skill performance and learning might be expected if warm-up exercise is potentially fatiguing and involves at the same time the main effectors that are important for the execution of the skill to be practiced in succession.

Acute Exercise After Learning

Immediately after practicing a motor skill, the motor memory trace is thought to be in a fragile state and, thus, susceptible to interference. The practice-induced skill improvements need to be transformed into a persistent and stable state during the post-practice period (McGaugh 2000; Robertson, Pascual-Leone, & Miall 2004). This applies for both declarative and procedural memories (Mayford, Siegelbaum, & Kandel 2012) and for the latter, incremental learning can be viewed as an ongoing cycle of consolidating fragile memory traces (Trempe & Proteau 2012). This is relevant for the entire motor learning period because already stabilized memories may become labile through reactivation in a subsequent practice session, and thus need to be re-stabilized again (Alberini 2005; Dudai 2012). The question how the consolidation of motor memory can be optimized received increased perception in the latest sport science literature (Yan, Abernethy, & Thomas 2008; Trempe & Proteau 2012).

While one promising possibility to facilitate consolidation is sleep, another lately discussed option might be a bout of endurance exercise immediately after practising a motor skill. The theoretical basis of this strategy is that the neurobiological machinery of memory formation remains active after the termination of motor practice. In the first hours after practicing a motor skill, molecular (Kleim et al. 2003) or electromagnetic (Muellbacher et al. 2002) blockades of movement representations in M1 or learning a motor interference task (Brashers-Krug, Shadmehr, & Bizzi 1996) can disrupt motor memory consolidation to a significant degree (reviewed in Robertson et al. 2004; Krakauer & Shadmehr 2006). With the passage of time after initial practice cessation, the susceptibility to interferences gradually descends (Krakauer & Shadmehr 2006).

From a neurobiological point of view, the persistence of LTP and its resistance against interfering stimuli could be the crucial mechanism allowing for proper skill consolidation (Cantarero, Tang, O'Malley, Salas, & Celnik 2013). Intact BDNF release and function of its receptor TrkB are important for the persistence of LTP (Bekinschtein et al. 2008). Therefore, the exercise-induced elevation of neurotrophins like BDNF and of catecholamines like norepinephrine (Skriver et al. 2014; Segal et al. 2012) might contribute to enhance between-session improvements and/or to minimize the effects of interfering stimuli in the consolidation time window.

Studies assessing motor memory consolidation typically comprise delayed retention tests that allow temporary performance effects to fade away (Schmidt & Lee 2011; Kantak & Winstein 2012). In succession, we will present one study using an off-line learning paradigm, thus allowing the learned skill to consolidate unhindered (Roig et al. 2012), and one study using a memory stabilization paradigm, including an interfering stimulus in the consolidation time window (Rhee et al. 2015).

The aforementioned study of Roig and co-workers showed that acute high-intensity exercise immediately after skill acquisition facilitates long-term motor memory (Roig et al. 2012). When directly comparing the two intervention groups (exercised before [PRE] or after skill acquisition [POST]) it was shown that the group that exercised after practice outperformed the group that exercised before practice in the retention test 7 d after skill acquisition. Hence, this study provided first evidence that a single bout of exercise after practicing a motor skill can enhance off-line learning.

But does post-learning exercise also protect against task interference within the consolidation window? To account for this question, Rhee et al. asked undergraduate subjects to learn a motor sequence task (Rhee et al. 2015). Three experimental groups practiced according to the classical memory stabilization paradigm: acquisition of the target sequence, followed by practicing an alternative (interfering) sequence 2 hours later and a retention test consisting of 3 trials of the target sequence 24 h after the first practice session. While one of these groups rested between acquisition of the target and alternative sequences (ALT), the experimental groups conducted an acute bout of exercise either immediately after the target sequence (IMM) or immediately before the alternative sequence (END). The authors found that exercise contributed to the emergence of an off-line performance gain in the retention test session despite of task interference. But this was only true for the first retention test trial in the END condition.

Long-Term Exercise

Regular exercise training conducted over months or even years leads to numerous epigenetic adaptations in different organ systems and tissues including skeletal and cardiac muscle cells or in the brain (Hawley, Hargreaves, Joyner, & Zierath 2014; Heinonen et al. 2014). Regarding the latter, Juvenal's (c. 60–c. 130) winged words *mens sana in corpore sano* have been scrutinized and experimentally confirmed in numerous studies (reviewed by Hillman et al. 2008; Prakash, Voss, Erickson, & Kramer 2015). Recently, the use of long-term endurance exercise to prime the molecular machinery for subsequent learning is increasingly recognized by scientists from basic research (Berchtold, Chinn, Chou, Kesslak, & Cotman 2005; Berchtold, Castello, & Cotman

2010; Korol, Gold, & Scavuzzo 2013). In line with this, priming exercise is assumed to be a promising intervention strategy especially for motor rehabilitation (Mang, Campbell, Ross, & Boyd 2013; Petzinger et al. 2013; Stoykov & Madhavan 2015) suggesting a general positive transfer effect of endurance exercise on motor skill learning (Kleim & Jones 2008) that has already been proved empirically (Quaney et al. 2009; Wang, Lin, & Lin 2015). However, there is a lack of studies examining the effects of long-term exercise on motor learning and function so that this area of research must be considered as largely underexplored to date (Stoykov & Madhavan 2015).

Nonetheless, PAS-induced synaptic plasticity is improved in subjects with high levels of aerobic physical activity compared with a sedentary group (Cirillo et al. 2009) providing indirect evidence for possible beneficial effects of regular exercise on motor learning. A first pilot study assessing the role of long-term physical activity on motor skill learning was conducted with 10 elderly subjects (age range 72–91 years) divided into two groups using a within-subjects design (Bakken et al. 2001). The exercise group passed through a physical activity program including calisthenics, stationary cycling, and walking over 8 weeks (3 training sessions/week), whereas controls rested. A finger-movement tracking task was tested before and after the 8 weeks. The exercise group showed a significant positive development in the accuracy index of a finger-movement tracking task from pre- to post-intervention compared with controls, whose performance worsened over time. However, the small sample size and the between-group differences especially regarding resting heart rate and blood pressure makes a generalization of the results difficult even for this age group.

In a more recent animal study, Buitrago et al. introduced the rotarod motor learning paradigm (balancing on an accelerated stick) and provided 5 rats daily access to a closed running wheel for a period of 7 days (Buitrago, Schulz, Dichgans, & Luft 2004). The rats were kept in the wheel until they ran a predetermined distance of 100 m (except for day 1). Wheel-running was followed by 8 days of rotarod training. In the control condition, 5 rats exclusively practiced the rotarod task. Interestingly, the exercise group showed higher initial levels of rotarod performance and this advantage remained stable until the end of the rotarod training period. The authors interpreted this finding as a positive transfer effect of wheel-running movements to the rotarod task by means of an improved motor control through placement of steps to maintain balance and speed. However, one might counter the assumption that wheel running led to a specific transfer effect (for example, on balance ability) since running is considered to be a simple, well-practiced, automated and therefore hardly challenging movement skill for mice (Black et al. 1990). In line with this assumption, prior studies failed to observe synaptogenesis in response to wheel running (Black et al. 1990). The occurrence of a general positive transfer effect evoked by long-term exercise (Adkins et al. 2006; Kleim & Jones 2008) should at least be considered as an alternative hypothesis to the assumption of a specific transfer of wheel running on locomotion-related abilities like balance.

Notably, the duration of endurance training in the reported intervention studies (Bakken et al. 2001; Buitrago et al. 2004) make the occurrence of protracted exercise-related epigenetic adaptations of heart or skeletal muscle quite unlikely (Ling & Rönn 2014). Consequently, the question of whether and how endurance ability in terms of maximal oxygen consumption (VO_2max) or individual anaerobic thresholds (IAT) impacts motor function and learning remains to be determined. Anyway, a recent artificial selection study with rats suggests that the inborn aerobic capacity might have a positive influence on motor learning (Wikgren et al. 2012). Just like the Buitragos study, the authors observed a better initial performance and a general upward shift of the learning curve, but in the absence of differences in the learning rate (Wikgren et al. 2012). On the contrary, the (sparse) existing evidence suggests that even comparably short periods of exercise are sufficient to prime the underlying neurobiological substrates for motor learning. Whether regular exercise over several months or years reveals additional benefits for motor learning is purely speculative to date. While a minimum amount of exercise is required to prime the molecular machinery for learning (Berchtold et al. 2005), the sustainability of exercise-induced adaptations is likely higher in the case of long-term compared with short-term exercise periods (Hötting & Röder 2013).

Hypothetical Mechanisms for Exercise-Induced Improvement in Motor Learning

A mechanistic understanding of the link between exercise-induced changes in peripheral biomarkers and their influence on the brain is crucial to design exercise interventions that enhance motor skill learning. Our working hypothesis is that endurance exercise improves motor learning through facilitation of motor learning-related neuroplasticity (see Figure 8.1). Undoubtedly, peripheral tissues, especially the skeletal muscles, can act as endocrine organs capable of secreting molecules relevant for neuroplasticity (Phillips et al. 2014; Lucas, Cotter, Brassard, & Bailey 2015). With respect to this, solid correlations between brain tissue and peripheral BDNF levels were found (Karege, Schwald, & Cisse 2002; Klein et al. 2011).

A possible way by which exercise increases BDNF could be the transport of peripheral-derived BDNF to the brain via the blood–brain barrier (Pan, Banks, Fasold, Bluth, & Kastin 1998). However, Matthews et al. (2009) showed that BDNF mRNA and protein are increased in skeletal muscles after exercise, but the increased BDNF seems not to be released into circulation. Analyses of blood samples from the radial artery and the internal jugular vein under resting and exercise conditions indicate that the brain itself may account for 70–80% of the BDNF levels circulating in peripheral blood vessels (Rasmussen et al. 2009). Therefore, changes in peripheral BDNF levels seem to be mainly caused by alterations in brain BDNF release into circulation.

Another biomarker of potential interest is lactate, which is known to modulate several brain functions (for overview see Barros 2013) such as the survival

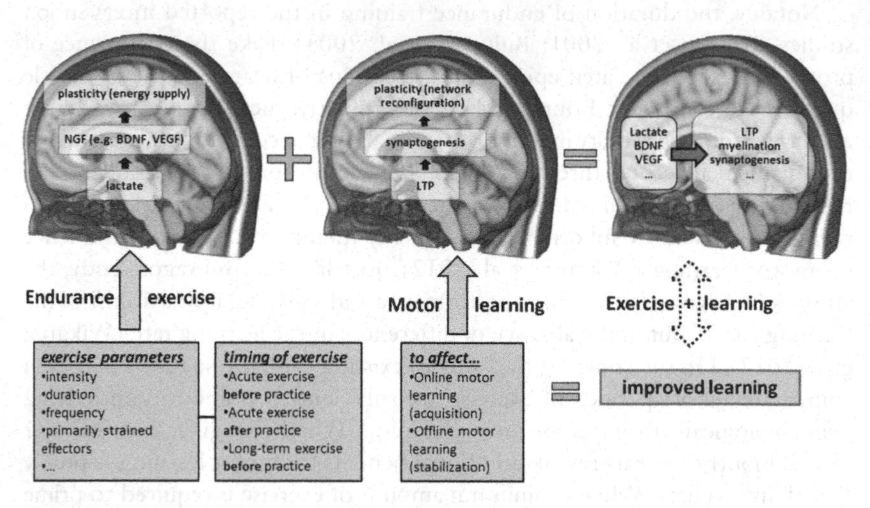

Figure 8.1 Schematic overview of candidate neurobiological correlates and contributing factors (exercise parameters and the timing of exercise sessions with respect to motor practice) of exercise-induced improvement in motor learning. NGF, nerve-growth factors; LTP, long-term potentiation; BDNF, brain-derived neurotrophic factor; VEGF, vascular-endothelial growth factor

of neurons (Fünfschilling et al. 2012; Lee et al. 2012) and axonal myelination (Rinholm et al. 2011). Furthermore, lactate is assumed to play a major role in the exercise-induced elevation of neural growth factors due to its well-known role as metabolite of endurance exercise. As we have outlined in the previous section, peripheral-derived lactate contributes significantly to brain metabolism under the conditions of physical exercise (Boumezbeur et al. 2010; van Hall et al. 2009). Current results from fundamental research, where endurance exercise is mimicked by sodium lactate treatments, generally underpin the notion of a connection between lactate and nerval growth factors. For example, Coco et al. (2013) treated cultures of astrocytes and SH-SY5Y (a cell line used as a model for neurons) in vitro for a period of 4 or 24 h with sodium lactate concentrations ranging from 5 to 25 mmol*l^{-1}. The results show that BDNF mRNA in the treated cultures is markedly increased in comparison to control cultures. When lactate was applied for 4 h, the BDNF mRNA increase was positively related with the concentration of sodium lactate in both cultures. This applied also for astrocytes after the 24 h treatment but not for the SH-SY5Y cells, where BDNF mRNA levels after 24 h returned to baseline. However, the exact mechanisms by which lactate increases BDNF mRNA remain to be clarified (Bergersen 2015). In a quite similar *in vivo* study, Lezi, Lu, Selfridge, Burns, and Swerdlow (2013) reproduced certain endurance exercise-related effects by infusing sodium lactate in resting mice. One of the main findings of this study is an elevation of VEGF levels, another neuroplasticity-related growth factor, in the brain. Importantly, Schiffer and colleagues (2011) recently showed that the peripheral infusion of sodium lactate leads to enhanced levels of serum BDNF in resting humans. Since sodium lactate has a basic pH-value, it is likely that increasing lactate concentrations instead of acidosis are causally linked with the observed changes in BDNF. In line with this, pH buffering via bicarbonate infusion during high-intensity cycling does not abolish the BDNF response, providing additional evidence that the exercise-induced elevation in BDNF levels is indeed due to increased lactatemia (Rojas Vega et al. 2012). Furthermore, it was found that lactate stimulates the expression of genes required for long-term memory in vitro and in vivo (Yang et al. 2014). To sum up, especially the existing studies conducted in laboratory settings indicate a positive relationship between lactate levels and the concentration of neurotrophic factors (especially BDNF), with strong evidence that this relationship may be causal in nature (Coco et al. 2013; Lezi et al. 2013; Schiffer et al. 2011). However, the results in applied settings are not unequivocal in this respect. Ferris, Williams, and Shen (2007) report that exercise-induced changes in serum BDNF correlate with changes in lactate ($r = 0.57$) as measured immediately after the cessation of exercise. On the other hand, Saucedo Marquez, Vanaudenaerde, Troosters, and Wenderoth (2015) recently did not observe any significant correlation between either cortisol or lactate and serum BDNF levels after the cessation of 20 min intense continuous or interval endurance exercise. In the latter study, only the subjective fatigue evaluation of the participants (Borg's CR-10 scale) was able to explain a significant proportion of the variance in the serum BDNF levels ($R^2 = 0.199$). Despite the absence of a correlation between exercise-induced elevations of lactate and BDNF levels after cessation of exercise, Skriver et al. (2014) showed that both biomarkers per se were highly associated with successful motor skill learning.

The reasons for these apparently contradictory findings are elusive to date and factors moderating the supposed relationship between lactate and BDNF must be identified. One possibility that might affect correlations between lactate and serum BDNF is the use of absolute or baseline-normalized lactate values. Based on the study of Kemppainen et al. (2005) who observed a negative correlation of serum lactate concentrations and brain glucose uptake, Singh and Staines (2015) suggest that cerebral metabolism is altered when exercise intensity is above the lactate threshold. If this holds, it could be useful for correlation analyses to express lactate values in relation to the individual anaerobic threshold of the participants. This is important because even individuals with similar maximum oxygen uptake may differ remarkably regarding their individual lactate thresholds (Coyle, Coggan, Hopper, & Walters 1988). Therefore, the individual strain of the anaerobic-lactic system might vary markedly even when subjects of similar endurance ability (as expressed by VO_2max) exercise at a given intensity in relation to VO_2max or maximum work rate. This in turn might affect the lactate uptake of the brain and therefore potentially the release of neurotrophic factors at the same place. Regarding BDNF, the respective gene polymorphism might play a role in mediating the effects of exercise on BDNF release. These and other factors (muscle fibre composition, activity of glycolytic or aerobic enzymes, methodological considerations with respect to the measurement of BDNF) could have a potential influence on the lactate–BDNF relationship in exercising humans (Karlsson, Sjödin, Jacobs, & Kaiser 1981; Klapcińska, Iskra, Poprzecki, & Grzesiok 2001; Pareja-Galeano et al. 2015).

How can exercise regimens be improved to optimize neuroplasticity? The aforementioned studies indicate the importance of high exercise intensities for a high BDNF response (Knaepen et al. 2010; Huang, Larsen, Ried-Larsen, Møller, & Andersen 2014), which may be mediated by an exercise-induced increase of lactate levels. Beyond that, high exercise intensities are proposed to increase cardiovascular health (Lucas et al. 2015) and showed beneficial effects on various cognitive functions (Angevaren et al. 2007; Ferris et al. 2007; Winter et al. 2007) and motor learning (Roig et al. 2012; Mang et al. 2014).

Exercise interventions that elevate peripheral BDNF levels include ramp or graded exercise tests to exhaustion (Rojas Vega, Hollmann, Vera Wahrmann, & Strüder 2012; Rojas Vega et al. 2006), continuous exercise of moderate to high intensities (Gold et al. 2003; Ferris et al. 2007; Schmidt-Kassow et al. 2012; Schmolesky, Webb, & Hansen 2013) and high-intensity interval (HIIT) as well as sprint interval training (Skriver et al. 2014; Mang et al. 2014; Winter et al. 2007). In contrast to ramp exercise and continuous exercise, interval training consists of repeated bouts of exercise interspersed with recovery periods that comprise light exercise or rest (Billat 2001) and is considered as an effective training method to improve endurance ability (Milanović, Sporiš, & Weston 2015). Moreover, as shown in animal research, 6 weeks of endurance training (6 times weekly) with either HIIT (95–100% VO_2max) or continuous exercise (80% VO_2max) elevated BDNF and GDNF (glial cell line-derived neurotrophic factor) in rat brain tissue significantly in comparison to a resting control group (Afzalpour et al. 2015). Moreover, the HIIT condition led to significantly higher BDNF and GDNF levels compared with the continuous condition (Afzalpour, Chadorneshin, Foadoddini, & Eivari 2015). The same holds for a direct comparison of the acute effects of HIIT (cycling at 90% of maximum work rate for 1 min, interspersed by 1 min rest periods) and continuous intensive exercise (70% of maximal work rate) of the same total duration (20 min) in humans as recently conducted by Saucedo Marquez et al. (2015). In their study it was found that both protocols significantly increased BDNF levels compared with a rest condition, with HIIT reaching higher BDNF levels than continuous exercise. Importantly, 73% of the participants preferred HIIT over continuous exercise (Saucedo Marquez et al. 2015).

The reason for the superiority of HIIT might be based on the very fact that exercise training scheduled that way can be performed at velocities above the IAT (Billat 2001), therefore allowing to subsequently accumulate considerable levels of lactate (Buchheit & Laursen 2013). On the contrary, continuous endurance exercise over longer durations have to be performed at intensities low enough *not* to induce lactate accumulation above the IAT (Rojas Vega et al. 2012) to avoid fatigue.

However, an important and unresolved issue to date is whether an exercise intervention should affect either the peak BDNF level at a fixed time point, for example after the cessation of exercise, or the total volume of circulating BDNF over time (Schmolesky et al. 2013). To make matters worse, the kinetics of exercise-induced BDNF changes during exercise are largely unknown to date, but existing data suggest that BDNF values reach their maximum after approximately 10–20 min of moderate or intensive exercise and show a slight decrease thereafter (Schmidt-Kassow et al. 2012; Saucedo Marquez et al. 2015). Nonetheless, long-term exercise interventions aiming at priming the molecular machinery of motor skill acquisition and stabilization might be most effective when conducted with high intensities.

Notwithstanding, recommendations regarding optimal exercise regimens are even more difficult to provide if motor skill learning should be affected by an acute bout of exercise. Even though some studies present evidence for a beneficial effect of HIIT for motor skill learning (Roig et al. 2012; Mang et al. 2014), this benefit might not apply for complex motor skill learning (McMorris et al. 2015). In the case of acute exercise prior to motor skill practice, this might be due to temporary peripheral and/or central fatigue effects (Taylor 2012), especially relevant if the pre-strained effectors are at the same time critically involved in the performance of the motor skill. Besides increasing intracortical inhibition of pre-strained muscles (Sidhu et al. 2013), high exercise intensities are also known to enhance cortisol levels (Rojas Vega et al. 2006) which might in turn diminish the potential for subsequent neuroplastic changes (Sale, Ridding, & Nordstrom 2008; McDonnell et al. 2013). Since low-to-moderate exercise intensities mainly revealed facilitating effects on various neuroplasticity indices and behavior (see previous sections), these intensities can be recommended for applied settings at the moment. However, high-intensity exercise might be useful if part-body movements of the upper limb should be facilitated by lower limb exercise (Roig et al. 2012; Mang et al. 2014) and maybe vice versa.

On the contrary, temporary fatigue effects theoretically should not be of disadvantage if exercise is conducted after practicing motor skills (Roig et al. 2012). However, further research is needed because the neuronal mechanisms that mediate motor memory consolidation in the time window after practice consolidation are not yet well known (Berghuis et al. 2015), let alone their potential interaction with a post-practice bout of exercise.

In sum, considerable knowledge gaps remain regarding the optimal type, intensity, duration and, if applicable, frequency of exercise to promote motor learning related neuroplasticity (van Praag, Fleshner, Schwartz, & Mattson 2014). However, especially the results obtained from basic research lay the foundation for more applied studies to be conducted in the future. In our view, properly scheduled endurance exercise protocols potentially reflect a promising intervention strategy to affect motor learning.

Future Directions

With respect to the practical applicability of the existing evidence, several important issues are still unresolved. First, the mechanisms by which exercise impacts motor learning are not fully understood. In our view, the field will profit from longitudinal studies that combine exercise and motor learning with the assessment of exercise-induced and learning-related biomarkers (Voss et al. 2013b) as was done by Skriver et al. (2014) or Mang et al. (2014). Besides humoral biomarkers, especially MRI or TMS parameters should be promising in this respect.

Strikingly, most of the existing studies examining exercise and motor learning took place in laboratory settings which manifests in the use of comparably simple motor skills practiced in blocked schedules (Roig et al. 2012; Mang et al. 2014; Rhee et al. 2015; Bakken et al. 2001). In order to prove the generalizability of the existent findings for applied settings, future studies should use more complex skills with higher ecological validity (Wulf & Shea 2002). In addition, most of the studies examining exercise and motor learning in humans used continuous skills like tracking (Schmidt & Lee 2011). Importantly, continuous skills are assumed to be generally better memorized than discrete skills (Schmidt & Lee 2011). Therefore, the facilitatory effect of exercise on discrete skill learning still needs to be demonstrated in future studies. In addition, most intervention studies to date examined motor learning for a rather short period of time and, therefore, primarily focus on the initial phases of motor skill learning (fast learning stage). However, motor skill practice at advanced levels of proficiency likely involves disparate neuronal circuits than in the initial phase of practicing (Lohse, Wadden, Boyd, & Hodges 2014). In other words, novice musicians or novice athletes likely recruit different brain networks and internal skill representations than the respective experts.

Another promising intervention strategy could be the combination of endurance exercise and coordinative-demanding training. For example, rats allowed to train very different motor tasks exhibit considerable synaptogenesis (Black et al. 1990; Kleim et al. 1996) and athletes performing coordinative demanding sports exhibit improved neuroplasticity (Kumpulainen et al. 2015). In line with this, it was shown that gymnasts outperform controls in acquiring new motor tasks (sequence learning, reaction time task, balance task, visual-manual tasks), an effect that might be based on positive transfer from the coordinative-demanding gymnastics domain (Pereira, Abreu, & Castro-Caldas 2013). Furthermore, non-specific coordination training (Hirtz & Wellnitz 1985; Nicklisch & Zimmermann 1981) or the learning of several unrelated motor tasks (Seidler 2004) prior to the acquisition of specific criterion tasks revealed faster learning of the latter, thus indicating higher motor adaptability (cf. the learning-to-learn-hypothesis, Seidler 2010). Given the partially distinct adaptations of the CNS in response to exercise and motor learning, a purposeful combination of varied motor experiences as previously outlined with endurance exercise potentially contributes to optimized neuroplasticity in the brain (Kempermann et al. 2010; Thomas et al. 2012) ensuring high adaptability to new environmental (learning) stimuli and high sustainability of the neuroplastic adaptations. In line with this assumption, skilled aerobic exercise training in a running wheel with irregularly spaced rungs enhanced brain reorganization in comparison to the classical endurance exercise paradigm (normal running wheel) in Parkinsonian rats (Wang, Guo, Myers, Heintz, & Holschneider 2015a). This highlights the potential of including coordinative-demanding tasks (for example requiring balance, rhythmicity, reaction, or kinesthetic differentiation) in physical exercise programs.

References

Adkins, D. L., J. Boychuk, M.S. Remple, & J.A. Kleim. 2006. "Motor training induces experience-specific patterns of plasticity across motor cortex and spinal cord." *J Appl Physiol 101*: 1776–82.

Afzalpour, M. E., H.T. Chadorneshin, M. Foadoddini, & H.A. Eivari. 2015. "Comparing interval and continuous exercise training regimens on neurotrophic factors in rat brain." *Physiol Behav 147*: 78–83.

Alberini, C. M. 2005. "Mechanisms of memory stabilization: Are consolidation and reconsolidation similar or distinct processes?" *Trends Neurosci 28*: 51–6.

Alonso, M., M. R. M. Vianna, A.M. Depino, T. M. e Souza, G. Pereira, H. Szapiro,... J.H. Medina. 2002. "BDNF-triggered events in the rat hippocampus are required for both short- and long-term memory formation." *Hippocampus 12*: 551–60.

Anderson, B. J., P. B. Eckburg, & K. I. Relucio. 2002. "Alterations in the thickness of motor cortical subregions after motor-skill learning and exercise." *Learn Mem 9*: 1–9.

Angevaren, M., L. Vanhees, W. Wendel-Vos, H. J. J. Verhaar, G. Aufdemkampe, A. Aleman, & W. M. M. Verschuren. 2007. "Intensity, but not duration, of physical activities is related to cognitive function." *Eur J Cardiovasc Prev Rehabil 14*: 825–30.

Anshel, M. H., & J. Novak. 1989. "Effects of different intensities of fatigue on performing a sport skill requiring explosive muscular effort: A test of the specificity of practice principle." *Percept Mot Skills 69*: 1379–89.

Antonov, I., I. Antonova, E. R. Kandel, & R. D. Hawkins. 2003. "Activity-dependent presynaptic facilitation and hebbian LTP are both required and interact during classical conditioning in Aplysia." *Neuron 37*: 135–47.

Baker, L. D., L. L. Frank, K. Foster-Schubert, P. S. Green, C. W. Wilkinson, A. McTiernan,...S.R. Plymate. 2010. "Effects of aerobic exercise on mild cognitive impairment: A controlled trial." *Arch Neurol 67*: 71–9.

Bakken, R. C., J. R. Carey, R. P. Di Fabio, T. J. Erlandson, J. L. Hake, & T. W. Intihar. 2001. "Effect of aerobic exercise on tracking performance in elderly people: a pilot study." *Phys Ther 81*: 1870–9.

Baldassarre, A., C. M. Lewis, G. Committeri, A. Z. Snyder, G. L. Romani, & M. Corbetta. 2012. "Individual variability in functional connectivity predicts performance of a perceptual task." *Proc Natl Acad Sci USA 109*: 3516–21.

Barros, L. F. 2013. "Metabolic signaling by lactate in the brain." *Trends Neurosci 36*: 396–404.

Bekinschtein, P., M. Cammarota, C. Katche, L. Slipczuk, J. I. Rossato, A. Goldin,... J.H. Medina. 2008. "BDNF is essential to promote persistence of long-term memory storage." *Proc Natl Acad Sci USA 105*: 2711–16.

Bekinschtein, P., M. Cammarota, & J. H. Medina. 2014. "BDNF and memory processing." *Neuropharmacology 76*: 677–83.

Bennett, E. L., M. C. Diamond, D. Krech, & M. R. Rosenzweig. 1964. "Chemical and anatomical plasticity of brain: Changes in brain through experience, demanded by learning theories, are found in experiments with rats." *Science 146*: 610–19.

Berchtold, N. C., N. Castello, & C. W. Cotman. 2010. "Exercise and time-dependent benefits to learning and memory." *Neuroscience 167*: 588–97.

Berchtold, N. C., G. Chinn, M. Chou, J. P. Kesslak, & C. W. Cotman. 2005. "Exercise primes a molecular memory for brain-derived neurotrophic factor protein induction in the rat hippocampus." *Neuroscience 133*: 853–61.

Bergersen, L. H. 2015. "Lactate transport and signaling in the brain: Potential therapeutic targets and roles in body-brain interaction." *J Cereb Blood Flow Metab 35*: 176–85.

Berghuis, K. M. M., M. P. Veldman, S. Solnik, G. Koch, I. Zijdewind, & T. Hortobágyi. 2015. "Neuronal mechanisms of motor learning and motor memory consolidation in healthy old adults." *Age (Dordr) 37*: 53.

Billat, L. V. 2001. "Interval training for performance: A scientific and empirical practice: Special recommendations for middle- and long distance running. Part I: Aerobic interval training." *Sports Med 31*: 13–31.

Black, J. E., K. R. Isaacs, B. J. Anderson, A. A. Alcantara, & W. T. Greenough. 1990. "Learning causes synaptogenesis, whereas motor activity causes angiogenesis, in cerebellar cortex of adult rats." *Proc Natl Acad Sci USA 87*: 5568–72.

Boumezbeur, F., K. F. Petersen, G. W. Cline, G. F. Mason, K. L. Behar, G. I. Shulman, & D. L. Rothman. 2010. "The contribution of blood lactate to brain energy metabolism in humans measured by dynamic 13C nuclear magnetic resonance spectroscopy." *J Neurosci 30*: 13983–91.

Brashers-Krug, T., R. Shadmehr, & E. Bizzi. 1996. "Consolidation in human motor memory." *Nature 382*: 252–5.

Buchheit, M., & P. B. Laursen. 2013. "High-intensity interval training, solutions to the programming puzzle: Part I: Cardiopulmonary emphasis." *Sports Med 43*: 313–38.

Buitrago, M. M., J. B. Schulz, J. Dichgans, & A. R. Luft. 2004. "Short and long-term motor skill learning in an accelerated rotarod training paradigm." *Neurobiol Learn Mem 81*: 211–16.

Butefisch, C. M., B. C. Davis, S. P. Wise, L. Sawaki, L. Kopylev, J. Classen, & L. G. Cohen. 2000. "Mechanisms of use-dependent plasticity in the human motor cortex." *Proc Natl Acad Sci USA 97*: 3661–5.

Cahill, L. S., P. E. Steadman, C. E. Jones, C. L. Laliberté, J. Dazai, J. P. Lerch,...J. G. Sled. 2015. "MRI-detectable changes in mouse brain structure induced by voluntary exercise." *NeuroImage 113*: 175–83.

Cantarero, G., B. Tang, R. O'Malley, R. Salas, & P. Celnik. 2013. "Motor learning interference is proportional to occlusion of LTP-like plasticity." *J Neurosci 33*: 4634–41.

Caspersen, C. J., K. E. Powell, & G. M. Christenson. 1985. "Physical activity, exercise, and physical fitness: definitions and distinctions for health-related research." *Public Health Rep 100*: 126–31.

Chaddock-Heyman, L., K. I. Erickson, J. L. Holtrop, M. W. Voss, M. B. Pontifex, L.B. Raine,...A.F. Kramer. 2014. "Aerobic fitness is associated with greater white matter integrity in children." *Front Hum Neurosci 8*: 584.

Chang, Y. K., J. D. Labban, J. I. Gapin, & J. L. Etnier. 2012. "The effects of acute exercise on cognitive performance: a meta-analysis." *Brain Res 1453*: 87–101.

Churchill, J. D., R. Galvez, S. Colcombe, R. A. Swain, A. F. Kramer, & W. T. Greenough. 2002. "Exercise, experience and the aging brain." *Neurobiol Aging 23*: 941–55.

Cirillo, J., A. P. Lavender, M. C. Ridding, & J. G. Semmler. 2009. "Motor cortex plasticity induced by paired associative stimulation is enhanced in physically active individuals." *J Physiol (Lond) 587*: 5831–42.

Coco, M., G. Alagona, G. Rapisarda, E. Costanzo, R.A. Calogero, V. Perciavalle, & V. Perciavalle. 2010. "Elevated blood lactate is associated with increased motor cortex excitability." *Somatosens Mot Res 27*: 1–8.

Coco, M., S. Caggia, G. Musumeci, V. Perciavalle, A.C. E. Graziano, G. Pannuzzo, & V. Cardile. 2013. "Sodium L-lactate differently affects brain-derived neurothrophic factor, inducible nitric oxide synthase, and heat shock protein 70 kDa production in human astrocytes and SH-SY5Y cultures." *J Neurosci Res 91*: 313–20.

Coyle, E. F., A. R. Coggan, M. K. Hopper, & T. J. Walters. 1988. "Determinants of endurance in well-trained cyclists." *J Appl Physiol 64*: 2622–30.

Currie, J., R. Ramsbottom, H. Ludlow, A. Nevill, & M. Gilder. 2009. "Cardio-respiratory fitness, habitual physical activity and serum brain-derived neurotrophic factor (BDNF) in men and women." *Neurosci Lett 451*: 152–5.

Dalsgaard, M. K., B. Quistorff, E. R. Danielsen, C. Selmer, T. Vogelsang, & N. H. Secher. 2004. "A reduced cerebral metabolic ratio in exercise reflects metabolism and not accumulation of lactate within the human brain." *J Physiol (Lond) 554*: 571–8.

Dayan, E., & L. G. Cohen. 2011. "Neuroplasticity subserving motor skill learning." *Neuron 72*: 443–54.

Dennis, A., A. G. Thomas, N. B. Rawlings, J. Near, T. E. Nichols, S. Clare,...C.J. Stagg. 2015. "An ultra-high field magnetic resonance spectroscopy study of post exercise lactate, glutamate and glutamine change in the human brain." *Front Physiol 6*: 351.

Dietrich, A., & M. Audiffren. 2011. "The reticular-activating hypofrontality (RAH) model of acute exercise." *Neurosci Biobehav Rev 35*: 1305–25.

Donchin, O., L. Sawaki, G. Madupu, L. G. Cohen, & R. Shadmehr. 2002. "Mechanisms influencing acquisition and recall of motor memories." *J Neurophysiol 88*: 2114–23.

Dudai, Y. 2012. "The restless engram: Consolidations never end." *Annu Rev Neurosci 35*: 227–47.

Ferris, L. T., J. S. Williams, & C.-L. Shen. 2007. "The effect of acute exercise on serum brain-derived neurotrophic factor levels and cognitive function." *Med Sci Sports Exerc 39*: 728–34.

Fields, R. D. 2008. "White matter in learning, cognition and psychiatric disorders." *Trends Neurosci 31*: 361–70.

Fritsch, B., J. Reis, K. Martinowich, H. M. Schambra, Y. Ji, L. G. Cohen, & B. Lu. 2010. "Direct current stimulation promotes BDNF-dependent synaptic plasticity: potential implications for motor learning." *Neuron 66*: 198–204.

Fumagalli, F., G. Racagni, & M. A. Riva. 2006. "The expanding role of BDNF: A therapeutic target for Alzheimer's disease?" *Pharmacogenomics J 6*: 8–15.

Fünfschilling, U., L. M. Supplie, D. Mahad, S. Boretius, A.S. Saab, J. Edgar,...B.G. Brinkmann. 2012. "Glycolytic oligodendrocytes maintain myelin and long-term axonal integrity." *Nature 485*: 517–21.

Gabrieli, J. D. E., S.S. Ghosh, & S. Whitfield-Gabrieli. 2015. "Prediction as a humanitarian and pragmatic contribution from human cognitive neuroscience." *Neuron 85*: 11–26.

Gold, S. M., K.-H. Schulz, S. Hartmann, M. Mladek, U. E. Lang, R. Hellweg,...C. Heesen. 2003. "Basal serum levels and reactivity of nerve growth factor and brain-derived neurotrophic factor to standardized acute exercise in multiple sclerosis and controls." *J Neuroimmunol 138*: 99–105.

Gómez-Pinilla, F., & C. Feng. 2012. "Molecular mechanisms for the ability of exercise supporting cognitive abilities and counteracting neurological disorders." In H. Boecker, C. H. Hillman, L. Scheef, & H. K. Strüder (Eds.), *Functional Neuroimaging in Exercise and Sport Sciences* (pp. 25–43). New York: Springer.

Harms, K. J., M. S. Rioult-Pedotti, D. R. Carter, & A. Dunaevsky. 2008. "Transient spine expansion and learning-induced plasticity in layer 1 primary motor cortex." *J Neurosci 28*: 5686–90.

Hawley, J. A., M. Hargreaves, M. J. Joyner, & J. R. Zierath. 2014. "Integrative biology of exercise." *Cell 159*: 738–49.

He, Y.-Y., X.-Y. Zhang, W.-H. Yung, J.-N. Zhu, & J.-J. Wang. 2013. "Role of BDNF in central motor structures and motor diseases." *Mol Neurobiol 48*: 783–93.

Hebb, D. O. 1949. *The Organization of Behavior: A Neuropsychological Theory.* New York: John Wiley & Sons.

Heinonen, I., K. K. Kalliokoski, J. C. Hannukainen, D. J. Duncker, P. Nuutila, & J. Knuuti. 2014. "Organ-specific physiological responses to acute physical exercise and long-term training in humans." *Physiology (Bethesda) 29*: 421–36.

Herting, M. M., J. B. Colby, E. R. Sowell, & B. J. Nagel. 2014. "White matter connectivity and aerobic fitness in male adolescents." *Dev Cogn Neurosci 7*: 65–75.

Hillman, C. H., K. I. Erickson, & A. F. Kramer. 2008. "Be smart, exercise your heart: Exercise effects on brain and cognition." *Nat Rev Neurosci 9*: 58–65.

Hirtz, P., & I. Wellnitz. 1985. "Hohes Niveau koordinativer Fähigkeiten führt zu besseren Ergebnissen im motorischen Lernen." *Körpererziehung 35*: 151–4.

Hötting, K., & B. Röder. 2013. "Beneficial effects of physical exercise on neuroplasticity and cognition." *Neurosci Biobehav Rev 37*: 2243–57.

Huang, T., K. T. Larsen, M. Ried-Larsen, N. C. Møller, & L. B. Andersen. 2014. "The effects of physical activity and exercise on brain-derived neurotrophic factor in healthy humans: A review." *Scand J Med Sci Sports 24*: 1–10.

Isaacs, K. R., B. J. Anderson, A. A. Alcantara, J. E. Black, & W. T. Greenough. 1992. "Exercise and the brain: Angiogenesis in the adult rat cerebellum after vigorous physical activity and motor skill learning." *J Cereb Blood Flow Metab 12*: 110–19.

Jensen, J. L., P. C. D. Marstrand, & J. B. Nielsen. 2005. "Motor skill training and strength training are associated with different plastic changes in the central nervous system." *J Appl Physiol 99*: 1558–68.

Jones, A. M., & D. C. Poole. 2009. "Physiological demands of endurance exercise." In R. J. Maughan (Ed.), *Olympic Textbook of Science in Sport* (pp. 43–55). Chichester, UK: Wiley-Blackwell.

Kantak, S. S., & C. J. Winstein. 2012. "Learning–performance distinction and memory processes for motor skills: A focused review and perspective." *Behav Brain Res 228*: 219–31.

Karege, F., M. Schwald, & M. Cisse. 2002. "Postnatal developmental profile of brain-derived neurotrophic factor in rat brain and platelets." *Neurosci Lett 328*: 261–4.

Karlsson, J., B. Sjödin, I. Jacobs, & P. Kaiser. 1981. "Relevance of muscle fibre type to fatigue in short intense and prolonged exercise in man." *Ciba Found Symp 82*: 59–74.

Kempermann, G., K. Fabel, D. Ehninger, H. Babu, P. Leal-Galicia, A. Garthe, & S.A. Wolf. 2010. "Why and how physical activity promotes experience-induced brain plasticity." *Front Neurosci 4*: 189.

Kemppainen, J., S. Aalto, T. Fujimoto, K. K. Kalliokoski, J. Långsjö, V. Oikonen,...J. Knuuti. 2005. "High intensity exercise decreases global brain glucose uptake in humans." *J Physiol (Lond) 568*: 323–32.

Kim, S., M. C. Stephenson, P. G. Morris, & S. R. Jackson. 2014. "tDCS-induced alterations in GABA concentration within primary motor cortex predict motor learning and motor memory: A 7 T magnetic resonance spectroscopy study." *NeuroImage 99*: 237–43.

Klapcińska, B., J. Iskra, S. Poprzecki, & K. Grzesiok. 2001. "The effects of sprint (300 m) running on plasma lactate, uric acid, creatine kinase and lactate dehydrogenase in competitive hurdlers and untrained men." *J Sports Med Phys Fitness 41*: 306–11.

Kleim, J. A., S. Barbay, N. R. Cooper, T. M. Hogg, C. N. Reidel, M. S. Remple, & R. J. Nudo. 2002. "Motor learning-dependent synaptogenesis is localized to functionally reorganized motor cortex." *Neurobiol Learn Mem 77*: 63–77.

Kleim, J. A., R. Bruneau, K. Calder, D. Pocock, P.M. VandenBerg, E. MacDonald,...K. Nader. 2003. "Functional organization of adult motor cortex is dependent upon continued protein synthesis." *Neuron 40*: 167–76.

Kleim, J. A., S. Chan, E. Pringle, K. Schallert, V. Procaccio, R. Jimenez, & S. C. Cramer. 2006. "BDNF val66met polymorphism is associated with modified experience-dependent plasticity in human motor cortex." *Nat Neurosci 9*: 735–7.

Kleim, J. A., N. R. Cooper, & P. M. VandenBerg. 2002. "Exercise induces angiogenesis but does not alter movement representations within rat motor cortex." *Brain Res 934*: 1–6.

Kleim, J. A., T. M. Hogg, P. M. VandenBerg, N. R. Cooper, R. Bruneau, & M. Remple. 2004. "Cortical synaptogenesis and motor map reorganization occur during late, but not early, phase of motor skill learning." *J Neurosci 24*: 628–33.

Kleim, J. A., & T. A. Jones. 2008. "Principles of experience-dependent neural plasticity: Implications for rehabilitation after brain damage." *J Speech Lang Hear Res 51*: S225–S239.

Kleim, J. A., E. Lussnig, E. R. Schwarz, T. A. Comery, & W. T. Greenough. 1996. "Synaptogenesis and Fos expression in the motor cortex of the adult rat after motor skill learning." *J Neurosci 16*: 4529–35.

Klein, A. B., R. Williamson, M.A. Santini, C. Clemmensen, A. Ettrup, M. Rios,...S. Aznar. 2011. "Blood BDNF concentrations reflect brain-tissue BDNF levels across species." *Int J Neuropsychopharmacol 14*: 347–53.

Klintsova, A. Y., E. Dickson, R. Yoshida, & W. T. Greenough. 2004. "Altered expression of BDNF and its high-affinity receptor TrkB in response to complex motor learning and moderate exercise." *Brain Res 1028*: 92–104.

Knaepen, K., M. Goekint, E. M. Heyman, & R. Meeusen. 2010. "Neuroplasticity – Exercise-induced response of peripheral Brain-Derived Neurotrophic Factor." *Sports Med 40*: 765–801.

Koppelmans, V., S. Hirsiger, S. Mérillat, L. Jäncke, & R. D. Seidler. 2015. "Cerebellar gray and white matter volume and their relation with age and manual motor performance in healthy older adults." *Hum Brain Mapp 36*: 2352–63.

Korol, D. L., P. E. Gold, & C. J. Scavuzzo. 2013. "Use it and boost it with physical and mental activity." *Hippocampus 23*: 1125–35.

Korte, M., P. Carroll, E. Wolf, G. Brem, H. Thoenen, & T. Bonhoeffer. 1995. "Hippocampal long-term potentiation is impaired in mice lacking brain-derived neurotrophic factor." *Proc Natl Acad Sci USA 92*: 8856–60.

Krakauer, J. W., & R. Shadmehr. 2006. "Consolidation of motor memory." *Trends Neurosci 29*: 58–64.

Kumpulainen, S., J. Avela, M. Gruber, J. Bergmann, M. Voigt, V. Linnamo, & N. Mrachacz-Kersting. 2015. "Differential modulation of motor cortex plasticity in skill- and endurance-trained athletes." *Eur J Appl Physiol 115*: 1107–15.

Lee, K. J., J. Im Rhyu, & D. T. S. Pak. 2014. "Synapses need coordination to learn motor skills." *Rev Neurosci 25*: 223–30.

Lee, Y., B. M. Morrison, Y. Li, S. Lengacher, M. H. Farah, P. N. Hoffman,...Y. Liu. 2012. "Oligodendroglia metabolically support axons and contribute to neurodegeneration." *Nature 487*: 443–8.

Lezi, E., J. Lu, J. E. Selfridge, J. M. Burns, & R. H. Swerdlow. 2013. "Lactate administration reproduces specific brain and liver exercise-related changes." *J Neurochem 127*: 91–100.

Ling, C., & T. Rönn. 2014. "Epigenetic adaptation to regular exercise in humans." *Drug Discov Today 19*: 1015–18.

Lohse, K. R., K. Wadden, L. A. Boyd, & N. J. Hodges. 2014. "Motor skill acquisition across short and long time scales: A meta-analysis of neuroimaging data." *Neuropsychologia 59*: 130–41.

Lu, Y., K. Christian, & B. Lu. 2008. "BDNF: A key regulator for protein synthesis-dependent LTP and long-term memory?" *Neurobiol Learn Mem 89*: 312–23.

Lucas, S. J. E., J. D. Cotter, P. Brassard, & D. M. Bailey. 2015. "High-intensity interval exercise and cerebrovascular health: Curiosity, cause, and consequence." *J Cereb Blood Flow Metab 35*: 902–11.

Maddock, R. J., G. A. Casazza, D. H. Fernandez, & M. I. Maddock. 2016. "Acute modulation of cortical glutamate and GABA content by physical activity." *J Neurosci 36*: 2449–57.

Magill, R. A. 2011. *Motor learning and control: Concepts and applications*. 9th ed. New York: McGraw-Hill.

Mang, C. S., K. L. Campbell, C. J. D. Ross, & L. A. Boyd. 2013. "Promoting neuroplasticity for motor rehabilitation after stroke: Considering the effects of aerobic exercise and genetic variation on brain-derived neurotrophic factor." *Phys Ther 93*: 1707–16.

Mang, C. S., N. J. Snow, K. L. Campbell, C. J. D. Ross, & L. A. Boyd. 2014. "A single bout of high-intensity aerobic exercise facilitates response to paired associative stimulation and promotes sequence-specific implicit motor learning." *J Appl Physiol 117*: 1325–36.

Markham, J. A., & W. T. Greenough. 2004. "Experience-driven brain plasticity: Beyond the synapse." *Neuron Glia Biol 1*: 351–63.

Matthews, V. B., M.-B. Aström, M. H. Chan, C. R. Bruce, K. S. Krabbe, O. Prelovsek,...T. Akerström. 2009. "Brain-derived neurotrophic factor is produced by skeletal muscle cells in response to contraction and enhances fat oxidation via activation of AMP-activated protein kinase." *Diabetologia 52*: 1409–18.

Mayford, M., S. A. Siegelbaum, & E. R. Kandel. 2012. "Synapses and memory storage." *Cold Spring Harb Perspect Biol 4*: a005751.

McDonnell, M. N., J. D. Buckley, G. M. Opie, M. C. Ridding, & J. G. Semmler. 2013. "A single bout of aerobic exercise promotes motor cortical neuroplasticity." *J Appl Physiol 114*: 1174–82.

McGaugh, J. L. 2000. "Memory – a Century of Consolidation." *Science 287*: 248–51.

McHughen, S. A., P. F. Rodriguez, J. A. Kleim, E. D. Kleim, L. M. Crespo, V. Procaccio, & S. C. Cramer. 2010. "BDNF Val66Met Polymorphism influences motor system function in the human brain." *Cereb Cortex 20*: 1254–62.

McKenzie, I. A., D. Ohayon, H. Li, J. P. de Faria, B. Emery, K. Tohyama, & W. D. Richardson. 2014. "Motor skill learning requires active central myelination." *Science 346*: 318–22.

McMorris, T., C. Gibbs, J. Palmer, A. Payne, & N. Torpey. 1994. "Exercise and performance of a motor skill." *Brit J Phys Educ Res Suppl 15*: 23–7.

McMorris, T., B. J. Hale, J. Corbett, K. Robertson, & C. I. Hodgson. 2015. "Does acute exercise affect the performance of whole-body, psychomotor skills in an inverted-U fashion? A meta-analytic investigation." *Physiol Behav 141*: 180–9.

Milanović, Z., G. Sporiš, & M. Weston. 2015. "Effectiveness of High-Intensity Interval Training (HIT) and continuous endurance training for VO2max improvements: A systematic review and meta-analysis of controlled trials." *Sports Med.* doi:10.1007/s40279-015-0365-0

Molteni, R., Z. Ying, & F. Gomez-Pinilla. 2002. "Differential effects of acute and chronic exercise on plasticity-related genes in the rat hippocampus revealed by microarray." *Eur J Neurosci 16*: 1107–16.

Muellbacher, W., U. Ziemann, J. Wissel, N. Dang, M. Kofler, S. Facchini,...M. Hallett. 2002. "Early consolidation in human primary motor cortex." *Nature 415*: 640–4.

Neeper, S. A., F. Gómez-Pinilla, J. Choi, & C. Cotman. 1995. "Exercise and brain neurotrophins." *Nature 373*: 109.

Nicklisch, R., & K. Zimmermann. 1981. "Die Ausbildung koordinativer Fähigkeiten und ihre Bedeutung für die technische beziehungsweise technisch-taktische Leistungsfähigkeit der Sportler." *Theorie und Praxis der Körperkultur 30*: 764–8.

Nudo, R. J. 2008. "Neurophysiology of motor skill learning." In H. Eichenbaum (Ed.), *Learning and Memory, Volume 3: Memory Systems* (pp. 403–21). Oxford: Elsevier.

Ogoh, S., & P. N. Ainslie. 2009. "Cerebral blood flow during exercise: Mechanisms of regulation." *J Appl Physiol 107*: 1370–80.

Pan, W., W. A. Banks, M. B. Fasold, J. Bluth, & A. J. Kastin. 1998. "Transport of brain-derived neurotrophic factor across the blood–brain barrier." *Neuropharmacology 37*: 1553–61.

Pareja-Galeano, H., R. Alis, F. Sanchis-Gomar, H. Cabo, J. Cortell-Ballester, M.C. Gomez-Cabrera,...J. Viña. 2015. "Methodological considerations to determine the effect of exercise on brain-derived neurotrophic factor levels." *Clin Biochem 48*: 162–6.

Pascual-Leone, A., A. Amedi, F. Fregni, & L. B. Merabet. 2005. "The plastic human brain cortex." *Annu Rev Neurosci 28*: 377–401.

Pereira, T., A. M. Abreu, & A. Castro-Caldas. 2013. "Understanding task- and expertise specific motor acquisition and motor memory formation and consolidation." *Percept Mot Skills 117*: 108–29.

Petzinger, G. M., B. E. Fisher, S. McEwen, J. A. Beeler, J. P. Walsh, & M. W. Jakowec. 2013. "Exercise-enhanced neuroplasticity targeting motor and cognitive circuitry in Parkinson's disease." *Lancet Neurol 12*: 716–26.

Phillips, C., M. A. Baktir, M. Srivatsan, & A. Salehi. 2014. "Neuroprotective effects of physical activity on the brain: a closer look at trophic factor signaling." *Front Cell Neurosci 8*: 170.

Plautz, E. J., G. W. Milliken, & R. J. Nudo. 2000. "Effects of repetitive motor training on movement representations in adult squirrel monkeys: Role of use versus learning." *Neurobiol Learn Mem 74*: 27–55.

Prakash, R. S., M. W. Voss, K. I. Erickson, & A. F. Kramer. 2015. "Physical activity and cognitive vitality." *Annu Rev Psychol 66*: 769–97.

Quaney, B. M., L. A. Boyd, J. M. McDowd, L. H. Zahner, J. He, M. S. Mayo, & R. F. Macko. 2009. "Aerobic exercise improves cognition and motor function poststroke." *Neurorehabil Neural Repair 23*: 879–85.

Rajab, A. S., D. E. Crane, L. E. Middleton, A. D. Robertson, M. Hampson, & B. J. MacIntosh. 2014. "A single session of exercise increases connectivity in sensorimotor-related brain networks: A resting-state fMRI study in young healthy adults." *Front Hum Neurosci 8*: 625.

Rasmussen, P., P. Brassard, H. Adser, M. V. Pedersen, L. Leick, E. Hart,...H. Pilegaard. 2009. "Evidence for a release of brain-derived neurotrophic factor from the brain during exercise." *Exp Physiol 94*: 1062–9.

Remple, M. S., R. M. Bruneau, P. M. VandenBerg, C. Goertzen, & J. A. Kleim. 2001. "Sensitivity of cortical movement representations to motor experience: Evidence that skill learning but not strength training induces cortical reorganization." *Behav Brain Res 123*: 133–41.

Rhee, J., J. Chen, S. M. Riechman, A. Handa, S. Bhatia, & D. L. Wright. 2015. "An acute bout of aerobic exercise can protect immediate offline motor sequence gains." *Psychol Res.* doi:10.1007/s00426-015-0682-9

Rinholm, J. E., N. B. Hamilton, N. Kessaris, W. D. Richardson, L. H. Bergersen, & D. Attwell. 2011. "Regulation of oligodendrocyte development and myelination by glucose and lactate." *J Neurosci 31*: 538–48.

Rioult-Pedotti, M. S., D. Friedman, & J. P. Donoghue. 2000. "Learning-induced LTP in Neocortex." *Science 290*: 533–6.

Robergs, R. A., F. Ghiasvand, & D. Parker. 2004. "Biochemistry of exercise-induced metabolic acidosis." *Am J Physiol Regul Integr Comp Physiol 287*: R502–R516.

Robertson, E. M., A. Pascual-Leone, & R. C. Miall. 2004. "Opinion: Current concepts in procedural consolidation." *Nat Rev Neurosci 5*: 576–82.

Roig, M., S. Nordbrandt, S. S. Geertsen, & J. B. Nielsen. 2013. "The effects of cardiovascular exercise on human memory: A review with meta-analysis." *Neurosci Biobehav Rev 37*: 1645–66.

Roig, M., K. Skriver, J. Lundbye-Jensen, B. Kiens, & J. B. Nielsen. 2012. "A single bout of exercise improves motor memory." *PLoS ONE 7*: e44594.

Rojas Vega, S., W. Hollmann, & H. K. Strüder. 2012. "Humoral factors in humans participating in different types of exercise and training." In H. Boecker, C. H. Hillman, L. Scheef, & H. K. Strüder (Eds.), *Functional Neuroimaging in Exercise and Sport Sciences* (pp. 169–96). New York: Springer.

Rojas Vega, S., W. Hollmann, B. Vera Wahrmann, & H. Strüder. 2012. "pH buffering does not influence BDNF responses to exercise." *Int J Sports Med 33*: 8–12.

Rojas Vega, S., H. K. Strüder, B. Vera Wahrmann, A. Schmidt, W. Bloch, & W. Hollmann. 2006. "Acute BDNF and cortisol response to low intensity exercise and following ramp incremental exercise to exhaustion in humans." *Brain Res 1121*: 59–65.

Rosenzweig, M. R., & E. L. Bennett. 1996. "Psychobiology of plasticity: Effects of training and experience on brain and behavior." *Behav Brain Res 78*: 57–65.

Rottensteiner, M., T. Leskinen, E. Niskanen, S. Aaltonen, S. Mutikainen, J. Wikgren,...K. Heikkilä. 2015. "Physical activity, fitness, glucose homeostasis, and brain morphology in twins." *Med Sci Sports Exerc 47*: 509–18.

Sale, M. V., M. C. Ridding, & M. A. Nordstrom. 2008. "Cortisol inhibits neuroplasticity induction in human motor cortex." *J Neurosci 28*: 8285–93.

Sampaio-Baptista, C., J. Scholz, M. Jenkinson, A. G. Thomas, N. Filippini, G. Smit,... H. Johansen-Berg. 2014. "Gray matter volume is associated with rate of subsequent skill learning after a long term training intervention." *NeuroImage 96*: 158–66.

Sanes, J. N., & J. P. Donoghue. 2000. "Plasticity and primary motor cortex." *Annu Rev Neurosci 23*: 393–415.

Saucedo Marquez, C. M., B. Vanaudenaerde, T. Troosters, & N. Wenderoth. 2015. "High-intensity interval training evokes larger serum BDNF levels compared with intense continuous exercise." *J Appl Physiol 119*: 1363–73.

Schlaffke, L., S. Lissek, M. Lenz, M. Brüne, G. Juckel, T. Hinrichs,...T. Schmidt-Wilcke. 2014. "Sports and brain morphology – a voxel-based morphometry study with endurance athletes and martial artists." *Neuroscience 259*: 35–42.

Schmidt, R. A., & T. D. Lee. 2011. *Motor control and learning: A behavioral emphasis*. 5th Ed. Champaign, IL: Human Kinetics.

Schmidt, R. A., & T. D. Lee. 2014. *Motor learning and performance: From principles to application*. 5th ed. Champaign, IL: Human Kinetics.

Schmidt-Kassow, M., S. Schädle, S. Otterbein, C. Thiel, A. Doehring, J. Lötsch, & J. Kaiser. 2012. "Kinetics of serum brain-derived neurotrophic factor following low-intensity versus high-intensity exercise in men and women." *Neuroreport 23*: 889–93.

Schmolesky, M. T., D. L. Webb, & R. A. Hansen. 2013. "The effects of aerobic exercise intensity and duration on levels of Brain-Derived Neurotrophic Factor in healthy men." *J Sports Sci Med 12*: 502–11.

Scholz, J., M. C. Klein, T. E. J. Behrens, & H. Johansen-Berg. 2009. "Training induces changes in white-matter architecture." *Nat Neurosci 12*: 1370–1.

Segal, S. K., C. W. Cotman, & L. F. Cahill. 2012. "Exercise-induced noradrenergic activation enhances memory consolidation in both normal aging and patients with amnestic mild cognitive impairment." *J Alzheimers Dis 32*: 1011–18.

Seidler, R. D. 2004. "Multiple motor learning experiences enhance motor adaptability." *J Cogn Neurosci 16*: 65–73.

Seidler, R. D. 2010. "Neural correlates of motor learning, transfer of learning, and learning to learn." *Exerc Sport Sci Rev 38*: 3–9.

Shellock, F. G., & W. E. Prentice. 1985. "Warming-up and stretching for improved physical performance and prevention of sports-related injuries." *Sports Med 2*: 267–78.

Sidhu, S. K., B. Lauber, A. G. Cresswell, & T. J. Carroll. 2013. "Sustained cycling exercise increases intracortical inhibition." *Med Sci Sports Exerc 45*: 654–62.

Singh, A. M., R. E. Duncan, J. L. Neva, & W. R. Staines. 2014. "Aerobic exercise modulates intracortical inhibition and facilitation in a nonexercised upper limb muscle." *BMC Sports Sci Med Rehabil 6*: 23.

Singh, A. M., J. L. Neva, & W. R. Staines. 2014. "Acute exercise enhances the response to paired associative stimulation-induced plasticity in the primary motor cortex." *Exp Brain Res 232*: 3675–85.

Singh, A. M., J. L. Neva, & W. R. Staines. 2016. "Aerobic exercise enhances neural correlates of motor skill learning." *Behav Brain Res 301*: 19–26.

Singh, A. M., & W. R. Staines. 2015. "The effects of acute aerobic exercise on the primary motor cortex." *J Mot Behav 47*: 328–39.

Skriver, K., M. Roig, J. Lundbye-Jensen, J. Pingel, J. W. Helge, B. Kiens, & J. B. Nielsen. 2014. "Acute exercise improves motor memory: Exploring potential biomarkers." *Neurobiol Learn Mem 116*: 46–58.

Smith, A. E., M. R. Goldsworthy, T. Garside, F. M. Wood, & M. C. Ridding. 2014. "The influence of a single bout of aerobic exercise on short-interval intracortical excitability." *Exp Brain Res 232*: 1875–82.

Smith, P. J., J. A. Blumenthal, B. M. Hoffman, H. Cooper, T. A. Strauman, K. Welsh-Bohmer,...A. Sherwood. 2010. "Aerobic exercise and neurocognitive performance: A meta-analytic review of randomized controlled trials." *Psychosom Med 72*: 239–52.

Snigdha, S., C. de Rivera, N. W. Milgram, & C. W. Cotman. 2014. "Exercise enhances memory consolidation in the aging brain." *Front Aging Neurosci 6*: Article 3.

Snow, N. J., C. S. Mang, M. Roig, M. N. McDonnell, K. L. Campbell, & L. A. Boyd. 2016. "The effect of an acute bout of moderate-intensity aerobic exercise on motor learning of a continuous tracking task." *PLoS ONE 11*: e0150039.

Schiffer, T., S. Schulte, B. Sperlich, S. Achtzehn, H. Fricke, & H. K. Strüder. 2011. "Lactate infusion at rest increases BDNF blood concentration in humans." *Neurosci Lett 488*: 234–7.

Stagg, C. J., V. Bachtiar, & H. Johansen-Berg. 2011. "The role of GABA in human motor learning." *Curr Biol 21*: 480–4.

Statton, M. A., M. Encarnacion, P. Celnik, & A. J. Bastian. 2015. "A single bout of moderate aerobic exercise improves motor skill acquisition." *PLoS ONE 10*: e0141393.

Stoykov, M. E., & S. Madhavan. 2015. "Motor priming in neurorehabilitation." *J Neurol Phys Ther 39*: 33–42.

Sumiyoshi, A., Y. Taki, H. Nonaka, H. Takeuchi, & R. Kawashima. 2014. "Regional gray matter volume increases following 7 days of voluntary wheel running exercise: A longitudinal VBM study in rats." *NeuroImage 98*: 82–90.

Suzuki, A., S. A. Stern, O. Bozdagi, G. W. Huntley, R. H. Walker, P. J. Magistretti, & C. M. Alberini. 2011. "Astrocyte-neuron lactate transport is required for long-term memory formation." *Cell 144*: 810–23.

Svatkova, A., R. C. W. Mandl, T. W. Scheewe, W. Cahn, R. S. Kahn, & H. E. Hulshoff Pol. 2015. "Physical exercise keeps the brain connected: biking increases white matter integrity in patients with schizophrenia and healthy controls." *Schizophr Bull 41*: 869–78.

Swain, R. A., A. Harris, E. Wiener, M. Dutka, H. Morris, B. Theien,...W. T. Greenough. 2003. "Prolonged exercise induces angiogenesis and increases cerebral blood volume in primary motor cortex of the rat." *Neuroscience 117*: 1037–46.

Szuhany, K. L., M. Bugatti, & M. W. Otto. 2015. "A meta-analytic review of the effects of exercise on brain-derived neurotrophic factor." *J Psychiatr Res 60*: 56–64.

Takahashi, K., A. Maruyama, K. Hirakoba, M. Maeda, S. Etoh, K. Kawahira, & J. C. Rothwell. 2011. "Fatiguing intermittent lower limb exercise influences corticospinal and corticocortical excitability in the nonexercised upper limb." *Brain Stimul 4*: 90–6.

Takimoto, M., & T. Hamada. 2014. "Acute exercise increases brain region-specific expression of MCT1, MCT2, MCT4, GLUT1, and COX IV proteins." *J Appl Physiol 116*: 1238–50.

Taubert, M., B. Draganski, A. Anwander, K. Müller, A. Horstmann, A. Villringer, & P. Ragert. 2010. "Dynamic properties of human brain structure: Learning-related changes in cortical areas and associated fiber connections." *J Neurosci 30*: 11670–7.

Taylor, J. L. 2012. "Motor control and motor learning under fatigue conditions." In A. Gollhofer, W. Taube, & J. B. Nielsen (Eds.), *Routledge handbook of motor control and motor learning* (pp. 353–83). New York: Routledge.

Teixeira, A. L., I. G. Barbosa, B. S. Diniz, & A. Kummer. 2010. "Circulating levels of Brain-Derived Neurotrophic Factor: Correlation with mood, cognition and motor function." *Biomarkers Med 4*: 871–87.

Thomas, A. G., A. Dennis, P. A. Bandettini, & H. Johansen-Berg. 2012. "The effects of aerobic activity on brain structure." *Front Psychol 3*: 86.

Toni, N., P. A. Buchs, I. Nikonenko, C. R. Bron, & D. Muller. 1999. "LTP promotes formation of multiple spine synapses between a single axon terminal and a dendrite." *Nature 402*: 421–5.

Tononi, G., & C. Cirelli. 2014. "Sleep and the price of plasticity: From synaptic and cellular homeostasis to memory consolidation and integration." *Neuron 81*: 12–34.

Trempe, M., & L. Proteau. 2012. "Motor skill consolidation." In Nicola J. Hodges & A. Mark Williams (Eds.), *Skill acquisition in sport: Research theory and practice* (pp. 192–210). 2nd ed. London: Routledge.

Tyč, F., A. Boyadjian, & H. Devanne. 2005. "Motor cortex plasticity induced by extensive training revealed by transcranial magnetic stimulation in human." *Eur J Neurosci 21*: 259–66.

van Hall, G., M. Strømstad, P. Rasmussen, O. Jans, M. Zaar, C. Gam,...H. B. Nielsen. 2009. "Blood lactate is an important energy source for the human brain." *J Cereb Blood Flow Metab 29*: 1121–9.

van Praag, H., M. Fleshner, M. W. Schwartz, & M. P. Mattson. 2014. "Exercise, energy intake, glucose homeostasis, and the brain." *J Neurosci 34*: 15139–49.

Voelcker-Rehage, C., & C. Niemann. 2013. "Structural and functional brain changes related to different types of physical activity across the life span." *Neurosci Biobehav Rev 37*: 2268–95.

Voss, M. W., S. Heo, R. S. Prakash, K. I. Erickson, H. Alves, L. Chaddock,...A. N. Szabo. 2013a. "The influence of aerobic fitness on cerebral white matter integrity and cognitive function in older adults: Results of a one-year exercise intervention." *Hum Brain Mapp 34*: 2972–85.

Voss, M. W., C. Vivar, A. F. Kramer, & H. van Praag. 2013b. "Bridging animal and human models of exercise-induced brain plasticity." *Trends Cogn Sci 17*: 525–44.

Wang, D.-C., Y.-Y. Lin, & H.-T. Lin. 2015. "Recovery of motor coordination after exercise is correlated to enhancement of brain-derived neurotrophic factor in lactational vanadium-exposed rats." *Neurosci Lett 600*: 232–7.

Wang, Z., Y. Guo, K. G. Myers, R. Heintz, & D. P. Holschneider. 2015a. "Recruitment of the prefrontal cortex and cerebellum in Parkinsonian rats following skilled aerobic exercise." *Neurobiol Dis 77*: 71–87.

Wang, Z., Y. Guo, K. G. Myers, R. Heintz, Y.-H. Peng, J.-M.I. Maarek, & D. P. Holschneider. 2015b. "Exercise alters resting-state functional connectivity of motor circuits in parkinsonian rats." *Neurobiol Aging 36*: 536–44.

Wikgren, J., G. G. Mertikas, P. Raussi, R. Tirkkonen, L. Äyräväinen, M. Pelto-Huikko,...H. Kainulainen. 2012. "Selective breeding for endurance running capacity affects cognitive but not motor learning in rats." *Physiol Behav 106*: 95–100.

Will, B., J. Dalrymple-Alford, M. Wolff, & J.-C. Cassel. 2008. "The concept of brain plasticity-Paillard's systemic analysis and emphasis on structure and function (followed by the translation of a seminal paper by Paillard on plasticity)." *Behav Brain Res 192*: 2–7.

Winter, B., C. Breitenstein, F. C. Mooren, K. Voelker, M. Fobker, A. Lechtermann,...K. Krueger. 2007. "High impact running improves learning." *Neurobiol Learn Mem 87*: 597–609.

Wulf, G., & C. H. Shea. 2002. "Principles derived from the study of simple skills do not generalize to complex skill learning." *Psychon Bull Rev 9*: 185–211.

Xu, T., X. Yu, A. J. Perlik, W. F. Tobin, J. A. Zweig, K. Tennant,...Y. Zuo. 2009. "Rapid formation and selective stabilization of synapses for enduring motor memories." *Nature 462*: 915–19.

Yamaguchi, T., T. Fujiwara, W. Liu, & M. Liu. 2012. "Effects of pedaling exercise on the intracortical inhibition of cortical leg area." *Exp Brain Res 218*: 401–6.

Yan, J. H., B. Abernethy, & J. R. Thomas. 2008. "Developmental and biomechanical characteristics of motor skill learning." In Y. Hong & R. Bartlett (Eds.), *Routledge handbook of biomechanics and human movement science* (pp. 565–80). London, New York: Routledge.

Yang, J., E. Ruchti, J.-M. Petit, P. Jourdain, G. Grenningloh, I. Allaman, & P. J. Magistretti. 2014. "Lactate promotes plasticity gene expression by potentiating NMDA signaling in neurons." *Proc Natl Acad Sci USA 111*: 12228–33.

Zatorre, R. J., R. D. Fields, & H. Johansen-Berg. 2012. "Plasticity in gray and white: Neuroimaging changes in brain structure during learning." *Nat Neurosci 15*: 528–36.

Zatorre, R. J. 2013. "Predispositions and plasticity in music and speech learning: Neural correlates and implications." *Science 342*: 585–9.

Acknowledgments

This work was supported by the Federal Institute of Sport Science (Bundesinstitut für Sportwissenschaft, ZMVI1-070610/14–16 [MT]) and the Max-Planck Institute for Human Cognitive and Brain Sciences.

9 Exercise Effects in Cognition and Motor Learning

Megan Herting

Introduction

When thinking about mental benefits of exercise, stress reduction and mood usually come to mind. However, beyond these benefits, exercise has been linked with a number of improvements in cognitive abilities. In fact, both animal and human studies suggest that acute exercise (going for a run) or more long-term habits regarding exercise may both influence the way in which we learn and remember information. In fact, exercise has been shown to influence the cell composition and function in the very brain regions that underlie learning (and subsequent), including the hippocampus and prefrontal cortex. Here we describe the way in which we learn and remember information as well as the ways in which exercise has been found to influence those abilities at the levels of both the observed behavior and the structure and function of the brain.

Learning and Memory

Learning is a long-term change in behavior that results from previous experience(s). *Memory* is the representation of this learned behavior, whereas *retrieval* is the recovery of information from a stored memory (Domjan 2003). However, memory can further be broken down into short-term versus long-term memory (Cowan 2008) (Figure 9.1). *Short-term memory* is the

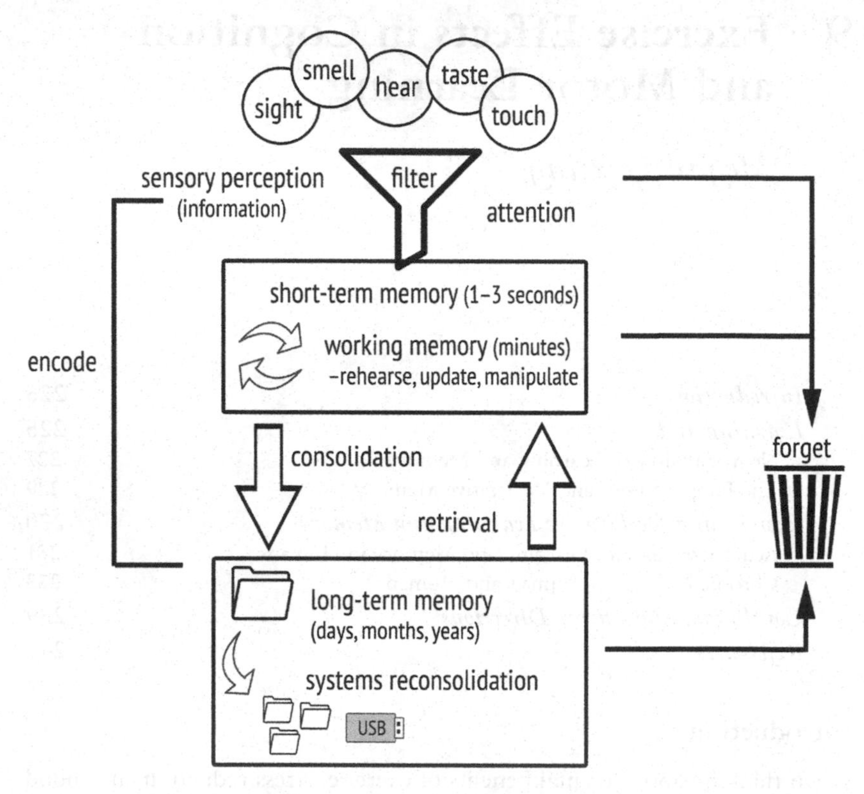

Figure 9.1 The concepts of learning and memory include a number of distinct processes. Once sensory information is captured by our attention, we can encode that information by either holding that information in **short-term memory** for a few seconds, or we can engage our **working memory** to actively rehearse, update, or apply that information. From there, information is thought to undergo **consolidation** in order to be transferred to **long-term memory** where it can be kept for much longer durations. Later on, long-term memories may be integrated and undergo further **system consolidation** for more permanent storage. These memories can then be accessed from long-term storage through the process of **retrieval**. Of course, it is also possible that the information may be lost through **forgetting**

ability to hold a limited capacity of information active in the mind and lasts only a few seconds. A distinct form of short-term memory has been given the term "working memory." As proposed by Alan Baddeley and Graham Hitch in 1974, *working memory* is a rehearsal-based system that allows for managing, updating, and manipulating recently acquired information (Baddeley & Hitch 1974). Unlike short-term memory, *long-term memory* reflects the retention of information that lasts minutes, days, weeks, months, or even years. The transfer of information from short to long-term memory is known as *memory consolidation* (Domjan 2003). Through consolidation, lasting memories may

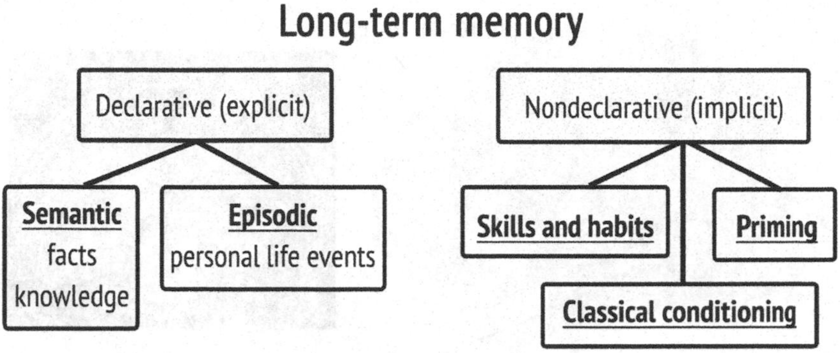

Long-term memory

Declarative (explicit)

Semantic
facts
knowledge

Episodic
personal life events

Nondeclarative (implicit)

Skills and habits

Priming

Classical conditioning

Figure 9.2 Multiple sub-categories of long-term memories

be reorganized, and even integrated, and then moved to a more permanent storage. This type of reconsolidation is thought to take years in humans (Roediger, Dudai, & Fitzpatrick 2007).

As shown in Figure 9.2, multiple forms of long-term memory have also been identified. These include declarative and non-declarative memory (Cohen & Squire 1980). *Declarative memory* (also referred to as explicit memory) is that of conscious recollection of information. This type of memory can be further broken down into semantic memory or episodic memory (Tulving 1972). *Semantic memory* includes conscious recollection of facts and knowledge, such as the capital city of Brazil. On the other hand, *episodic memory* is that of autobiographical life events, such as the gifts you received on your tenth birthday. Declarative memories are often assessed by conventional tests of conscious retrieval, such as recall or recognition. In contrast, *non-declarative memory*, also known as implicit memory, includes processes that do not require conscious retrieval (Cohen & Squire 1980). This includes *procedural memory*, which reflects skill-based (motor, perceptual, etc.) learning and habits, as well as priming and simple classical conditioning. Non-declarative memories are therefore assessed through performance of a behavioral test such as riding a bicycle or writing with a pencil.

Neurocircuitry of Learning and Memory

A large body of evidence has shown that distinct brain regions, including the prefrontal cortex and the hippocampus, are important for learning and memory processes (Figure 9.3). The dorsolateral prefrontal cortex plays an important role in both short-term and working memory (Gabrieli 1998; Owen 1997). Distinct brain systems have also been shown to subserve the different types of long-term memory. Studies in animals and humans have shown that the hippocampus plays a vital role in declarative learning and memory formation (Milner 1972; Squire 1992, 1982, 2004) (see below).

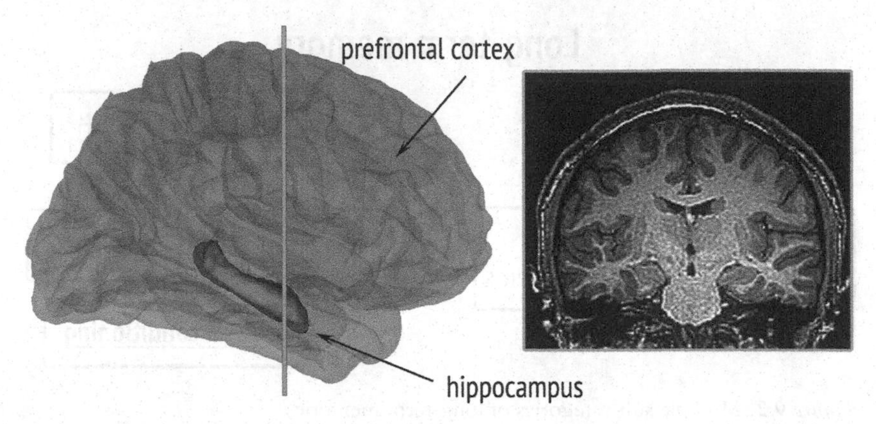

Figure 9.3 Brain regions found to be especially important for learning and memory processes, including the prefrontal cortex and the hippocampus. The image on the right highlights a coronal view of the left and right hippocampus, which corresponds to the vertical line on the left image

Along with the hippocampus, a number of additional cortical structures are also important for learning new types of declarative memories (Eichenbaum 2000). For example, sensory cortices and association areas process stimuli to be encoded by the hippocampus. The prefrontal and parietal cortices further contribute to declarative memories, through regulating attention, organization, and use of various strategies needed to successfully encode and retrieve information. Lastly, other limbic brain regions, such as the amygdala and the ventral striatum (including the nucleus accumbens), likewise influence learning and memory behavior through arousal and motivation (McDonald & White 1993). For example, the ventral striatum is important in outcome prediction learning, and promotes motivational processing for goal-directed behaviors, such as learning new information (for review see Pennartz, Ito, Verschure, Battaglia, & Robbins 2011).

While cortical and hippocampal circuits are vital to declarative memory encoding and retrieval, non-declarative memories do not rely on the hippocampus (Squire 1992). Rather, procedural memories, such as skill learning and habits, involve interactions between cortical areas and the dorsal striatum, including the caudate nucleus and putamen (for review see Packard & Knowlton 2002). Specifically, projection loops between the frontal, cortical, and striatum brain regions, known as fronto-cortico-striatal loops, have been identified and deemed central for motor and stimulus-response learning. In addition, the cerebellum has also been shown to be important for motor learning for both procedural tasks (for review see Saywell & Taylor 2008) as well as eyeblink and other simple forms of classical conditioning (Freeman & Steinmetz 2011; Thompson & Steinmetz 2009).

The Hippocampus and Declarative Memory

The hippocampal formation is located deep in the medial temporal cortices, and has been widely studied for its vital role in learning and memory (Figure 9.3). Hippocampal lesions in humans have been shown to result in memory impairments. Damage to the hippocampus has been shown to result in both the loss of access to memories in the past, known as **retrograde amnesia**, as well as the inability to create new memories, or **anterograde amnesia** (Milner 1972; Rempel-Clower, Zola, Squire, & Amaral 1996). A well-known medical case of a patient called "HM" was vital in illuminating the role of the hippocampus in declarative memory. In an attempt to reduce severe epilepsy, HM had surgery to remove large portions of his temporal lobes, which included the hippocampus. Following this procedure, HM was unable to create new declarative memories (i.e. anterograde amnesia), but was fully capable of learning non-declarative motor memory tasks (Milner 1962). He also displayed severe retrograde amnesia, as he could not remember information from the time of his 15th birthday to the day of his surgery (Milner 1962). However, he was able to recall information prior to his 15th birthday. Given that HM could retrieve very old memories of being a child, but not more recent memories as a young adult, it seems that the hippocampus may have a role in memory storage but its role has a finite time-limit. Specifically, it is thought that the hippocampus is where your most recent memories are initially stored, but then, over time, encoded memories are moved from the hippocampus to the cortex for more permanent storage (Squire 1992). While the discovery of HM's learning and memory impairments resulted in a number of our current cognitive theories, animal studies have been invaluable for clarifying and refining our understanding of which brain regions are important for learning and memory processes. For example, animal studies have since been performed to confirm the importance of the hippocampus in both acquiring as well as retrieving declarative memories. Removal of both the left and right hippocampus has been shown to impair learning and memory of spatial (Morris, Garrud, Rawlins, & O'Keefe 1982) and non-spatial (Squire 1992) information in animals. Furthermore, non-human primate studies have replicated that the hippocampus is important for initial memory storage, but ultimately reduces its role in memory storage over time (Zola-Morgan & Squire 1990).

In addition to being essential for declarative memories, the hippocampus is thought to be especially important for associative learning and memory. **Associative**, or relational, learning and memory involves forming and using representations among elements within an internal or external environment (Henke, Buck, Weber, & Wieser 1997). An example would be the nostalgic alphabet posters in grade school classrooms that pair a letter (such as A) with an item (like an Apple). By forming this association (A for Apple), children are able to learn the letters, their sounds, and how they may be used to begin to spell words. Beyond the alphabet, associative learning is important as it allows for binding information and is integral to learning within and outside the classroom. *In vivo* functional magnetic resonance imaging (fMRI) studies have provided further evidence that the hippocampus is engaged during encoding and retrieval of declarative information (Binder, Bellgowan, Hammeke, Possing, & Frost 2005; Chua, Schacter, Rand-Giovannetti, & Sperling 2007; Davachi & Wagner 2002; Giovanello, Schnyer, & Verfaellie 2004; Greicius et al. 2003; Karlsgodt, Shirinyan, van Erp, Cohen, & Cannon 2005; Reber, Wong, & Buxton 2002; Schacter & Wagner 1999), and is especially so during associative learning (Giovanello et al. 2004). Additionally, the hippocampus is activated alongside the prefrontal cortex during memory retrieval; although this pattern is moderated by type of retrieval being performed (e.g. recall versus recognition) (Okada, Vilberg, & Rugg 2012).

Taken together, it has become increasingly clear from both animal studies, as well as patient case studies like HM, and human imaging studies, that the prefrontal cortex and hippocampus are important for learning as well as memory for declarative information. Moreover, while the hippocampus is engaged in learning and during retrieval, it is thought to be especially involved in learning associations.

Exercise and Declarative Learning and Memory

Exercise is often described in terms of its type and duration. Types of exercise include aerobic and anaerobic. **Aerobic exercise** is defined as sustained activity that stimulates heart and lung function, resulting in improved oxygen transport to the body's cells, and includes activities such as running, walking, swimming, and cycling (Armstrong & Welsman 2007). On the other hand, **anaerobic exercise** is performed in the absence of oxygen, which includes both sprints and resistance and strength training. In research, typical exercise durations studied in humans include acute and chronic exercise. **Acute** exercise studies utilize a single session of aerobic exercise that occurs immediately preceding the measured outcome. Studies of **chronic** effects, on the other hand, examine how repeated, regular, long-term aerobic exercise activity relates to the dependent variables of interest (Coles & Tomporowski 2008; Hillman et al. 2009; Tomporowski, Davis, Lambourne, Gregoski, & Tkacz 2008; Hillman, Erickson, & Kramer 2008; van Praag 2009).

Similar to the field of learning and memory, animal studies have also been useful in providing evidence as to what effect exercise has on learning and memory performance. While the types of exercise (aerobic vs. anaerobic exercise, such as strength training) are similarly studied in animals as they are in humans, very few animal studies use an acute timescale as seen in human research. Rather, animal studies usually allow animals to exercise daily for longer periods, such as a few days or up to several months. Furthermore, while a few studies have closely examined anaerobic exercise, to date aerobic exercise has predominately been studied. Thus, below I summarize the influence of exercise on learning and memory with a strong focus on aerobic exercise, including "acute" aerobic exercise studies in humans and "chronic" aerobic exercise studies in both animals and humans.

Acute Exercise and Learning and Memory in Humans

Compared to chronic exercise, less research has been performed on the topic of acute exercise and how it relates to learning memory (Hillman et al. 2008; Coles & Tomporowski 2008). This paucity in the literature may be due to the fact that acute exercise studies cannot readily disentangle if there is a specific effect of aerobic exertion on learning and memory processes (such as encoding or retrieval), or if the relationship is due to general increases in arousal, or attention, that accompany aerobic exercise (Hopkins, Davis, Vantieghem, Whalen, & Bucci 2012). In other words, mere increases in arousal may help the individual attune to the information to be encoded, which may result in encoding/recall of that information. However, if this were the case, the influence of acute exercise would not necessarily be due to actual changes in learning and memory processes per say, such as the ability to encode or remember information, rather it would be due to changes in attention. Nonetheless, a few studies have shown acute exercise to relate to better memory performance on tasks of short-term and long-term memory, as well as working memory.

A number of these acute exercise studies have used a simple word list task to show an effect of a single bout of exercise on short and long-term memory. In the word list task, individuals are presented with a list of words (Figure 9.4) and then asked to recall as many words as possible immediately after seeing the list (i.e. immediate recall) as well as after a given delay (e.g. a few minutes to hours) (i.e. delayed recall). Using this test, most individuals show what is known as a "serial position effect" (Figure 9.4), where they tend to remember words at the beginning and the end of the list, but are more likely to forget the words that fell in the middle. The recall of the first few words is referred to as **primacy memory**, because they occurred first; whereas the recall of the last few words in the list is referred to as **recency memory**, because they were the items that were seen most recently. Using the word list task, a few studies have shown that acute exercise relates to better memory immediately after encoding as well as after the longer delay. For example, one acute exercise study asked children (ages 11–12 years) to learn 20 words after acute exercise (i.e. physical education class of team games or circuit training) or after a baseline condition (i.e. no acute exercise). Acute exercise via team games led to the children remembering more words immediately after studying the list of words, including better primacy as well as recency memory (Pesce, Crova, Cereatti, Casella, & Bellucci 2009). Both team games and circuit training led to better memory after the delay (e.g. delayed recall) as compared to when the children did not participate in exercise prior to the word list task. Similarly, young adults subjected to a 40-minute rowing session also demonstrated improved primacy and recency memory for a word list during immediate and delayed recall sessions.

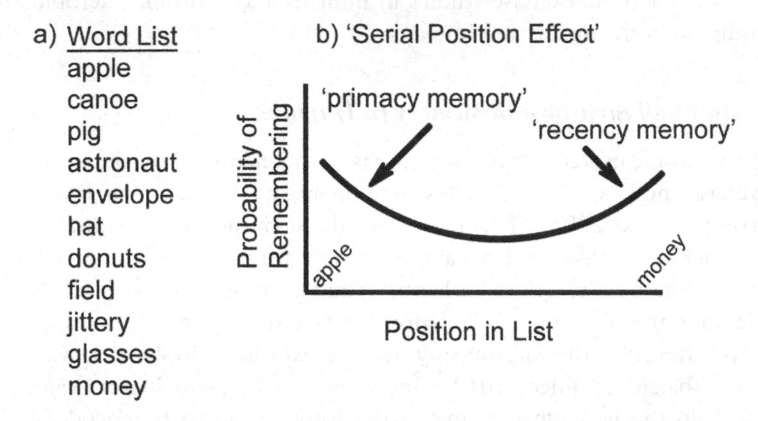

Figure 9.4 Example of a simple word list test used to examine short-term and long-term memory. A) Individuals are asked to view a word list and then asked to recall the words either immediately after seeing them (short-term memory) or after a delay (long-term memory). B) Most individuals are more likely to remember the first few words (primacy effect) and the last few words of the list (recency effect), known as the serial position effect. Using these measurements of learning and memory, a few studies have found that both primacy and recency effects are large after a single session of aerobic exercise

Beyond aerobic exercise, a study also showed that acute anaerobic exercise (duration >2 minutes), leads to better short-term memory (Davey 1973). Albeit others have not been able to replicate such immediate beneficial effects of exercise on memory after anaerobic exercise in college-age students (e.g. a treadmill run to voluntary exhaustion) (Tomporowski, Ellis, & Stephens 1987), suggesting the type of exercise (aerobic versus anaerobic) may be important in reaping the memory benefits of acute exercise. In fact, Winter and colleagues (Winter et al. 2007) sought to directly examine how the type of acute exercise influences learning speed and long-term memory retention in adult men (ages 19–27). Specifically, participants were asked to encode novel words (akin to language learning) via associative learning following different exercise conditions. The conditions were performed on different days and separated 1 week apart. The exercise conditions included either: 1) 3 minutes of high-impact anaerobic sprints, 2) 40 minutes of low-impact aerobic running, or 3) 15 minutes resting (no exercise). In addition to learning the new words following exercise or rest, the men also performed a transfer test in which they tried to use their new memories immediately after learning the words. They repeated the transfer test after one week and 6–8 months later. The results showed that vocabulary learning was 20 percent faster following anaerobic exercise as compared to either the aerobic or the relaxed non-exercise conditions. However, no differences were found between the exercise conditions in the immediate or delayed (1 week or 6–8 month) memory transfer tests. Interestingly, however, they also found that the individuals with the most robust increases in adrenaline (an arousal hormone and neurotransmitter) as a function of the anaerobic sprints, had better long-term retention of the new vocabulary set at the 6–8 month follow-up test. While more research is needed to determine how the type of acute exercise can influence learning and memory, these latter findings by Winter and colleagues support the idea that increased memory benefits may be closely tied to the arousal level at the time of learning.

In addition to its impact on short- and long-term memory, more recent research suggests that acute exercise may also influence working memory abilities. The Sternberg Task, which highlights speed and accuracy in completing a working memory task, was performed on 21 college-age students either before, immediately after, or 30 minutes after each of the following conditions: 1) 30 minutes of aerobic exercise, 2) 30 minutes of resistance training, or 3) 30 minutes of rest (Pontifex, Hillman, Fernhall, Thompson, & Valentini 2009). While exercise type did not change accuracy on the task, students subjected to aerobic exercise had faster responses as compared to the resistance training and rest conditions. In fact, a recent meta-analysis found that the benefit of acute exercise of intermediate intensity may be limited to working memory response time and not working memory accuracy (McMorris, Sproule, Turner, & Hale 2011). However, fMRI studies also suggest that, while working memory accuracy may not have changed, a single bout of acute exercise may still influence how the brain tackles a working memory challenge. Li and colleagues (Li et al. 2014) had 15 young women (mean age=19.5) complete a working memory task (N-back 2 versus rest) 15 minutes following either 1) a 20 minute moderate intensity aerobic exercise intervention, or 2) 20 minutes of rest. Although working memory performance did not improve after the 20 minutes, there were changes in brain function during the working memory task, including recruitment of prefrontal and occipital cortices, and deactivation of the anterior cingulate and left frontal brain regions. Again, this suggests that a single episode of aerobic exercise may change how the brain processes learning and memory information.

In summary, acute exercise has been linked with improvements in various components on short-term learning as well as long-term and working-memory tests. Moreover, it seems that the type of acute exercise may especially be important in terms of whether you might expect to see benefits in learning versus the various types of memory sub-processes. In addition, it seems that the acute physiological changes from the exercise, such as changes in brain activity after exercise as well as the exercise-induced increases in arousal, may ultimately contribute to how we process information. Taken together, these changes could contribute to better performance on tests of learning and memory.

Chronic Exercise in Animals and Humans

The strong interest in understanding the effects of chronic exercise on learning and memory first appeared within the context of aging. That is, normal aging is characterized by a decline in cognitive functions, including episodic memory loss. Furthermore, dementia, or a disabling decline in intellectual functioning, including memory impairments, can also develop with aging. Thus, the field of aging was especially interested to determine if chronic aerobic exercise might improve, or at least slow down, some of those age-related changes in memory loss. Research in this field has found that chronic aerobic exercise enhances learning and memory in both aging rodents and humans (van Praag, Kempermann, & Gage 1999; Hillman et al. 2008).

In addition to age-related cognitive benefits throughout the lifetime, chronic exercise has been found to benefit the process of learning about one's environment. In animals, such as rats and mice, running on a running wheel has been shown to increase learning and memory on a number of behavioral tasks that require the hippocampus (Hopkins, Nitecki, & Bucci 2011). One such task used with rats and mice is the Morris Water Maze (Figure 9.5). This task consists of a large circular pool with a submerged platform under the surface of the water. Surrounding the pool are various visual spatial cues throughout the room. Given that rodents dislike swimming, they struggle to find footing upon placement in the pool. After only a few times of being placed into the pool, both rats and mice show a learned association of the visual cues and the location of the hidden platform. That is, over time, a rodent placed in different starting locations will learn to take the quickest escape route by using the visual cues in the room. Over multiple trials, the animal shows a faster time and/or shorter distance to the platform (as compared to the first time they were placed into the pool), reflecting that they have in fact learned where the platform is located (Figure 9.5a). After the animal has learned the platform location, the hidden platform is removed from the pool and a "probe test" is utilized to assess memory after a delay (e.g. hours or days later) from the last learning trial. The animal is then placed into the pool for one minute, and the time spent swimming near the previous platform location is used as a test of memory recall (Figure 9.5a). Using this task, Van Praag and colleagues (van Praag et al. 1999) have found that volunteer wheel running in adult mice leads to significantly faster learning of the hidden platform location. Moreover, the mice that were runners also spent more time in the platform area during the "probe test" as compared to non-runners, suggesting better spatial memory of where the platform was previously located after a delay. Additional studies have also reported wheel running to improve learning and memory using other tasks, such as the T-maze, and Y-maze (van Praag, Shubert, Zhao, & Gage 2005; van Praag et al. 1999; Van der Borght, Havekes, Bos, Eggen, & Van der Zee 2007; Uysal et al. 2005).

Moreover, chronic aerobic exercise related improvements in spatial learning and memory in rodents have been linked with changes in hippocampal structure (Van der Borght et al. 2007; Uysal et al. 2005; van Praag et al. 2005; van Praag et al. 1999). For example, wheel running has also been shown to lead to a 3- to 4-fold increase in new brain cell growth, known as neurogenesis, in the dentate gyrus sub-region of the hippocampus (van Praag 2008; Kim

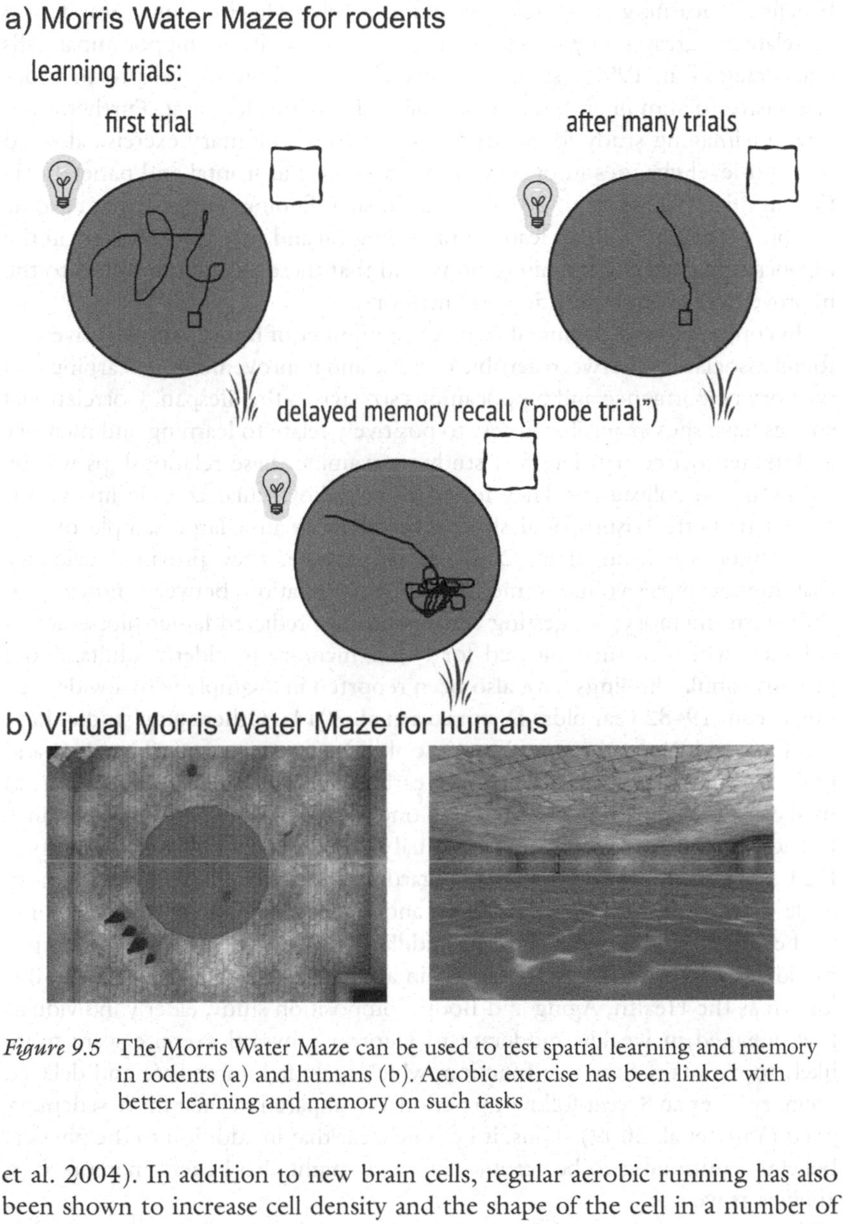

Figure 9.5 The Morris Water Maze can be used to test spatial learning and memory in rodents (a) and humans (b). Aerobic exercise has been linked with better learning and memory on such tasks

et al. 2004). In addition to new brain cells, regular aerobic running has also been shown to increase cell density and the shape of the cell in a number of hippocampal regions (Uysal et al. 2005; Vaynman, Ying, & Gomez-Pinilla 2004; Redila & Christie 2006). These changes in brain cell composition, such as increases in new cell growth, have also been shown to relate to greater volumes of sub-regions of the hippocampus following exercise (Clark et al. 2008), as well as total size of the hippocampus as seen with MRI (Biedermann et al. 2012), following regular wheel running in mice. A number of the other physiological changes are likely to contribute to the chronic exercise-induced benefits in learning and memory. For example, wheel running was found to relate to greater physiological markers of plasticity in hippocampal cells (van Praag et al. 1999), suggesting exercise may allow for cells to be more responsive to stimuli and, ultimately, allow for better learning. Furthermore, a recent imaging study found that four weeks of voluntary exercise also led to trend-level changes in brain volume expansion in frontal and parietal cortices in mice (Cahill et al. 2015). Overall, such findings suggest that chronic aerobic exercise in animals leads to physiological and structural changes in the hippocampus and other brain regions, and that these changes may lead to the improvements seen in learning and memory.

In conjunction with animal research, a number of human studies have also found associations between aerobic exercise and improvements in learning and memory performance and hippocampus size across the lifespan. Correlational studies have shown aerobic fitness to positively relate to learning and memory and its neurocircuitry. The first study to examine these relationships was by Erickson and colleagues. They found that cardiovascular aerobic fitness was related to better visuospatial short-term memory in a large sample of elderly adults (Erickson et al. 2009). Furthermore, they provided evidence that hippocampal volumes mediated the correlation between fitness and short-term memory; suggesting aerobic fitness predicted larger hippocampal volumes, which in turn allowed for better memory in elderly adults. More recently, similar findings have also been reported in a sample with a wider age range from 19–82 year olds (Demirakca et al. 2014). Other aging studies have also found a beneficial effect of exercise abilities on memory performance and brain size in elderly adults who have early signs of dementia, also known as mild cognitive impairments (MCI). In one study, a 6-minute walking distance test was found to relate to better visual memory (Rey-Osterrieth Complex Figure Tests) and logical memory regarding storytelling (Wechsler Memory Scale-Revised Logical Memory Tests) and larger cerebral gray matter volume in the left middle temporal gyrus, middle occipital gyrus, and hippocampus in older adults with MCI. Moreover, in a large prospective study (N=2,509) known as the Health, Aging and Body Composition study, elderly individuals that engaged in weekly, moderate to vigorous physical exercise were more likely to maintain cognitive function, which included immediate and delayed memory, over an 8-year follow-up period as compared to their more sedentary peers (Yaffe et al. 2009). Thus, it became clear that in addition to the physical benefits, continuing to be aerobically active might also benefit memory capabilities as we age.

From the aerobic exercise findings in adults and elderly, the question then arose if aerobic exercise might also benefit learning and memory during development. Specifically, do aerobic exercise behaviors also relate to how healthy children and adolescents learn and retain information? In this regard, associations between aerobic fitness, larger hippocampal volumes, and better memory performance were found in children (Chaddock et al. 2010;

Chaddock, Hillman, Buck, & Cohen 2011a) as well as adolescents (Herting & Nagel 2012). Moreover, these studies also expanded on what types of learning and memory may be influenced by exercise in children and adolescents. For example, in one of these studies, 9 to 10-year-old high- and low-fit were asked to complete two different types of learning and memory tasks. In one learning and memory task they were asked to learn pictures in relation to one another (e.g. a relational memory task), whereas in the other task the children were asked to learn pictures but in a non-relational item memory task. The results showed that high-fit children displayed better performance on relational learning and memory as compared to the low-fit children, but the groups were not different on the non-relational memory task (Chaddock et al. 2010; Chaddock et al. 2011a). Because the hippocampus is especially vital for associative learning, the specificity of improved memory performance on a relational as compared to a non-relational memory task suggests that fitness may be improving performance via the hippocampus. This idea was later further supported by findings that in fact larger hippocampal volumes were found to mediate the relationship between fitness and relational memory performance in children (Chaddock et al. 2010), as well as found to predict better performance on another spatial memory task a year later (Chaddock, Pontifex, Hillman, & Kramer 2011b).

In addition to the studies in children, we have also examined the association between aerobic fitness and learning and memory in teenagers. Specifically, adolescent males (ages 15 to 18 years) with greater aerobic fitness were found to have better visuospatial associative learning on a virtual human analogue of the Morris Water Task (Figure 9.5b) (Herting & Nagel 2012). Similar to the animal version, the virtual Morris Water Task is performed on a computer. Participants complete six trials where they are placed into a pool of water with visual cues around the pool and they are instructed to find a hidden platform as fast as possible. Over six trials, both high and low-fit teenage boys learned to spend more time searching in the correct location of the hidden platform in the pool, but those with greater aerobic fitness showed the most improvement in learning between the first and last trial of the task. As seen in adults and children, the higher-fit adolescent boys also had larger hippocampal volumes (Herting & Nagel 2012). Furthermore, using the same high and low-fit adolescent boys, we also examined how aerobic fitness related to brain activity during the encoding of verbal associations (Herting & Nagel 2013). Specifically, subjects were asked to learn random word pairs (e.g. "chocolate" and "nun") presented on the screen during an fMRI scan. While high and low-fit teen boys remembered a similar number of word-pairs, low-fit youth showed greater hippocampal activation compared to high-fit youth while encoding word-pairs they later remembered. These findings may suggest that higher-fit adolescents may have increased hippocampal efficiency when encoding new verbal associative information compared to their low-fit peers. Alternatively, these findings could also suggest that low-fit teens may have to recruit additional brain resources to achieve a similar level of learning as compared to more aerobically fit boys (Herting & Nagel 2013). These fMRI findings suggest that exercise may not only relate to changes in structural composition or size of memory-related brain regions, but may also impact how the brain functions when learning new information.

While these aforementioned studies in children, teens, and young adults have linked exercise to memory encoding and retrieval, the use of correlational (e.g. cross-sectional) study designs does not allow for conclusions to be made regarding the cause and effect of exercise on learning and memory. Rather, exercise intervention studies, in which individuals perform learning and memory tests before and after they participate in an exercise program, are the gold standard in determining cause-and-effect relationships. In this regard, the few randomized controlled exercise intervention studies that have been implemented in children, adults, and elderly are extremely valuable. Across the board, intervention studies have largely shown that repeated exercise does indeed lead to changes in learning and memory at both the level of the brain and in behavior. In one physical activity and memory intervention study, increases in total low or medium-intensity physical activity over six months led to increases in memory performance on an auditory verbal learning test as well as local gray matter volume in prefrontal and cingulate cortex regions in 62 healthy elderly adults (Ruscheweyh et al. 2011). A similar result was also seen in another 6-month randomized intervention study in 86 healthy elderly women (ages 70 to 80 years). In this case, aerobic training twice per week led to better verbal memory on the Rey Auditory Verbal Learning Test as compared to a balance and toning control condition (Nagamatsu et al. 2013). In terms of brain changes, one intervention study found that elderly individuals who were asked to engage in aerobic exercise three days a week for one year had an approximately 2 % increase in the volume of the anterior hippocampus, as well as improvements in spatial memory, whereas a stretching control group displayed over a 1% decline in volume over this same one-year interval (Erickson et al. 2011). Another study found similar results with increases in bilateral hippocampus volume following 10 weeks of an exercise intervention program (Parker et al. 2011).

Beyond brain size, a few studies also suggest changes in brain function as a direct result of aerobic exercise. For example, an aerobic intervention in adults (ages 21–45 years) resulted in increased cerebral blood volume restricted to the dentate gyrus region of the hippocampus, which correlated with improved learning (Pereira et al. 2007). In addition, a 6-month joint exercise and spatial training intervention in middle-aged men and women (40–55 years of age) led to fitness-related modulation of brain activity in prefrontal and hippocampal regions during successful spatial learning of a virtual town (Holzschneider, Wolbers, Roder, & Hotting 2012). In a separate 12-week intervention study in adults over 60, Nishiguchi and colleagues (Nishiguchi et al. 2015) also examined how an intervention that consisted of class education and pedometer-based walking impacted working memory performance and brain activity. Despite no overt behavioral effect on working memory performance, the exercise intervention led to decreases in superior frontal brain activation on a common fMRI working memory tasks (i.e. n-back 1 versus 0 task) compared to the control group. These findings again highlight that exercise may directly improve brain processing in neural regions essential for learning and various types of memory.

Unlike the adult and aging study, even fewer exercise intervention studies in children and adolescents have been published. One exception is the Fitness Improves Thinking (FITKids) after-school intervention program (Castelli, Hillman, Hirsch, Hirsch, & Drollette 2011; Kamijo et al. 2011). Briefly, this important study randomly assigned children, ages 8 to 10 years, to either participate in an after-school fitness program or be placed on a waitlist for the program. The fitness program ran for 9 months and included 2 hours everyday after school where children were encouraged to achieve 70 minutes of moderate to vigorous physical activity (as confirmed by wearing a heart-rate monitor). While a wonderful design, the cognitive tests that were examined both prior to and after the intervention, unfortunately, did not include memory. Rather, relational and item memory testing of pictures of faces and scenes was only collected on the children after they had completed the FITKids program (Monti, Hillman, & Cohen 2012). While both pre- and post-memory testing is still needed to truly rule out any pre-existing differences in those randomly assigned to the aerobic fitness after-school program versus those children put on the waitlist, it was found that the children that completed the aerobic fitness intervention displayed greater eye fixation on the faces that were previously paired with scenes during the learning test. The authors conclude that this difference in eye fixation for previously paired faces is representative of better relational memory following the aerobic exercise intervention (Monti et al. 2012). Interestingly, no difference was found in the accuracy of relational or item memory between the aerobic fitness intervention and waitlist groups of children. Again this study strengthens the link between exercise and memory in children, but also highlights that additional exercise intervention studies in children and adolescents are required to establish a casual effect of aerobic exercise on learning and memory across development.

Overall, previous research shows a strong connection between aerobic fitness and the neurobiology and behavior of various types of learning and memory processes. In fact a number of adult and elderly studies have shown a causal relationship between engaging in aerobic exercise and changes in brain structures, encoding of new information, and later retrieval of memories. Furthermore, a lifestyle that includes aerobic exercise is linked with improvement in learning and the hippocampus across the lifespan, as well as establishing a buffer against common age-related decline often seen in memory capabilities.

Conclusions and Future Directions

Taken together, the literature suggests that acute and repeated aerobic exercise is associated with distinct benefits in learning and memory. Despite these initial findings, there are a number of theoretical and methodological developments that will help us to continue to clarify the relationship between exercise and learning and memory. One of the biggest questions that needs to be addressed is determining if there is an optimal effect of exercise on learning and memory. That is, the type of exercise, the amount or dose, and the duration needed to achieve the maximum benefits in learning and memory are still unknown. While the literature has focused on aerobic exercise, very few studies have examined how other types of exercise, such as strength training, anaerobic sprints, or stretching affect our brains and behaviors. The acute aerobic studies have especially highlighted how exercise of varying types and intensities may have different effects on short and long-term memory retention. Furthermore, while chronic studies have focused on aerobic exercise, animal research has also found that strength training in rats relates to increases in learning and memory via a different physiological pathway as compared to aerobic exercise (Cassilhas et al. 2012). A more recent study has also reported a correlation between the amount, duration, and frequency of total daily walking activity and hippocampus size among older women, albeit this was not seen among men (Varma, Chuang, Harris, Tan, & Carlson 2015). Thus, more studies are needed to determine what combinations of aerobic and non-aerobic exercise of varying intensity (vigorous or low/moderate), and for how many minutes per day and/or week, is required to improve learning and memory.

While exercise seems to similarly relate to learning and memory processes in children, adolescents, and young and old adults, it also remains to be determined if exercise may be more beneficial at certain time frames across the lifespan. Studies that examine wider age ranges may help to determine if there are windows of time in which exercise may have a bigger effect, such as on the young brains of children and adolescents or the aging brains of the elderly. At least for the hippocampus, this hypothesis is feasible as more robust hippocampal changes have been noted in adolescents and young rats as compared to much older rats following exercise (Kim et al. 2004). Moreover, larger studies that aim to characterize exercise habits in the same individuals over their lives are required to determine if the effects of exercise on learning and memory are long lasting. That is, if an individual chooses to stop engaging in exercise, does the benefit also taper out over time? Or, does exercising during childhood and adolescence lead to long-lasting changes in learning and memory circuitry that remain stable despite physical activity regimens performed later in adulthood? Alternatively, are the effects of exercise on learning and memory additive? On the other end of the age spectrum, is exercise needed earlier in adulthood or only at the time of aging in order to mitigate memory loss?

From the literature it is also clear that there are quite a few mixed findings as to which types of learning and memory are influenced by exercise. Some studies used short or long-term memory protocols, whereas in other cases the "memory" tasks also tap more executive processes, such as working memory.

While hippocampal-dependent learning and memory, such as associative declarative memories, are repeatedly linked to exercise, other links between exercise and learning and memory processes are less clear. The differences in learning modality (learning new visual versus verbal information) as well as the length of memory retention studied (ranging from immediately after encoding versus a few hours or days later) all seem to be important to consider in further clarifying how exercise influences what we can learn as well as what we can later remember. The similarity of the results across species and the possibility of employing readily available translational learning and memory tests, such as the Morris Water Maze (Figure 9.5), suggest that both rodent and human research can be useful resources in pursuing these avenues of research. When interpreting our research findings across all species, however, it is important that we consider an evolutionary perspective. In the case of rodent studies, rodents that are used in these studies are housed in a cage with or without a wheel. In the corresponding human studies, we often compare high and low-fit peers. In both cases, from an evolutionary perspective, the running rodents and the high-active human beings are the norm; both animals and humans were made to move. It follows that physical activity and better aerobic fitness would likely allow for learning and remembering the necessary knowledge needed to survive in the environment, as well as to catch prey and avoid predators. Thus, rather than aiming to enhance our learning and memory capabilities with exercise, the parallel findings between animals and humans may remind us that learning and memory abilities seen with aerobic exercise may be our evolutionary norm, and sedentary behavior may actually be harmful, or even detrimental to our brains' ability to both learn and remember.

In conclusion, beyond the numerous physical existing benefits, exercise may directly influence how and what we learn and remember. Nonetheless, more research is needed, especially as the answers to these pressing questions will inform us in determining if the exercise regimens often suggested to benefit physical health (such as "30 minutes a day for 3 times a week") are the same or differ from those that may help us to maximize our ability to learn, retain, and retrieve new information.

References

Armstrong, N., & J. R. Welsman. 2007. "Aerobic fitness: What are we measuring?" *Med Sport Sci 50*: 5–25.

Baddeley, A., & G. J. Hitch. 1974. "Recent advances in learning and motivation." In G. A. Bower (Ed.), *Working Memory* (pp. 47–90). New York: Academic Press.

Biedermann, S., J. Fuss, L. Zheng, A. Sartorius, C. Falfan-Melgoza, T. Demirakca,… W. Weber-Fahr. 2012. "In vivo voxel based morphometry: Detection of increased hippocampal volume and decreased glutamate levels in exercising mice." *Neuroimage 61* (4): 1206–12. doi:10.1016/j.neuroimage.2012.04.010

Binder, J. R., P. S. Bellgowan, T. A. Hammeke, E. T. Possing, & J. A. Frost. 2005. "A comparison of two FMRI protocols for eliciting hippocampal activation." *Epilepsia 46* (7): 1061–70.

Cahill, L. S., P. E. Steadman, C. E. Jones, C. L. Laliberte, J. Dazai, J. P. Lerch,...J. G. Sled. 2015. "MRI-detectable changes in mouse brain structure induced by voluntary exercise." *Neuroimage 113*: 175–83. doi:10.1016/j.neuroimage.2015.03.036

Cassilhas, R. C., K. S. Lee, J. Fernandes, M. G. Oliveira, S. Tufik, R. Meeusen,...T. de Mello. 2012. "Spatial memory is improved by aerobic and resistance exercise through divergent molecular mechanisms." *Neuroscience 202*: 309–17. doi:10.1016/j.neuroscience.2011.11.029

Castelli, D. M., C. H. Hillman, J. Hirsch, A. Hirsch, & E. Drollette. 2011. "FIT Kids: Time in target heart zone and cognitive performance." *Prev Med 52* Suppl 1: S55–9. doi:10.1016/j.ypmed.2011.01.019

Chaddock, L., K. I. Erickson, R. S. Prakash, J. S. Kim, M. W. Voss, M. Vanpatter,...& A. F. Kramer. 2010. "A neuroimaging investigation of the association between aerobic fitness, hippocampal volume, and memory performance in preadolescent children." *Brain Res 1358*: 172–83.

Chaddock, L., C. H. Hillman, S. M. Buck, & N. J. Cohen. 2011. "Aerobic fitness and executive control of relational memory in preadolescent children." *Med Sci Sports Exerc 43* (2): 344–9.

Chaddock, L., M. B. Pontifex, C. H. Hillman, & A. F. Kramer. 2011. "A review of the relation of aerobic fitness and physical activity to brain structure and function in children." *J Int Neuropsychol Soc 17* (6): 975–85.

Chua, E. F., D. L. Schacter, E. Rand-Giovannetti, & R. A. Sperling. 2007. "Evidence for a specific role of the anterior hippocampal region in successful associative encoding." *Hippocampus 17* (11): 1071–80.

Clark, P. J., W. J. Brzezinska, M. W. Thomas, N. A. Ryzhenko, S. A. Toshkov, & J. S. Rhodes. 2008. "Intact neurogenesis is required for benefits of exercise on spatial memory but not motor performance or contextual fear conditioning in C57BL/6J mice." *Neuroscience 155* (4): 1048–58.

Cohen, N. J., & L. R. Squire. 1980. "Preserved learning and retention of pattern-analyzing skill in amnesia: Dissociation of knowing how and knowing that." *Science 210* (4466): 207–10.

Coles, K., & P. D. Tomporowski. 2008. "Effects of acute exercise on executive processing, short-term and long-term memory." *J Sports Sci 26* (3): 333–44.

Cowan, N. 2008. "What are the differences between long-term, short-term, and working memory?" *Prog Brain Res 169*: 323–38. doi:10.1016/S0079-6123(07)00020-9

Davachi, L., & A. D. Wagner. 2002. "Hippocampal contributions to episodic encoding: insights from relational and item-based learning." *Journal of Neurophysiology 88* (2): 982–90.

Davey, C. P. 1973. "Physical exertion and mental performance." *2007 16* (5): 595–9.

Demirakca, T., W. Brusniak, N. Tunc-Skarka, I. Wolf, S. Meier, F. Matthaus,...C. Diener. 2014. "Does body shaping influence brain shape? Habitual physical activity is linked to brain morphology independent of age." *World J Biol Psychiatry 15* (5): –96. doi:10.3109/15622975.2013.803600

Domjan, M. 2003. *The principles of learning and behavior, 5th edition*. Belmont, CA: Wadsworth.

Eichenbaum, H. 2000. "A cortical-hippocampal system for declarative memory." *Nat Rev Neurosci 1* (1): 41–50. doi:10.1038/35036213

Erickson, K. I., R. S. Prakash, M. W. Voss, L. Chaddock, L. Hu, K. S. Morris,...A. F. Kramer. 2009. "Aerobic fitness is associated with hippocampal volume in elderly humans." *Hippocampus 19* (10): 1030–9.

Erickson, K. I., M. W. Voss, R. S. Prakash, C. Basak, A. Szabo, L. Chaddock,...A. F. Kramer. 2011. "Exercise training increases size of hippocampus and improves memory." *Proc Natl Acad Sci U S A 108* (7): 3017–22.

Freeman, J. H., & A. B. Steinmetz. 2011. "Neural circuitry and plasticity mechanisms underlying delay eyeblink conditioning." *Learn Mem 18* (10): 666–77. doi:10.1101/lm.2023011

Gabrieli, J. D. 1998. "Cognitive neuroscience of human memory." *Annual Review of Psychology 49*: 87–115.

Giovanello, K. S., D. M. Schnyer, & M. Verfaellie. 2004. "A critical role for the anterior hippocampus in relational memory: Evidence from an fMRI study comparing associative and item recognition." *Hippocampus 14* (1): 5–8.

Greicius, M. D., B. Krasnow, J. M. Boyett-Anderson, S. Eliez, A. F. Schatzberg, A. L. Reiss, & V. Menon. 2003. "Regional analysis of hippocampal activation during memory encoding and retrieval: fMRI study." *Hippocampus 13* (1): 164–74.

Henke, K., A. Buck, B. Weber, & H. G. Wieser. 1997. "Human hippocampus establishes associations in memory." *Hippocampus 7* (3): 249–56.

Herting, M. M., & B. J. Nagel. 2012. "Aerobic fitness relates to learning on a virtual Morris Water Task and hippocampal volume in adolescents." *Behavioural Brain Research 233* (2): 517–25. doi:10.1016/j.bbr.2012.05.012

Herting, M. M., & B. J. Nagel. 2013. "Differences in brain activity during a verbal associative memory encoding task in high- and low-fit adolescents." *Journal of Cognitive Neuroscience 25* (4): 595–612. doi:10.1162/jocn_a_00344

Hillman, C. H., K. I. Erickson, & A. F. Kramer. 2008. "Be smart, exercise your heart: Exercise effects on brain and cognition." *Nat Rev Neurosci 9* (1): 58–65.

Hillman, C. H., M. B. Pontifex, L. B. Raine, D. M. Castelli, E. E. Hall, & A. F. Kramer. 2009. "The effect of acute treadmill walking on cognitive control and academic achievement in preadolescent children." *Neuroscience 159* (3): 1044–54.

Holzschneider, K., T. Wolbers, B. Roder, & K. Hotting. 2012. "Cardiovascular fitness modulates brain activation associated with spatial learning." *Neuroimage 59* (3): 3003–14. doi:10.1016/j.neuroimage.2011.10.021

Hopkins, M. E., F. C. Davis, M. R. Vantieghem, P. J. Whalen, & D. J. Bucci. 2012. "Differential effects of acute and regular physical exercise on cognition and affect." *Neuroscience*. doi:10.1016/j.neuroscience.2012.04.056

Hopkins, M. E., R. Nitecki, & D. J. Bucci. 2011. "Physical exercise during adolescence versus adulthood: Differential effects on object recognition memory and brain-derived neurotrophic factor levels." *Neuroscience 194*: 84–94. doi:10.1016/j.neuroscience.2011.07.071

Kamijo, K., M. B. Pontifex, K. C. O'Leary, M. R. Scudder, C. T. Wu, D. M. Castelli, & C. H. Hillman. 2011. "The effects of an afterschool physical activity program on working memory in preadolescent children." *Dev Sci 14* (5): 1046–58. doi:10.1111/j.1467-7687.2011.01054.x

Karlsgodt, K. H., D. Shirinyan, T. G. van Erp, M. S. Cohen, & T. D. Cannon. 2005. "Hippocampal activations during encoding and retrieval in a verbal working memory paradigm." *Neuroimage 25* (4): 1224–31.

Kim, Y. P., H. Kim, M. S. Shin, H. K. Chang, M. H. Jang, M. C. Shin,...C. J. Kim. 2004. "Age-dependence of the effect of treadmill exercise on cell proliferation in the dentate gyrus of rats." *Neurosci Lett 355* (1–2): 152–4.

Li, L., W. W. Men, Y. K. Chang, M. X. Fan, L. Ji, & G. X. Wei. 2014. "Acute aerobic exercise increases cortical activity during working memory: A functional MRI study in female college students." *PLoS One 9* (6): e99222. doi:10.1371/journal.pone.0099222.

McDonald, R. J., & N. M. White. 1993. "A triple dissociation of memory systems: Hippocampus, amygdala, and dorsal striatum." *Behav Neurosci 107* (1): 3–22.

McMorris, T., J. Sproule, A. Turner, & B. J. Hale. 2011. "Acute, intermediate intensity exercise, and speed and accuracy in working memory tasks: A meta-analytical comparison of effects." *Physiol Behav 102* (3–4): 421–8. doi:10.1016/j.physbeh.2010.12.007

Milner, B. 1962. "Les troubles de la memoire accompagnant des lesions hippocampiques bilaterales." In P. Passouant (Ed.), *Physiologie de l'hippocampe* (pp. 257–72). Paris: Centre National de la Recherche Scientifique.

Milner, B. 1972. "Disorders of learning and memory after temporal lobe lesions in man." *Clin Neurosurg 19*: 421–46.

Monti, J. M., C. H. Hillman, & N. J. Cohen. 2012. "Aerobic fitness enhances relational memory in preadolescent children: The FITKids randomized control trial." *Hippocampus 22* (9): 1876–82. doi:10.1002/hipo.22023

Morris, R. G., P. Garrud, J. N. Rawlins, & J. O'Keefe. 1982. "Place navigation impaired in rats with hippocampal lesions." *Nature 297* (5868): 681–3.

Nagamatsu, L. S., A. Chan, J. C. Davis, B. L. Beattie, P. Graf, M. W. Voss,...T. Liu-Ambrose. 2013. "Physical activity improves verbal and spatial memory in older adults with probable mild cognitive impairment: A 6-month randomized controlled trial." *J Aging Res 2013*: 861893. doi:10.1155/2013/861893

Nishiguchi, S., M. Yamada, T. Tanigawa, K. Sekiyama, T. Kawagoe, M. Suzuki,...T. Tsuboyama. 2015. "A 12-week physical and cognitive exercise program can improve cognitive function and neural efficiency in community-dwelling older adults: A randomized controlled trial." *J Am Geriatr Soc 63* (7): 1355–63. doi:10.1111/jgs.13481

Okada, K., K. L. Vilberg, & M. D. Rugg. 2012. "Comparison of the neural correlates of retrieval success in tests of cued recall and recognition memory." *Hum Brain Mapp 33* (3): 523–33. doi:10.1002/hbm.21229

Owen, A. M. 1997. "The functional organization of working memory processes within human lateral frontal cortex: The contribution of functional neuroimaging." *European Journal of Neuroscience 9* (7): 1329–39.

Packard, M. G., & B. J. Knowlton. 2002. "Learning and memory functions of the Basal Ganglia." *Annu Rev Neurosci 25*: 563–93. doi:10.1146/annurev.neuro.25.112701.142937

Parker, B. A., P. D. Thompson, K. C. Jordan, A. S. Grimaldi, M. Assaf, K. Jagannathan, & G. D. Pearlson. 2011. "Effect of exercise training on hippocampal volume in humans: A pilot study." *Res Q Exerc Sport 82* (3): 585–91. doi:10.1080/02701367.2011.10599793

Pennartz, C. M., R. Ito, P. F. Verschure, F. P. Battaglia, & T. W. Robbins. 2011. "The hippocampal-striatal axis in learning, prediction and goal-directed behavior." *Trends Neurosci 34* (10): 548–59. doi:10.1016/j.tins.2011.08.001

Pereira, A. C., D. E. Huddleston, A. M. Brickman, A. A. Sosunov, R. Hen, G. M. McKhann,...S. A. Small. 2007. "An in vivo correlate of exercise-induced neurogenesis in the adult dentate gyrus." *Proc Natl Acad Sci U S A 104* (13): 5638–43.

Pesce, C., C. Crova, L. Cereatti, R. Casella, & M. Bellucci. 2009. "Physical activity and mental performance in preadolescents: Effects of acute exercise on free-recall memory." *Mental Health and Physical Activity 2* (1): 16–22.

Pontifex, M. B., C. H. Hillman, B. Fernhall, K. M. Thompson, & T. A. Valentini. 2009. "The effect of acute aerobic and resistance exercise on working memory." *Med Sci Sports Exerc 41* (4): 927–34. doi:10.1249/MSS.0b013e3181907d69

Reber, P. J., E. C. Wong, & R. B. Buxton. 2002. "Encoding activity in the medial temporal lobe examined with anatomically constrained fMRI analysis." *Hippocampus 12* (3): 363–76.

Redila, V. A., & B. R. Christie. 2006. "Exercise-induced changes in dendritic structure and complexity in the adult hippocampal dentate gyrus." *Neuroscience 137* (4) :1299–307. doi:10.1016/j.neuroscience.2005.10.050

Rempel-Clower, N. L., S. M. Zola, L. R. Squire, & D. G. Amaral. 1996. "Three cases of enduring memory impairment after bilateral damage limited to the hippocampal formation." *J Neurosci 16* (16): 5233–55.

Roediger, H. L., Y. Dudai, & S. M. Fitzpatrick. 2007. *Science of memory: Concepts.* New York: Oxford University Press.

Ruscheweyh, R., C. Willemer, K. Kruger, T. Duning, T. Warnecke, J. Sommer,...A. Floel. 2011. "Physical activity and memory functions: An interventional study." *Neurobiol Aging 32* (7): 1304–19. doi:10.1016/j.neurobiolaging.2009.08.001

Saywell, N., & D. Taylor. 2008. "The role of the cerebellum in procedural learning -are there implications for physiotherapists' clinical practice?" *Physiother Theory Pract 24* (5): 321–8. doi:10.1080/09593980701884832

Schacter, D. L., & A. D. Wagner. 1999. "Medial temporal lobe activations in fMRI and PET studies of episodic encoding and retrieval." *Hippocampus 9* (1): 7–24.

Squire, L. R. 1982. "The neuropsychology of human memory." *Annu Rev Neurosci 5*: 241–73.

Squire, L. R. 1992. "Memory and the hippocampus: A synthesis from findings with rats, monkeys, and humans." *Psychological Review 99* (2): 195–231.

Squire, L. R. 2004. "Memory systems of the brain: A brief history and current perspective." *Neurobiol Learn Mem 82* (3): 171–7.

Thompson, R. F., & J. E. Steinmetz. 2009. "The role of the cerebellum in classical conditioning of discrete behavioral responses." *Neuroscience 162* (3): 732–55. doi:10.1016/j.neuroscience.2009.01.041

Tomporowski, P. D., C. L. Davis, K. Lambourne, M. Gregoski, & J. Tkacz. 2008. "Task switching in overweight children: Effects of acute exercise and age." *J Sport Exerc Psychol 30* (5): 497–511.

Tomporowski, P. D., N. R. Ellis, & R. Stephens. 1987. "The immediate effects of strenuous exercise on free-recall memory." *Ergonomics 30* (1): 121–9. doi:10.1080/00140138708969682

Tulving, E. 1972. "Episodic and semantic memory." In E. Tulving & W. Donaldson (Eds.), *Organization of memory* (pp. 381–403). New York: Academic Press.

Uysal, N., K. Tugyan, B. M. Kayatekin, O. Acikgoz, H. A. Bagriyanik, S. Gonenc,...I. Semin. 2005. "The effects of regular aerobic exercise in adolescent period on hippocampal neuron density, apoptosis and spatial memory." *Neurosci Lett 383* (3): 241–5.

Van der Borght, K., R. Havekes, T. Bos, B. J. Eggen, & E. A. Van der Zee. 2007. "Exercise improves memory acquisition and retrieval in the Y-maze task: Relationship with hippocampal neurogenesis." *Behav Neurosci 121* (2): 324–34.

van Praag, H. 2008. "Neurogenesis and exercise: Past and future directions." *Neuromolecular Med 10* (2): 128–40.

van Praag, H. 2009. "Exercise and the brain: Something to chew on." *Trends Neurosci 32* (5): 283–90.

van Praag, H., G. Kempermann, & F. H. Gage. 1999. "Running increases cell proliferation and neurogenesis in the adult mouse dentate gyrus." *Nat Neurosci 2* (3): 266–70.

van Praag, H., T. Shubert, C. Zhao, & F. H. Gage. 2005. "Exercise enhances learning and hippocampal neurogenesis in aged mice." *J Neurosci 25* (38): 8680–5.

Varma, V. R., Y. F. Chuang, G. C. Harris, E. J. Tan, & M. C. Carlson. 2015. "Low-intensity daily walking activity is associated with hippocampal volume in older adults." *Hippocampus 25* (5): 605–15. doi:10.1002/hipo.22397

Vaynman, S., Z. Ying, & F. Gomez-Pinilla. 2004. "Exercise induces BDNF and synapsin I to specific hippocampal subfields." *J Neurosci Res 76* (3): 356–62.

Winter, B., C. Breitenstein, F. C. Mooren, K. Voelker, M. Fobker, A. Lechtermann,… S. Knecht. 2007. "High impact running improves learning." *Neurobiol Learn Mem 87* (4): 597–609.

Yaffe, K., A. J. Fiocco, K. Lindquist, E. Vittinghoff, E. M. Simonsick, A. B. Newman,… T. B. Harris. 2009. "Predictors of maintaining cognitive function in older adults: The Health ABC study." *Neurology 72* (23): 2029–35.

Zola-Morgan, S. M., & L. R. Squire. 1990. "The primate hippocampal formation: Evidence for a time-limited role in memory storage." *Science 250* (4978): 288–90.

Section 4

Sport vs. Exercise and Their Effects on Emotions and Psychological Diseases

10 Exercise in the Prevention, Treatment, and Management of Addictive Substance Use

Adrian Taylor and Tom Thompson

Introduction

Prevalence, Health Consequences, and Treatment Options for Cigarette Smoking

Globally, tobacco causes about 5 million deaths annually (World Health Organization 2011b). Due to the insidious nature of nicotine addiction, smoking cessation is a challenge with unaided quit attempts resulting in abstinence after 6–12 months of only 3–5%. With a combination of behavioral counselling and nicotine replacement therapy (NRT), bupropion or varenicline, long-term abstinence can improve success rates by around 7–9% (Cahill, Stead, & Lancaster 2011; Hughes, Stead, & Lancaster 2007; Stead et al. 2012). There is, therefore, an ongoing search for a better understanding of what may work, and how, in an effort to identify new therapies. Relapse after quitting is often caused by reasons such as elevated cravings, low mood and weight gain, so behavioral and pharmacological therapies that target these causes may be effective. Physical activity may be targeted at helping people to overcome some of the reasons why smokers fail to quit.

Prevalence, Health Consequences, and Treatment Options for Alcohol and Substance Use Disorders

Alcohol and substance use disorders (AUD; SUDs) are common: globally, 5.9% and 1% of deaths are attributable to alcohol and illicit drug use, respectively (World Health Organization 2014; United Nations Office on Drugs and Crime 2012). In the UK, alcohol is attributed to over 1 in 5 deaths of men aged 16–54 years old (NorthWest Public Health Observatory 2008), and alcohol abuse is associated with an economic annual cost of around £21bn (The Centre for Social Justice 2013) (£3.5bn in healthcare (The National Treatment Agency for Substance Misuse 2013)). Illicit drug abuse in the UK has an economic cost of around £15bn (The Centre for Social Justice 2013) (£488m through healthcare (The National Treatment Agency for Substance Misuse 2012)) with nearly 1 in 10 adults aged 16–59 in England and Wales having used illicit drugs in the past year (Public Health England 2014). Worldwide, alcohol attributable deaths increased from 3.8% in 2004 (World Health Organization 2011a) to 5.9% in 2012 (World Health Organization 2014), and illicit drug use levels have failed to improve between 2005 and 2010 (United Nations Office on Drugs and Crime 2012), with a slight increase in the UK in recent years (Public Health England 2014). Alcohol and drug use disorders are commonly co-existent with common and severe mental health problems (Perry et al. 2015), as well as a number of negative health-related consequences such as engagement in violent acts (Stuart, O'Farrell, & Temple 2009), driving while under the influence (Cherpitel et al. 2010), and suicide (Pitman, Krysinska, Osborn, & King 2012).

Pharmacological interventions for AUD and SUDs have been well researched and reported on for addressing withdrawal, dependence and abstinence maintenance. The Cochrane Drug and Alcohol Group (Cochrane Drug and Alcohol Group 2015) have 11 and 30 published reviews for pharmacological interventions for AUDs and SUDs, respectively. Psychosocial interventions (e.g. brief interventions and motivational interviewing) are less well reported, with 6 and 8 published reviews for AUDs and SUDs, respectively. Preventive interventions only have five reviews for AUDs, and three reviews for SUDs. Due to the heterogeneity of the types of drugs used and style of intervention, it is hard to meaningfully summarize the available data of existing interventions. However, with relapse rates as high as 60% one year after treatment for SUDs (McLellan, Lewis, O'Brien, & Kleber 2000; Ramo & Brown 2008; Connors, Maisto, & Donovan 1996) and 60–90% for AUDs (Brownell, Marlatt, Lichtenstein, & Wilson 1986; Maisto, Connors, & Zywiak 2000; Miller, Walters, & Bennett 2001; Xie, McHugo, Fox, & Drake 2005) and common drug replacement therapies being associated with innate complications (Fareed, Casarella, Amar, Vayalapalli, & Drexler 2009; Wapf et al. 2008; Fischer, Rehm, Kim, & Kirst 2005; Maruyama, Macdonald, Borycki, & Zhao 2013), there is a need for evidence for new treatments and preventive interventions to help address the burden of AUDs and SUDs.

Other Possible Addictions of Relevance to the Chapter

Other addictions to be considered later in the present chapter may arguably have a generally lower level of acceptance as a major concern for public health, morbidity, and mortality. However, there is increasing interest in the role of foods such as high energy snacks, fat and caffeine which can be consumed and contribute to similar behaviors, cognitions, and symptoms to other addictions (Meule & Gearhardt 2014; Schulte, Joyner, Potenza, Grilo, & Gearhardt 2015; Gearhardt, Grilo, DiLeone, Brownell, & Potenza 2011). The major contribution of Western diets to obesity and poor health makes this a particularly important issue, and we will consider the growing research on exercise and food addictions.

There are other possible behaviors which may take on addictive properties. This review will not explore the role of exercise in connection with these behaviors, although we have identified studies which have been reported as a therapeutic option for smartphone use (Kim 2013), and gambling (Angelo, Tavares, & Zilberman 2013). Given that engagement in each addictive behavior may have some similarities but also unique differences, there is scope to explore in future research if exercise has common mechanisms in its possible influence. Finally, it is important to clarify that while much has been written about exercise addiction, this chapter will not consider this but will instead focus on the role of exercise on other addictive behaviors.

A Model for Understanding the Role of Exercise as a Treatment Option for Addictions

Before we start to consider the evidence for the relationship between physical activity and addictions, it may be useful to lay out a framework for understanding this link and what the implications may be. Figure 10.1 provides a framework for understanding the literature concerned with physical activity, sport, exercise, and sedentary behavior and addictions, with numbered links to simplify things.

The questions that emerge are the following: (1) Does physical activity (in whatever form), and sedentary behavior, have a causal effect on addictions? (2) If so then is there a dose response and (3) what mechanisms are involved? (4) If there is evidence of a causal effect, what support and interventions can help people change their behavior, and (5) what psychological constructs should be targeted to help people change behavior? Finally, (6) how much support is needed (dose of intervention) to maximize the likelihood of behavior changing?

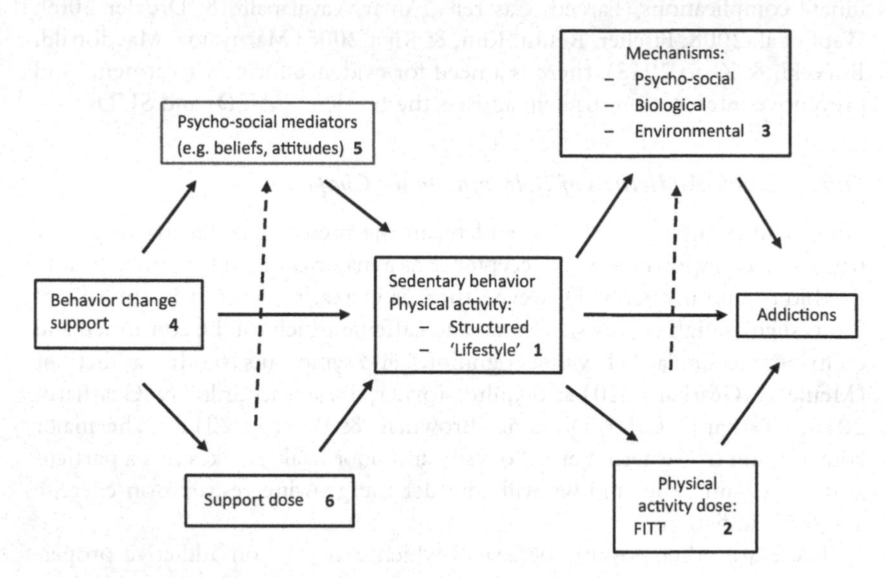

Figure 10.1 A framework for understanding physical activity and substance and behavioral addictions relationships

The framework encompasses both efficacy and effectiveness questions. When physical activity behavior is carefully monitored and recorded in humans and animals, what effects do different amounts have (in terms of frequency, intensity, duration, type, and timing) on addictive behaviors, and what do we know about how it works on addictive processes under these controlled situations. Knowledge about some of these questions can help to inform what and how physical activity should or could be promoted: in other words, what do we know about the translational science in the context of physical activity and addiction prevention, treatment, relapse prevention and for harm reduction?

Exercise and Addictions: What is the Evidence?

Physical Activity, Smoking, and Nicotine Use

Kaczynski and colleagues (Kaczynski, Manske, Mannell, & Grewal 2008) reviewed 50 cross-sectional studies and reported that physically active people were less likely to smoke (and smokers were less likely to be active). Prospectively, some studies (Rodriguez, Dunton, Tscherne, & Sass 2008) have reported that being substantially involved in sport and exercise somehow protects adolescents from becoming smokers. Of course these associations do not imply causality, and simply becoming physically active does not necessarily reduce smoking. But there has been a considerable growth in the number of studies over the past 15 years which have examined the effects of exercise on smoking outcomes, which can be classified as those targeting smokers who wish to reduce but not quit and those involving those who do want to quit.

Interventions targeted at harm reduction have only recently become more common, because successes in reduction can be an important step towards quitting, and there has been a recent global interest in harm reduction using e-cigarettes and controlled nicotine products. In a pilot randomized controlled trial (RCT) (Taylor et al. 2014), Thompson and colleagues (Thompson et al. 2015) reported that a behavioral support intervention called Exercise Assisted Reduction then Stop (EARS), targeted at disadvantaged smokers who did not want to quit but did want to reduce smoking, led to 36% in the intervention vs. 10% in the usual care control who made a quit attempt. Further findings were 23% (intervention) vs. 6% (control) had expired air carbon monoxide (CO) confirmed abstinence 4–8 weeks post quit, at 16 weeks 10% (intervention) vs. 4% (control) achieved point prevalence abstinence (i.e. a specified abstinence for a relatively brief period), and 63% (intervention) vs. 32% (control) achieved at least a 50% reduction in the number of cigarettes smoked daily. The study was not powered to detect significant differences but these findings provide good evidence of the potential of such a physical activity intervention. Given the intervention also provides pragmatic support for smoking reduction and increasing physical activity, further research is needed to understand how much the change in smoking was due to increasing physical activity. The study did provide a case study of how exercise had interacted with changes in smoking (Taylor et al. 2014).

In the latest Cochrane Systematic Review of the effects of exercise on smoking cessation at least 6 months post baseline, 20 trials were identified with a total of 5,870 participants (Ussher, Taylor, & Faulkner 2014). The studies ranged from one with 2,318 participants to eight trials with fewer than 30 people in each treatment arm. There was considerable heterogeneity in study design, type of intervention and control group, and methods used to assess the main smoking outcomes, and physical activity, and overall only one study was judged to have low risk of bias across the domains assessed. Significantly higher abstinence rates in a physically active group versus a control group at the end of treatment were shown in four studies, with only one showing a benefit from exercise at subsequent 3 and (borderline significance) 12 month assessments (Marcus et al. 1999): 11.9% v 5.4% were still abstinent. Another study reported significantly higher abstinence rates at 6-month follow-up for a combined exercise and smoking cessation programme compared with brief smoking cessation advice (Horn et al. 2011). One study showed significantly higher abstinence rates for the exercise group versus a control group at the 3-month follow-up but not at the end of treatment or 12-month follow-up (Marcus et al. 2005). The other studies showed no significant effect for exercise on abstinence, though greater physical activity intervention adherence appeared to strengthen any effects.

Two further notable trials since the Cochrane Review have been published: Ussher and colleagues (Ussher et al. 2015) reported no effects of a physical activity intervention on smoking cessation among 785 pregnant smokers despite increases in physical activity. Smits and colleagues (Smits et al. 2015) reported a significant effect of a physical activity intervention (vs. wellness education control), when augmenting standard smoking cessation support, at 4 and 6 months post quit date, among smokers with high anxiety sensitivity.

In summary, there is some limited evidence for a beneficial effect of exercise on long-term smoking abstinence, but further research is needed. With reference to Figure 10.1, the chronic intervention studies included both efficacy and effectiveness studies. For example, the studies by Marcus (1999) and Smits (2015) had strong control over what was delivered in the physical activity intervention and adherence was excellent, whereas studies by Ussher and colleagues (2003; 2015) involved a physical activity counselling intervention which could easily be implemented into treatment services, but involved rather mixed participant engagement in the intervention. So while there is some encouraging efficacy evidence, we need to know more about how a physical activity intervention could be implemented to ensure it is feasible and acceptable (effective) in the contexts where support is provided. A wide range of additional literature not included in the Cochrane Review has provided valuable information on how best to support adherence to, and maintenance of physical activity in the short and long term, to increase smoking abstinence.

Systematic reviews (Ussher et al. 2014; Ledochowski et al. 2013) and meta-analysis (Haasova et al. 2013) on the acute effects of a single session of PA on urges to smoke have shown consistent positive effects; only one or two of over 30 studies have shown no beneficial effect. A few studies have reported the effects during an actual attempt to quit, but it is difficult to standardize at what initial level of craving they begin with. Studies have typically therefore required participants to refrain from smoking for 2–24 hours to elevate urges to smoke. A variety of doses of PA are effective, including seated MP3-guided isometric exercises, yoga and aerobic activity (from 5–30 minutes of brisk walking or running). Vigorous PA is no more effective than moderate intensity exercise, with both reducing cravings by about 30%, while light exercise reduces them by about 10% compared with a passive control condition (Haasova et al. 2014).

Other smoking-related outcomes have also been reported. Exercise can attenuate cue-elicited cravings (when faced with a lit cigarette or smoking-related images) (Taylor & Katomeri 2007); reduce attentional bias to smoking images (assessed using eye-tracking technology or the dot probe task) (Van Rensburg, Taylor, & Hodgson 2009); and alter regional brain activation (measured with fMRI technology) in response to smoking images (Van Rensburg, Taylor, Hodgson, & Benattayallah 2009; Van Rensburg, Taylor, Benattayallah, & Hodgson 2012). A number of studies (e.g. Taylor & Katomeri 2007; Faulkner, Arbour-Nicitopoulos, & Hsin 2010) have also shown that exercise delays smoking and reduces time spent, and number of puffs, smoking the next cigarette. In summary, in acute experimental studies there is strong evidence that a single session of exercise can help a smoker avoid smoking in the short term, compared with being passive. This finding can be translated with confidence into interventions to help smokers who wish to cut down or quit.

Physical Activity, Alcohol Use, and Alcohol Use Disorders

Compared to the study of the effects of physical activity for smoking reduction and cessation, the role of physical activity in targeting alcohol use is, at present, relatively under researched. Whilst smoking is universally accepted to have negative health consequences at any level, alcohol consumption often comes with recommended safe consumption guidelines (e.g. Nutt & Rehm 2014). There is uncertainty over what constitutes "safe" alcohol consumption illustrated by the variation in guidelines from country to country, a mixture of classification systems (alcohol units vs grams of alcohol), what constitutes a "unit," and differences in recommended daily or weekly consumption patterns (Furtwaengler & de Visser 2013; International Centre for Alcohol Policies 2003). There is also a lack of common rubric and nomenclature when classifying drinking levels, and researchers will often arbitrarily treat data as ordinal, and often exclude participants with alcohol use disorders (AUDs) and dependence who may differ from simply being a "heavy drinker." For these reasons, the landscape of evidence for the relationship between physical activity and alcohol consumption and AUDs is complex, uncertain, and often not as expected. Similarly, the mechanisms for how physical activity may be related to alcohol behavior can work in very different ways and hence lead to very different relationships.

In the context of this chapter, we will define "alcohol use" as being the amount of alcohol consumed within a certain timeframe (usually on a daily basis), and AUDs as disorders which describe "a problematic pattern of using alcohol [...] that results in impairment in daily life or noticeable distress" (American Psychiatric Association 2013) usually accompanied by clinical diagnoses and treatment.

While more physically active individuals are less likely to smoke (Kaczynski et al. 2008), due to typically observed clustering of other poor health behaviors, one might expect levels of alcohol to be also negatively associated with physical activity (Coups, Gaba, & Orleans 2004; Fine, Philogene, Gramling, Coups, & Sinha 2004; McAloney, Graham, Law, & Platt 2013; Poortinga 2007; Pronk et al. 2004). However, this is predominantly shown not to be the case (Noble, Paul, Turon, & Oldmeadow 2015), and the relationship between alcohol use and physical activity appears, to some extent, to be influenced by the type of physical activity. In a recent review of 17 longitudinal studies of the relationship between sports participation and alcohol use in adolescents and young adults, over periods from 6 months to 16 years, Kwan and colleagues (2014) revealed that sports participation was associated with greater alcohol use in 14 studies, and no clear relationship in three studies. One study (Wichstrøm

& Wichstrøm 2009) went further and revealed those participating in team or technical sports consumed more alcohol than those participating in individual or endurance sports. So it appears that while some have suggested for some time that an active lifestyle may reduce risk-taking behaviors (Shephard 1989), some sports involve processes that could foster a culture of alcohol use, or alternatively attract young people who are more prone to risk taking across a range of health behaviors.

In terms of physical activity more generally (rather than sports participation) the relationship to alcohol consumption reveals contrasting findings. One rigorous study in Finland (Korhonen, Kujala, Rose, & Kaprio 2009) followed 1,870 twin pairs from the age of 16 to 27 and concluded that low physical activity levels increased the risk of weekly alcohol intoxication and problems due to alcohol use (although not related to weekly alcohol consumption). Other studies have also shown that drinking at hazardous levels was associated with lower physical activity levels (Liangpunsakul, Crabb, & Qi 2010; Berrigan, Dodd, Troiano, Krebs-Smith, & Barbash 2003), but the way that "hazardous" drinking was defined may have influenced the comparability with other studies which reported different relationships. For example, French and colleagues (2009) when analysing data from a large US population survey (n = 230,856) found that alcohol use and PA were positively associated even at heavy drinking levels with heavy drinkers completing 10 minutes more PA a week than moderate drinkers, and 20 minutes more per week than abstainers: PA intensity was not related to alcohol. However, the researchers did not make reference to "hazardous" or dependent drinking and classified people into one of four categories (abstainer, light, moderate, or heavy) for alcohol consumption. In contrast, Berrigan and colleagues (2003) only had two categories of alcohol consumption: those who drank under the recommended guidelines of the country the research took place, and those drinking above it were classified as hazardous, which makes interpretation of the findings challenging.

In a study which looked at PA and AUDs, physical activity has been shown to be positively associated with less severe forms of alcohol use disorders (alcohol use) but not with more severe forms (dependence) (Lisha, Sussman, Fapa, & Leventhal 2013), suggesting an inverted "J" relationship between alcohol consumption and physical activity levels. This type of relationship has been shown in other population research in the United States (Smothers & Bertolucci 2001) based on a sample of 41,104 participants. They reported the odds ratios of having a physically active lifestyle increased from alcohol abstinence (1.00) to moderate drinking (1.84) then declined at heavier consumption (1.61). Interestingly they also report this relationship was evident among current smokers, the elderly, and those with cardiovascular risk conditions. They also examined the impact of PA intensity, and as the intensity increased the "J" shape peak became even more pronounced for moderate drinkers.

In summary, it would seem that an inverted "J" shape relationship is the most plausible and likely to explain the cross-sectional relationship between

PA and alcohol consumption: moderate alcohol consumers are likely to be the most physically active, and abstainers and heavy alcohol consumers and those with an AUD being less active. But clearly, research on the association between sport, exercise, physical activity and alcohol use is fraught with methodological issues and without appropriate statistical control and care with operationalising independent and dependent variables could lead to spurious findings and conclusions.

In terms of causal effects, recent reviews examining the role of physical activity and exercise in the treatment of AUDs highlight a lack of evidence. A recent review completed by Wang et al. (2014) only included three studies on physical activity and alcohol abuse in their meta analyses, but this was due to an incomplete search strategy omitting key databases. Their main analysis incorporated alcohol, other substances, and nicotine into one analysis which is problematic due to the inherent differences between the addiction profiles and treatments for different substances. With only three studies no meaningful analyses on PA and alcohol were presented.

A more comprehensive review (Zschucke, Heinz, & Ströhle 2012) revealed nine studies that examined the effects of exercise programs on abstinence, relapse rates, and/or different associated somatic, emotional, and psychological outcomes. In the reported studies treatment durations ranging from 4 weeks (Gary & Guthrie 1972; Palmer, Vacc, & Epstein 1988) to 4 months (Weber 1984) with training frequencies ranging from three to five times per week and all but one (Ermalinski, Hanson, Lubin, Thornby, & Nahormek 1997) being aerobic exercise based (Gary & Guthrie 1972; Frankel & Murphy 1974; Sinyor, Brown, Rostant, & Seraganian 1982; Weber 1984; Murphy, Pagano, & Marlatt 1986; Palmer et al. 1988; Donaghy 1997; Brown et al. 2009). Six studies reported drinking episodes, craving, or days of abstinence as alcohol-related outcomes, four of which reported significantly stronger improvements in the exercise group, with the other two reporting no differences. The evidence presented in this review should be viewed with caution, as only one of the included studies (a Ph.D., thesis) fulfilled the criteria for an RCT (Donaghy 1997), and the rest were methodologically flawed, especially in terms of inadequate statistical power, and with high risk of bias for interpreting the findings.

A more methodologically robust review of clinical exercise interventions in AUDs (Giesen, Deimel, & Bloch 2015) included 14 studies. Overall, they reported inconsistent effects but a trend towards positive effects of exercise on anxiety, mood management, craving, and drinking behavior. They rated the overall level of evidence as moderate. The review also provided some evidence for the feasibility and acceptability of exercise interventions as a complementary part of the therapy for AUDs. In summary the authors recommended that the results should be interpreted cautiously due to the numerous methodological flaws and heterogeneity of the interventions and measures, and called for further rigorous research.

Finally, we are aware of just two studies involving humans that have investigated the acute effects of exercise on alcohol-related outcomes such as cravings. Ussher and colleagues (Ussher, Sampuran, Doshi, West, & Drummond 2004) reported that alcohol urges were significantly reduced during, but not after, 10 minutes of moderate intensity exercise, compared with light intensity exercise, among participants undergoing alcohol detoxification at a psychiatric hospital. Taylor and colleagues (2013) reported that regular alcohol drinkers, who were temporarily abstinent, had reduced urges to drink and attentional bias to alcohol (vs. neutral images) after a 15 min brisk walk, compared with a passive control condition. These findings match those reported above for smoking-related outcomes and also suggest that exercise may acutely help to manage an urge to drink and interest in alcohol though further research is needed to understand the mechanisms, dose, and potential for extending the finding into natural settings as part of a chronic intervention.

Physical Activity and Other Substance Use Disorders

The DSM-5 (American Psychiatric Association 2013) refers to use of opioids, sedatives/hypnotics/anxiolytics, cocaine, cannabis, hallucinogens, inhalants, amphetamines, phencyclidine, and polysubstance use which contribute to substance use disorders (SUDs) (in addition to tobacco, alcohol, and caffeine use). This section will focus mainly on the use of cannabis, cocaine, and opiates (alone or in combination) as this is where the exercise literature has mainly been focused. Because these substances are illegal in most countries, the research focus has been on the role of exercise in prevention and abstinence (rather than reduction).

Kwan and colleagues (2014) reviewed longitudinal studies of sports participation and illicit drug use among adolescents and young adults. The review included 11 studies that assessed the use of illicit substances but identified a lack of common standardized measures which limited the scope to produce overall findings. Eight studies examined cannabis use, with four reporting negative associations between sports participation (Barber, Eccles, & Stone 2001; Darling 2005; Dawkins, Williams, & Guilbault 2005; Terry-McElrath & O'Malley 2011) and four no association (Aaron et al. 1995; Eccles & Barber 1999; Mahoney & Vest 2012; Wichstrøm & Wichstrøm 2009). Five other studies examined the use of other illicit drugs, four of which report a negative association (Barber, Eccles, & Stone 2001; Darling 2005; Eitle, Turner, & Eitle 2003; Terry-McElrath & O'Malley 2011) and one which found no significant relationship between sports participation and illicit drug use (Eccles & Barber 1999). In summary, they concluded that with 7 of 11 studies finding negative associations between sports participation and overall illicit drug use, sports participation may protect against drug use, with no differences between short- and long-term studies. Similar results were also reported in an earlier review of high school and college sports participation and illicit drug use, where nine of 16 studies reported an inverse relationship between sports participation and substance use (Lisha & Sussman 2010). They also report some studies which revealed a moderating effect of sport and gender (Ford 2007; Peretti-Watel, Beck, & Legleye 2002; Ewing 1998). Both reviews reported difficulty in accurately summarizing the findings of studies due to the lack of common outcome measures, and a tendency to analyze all varieties of drug use in one analysis.

In terms of physical activity more broadly, a cross-sectional study of 2,458 adolescents and young adults found that SUDs were less prevalent in those who were regularly physically active (Ströhle et al. 2007). Prospectively, Korhonen and colleagues (2009) followed 1,870 twin pairs for the age 16 to 27 years, and showed that drug use in adulthood was common among those who were persistently physically inactive in adolescence (OR=3.7, p=<0.001) in comparison to those who were persistently active. They also report that a sedentary lifestyle predicted illicit drug use even when controlling for familial factors.

In terms of causal effects, recent reviews examining the role of physical activity and exercise in the treatment of SUDs highlight a lack of evidence (Abrantes, Matsko, Wolfe, & Brown 2013). A review (Zschucke et al. 2012) identified eight studies, none of which were RCTs. The studies examined the effects of exercise-based interventions on recovery trajectories, physical fitness, and psychological variables. The studies included treatment durations ranging from 2 weeks (Buchowski et al. 2011) to 6 months (Roessler 2010), with training frequencies from several times a day (Li, Chen, & Mo 2002) to twice a week (Williams 2000). Six studies reported substance-related outcomes (e.g. craving, percentage of days abstinent, continuous days abstinent) (Collingwood, Reynolds, Kohl, Smith, & Sloan 1991; Burling, Seidner, Robbins-Sisco, Krinsky, & Hanser 1992; Li et al. 2002; Roessler 2010; Brown et al. 2010; Buchowski et al. 2011) which improved with exercise treatment in all studies. Within this review the authors acknowledge that all studies suffered severe methodological limitations which make firm conclusions difficult. Only two studies included a control group (Burling et al. 1992; Li et al. 2002) and two studies performed post hoc classifications between improvers and non-improvers (Collingwood et al. 1991; Williams 2000), and four studies had no control conditions at all (Palmer, Palmer, Michiels, & Thigpen 1995; Roessler 2010; Brown et al. 2010; Buchowski et al. 2011). Sample sizes were very small, with 50 or fewer participants in five of the studies (Palmer et al. 1995; Williams 2000; Roessler 2010; Brown et al. 2010; Buchowski et al. 2011), and one study contained unequal group sizes (Burling et al. 1992). Most studies conducted no intention to treat analyses to account for the high number of dropouts. Two studies included culture-specific interventions which were also not pure exercise interventions (Burling et al. 1992; Li, Chen, & Mo 2002) which greatly limits generalizability. In three studies group differences related to specific outcomes were already present at baseline, which partly explained group differences at the end of treatment (Collingwood et al. 1991; Palmer et al. 1995; Li et al.

2002). Finally, three studies did not chemically verify self-reported abstinence (Collingwood et al. 1991; Roessler 2010; Brown et al. 2010).

A more recent review (Wang et al. 2014) examining the impact of physical exercise on SUDs identified five studies which focussed solely on SUDs (in this case all heroin). The review included studies with participants who were diagnosed with a substance use disorder and only included experimental studies looking at chronic physical exercise. They report a positive effect of exercise on SUDs as part of a sub-group analysis including three of the five studies. Aside from the low number of studies included in the analyses, the studies included in the review present other reasons to cautiously interpret this result. Of the five studies, three did not conduct intention to treat analysis (Huang et al. 2000; Li et al. 2013; Zhuang, An, & Zhao 2013), three provided no details of blinding (Li et al. 2002; Li et al. 2013; Zhuang et al. 2013), and three provided no details of allocation concealment (Huang et al. 2000; Huang & Yang 2000; Li et al. 2002), suggesting those studies included were of low methodological quality and with a notable risk of bias. As with other reviews, the interventions were varied, from 10 days (Huang et al. 2000; Li et al. 2002) to 6 months (Huang & Yang 2000; Zhuang et al. 2013) with various frequency and intensity. Three studies included physical activity which could be widely generalizable in the form of brisk walking (Huang & Yang 2000), jogging (Huang et al. 2000), and yoga (Zhuang et al. 2013), and the remaining two involved activity heavily culture specific to Eastern culture (Li et al. 2002; Li et al. 2013). The literature has provided some evidence for the feasibility and acceptability of exercise interventions as a complementary part of the therapy for SUDs (Abrantes et al. 2011), setting the stage for more substantive research.

In summary the authors of reviews recommended that the risk of bias is high and that no clear effects of exercise interventions for SUDs have been demonstrated. Studies have included numerous methodological flaws and the heterogeneity of the interventions and measures clearly point to the need for further rigorous research. While any effects of exercise on addiction outcomes and processes have yet to be shown, there may be benefits for general health improvement during recovery (Smith & Lynch 2011; Zschucke et al. 2012). We are not aware of any studies involving human participants to investigate the acute effects of exercise on cravings and other outcomes associated with substance misuse.

Physical Activity and Uncontrolled Snacking

There is a very large literature on the role of exercise for weight management that focuses largely on the value of energy expenditure, and to a lesser extent on if and how exercise regulates appetite and hunger (e.g. via psychobiological compensatory mechanisms such as the regulation of hormones related to appetite and hunger such as grehlin and peptide YY (Li, Asakawa, Li, Cheng, & Inui 2011)). We are not, however, aware of any chronic studies that have explicitly developed an intervention in which physical activity (or reduced sedentary time) has been targeted at regular snackers who find it difficult to self-regulate behavior in a similar way to those with other substance use disorders. There is strong evidence though that cravings for specific foods (e.g. high-energy snack foods which may be associated with an energy boost or a sense of pleasure) are more common among regular snackers, and are influenced by a range of factors such as presence of salient cues, and emotional states such as stress and boredom, independent of hunger and appetite. But our interest here is in whether exercise (or interruptions in sedentary behavior) can reduce the urge to snack and uncontrolled snacking.

We are aware of six published studies that have examined the acute effects of exercise on food-related outcomes. An early study by Thayer and colleagues (1993) supported the idea that exercise could reduce snacking, and mood as a possible mediator of eating high-energy foods. The study was not well controlled though; participants were not regular snackers nor were they asked to abstain from snacking. Five studies have shown reduced snack food cravings following exercise, compared with a passive control condition (Ledochowski, Ruedl, Taylor, & Kopp 2015; Oh & Taylor 2012; Oh & Taylor 2013; Oh & Taylor 2014; Taylor & Oliver 2009) and attenuated cue-elicited cravings (Taylor & Oliver 2009) among abstinent snackers. Manipulated experimental conditions suggested the effects were also evident during simulated stress and among normal, overweight, and obese participants. Exercise also reduced chocolate consumption, and attentional bias to still chocolate images (Taylor, Oliver, & Van Rensburg 2009) and video clip images (Oh & Taylor 2014). The latter study involved smokers who were abstinent from snacking and smoking (given the propensity for smokers to comfort eat during abstinence and gain weight), and a comparison of moderate, vigorous, and passive conditions. Both moderate and vigorous intensity cycling reduced the desire to snack (and smoke), and attentional bias to snack food (and smoking).

In summary, there is consistent evidence that exercise can acutely reduce cravings, attentional bias and snack food consumption among regular snackers who are abstinent, especially in situations where snacking behavior may be most hard to self-regulate. Thus the literature appears to mimic that reported for cigarette smoking and to a lesser extent alcohol consumption.

Possible Mechanisms

Identifying one or more plausible mechanisms is a key criterion for demonstrating causality, and can also help us to understand if and how different doses (FITTT: frequency, intensity, time, type, and timing) impact on behavioral and other outcomes associated with an addiction, which may be important for identifying optimum guidance on using physical activity. Figure 10.1 identifies psycho-social, biological, and environmental factors as possible mediators, and highlights the importance of interdisciplinary research.

At a psycho-social level interest has been in cognitive processes associated with physical active (acutely and chronically) including the following: distraction (mindfulness); reduced cravings and urges to engage in the behavior during and in the absence of salient cues; reduced impulsivity/attentional bias; improved mood and affect; reduced anxiety sensitivity; changes in key positive beliefs and attitudes about the self, the behavior and others (e.g. self-esteem, a sense of competence to change behavior and overcome self-efficacy barriers, control and relatedness with others); an identity shift (e.g. from being "an addict" to "an exerciser"); and finally changes in or stable body-image (e.g. preventing weight gain normally associated with smoking cessation). Different mechanisms may mediate the relationship between reduced sedentary behavior and addictive behaviors, between structured exercise (e.g. 3 x 30 min of moderate-vigorous gym based activity), and between doing short regular short bouts of one or more activities daily.

Earlier in the chapter, reference was made to some literature on the positive acute effects of exercise on cravings and withdrawal symptoms. So the accumulation of these effects (from repeated bouts) may be important in breaking the cycle between the urge to smoking and reward and pleasure from the behavior, especially if it also breaks the link between environmental cues, cravings, and behavior. In the acute literature, there is no evidence that exercise mediates cravings simply by distraction (see Taylor & Ussher 2013 for a review) or that exercise increases positive affect and reduces negative affect which in turn reduces cravings (Haasova et al. 2014). The latter is the most rigorous attempt, to date, to quantitatively identify moderators and mediators using original individual data from many acute studies in a combined meta-analysis. Other interesting studies have also indirectly provided clues about mechanisms. For example, Tritter, Fitzgeorge, and Prapavessis (2015) reported that an acute bout of exercise provided additional craving relief to the nicotine lozenge in recently quit smokers. This suggests that exercise targets mechanisms other than those received from nicotine replacement therapy.

One notable study (Smits et al. 2015) revealed that a chronic exercise intervention, compared with a health promotion intervention, increased 6-month smoking abstinence but only for those with high anxiety sensitivity. While these moderator effects do not demonstrate a causal mechanism, the authors recommend further research to determine the mechanisms underlying the effects of exercise on smoking cessation among those with high anxiety sensitivity.

The range of possible psycho-social mechanisms (which may act for different people in different ways) may do little to help us understand causality but in designing complex interventions, in the form of counselling support for individuals, for example, helping individuals to be more aware of the implicit and explicit changes that occur as a result of physical activity may be very important. For example, client-centred approaches based on Self-Determination Theory (Deci & Ryan 1985) and/or motivational interviewing (Miller & Rollnick 2012) may allow a client to recognize how exercise helps to overcome or replace triggers that would otherwise have led to addictive behaviors. Qualitatively, a case study reported in our EARS study (Taylor et al. 2014) highlighted the complex psycho-social processes that can operate: changes in smoking influenced physical activity as well as exercise influenced decisions about smoking. The counsellor was simply there to help the client to understand these relationships and use the opportunity for positive behavioral changes that could be sustained because the client "owned" the change, rather than being given any prescriptive dose of exercise.

A few studies have examined the acceptability and feasibility of physical activity interventions and the associated mechanisms between changes in physical activity and addictive behaviors other than smoking, and intuitively some similar processes would seem plausible (Neale, Nettleton, & Pickering 2012; Beynon et al. 2013; Abrantes et al. 2011). Non-academic literature has widely reported how sport (e.g. boxing) has "saved" individuals from a life with AUDs and SUDs, but also there are notorious cases where "sport" has socialized individuals into AUDs and SUDs. Given this possibly quite diverse range of responses to "sport" it may be hard to believe that exercise per se can have a neuro-protective or biological effect which mediates addictive behavior, leading to the possibility that the psycho social impact of sport negates any neuro-protective effect of exercise.

Biological Mechanism

There has been an exponential increase in interest in the biological changes associated with exercise which could help lead to pharmacological therapies to manage addictive behaviors. These have largely involved animal studies which will be considered in a moment but first a few human studies have considered neuro-biological processes.

Substantial evidence suggests that specific areas of the brain are associated with how we process information: in particular, cues associated with an addictive behavior elicit greater blood flow from functional magnetic resonance imagery (fMRI) for whom the images are salient. Two studies have examined whether exercise acutely changes blood flow to regional areas of interest while viewing smoking related images, concurrent with self-reported reductions in cravings or urges to engage in a behavior. Van Rensburg (2009) reported that after rest, viewing smoking images was associated with significant activation in areas associated with reward (caudate nucleus), motivation (orbitofrontal cortex), and visuo-spatial attention (parietal lobe, parahippocampal and fusiform gyrus), whereas after exercise there was less activation in these areas and a concomitant shift of activation towards areas identified in the "brain default mode" (Broadmanns Area, BA, 10). In a second study (Van Rensburg et al.

2012) with smokers also experiencing withdrawal, significant areas of activation were found in areas of the limbic lobe and in areas associated with visual attention in response to smoking-related stimuli after both exercise and rest. Smokers showed increased activation to smoking images in areas associated with primary and secondary visual processing following rest, but not following a session of exercise.

In food-related research, Cornier and colleagues (2012) reported that after a 6-month exercise intervention the presentation of visual stimuli of foods of high hedonic value, as compared to neutral images, there was a reduction in the neuronal response to food, primarily in the posterior attention network and insula. Another study (Crabtree, Chambers, Hardwick, & Blannin 2014) examined the acute effects of high-intensity exercise on fMRI response to low- and high-calorific food images. They reported changes in neural responses to images of high-calorie foods with significantly increased dorsolateral prefrontal cortex activation and suppressed orbitofrontal cortex (OFC) and hippocampus activation after EX compared with rest. Both studies do not provide clear evidence that addictive processes, as observed through fMRI, are changed by exercise, since participants were not recruited on the basis of being regular (possibly compulsive or addicted) snackers, but they are selected here because they provide examples to open further research.

Given the co-existence of mental health problems with addictions, there has been growing interest in how the possible mechanism by which exercise acutely and chronically impacts on mental health could also play an important role in helping in the treatment of individuals with an addiction. Also, pharmacological therapies licensed for reducing anxiety or depression have been shown to have utility in treating addictions. There is a more detailed analysis of the literature on the biological pathways by which exercise is thought to enhance mental health elsewhere in this book.

Animal studies have provided strong evidence for the beneficial effects of the running wheel on substance use. Self-administered cocaine use was reduced in studies concerned with naive acquisition (Smith & Pitts 2011), maintenance (Smith, Schmidt, Iordanou, & Mustroph 2008), escalation (Smith, Walker, Cole, & Lang 2011), and reinstatement after abstinence (Smith, Pennock, Walker, & Lang 2012). Reduced use of alcohol (Ehringer, Hoft, & Zunhammer 2009), morphine (Hosseini, Alaei, Naderi, Sharifi, & Zahed 2009), and methamphetamine (Miller et al. 2012) have also been reported.

The review by Smith and Lynch (2013) provides an excellent summary of the possible mediating effects of dopamine, glutamate, norepinephrine, opioids, PKA, extracellular signal-regulated kinase, brain-derived neurotrophic factor, and other molecules, in the effects of exercise on prevention and treatment of substance use. But no clear single contender is currently evident, and it may be that genes interact with the effects of exercise on molecules such as brain-derived neurotrophic factor (BDNF) so that only those with a specific genetic polymorphism show reduced uptake and reduction in substance use in response to exercise.

Additional animal research conducted in the 1970s and 1980s by Bruce K. Alexander examined the effects of housing on self-administered morphine within rats (Alexander, Coambs, & Hadaway 1978; Alexander, Beyerstein, Hadaway, & Coambs 1981). It aimed to investigate the effect of environment on addiction to investigate the disease model of addiction. It found that rats kept in isolation would consume more morphine than water compared to rats housed in enriched colony environments (with space, other rats, exercise wheels) where it was thought consumption of morphine interfered with species-specific behavior. Even when morphine-addicted rats were moved from isolation to enriched environments, there was a tendency for the rats to wean themselves off the morphine solution. This points towards the powerful influence of environment in mediating addiction, and the important role it plays in addiction (Figure 10.1).

Future Research Questions

Gaps in the research have been identified throughout this chapter. Figure 10.1 provides a framework for a whole range of pre-clinical to pragmatic effectiveness research. The neural hormonal animal research is exciting in that it provides an opportunity to identify objective evidence that exercise does have causal effects on uptake and use of addictive substances. Future research may well identify if these effects are true for specific phenotypes, and what dose of exercise is most effective, which can be tested in human studies. In other studies to test psychosocial mechanisms 10 years or more of studies on the acute effects of exercise on nicotine addiction have both confirmed a moderate to strong positive effect but also helped to understand some mechanisms and dose-response questions. Further research is needed to understand which smokers most benefit from exercise and why. Only two acute exercise and alcohol studies have been conducted and further research is needed. Recognition that some eating behaviors (e.g. snacking irrespective of satiety) may have similarities to addictions opens the opportunity to conduct further studies and how exercise influences eating related processes and what is the effect of dose.

There is a need to learn much more about how to translate this growing research base into acceptable and feasible interventions. The delivery of exercise programmes or the promotion of sustainable physical activity through client-centered support both offer opportunities to augment existing therapies but also as standalone interventions targeted at vulnerable young people and for relapse prevention. It will be important to carefully describe the components of any intervention to ensure replication and also develop a clearer understanding of which components are most valued and effective within comprehensive process evaluations, and where possible through mediation analysis and qualitative investigation.

Finally, when developing ideas for a study it is important to understand the context in which an intervention may be implemented. The alcohol research is difficult to synthesize because of the heterogeneity in context (where, when, and who), intervention design, content and quality of reporting, and study methods (including outcome measures).

Practical Implications

Treatments for smoking cessation, and alcohol and substance use are dominated by pharmacological, and to a lesser extent psychological evidence-based interventions. Practitioners are well trained to understand how to administer treatments and their effects and possible side effects. It may take many years to build up similar evidence bases for the effects of exercise interventions. The extent to which the findings from pre-clinical animal studies can be translated into a human prescription is unclear, powerful as they are. In our early versions of our exercise and smoking cessation review, the Cochrane panel did not see the relevance of acute exercise and smoking studies (e.g. assessing cravings), but now they are included and provide important practical value. We know that even short moderate intensity bouts of exercise reduce cravings and are likely to be easily adopted as lifestyle physical activity. But it is important to note that the effects are only evident when cravings have been elevated through experimental manipulation. In other words, the value of exercise for self-management of smoking is unlikely to become learned if someone only ever exercises after smoking, when cravings are all but non-existent.

The focus of exercise research has largely been to identify support for it as a treatment for smoking cessation, and alcohol and substance use disorders. There is increasing interest in the role of physical activity for prevention, harm reduction and relapse prevention. Changing multiple health behaviors (exercise and substance use) may not be easy and can reduce success in all targeted behaviors. The timing of when to engage clients in physical activity may be critical and would most likely be sustainable when the person is involved in this decision and a range of other choices about frequency, type and duration, as well support provided by others. Practitioners can be trained to use motivational interviewing and client-centered approaches to support increases in PA as an aid for managing an addictive behavior.

References

Aaron, D. J., S. R. Dearwater, R. Anderson, T. Olsen, A. M. Kriska, & R. E. Laporte. 1995. "Physical activity and the initiation of high-risk health behaviors in adolescents." *Medicine and Science in Sports and Exercise 27* (12): 1639–45. www.ncbi.nlm.nih.gov/pubmed/8614320.

Abrantes, A. M., C. L. Battle, D. R Strong, E. Ing, M. E. Dubreuil, A. Gordon, & R. A. Brown. 2011. "Exercise preferences of patients in substance abuse treatment." *Mental Health and Physical Activity 4* (2): 79–87. doi:10.1016/j.mhpa.2011.08.002

Abrantes, A. M., S. Matsko, J. Wolfe, & R. A. Brown. 2013. "Physical activity and alcohol and drug use disorders." In P. Ekkekkakis (Ed.), *Routledge handbook of physical activity and mental health* (pp. 465–77). New York: Routledge.

Alexander, B. K., B. L. Beyerstein, P. F. Hadaway, & R. B. Coambs. 1981. "Effect of early and later colony housing on oral ingestion of morphine in rats." *Pharmacology, Biochemistry, and Behavior 15* (4): 571–6. www.ncbi.nlm.nih.gov/pubmed/7291261.

Alexander, B. K., R. B. Coambs, & P. F. Hadaway. 1978. "The effect of housing and gender on morphine self-administration in rats." *Psychopharmacology 58* (2): 175–9. doi:10.1007/BF00426903

American Psychiatric Association. 2013. *Diagnostic and Statistical Manual of Mental Disorders: DSM-5*. Washington DC: American Psychiatric Association.

Angelo, D. L., H. Tavares, & M. L. Zilberman. 2013. "Evaluation of a physical activity program for pathological gamblers in treatment." *Journal of Gambling Studies / Co-Sponsored by the National Council on Problem Gambling and Institute for the Study of Gambling and Commercial Gaming 29* (3): 589–99. doi:10.1007/s10899-012-9320-2

Barber, B. L., J. S. Eccles, & M. R. Stone. 2001. "Whatever happened to the Jock, the Brain, and the Princess? Young adult pathways linked to adolescent activity involvement and social identity." *Journal of Adolescent Research 16* (5): 429–55. doi:10.1177/0743558401165002

Berrigan, D., K. Dodd, R. P. Troiano, S. M. Krebs-Smith, & R. B. Barbash. 2003. "Patterns of health behavior in U.S. adults." *Preventive Medicine 36* (5): 615–23. www.ncbi.nlm.nih.gov/pubmed/12689807.

Beynon, C. M., A. Luxton, R. Whitaker, N. T. Cable, L. Frith, A. H. Taylor,... L. Zou. 2013. "Exercise referral for drug users aged 40 and over: Results of a pilot study in the UK." *BMJ Open 3* (5). doi:10.1136/bmjopen-2013-002619

Brown, R. A., A. M. Abrantes, J. P. Read, B. H. Marcus, J. Jakicic, D. R. Strong,...J. R. Oakley. 2009. "Aerobic exercise for alcohol recovery: Rationale, program description, and preliminary findings." *Behavior Modification 33* (2): 220–49. doi:10.1177/0145445508329112

Brown, R. A., A. M. Abrantes, J. P. Read, B. H. Marcus, J. Jakicic, D. R. Strong,...J. R. Oakley. 2010. "A pilot study of aerobic exercise as an adjunctive treatment for drug dependence." *Mental Health and Physical Activity 3* (1): 27–34. doi:10.1016/j.mhpa.2010.03.001

Brownell, K. D., G. A. Marlatt, E. Lichtenstein, & G. T. Wilson. 1986. "Understanding and preventing relapse." *American Psychologist 41* (7): 765–82. www.ncbi.nlm.nih.gov/pubmed/3527003.

Buchowski, M. S., N. N. Meade, E. Charboneau, S. Park, M. S. Dietrich, R. L. Cowan, & P. R. Martin. 2011. "Aerobic exercise training reduces cannabis craving and use in non-treatment seeking cannabis-dependent adults." *PLoS ONE 6* (3): e17465. doi:10.1371/journal.pone.0017465

Burling, T. A., A. L. Seidner, D. Robbins-Sisco, A. Krinsky, & S. B. Hanser. 1992. "Batter up! Relapse prevention for homeless veteran substance abusers via softball team participation." *Journal of Substance Abuse 4* (4): 407–13. www.ncbi.nlm.nih.gov/pubmed/1338187.

Cahill, K., L. F. Stead, & T. Lancaster. 2011. "Nicotine receptor partial agonists for smoking cessation." *Cochrane Database Syst Rev 2*: CD006103. doi:10.1002/14651858.CD006103.pub5

The Centre for Social Justice. 2013. "No quick fix: Exposing the depth of Britain's drugs and alcohol problem." London.

Cherpitel, C. J., Y. Ye, T. K. Greenfield, J. Bond, W. C. Kerr, & L. T. Midanik. 2010. "Alcohol-related injury and driving while intoxicated: A risk function analysis of two alcohol-related events in the 2000 and 2005 National Alcohol Surveys." *American Journal of Drug and Alcohol Abuse 36* (3): 168–74. doi:10.3109/00952991003793851

Cochrane Drug and Alcohol Group. 2015. "Cochrane Drug and Alcohol Group – Our Reviews." http://cdag.cochrane.org/our-reviews.

Collingwood, T. R., R. Reynolds, H. W. Kohl, W. Smith, & S. Sloan. 1991. "Physical fitness effects on substance abuse risk factors and use patterns." *Journal of Drug Education 21* (1): 73–84. www.ncbi.nlm.nih.gov/pubmed/2016666.

Connors, G. J., S. A. Maisto, & D. M. Donovan. 1996. "Conceptualizations of relapse: A summary of psychological and psychobiological models." *Addiction 91* Suppl (December): S5–13. www.ncbi.nlm.nih.gov/pubmed/8997777.

Cornier, M.-A., E. L. Melanson, A. K. Salzberg, J. L. Bechtell, & J. R. Tregellas. 2012. "The effects of exercise on the neuronal response to food cues." *Physiology and Behavior 105* (4): 1028–34. doi:10.1016/j.physbeh.2011.11.023

Coups, E. J., A. Gaba, & C. T. Orleans. 2004. "Physician screening for multiple behavioral health risk factors." *American Journal of Preventive Medicine 27* (2 Suppl): 34–41. doi:10.1016/j.amepre.2004.04.021

Crabtree, D. R., E. S. Chambers, R. M. Hardwick, & A. K. Blannin. 2014. "The effects of high-intensity exercise on neural responses to images of food." *American Journal of Clinical Nutrition 99* (2).

Darling, N. 2005. "Participation in extracurricular activities and adolescent adjustment: Cross-sectional and longitudinal findings." *Journal of Youth and Adolescence 34* (5): 493–505. doi:10.1007/s10964-005-7266-8

Dawkins, M. P., M. M. Williams, & M. Guilbault. 2005. "Participation in school sports: Risk or protective factor for drug use among black and white students?" *Journal of Negro Education 75* (1).

Deci, E. L., & R. M. Ryan. 1985. *Intrinsic motivation and self-determination in human behavior*. New York: Plenum Press.

Donaghy, M. E. 1997. "The investigation of exercise as an adjunct to the treatment and rehabilitation of the problem drinker." http://theses.gla.ac.uk/3250/1/1997donaghyphd.pdf.

Eccles, J. S., & B. L. Barber. 1999. "Student council, volunteering, basketball, or marching band: What kind of extracurricular involvement matters?" *Journal of Adolescent Research 14* (1): 10–43. doi:10.1177/0743558499141003

Ehringer, M. A., N. R. Hoft, & M. Zunhammer. 2009. "Reduced alcohol consumption in mice with access to a running wheel." *Alcohol 43* (6): 443–52. doi:10.1016/j.alcohol.2009.06.003

Eitle, D., R. J. Turner, & T. M. Eitle. 2003. "The deterrence hypothesis reexamined: Sports participation and substance use among young adults." *Journal of Drug Issues 33* (1): 193–221. doi:10.1177/002204260303300108

Ermalinski, R., P. G. Hanson, B. Lubin, J. I. Thornby, & P. A. Nahormek. 1997. "Impact of a body-mind treatment component on alcoholic inpatients." *Journal of Psychosocial Nursing and Mental Health Services 35* (7): 39–45. www.ncbi.nlm.nih.gov/pubmed/9243422.

Ewing, B. T. 1998. "High school athletes and marijuana use." *Journal of Drug Education 28* (2): 147–57. www.ncbi.nlm.nih.gov/pubmed/9673074.

Fareed, A., J. Casarella, R. Amar, S. Vayalapalli, & K. Drexler. 2009. "Benefits of retention in methadone maintenance and chronic medical conditions as risk factors for premature death among older heroin addicts." *Journal of Psychiatric Practice 15* (3): 227–34. doi:10.1097/01.pra.0000351884.83377.e2

Faulkner, G., K. P. Arbour-Nicitopoulos, & A. Hsin. 2010. "Cutting down one puff at a time: The acute effects of exercise on smoking behavior." *Journal of Smoking Cessation 5*: 130–5.

Fine, L. J., G. S. Philogene, R. Gramling, E. J. Coups, & S. Sinha. 2004. "Prevalence of multiple chronic disease risk factors: 2001 National Health Interview Survey." *American Journal of Preventive Medicine 27* (2 Suppl): 18–24. doi:10.1016/j.amepre.2004.04.017

Fischer, B., J. Rehm, G. Kim, & M. Kirst. 2005. "Eyes wide shut? – A conceptual and empirical critique of methadone maintenance treatment." *European Addiction Research 11* (1): 1–9; discussion 10–14. doi:10.1159/000081410

Ford, J. A. 2007. "Substance use among college athletes: A comparison based on sport/team affiliation." *Journal of American College Health 55* (6): 367–73. doi:10.3200/JACH.55.6.367-373.

Frankel, A., & J. Murphy. 1974. "Physical fitness and personality in alcoholism: Canonical analysis of measures before and after treatment." *Quarterly Journal of Studies on Alcohol 35* (4 Pt A): 1272–8. www.ncbi.nlm.nih.gov/pubmed/4155516.

French, M. T., I. Popovici, & J. C. Maclean. 2009. "Do alcohol consumers exercise more? Findings from a national survey." *American Journal of Health Promotion 24* (1): 2–10. doi:10.4278/ajhp.0801104

Furtwaengler, N. A. F. F., & R. O. de Visser. 2013. "Lack of international consensus in low-risk drinking guidelines." *Drug and Alcohol Review 32* (1): 11–18. doi:10.1111/j.1465-3362.2012.00475.x

Gary, V., & D. Guthrie. 1972. "The effect of jogging on physical fitness and self concept in hospitalized alcoholics." *Quarterly Journal of Studies on Alcohol 33* (4): 1073–8. www.ncbi.nlm.nih.gov/pubmed/4648626.

Gearhardt, A. N., C. M. Grilo, R. J. DiLeone, K. D. Brownell, & M. N. Potenza. 2011. "Can food be addictive? Public health and policy implications." *Addiction 106* (7): 1208–12. doi:10.1111/j.1360-0443.2010.03301.x

Giesen, E. S., H. Deimel, & W. Bloch. 2015. "Clinical exercise interventions in alcohol use disorders: A systematic review." *Journal of Substance Abuse Treatment 52* (May): 1–9. doi:10.1016/j.jsat.2014.12.001

Haasova, M., F. C. Warren, M. Ussher, K. J. Van Rensburg, G. Faulkner, M. Cropley,...A. H. Taylor. 2013. "The acute effects of physical activity on cigarette cravings: Systematic review and meta-analysis with Individual Participant Data." *Addiction 108* (1): 26–37. doi:10.1111/j.1360-0443.2012.04034.x

Haasova, M., F. C. Warren, M. Ussher, K. J. Van Rensburg, G. Faulkner, M. Cropley,...A. H. Taylor. 2014. "The acute effects of physical activity on cigarette cravings: Exploration of potential moderators, mediators and physical activity attributes using Individual Participant Data (IPD) meta-analyses." *Psychopharmacology 231* (7): 1267–75. doi:10.1007/s00213-014-3450-4

Horn, K., G. Dino, S. A. Branstetter, J. Zhang, N Noerachmanto, T. Jarrett, & M. Taylor. 2011. "Effects of physical activity on teen smoking cessation." *Pediatrics 128* (4): e801–11. doi:10.1542/peds.2010–2599

Hosseini, M., H. A. Alaei, A. Naderi, M. R. Sharifi, & R. Zahed. 2009. "Treadmill exercise reduces self-administration of morphine in male rats." *Pathophysiology: The*

Official Journal of the International Society for Pathophysiology 16 (1): 3–7. doi:10.1016/j.pathophys.2008.11.001

Huang, H., & F. Yang. 2000. "Exercise therapy adjuvant treatment of heroin dependence on clinical observation." *Chinese Journal of Drug Abuse Prevention and Treatment 26*: 30–1.

Huang, H., F. Yang, S. S. Yang, D. S. Xiao, & A. H. Nie. 2000. "Influence of aerobic training on recovery of heroin addicts." *Chinese Journal of Physical Therapy 23*: 267–70.

Hughes, J. R., L. F. Stead, & T. Lancaster. 2007. "Antidepressants for smoking cessation." *Cochrane Database Syst Rev 1*: CD000031. doi:10.1002/14651858.CD000031.pub3

International Centre for Alcohol Policies. 2003. "International Drinking Guidelines." www.icap.org/LinkClick.aspx?fileticket=KtXj8PGibT8%3D&tabid=75.

Kaczynski, A. T., S. R. Manske, R. C. Mannell, & K. Grewal. 2008. "Smoking and physical activity: A systematic review." *Am J Health Behav 32* (1): 93–110. doi:10.5555/ajhb.2008.32.1.93

Kim, H. 2013. "Exercise rehabilitation for smartphone addiction." *Journal of Exercise Rehabilitation 9* (6): 500–5. doi:10.12965/jer.130080

Korhonen, T., U. M. Kujala, R. J. Rose, & J. Kaprio. 2009. "Physical activity in adolescence as a predictor of alcohol and illicit drug use in early adulthood: A longitudinal population-based twin study." *Twin Research and Human Genetics: The Official Journal of the International Society for Twin Studies 12* (3): 261–8. doi:10.1375/twin.12.3.261

Kwan, M., S. Bobko, G. Faulkner, P. Donnelly, & J. Cairney. 2014. "Sport participation and alcohol and illicit drug use in adolescents and young adults: A systematic review of longitudinal studies." *Addictive Behaviors 39* (3): 497–506. doi:10.1016/j.addbeh.2013.11.006

Ledochowski, L., G. Ruedl, A. H. Taylor, & M. Kopp. 2015. "Acute effects of brisk walking on sugary snack cravings in overweight people, affect and responses to a manipulated stress situation and to a sugary snack cue: A crossover study." *PLOS ONE 10* (3): e0119278. doi:10.1371/journal.pone.0119278

Ledochowski, L., A. H. Taylor, M. Haasova, G. E. Faulkner, M. H. Ussher, & M. Kopp. 2013. "Unmittelbare Auswirkungen einzelner Bewegungseinheiten auf das Bedürfnis zu rauchen." *Zeitschrift für Gesundheitspsychologie 21* (3): 122–37. doi:10.1026/0943-8149/a000099

Li, D.-X., X.-Y. Zhuang, Y.-P. Zhang, H. Guo, Z. Wang, Q. Zhang,...Y.-G. Yao. 2013. "Effects of Tai Chi on the protracted abstinence syndrome: A time trial analysis." *American Journal of Chinese Medicine 41* (1): 43–57. doi:10.1142/S0192415X13500043

Li, J.-B., A. Asakawa, Y. Li, K. Cheng, & A. Inui. 2011. "Effects of exercise on the levels of Peptide YY and Ghrelin." *Experimental and Clinical Endocrinology and Diabetes: Official Journal, German Society of Endocrinology [and] German Diabetes Association 119* (3): 163–6. doi:10.1055/s-0030-1262790

Li, M., K. Chen, & Z. Mo. 2002. "Use of Qigong Therapy in the detoxification of heroin addicts." *Alternative Therapies in Health and Medicine 8* (1): 50–4, 56–9. www.ncbi.nlm.nih.gov/pubmed/11795622.

Liangpunsakul, S., D. W. Crabb, & R. Qi. 2010. "Relationship among alcohol intake, body fat, and physical activity: A population-based study." *Annals of Epidemiology 20* (9): 670–5. doi:10.1016/j.annepidem.2010.05.014

Lisha, N. E., & S. Sussman. 2010. "Relationship of high school and college sports participation with alcohol, tobacco, and illicit drug use: A review." *Addictive Behaviors 35* (5): 399–407. doi:10.1016/j.addbeh.2009.12.032

Lisha, N. E., S. Sussman, F. Fapa, & A. M. Leventhal. 2013. "Physical activity and alcohol use disorders." *American Journal of Drug and Alcohol Abuse 39* (2): 115–20. doi:10.3109/00952990.2012.713060

McAloney, K., H. Graham, C. Law, & L. Platt. 2013. "A scoping review of statistical approaches to the analysis of multiple health-related behaviours." *Preventive Medicine 56* (6): 365–71. doi:10.1016/j.ypmed.2013.03.002

McLellan, A. T., D. C. Lewis, C. P. O'Brien, & H. D. Kleber. 2000. "Drug dependence, a chronic medical illness: Implications for treatment, insurance, and outcomes evaluation." *JAMA 284* (13): 1689–95. www.ncbi.nlm.nih.gov/pubmed/11015800.

Mahoney, J. L., & A. E. Vest. 2012. "The over-scheduling hypothesis revisited: Intensity of organized activity participation during adolescence and young adult outcomes." *Journal of Research on Adolescence 22* (3): 409–18. doi:10.1111/j.1532-7795.2012.00808.x

Maisto, S. A., G. J. Connors, & W. H. Zywiak. 2000. "Alcohol treatment, changes in coping skills, self-efficacy, and levels of alcohol use and related problems 1 year following treatment initiation." *Psychology of Addictive Behaviors: Journal of the Society of Psychologists in Addictive Behaviors 14* (3): 257–66. www.ncbi.nlm.nih.gov/pubmed/10998951.

Marcus, B. H., A. E. Albrecht, T. K. King, A. F. Parisi, B. M. Pinto, M. Roberts,... D. B. V. Abrams. 1999. "The efficacy of exercise as an aid for smoking cessation in women: A randomized controlled trial." *Archives of Internal Medicine 159* (11): 1229–34. www.ncbi.nlm.nih.gov/pubmed/10371231.

Marcus, B. H., B. A. Lewis, J. Hogan, T. K. King, A. E. Albrecht, B. Bock,...D. B. Abrams. 2005. "The efficacy of moderate-intensity exercise as an aid for smoking cessation in women: A randomized controlled trial." *Nicotine and Tobacco Research: Official Journal of the Society for Research on Nicotine and Tobacco 7* (6): 871–80. doi:10.1080/14622200500266056

Maruyama, A., S. Macdonald, E. Borycki, & J. Zhao. 2013. "Hypertension, chronic obstructive pulmonary disease, diabetes and depression among older methadone maintenance patients in British Columbia." *Drug and Alcohol Review 32* (4): 412–18. doi:10.1111/dar.12031

Meule, A., & A. N. Gearhardt. 2014. "Food addiction in the light of DSM-5." *Nutrients 6* (9): 3653–71. doi:10.3390/nu6093653

Miller, M. L., B. D. Vaillancourt, M. J. Wright, S. M. Aarde, S. A. Vandewater, K. M. Creehan, & M. A. Taffe. 2012. "Reciprocal inhibitory effects of intravenous D-Methamphetamine self-administration and wheel activity in rats." *Drug and Alcohol Dependence 121* (1–2): 90–6. doi:10.1016/j.drugalcdep.2011.08.013

Miller, W. R., & S. Rollnick. 2012. "Meeting in the middle: Motivational interviewing and self-determination theory." *Int J Behav Nutr Phys Act 9*: 25. doi:1479-5868-9-25 10.1186/1479-5868-9-25

Miller, W. R., S. T. Walters, & M. E. Bennett. 2001. "How effective is alcoholism treatment in the United States?" *Journal of Studies on Alcohol 62* (2): 211–20. www.ncbi.nlm.nih.gov/pubmed/11327187.

Murphy, T. J., R. R. Pagano, & G. A. Marlatt. 1986. "Lifestyle modification with heavy alcohol drinkers: Effects of aerobic exercise and meditation." *Addictive Behaviors 11* (2): 175–86. www.ncbi.nlm.nih.gov/pubmed/3526824.

The National Treatment Agency for Substance Misuse. 2012. "Why invest? How drug treatment and recovery services work for individuals, communities and society." London. www.nta.nhs.uk/uploads/whyinvest2final.pdf.

The National Treatment Agency for Substance Misuse. 2013. "Alcohol Treatment in England 2011–2012." London. www.nta.nhs.uk/uploads/alcoholcommentary2013final.pdf.

Neale, J., S. Nettleton, & L. Pickering. 2012. "Heroin users' views and experiences of physical activity, sport and exercise." *International Journal on Drug Policy 23* (2): 120–7. doi:10.1016/j.drugpo.2011.06.004

Noble, N., C. Paul, H. Turon, & C. Oldmeadow. 2015. "Which modifiable health risk behaviours are related? A systematic review of the clustering of smoking, nutrition, alcohol and physical activity ('SNAP') health risk factors." *Preventive Medicine 81* (July): 16–41. doi:10.1016/j.ypmed.2015.07.003

NorthWest Public Health Observatory. 2008. "Alcohol-attributable fractions for England." Liverpool. www.nwph.net/nwpho/publications/alcoholattributablefractions.pdf.

Nutt, D. J., & J. Rehm. 2014. "Doing it by numbers: A simple approach to reducing the harms of alcohol." *Journal of Psychopharmacology 28* (1): 3–7. doi:10.1177/0269881113512038

Oh, H., & A. H. Taylor. 2012. "Brisk walking reduces ad libitum snacking in regular chocolate eaters during a workplace simulation." *Appetite 58* (1): 387–92. doi:10.1016/j.appet.2011.11.006

Oh, H., & A. H. Taylor. 2013. "Self-regulating smoking and snacking through physical activity." *Health Psychology, 33* (4), 349–59. doi:10.1037/a0032423

Oh, H., & A. H. Taylor. 2014. "Self-regulating smoking and snacking through physical activity." *Health Psychology: Official Journal of the Division of Health Psychology, American Psychological Association 33* (4): 349–59. doi:10.1037/a0032423

Palmer, J. A., L. K. Palmer, K. Michiels, & B. Thigpen. 1995. "Effects of type of exercise on depression in recovering substance abusers." *Perceptual and Motor Skills 80* (2): 523–30. doi:10.2466/pms.1995.80.2.523

Palmer, J., N. Vacc, & J. Epstein. 1988. "Adult inpatient alcoholics: Physical exercise as a treatment intervention." *Journal of Studies on Alcohol 49* (5): 418–21. www.ncbi.nlm.nih.gov/pubmed/3216644.

Peretti-Watel, P., F. Beck, & S. Legleye. 2002. "Beyond the U-Curve: The relationship between sport and alcohol, cigarette and cannabis use in adolescents." *Addiction 97* (6): 707–16. www.ncbi.nlm.nih.gov/pubmed/12084140.

Perry, A. E., M. Neilson, M. Martyn-St James, J. M. Glanville, R. Woodhouse, C. Godfrey, & C. Hewitt. 2015. "Interventions for drug-using offenders with co-occurring mental illness." *Cochrane Database of Systematic Reviews 6* (January): CD010901. doi:10.1002/14651858.CD010901.pub2

Pitman, A., K. Krysinska, D. Osborn, & M. King. 2012. "Suicide in young men." *Lancet 379* (9834): 2383–92. doi:10.1016/S0140-6736(12)60731-4

Poortinga, W. 2007. "The prevalence and clustering of four major lifestyle risk factors in an English adult population." *Preventive Medicine 44* (2): 124–8. doi:10.1016/j.ypmed.2006.10.006

Pronk, N. P., L. H. Anderson, A. L. Crain, B. C. Martinson, P. J. O'Connor, N. E. Sherwood, & R. R. Whitebird. 2004. "Meeting recommendations for multiple healthy lifestyle factors. prevalence, clustering, and predictors among adolescent, adult, and senior health plan members." *American Journal of Preventive Medicine 27* (2 Suppl): 25–33. doi:10.1016/j.amepre.2004.04.022

Public Health England. 2014. "Drug treatment in England 2013–2014." London. www.nta.nhs.uk/uploads/drug-treatment-in-england-2013-14-commentary.pdf.

Ramo, D. E., & S. A. Brown. 2008. "Classes of substance abuse relapse situations: A comparison of adolescents and adults." *Psychology of Addictive Behaviors: Journal of the Society of Psychologists in Addictive Behaviors 22* (3): 372–9. doi:10.1037/0893-164X.22.3.372

Rodriguez, D., G. F. Dunton, J. Tscherne, & J. Sass. 2008. "Physical activity and adolescent smoking: A moderated mediation model." *Mental Health and Physical Activity 1* (1): 17–25. doi:10.1016/j.mhpa.2008.04.001

Roessler, K. K. 2010. "Exercise treatment for drug abuse – A Danish pilot study." *Scandinavian Journal of Public Health 38* (6): 664–9. doi:10.1177/1403494810371249

Schulte, E. M., M. A. Joyner, M. N. Potenza, C. M. Grilo, & A. N. Gearhardt. 2015. "Current considerations regarding food addiction." *Current Psychiatry Reports 17* (4): 563. doi:10.1007/s11920-015-0563-3

Shephard, R. J. 1989. "Adolphe Abrahams Memorial Lecture, 1988. Exercise and lifestyle change." *British Journal of Sports Medicine 23* (1): 11–22. www.pubmedcentral.nih.gov/articlerender.fcgi?artid=1478656&tool=pmcentrez&rendertype=abstract.

Sinyor, D. T. Brown, L. Rostant, & P. Seraganian. 1982. "The role of a physical fitness program in the treatment of alcoholism." *Journal of Studies on Alcohol 43* (3): 380–6. www.ncbi.nlm.nih.gov/pubmed/7121004.

Smith, M. A., & W. J. Lynch. 2011. "Exercise as a potential treatment for drug abuse: Evidence from preclinical studies." *Frontiers in Psychiatry 2* (January): 82. doi:10.3389/fpsyt.2011.00082

Smith, M. A., & W. J. Lynch. 2013. "The neurobiology of exercise and drug-seeking behaviour." In P. Ekkekakis (Ed.), *Handbook on physical activity and mental health.* New York: Routledge.

Smith, M. A., M. M. Pennock, K. L. Walker, & K. C. Lang. 2012. "Access to a running wheel decreases cocaine-primed and cue-induced reinstatement in male and female rats." *Drug and Alcohol Dependence 121* (1–2): 54–61. doi:10.1016/j.drugalcdep.2011.08.006

Smith, M. A., & E. G. Pitts. 2011. "Access to a running wheel inhibits the acquisition of cocaine self-administration." *Pharmacology, Biochemistry, and Behavior 100* (2): 237–43. doi:10.1016/j.pbb.2011.08.025

Smith, M. A., K. T. Schmidt, J. C. Iordanou, & M. L. Mustroph. 2008. "Aerobic exercise decreases the positive-reinforcing effects of cocaine." *Drug and Alcohol Dependence 98* (1–2): 129–35. doi:10.1016/j.drugalcdep.2008.05.006

Smith, M. A., K. L. Walker, K. T. Cole, & K. C. Lang. 2011. "The effects of aerobic exercise on cocaine self-administration in male and female rats." *Psychopharmacology 218* (2): 357–69. doi:10.1007/s00213-011-2321-5

Smits, J. A. J., M. J. Zvolensky, M. L. Davis, D. Rosenfield, B. H. Marcus, T. S. Church,...M. B. Powers. 2015. "The efficacy of vigorous-intensity exercise as an aid to smoking cessation in adults with high anxiety sensitivity: A randomized controlled trial." *Psychosomatic Medicine* (October). doi:10.1097/PSY.0000000000000264

Smothers, B., & D. Bertolucci. 2001. "Alcohol consumption and health-promoting behavior in a U.S. household sample: Leisure-time physical activity." *Journal of Studies on Alcohol 62* (4): 467–76. www.ncbi.nlm.nih.gov/pubmed/11513224.

Stead, L. F., R. Perera, C. Bullen, D. Mant, J. Hartmann-Boyce, K. Cahill, & T. Lancaster. 2012. "Nicotine replacement therapy for smoking cessation." *Cochrane Database Syst Rev 11*: CD000146. doi:10.1002/14651858.CD000146.pub4

Ströhle, A., M. Höfler, H. Pfister, A.-G. Müller, J. Hoyer, H.-U. Wittchen, & R. Lieb. 2007. "Physical activity and prevalence and incidence of mental disorders in adolescents and young adults." *Psychological Medicine 37* (11): 1657–66. doi:10.1017/S003329170700089X

Stuart, G. L., T. J. O'Farrell, & J. R. Temple. 2009. "Review of the association between treatment for substance misuse and reductions in intimate partner violence." *Substance Use and Misuse 44* (9–10): 1298–1317. doi:10.1080/10826080902961385

Taylor, A. H., & M. Katomeri. 2007. "Walking reduces cue-elicited cigarette cravings and withdrawal symptoms, and delays ad libitum smoking." *Nicotine Tob Res 9* (11): 1183–90. doi:783697604 [pii]10.1080/14622200701648896

Taylor, A. H., H. Oh, & S. Cullen. 2013. "Acute effect of exercise on alcohol urges and attentional bias towards alcohol related images in high alcohol consumers." *Mental Health and Physical Activity 6* (3): 220–6. doi:10.1016/j.mhpa.2013.09.004

Taylor, A. H., & A. J. Oliver. 2009. "Acute effects of brisk walking on urges to eat chocolate, affect, and responses to a stressor and chocolate cue. An experimental study." *Appetite 52* (1): 155–60. doi:10.1016/j.appet.2008.09.004

Taylor, A. H., A. Oliver, & K. J. Van Rensburg. 2009. "The acute effects of exercise on visual attentional biases to chocolate-related images." Paper presented at International Society of Sport Psychology Conference.

Taylor, A. H., T. P. Thompson, C. J. Greaves, R. S. Taylor, C. Green, F. C. Warren,...R. Kandiyali. 2014. "A pilot randomised trial to assess the methods and procedures for evaluating the clinical effectiveness and cost-effectiveness of Exercise Assisted Reduction Then Stop (EARS) among disadvantaged smokers." *Health Technology Assessment 18* (4): 1–324. www.ncbi.nlm.nih.gov/pubmed/24433837.

Taylor, A. H., & M. Ussher. 2013. "Physical activity as an aid to smoking cessation." In P. Ekkekakis (Ed.), *Handbook on physical activity and mental health* (pp. 451–64). New York: Routledge.

Terry-McElrath, Y. M, & P. M. O'Malley. 2011. "Substance use and exercise participation among young adults: Parallel trajectories in a national cohort-sequential study." *Addiction 106* (10): 1855–65; discussion 1866–7. doi:10.1111/j.1360-0443.2011.03489.x

Thayer, R. E., D. P. Peters, P. J. Takahashi, & A. M. Birkhead-Flight. 1993. "Mood and behavior (smoking and sugar snacking) following moderate exercise: A partial test of self-regulation theory." *Personality and Individual Differences 14* (1): 97–104. doi:10.1016/0191-8869(93)90178-6

Thompson, T. P., C. J. Greaves, R. Ayres, P. Aveyard, F. C. Warren, R. Byng,...R. S. Taylor. 2015. "An exploratory analysis of the smoking and physical activity outcomes from a pilot randomized controlled trial of an exercise assisted reduction to stop smoking intervention in disadvantaged groups." *Nicotine and Tobacco Research: Official Journal of the Society for Research on Nicotine and Tobacco*, 2016 Mar; *18* (3): 289–97. doi: 10.1093/ntr/ntv099. Epub 2015 May 11.

Tritter, A., L. Fitzgeorge, & H. Prapavessis. 2015. "The effect of acute exercise on cigarette cravings while using a nicotine lozenge." *Psychopharmacology 232* (14): 2531–9. doi:10.1007/s00213-015-3887-0

United Nations Office on Drugs and Crime. 2012. "World Drug Report 2012." New York. www.unodc.org/documents/data-and-analysis/WDR2012/WDR_2012_web_small.pdf.

Ussher, M. H., S. Lewis, P. Aveyard, I. Manyonda, R. West, B. Lewis,...B. Marcus. 2015. "Physical activity for smoking cessation in pregnancy: Randomised controlled trial." *BMJ (Clinical Research Ed.) 350* (January): h2145. www.pubmedcentral.nih.gov/articlerender.fcgi?artid=4431606&tool=pmcentrez&rendertype=abstract.

Ussher, M., A. K. Sampuran, R. Doshi, R. West, & D. C. Drummond. 2004. "Acute effect of a brief bout of exercise on alcohol urges." *Addiction 99* (12): 1542–7.

Ussher, M. H., A. H. Taylor, & G. E. J. Faulkner. 2014. "Exercise interventions for smoking cessation." *Cochrane Database of Systematic Reviews 8* (January): CD002295. doi:10.1002/14651858.CD002295.pub5

Ussher, M. H., R. West, A. McEwen, A. Taylor, & A. Steptoe. 2003. "Efficacy of exercise counselling as an aid for smoking cessation: A randomized controlled trial." *Addiction 98* (4): 523–32. doi:346 [pii]

Van Rensburg, K. J., K. A. Taylor, A. Benattayallah, & T. Hodgson. 2012. "The effects of exercise on cigarette cravings and brain activation in response to smoking-related images." *Psychopharmacology 221* (4): 659–66. doi:10.1007/s00213-011-2610-z

Van Rensburg, K. J., A. Taylor, & T. Hodgson. 2009. "The effects of acute exercise on attentional bias towards smoking-related stimuli during temporary abstinence from smoking." *Addiction 104* (11): 1910–17. doi:ADD2692 [pii]10.1111/j.1360-0443.2009.02692.x

Van Rensburg, K. J., K. A. Taylor, T. Hodgson, & A. Benattayallah. 2009. "Acute exercise modulates cigarette cravings and brain activation in response to smoking-related images: An fMRI study." *Psychopharmacology 203* (3): 589–98. doi:10.1007/s00213-008-1405-3

Wang, D., Y. Wang, Y. Wang, R. Li, & C. Zhou. 2014. "Impact of physical exercise on substance use disorders: A meta-analysis." *PloS One 9* (10): e110728. doi:10.1371/journal.pone.0110728

Wapf, V., M. Schaub, B. Klaeusler, L. Boesch, R. Stohler, & D. Eich. 2008. "The barriers to smoking cessation in Swiss methadone and buprenorphine-maintained patients." *Harm Reduction Journal* 5 (January): *10.* doi:10.1186/1477-7517-5-10

Weber, A. 1984. "Laufen Als Behandlungskontrolle – Eine Experimentelle Untersuchung an Alkoholabhängigen in Der Klinik...Running as treatment for hospitalized alcoholics: an experimental approach." *Suchtgefahren 30* (3): 160–7.

Wichstrøm, T., & L. Wichstrøm. 2009. "Does sports participation during adolescence prevent later alcohol, tobacco and cannabis use?" *Addiction 104* (1): 138–49. doi:10.1111/j.1360-0443.2008.02422.x

Williams, D. J. 2000. "Exercise and substance abuse treatment: Predicting programme completion." *Corrections Compendium 25* (25): 4–7.

World Health Organization. 2011a. "Global Status Report on Alcohol and Health." Geneva. www.who.int/substance_abuse/publications/global_alcohol_report/msbgsruprofiles.pdf.

World Health Organization. 2011b. "WHO Report on the Global Tobacco Epidemic, 2011." World Health Organization. www.who.int/tobacco/global_report/2011/en/.

World Health Organization. 2014. "Global Status Report on Alcohol and Health 2014." Geneva. www.who.int/substance_abuse/publications/global_alcohol_report/msb_gsr_2014_1.pdf?ua=1.

Xie, H., G. J. McHugo, M. B. Fox, & R. E. Drake. 2005. "Substance abuse relapse in a ten-year prospective follow-up of clients with mental and substance use disorders." *Psychiatric Services 56* (10): 1282–7. doi:10.1176/appi.ps.56.10.1282

Zhuang, S., S. An, & Y. Zhao. 2013. "Yoga effects on mood and quality of life in Chinese women undergoing heroin detoxification: A randomized controlled trial." *Nursing Research 62* (4): 260–8. doi:10.1097/NNR.0b013e318292379b

Zschucke, E., A. Heinz, & A. Ströhle. 2012. "Exercise and physical activity in the therapy of substance use disorders." *Scientific World Journal* (January): 901741. doi:10.1100/2012/901741

11 Morbid Exercise Behavior

Addiction or Psychological Escape?*

Attila Szabo, Zsolt Demetrovics, and Mark D. Griffiths

Introduction: Definition and Terminology

Competitive athletes train hard and for long hours within a well-balanced schedule, having full control over their exercise behavior. Sometimes, the same exercise pattern may also be observed in leisure exercisers. However, a very small proportion of these people may lose control over their exercise behavior and experience the compulsive need to engage in the activity at unpredictable times, that is determined by the severity of an inner urge. This "inner urge" may occasionally be the attempt to conquer something behind an individual's ability (like running a marathon in a subjectively perceived possible timeframe that is in discord with the physical ability of the person), or more often may be *to escape* from something to cope with and/or escape something in their lives (e.g. job stress, relationship problems, etc.). Exercising to the point where an individual *loses control* over the behavior that becomes obligatory and leads to physical and mental damage is referred to as *exercise addiction* (Griffiths 1997; Thaxton 1982).

The same concept is also often described as *exercise dependence* by a number of scholars (e.g. Cockerill & Riddington 1996; Hausenblas & Downs 2002). Furthermore, scientists have often referred to the condition as *obligatory exercising* (e.g. Pasman & Thompson 1988), by stressing the *compulsive* aspect of the behavior. Indeed, in the mass media, exercise addiction is also frequently termed *compulsive exercise* (Eberle 2004), or as *exercise abuse* (Davis 2000). It is important to re-emphasize that all these seemingly synonymous terms are theoretically intended to label the *same* psychological condition (Figure 11.1). However, there are several reasons why alternating the terminology in naming the same phenomenon may be unproductive. There is a convincing argument for differentiating addiction from dependence (O'Brien, Volkow, & Li 2006). While the term *dependence* is often "carelessly" used as a synonym for addiction, the latter includes the former, and also includes *compulsion* (Goodman 1990). Accordingly, the general formula for addiction may be described as: *addiction* = *dependence* + *compulsion*. Consequently, by adopting the term "dependence," one misses a key component of the morbid exercise behavior that is the urge or *compulsion*, which is the main propelling force behind the disorder. Goodman highlights that not all dependencies and compulsions may be classified as addiction(s).

Primary and Secondary Exercise Addiction

In addition to terminology issues, there is a conceptually incorrect separation between two similar exaggerated forms of exercise behavior in the academic literature. The terms "primary exercise dependence" and "secondary exercise dependence" were used by De Coverley Veale (1987) to differentiate between compulsive exercise with no other comorbid behavior (addiction), or an obligatory form of exercise that surfaced as a comorbidity with other behaviors such as eating disorders (i.e. anorexia nervosa, bulimia, binge eating, etc.).

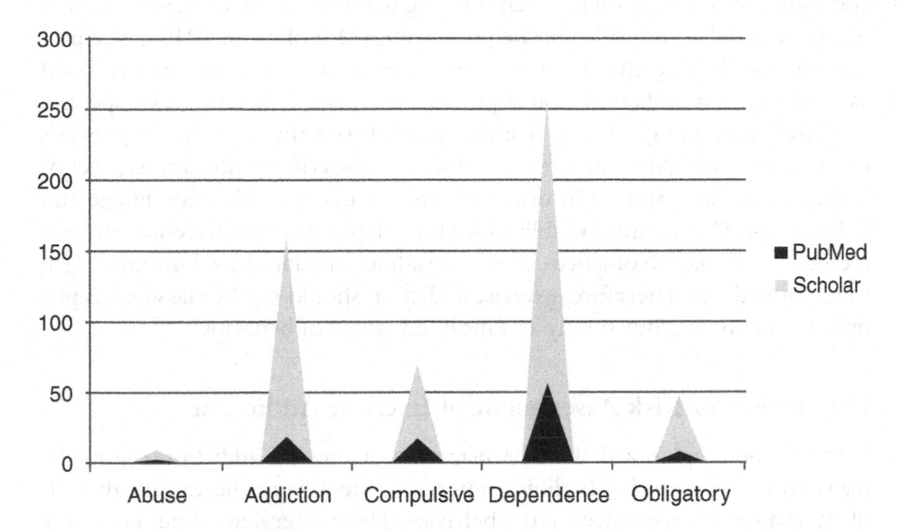

Figure 11.1 Various terms used for the morbid pattern of exercise behavior

Some authors (including one of the present authors) have rejected the concept of secondary exercise addiction (Szabo 2010), because the purpose of the exaggerated exercise in eating disorders is to lose or control one's weight. Consequently, exercise is an auxiliary means of achieving an end. In the "actual" (or primary) exercise addiction, the uncontrolled workouts represent a false expectation for solving or avoiding something importing to the person via the high volume (and also intensity in some cases) of exercise.

Indeed, exercise addiction has two distinctive features, that in addition to high risk scores on screening tools, should affirm the morbid nature of the behavior. The first is *loss of control* and the second is a *negative consequence* to the physical or social health of the person. A false belief, or the expectation that exercise is the solution (for example) to escape stress, is also a general feature of exercise addiction. This false belief, leading to (1) compulsive urges and tolerance, (2) frequent and intense exercise sessions at times when the urge becomes unbearable, and (3) exercise tolerance with short-duration relief after exercise, represent a morbid pattern of the behavior. In contrast to eating disorders, exercise addiction involves the loss of control, the uncontrolled urge to exercise, and a compulsion fuelled by false expectation(s). These are the morbid aspects of the exercise behavior that are not usually observed in eating disorder, which is a relatively common morbidity in contrast to primary exercise addiction which is much rarer. In fact, Bamber, Cockerill, and Carroll (2000) suggested that primary exercise addiction may not even warrant serious scholastic attention. In spite of the ongoing heated debate in acknowledging primary and/or secondary exercise addiction, Veale (1995) affirmed that primary exercise addiction exists but it is very rare. This assertion of rarity is corroborated by empirical research showing that the risk for exercise addiction is approximately 0.3%–0.5% of the population (Mónok et al. 2012). The present authors believe that high volumes of exercise could only be perceived as morbid when addiction is also present, that usually means an escape and temporary psychological relief from the pain of something else in the person's life. In most eating disorders, the morbidity is associated with various psychological problems resulting in distorted and an unacceptable body image and self-concept. One should be able to see the relatively large difference between escape and/or pain avoidance (in exercise addiction) and drive for thinness (in eating disorders). Therefore, exercise addiction should not be classified as primary or secondary, but simply as a morbid pattern of behavior.

Diagnosis and Risk Assessment of Exercise Addiction

It should be emphasized that the diagnosis of exercise addiction cannot be made without an in-depth clinical interview, identifying the causes, the etiology and the consequences of the behavior. These interviews should not only look for presence of symptoms, but also to the severity of the symptoms of addiction and the risk of damage to the physical and mental health of the individual. Any existing, or foreseeable, detrimental health effects, in parallel with the observance of the markers of other behavioral addictions (i.e. gambling disorder) in DSM-5 (Diagnostic and Statistical Manual for Mental Disorders) (APA 2013), should be interpreted as indices of the morbid exercise pattern.

The *risk*, or proneness, for morbid exercise behavior can be approximated with brief screening tools. Several screening measures have been developed for assessing exercise addiction. Most of them are based on the most common symptoms of behavioral and/or chemical addictions. In general, the frequency and intensity of the symptoms reported by the tested individuals are computed to generate a score that may reflect exercise addiction. However, these self-reported scores only mirror the *risk* for the morbidity, which by no means should be interpreted as a diagnosis.

The Obligatory Exercise Questionnaire (OEQ)

The OEQ was among the first instruments developed to assess exercise addiction and was modified from the original Obligatory Running Questionnaire (ORQ – Blumenthal, O'Toole, & Chang 1984). Later, the OEQ was modified to a version that was a more general measure of running and exercise activity (Thompson & Pasman 1991). The new version of the scale consists of 20 items relating to running or exercise habits, and are rated on a 4-point frequency scale: 1-never, 2-sometimes, 3-usually, 4-always. Two of the items are inversely rated during scoring. The psychometric properties of the tool have been well established (Coen & Ogles 1993). The internal reliability (Cronbach α) of the OEQ was reported to be α =. 96 and its concurrent validity was r =.96 (Thompson & Pasman 1991).

The Exercise Dependence Questionnaire (EDQ)

The EDQ was developed with a sample of 449 participants who exercised for more than four hours a week (Ogden, Veale, & Summers 1997). The scale consists of 29 items and includes eight subscales: 1) interference with social/family/work life, 2) positive reward, 3) withdrawal symptoms, 4) exercise for weight control, 5) insight into the problem, 6) exercise for social reasons, 7) exercise for health reasons, and 8) stereotyped behavior. The EDQ has moderate to good internal reliability, ranging from α =. 52 to α =. 84. Its concurrent validity with other instruments has not been reported. Furthermore, certain items assess attitudes and social practices rather than addiction.

The Exercise Dependence Scale (EDS)

The EDS is a popular and well-validated instrument developed by Hausenblas and Symons Downs (2002). The authors use the term exercise *dependence* rather than exercise addiction. Dependence is described as a craving for exercise that results in uncontrolled and excessive workouts and manifests in the form of physiological symptoms, psychological symptoms, or both (Hausenblas & Symons Downs 2002). The EDS was based on the earlier *Diagnostic and Statistical Manual of Mental Disorders* criteria for substance dependence (APA 1994). This measure differentiates between at-risk, non-dependent-symptomatic, and non-dependent-asymptomatic exercisers. It also specifies whether individuals may have a physiological dependence or non-physiological dependence. All the items on the EDS are rated on a six-point Likert scale, ranging from 1 (never) to 6 (always). Evaluation is made in reference to the DSM-IV criteria (APA 1994), screening for the presence of three or more of the following symptoms: 1) tolerance, 2) withdrawal, 3) intention effects (exercise is often taken in larger amounts or over a longer period than was intended), 4) loss of control, 5) time (too much time is spent in activities conducive to the obtainment of exercise), 6) conflict, and 7) continuance (exercise is continued in spite of recurrent physical or psychological problems caused by exaggerated exercise). The EDS provides a total score and subscale scores. The higher the score, the higher is the risk for *dependence*. It has been shown that the scale possesses good internal reliability (α =. 78 to α =. 92) and test–retest reliability (r = 0.92). Not long after the development of the original scale, an improved (fully revised) scale was released by the developers (Symons Downs, Hausenblas, and Nigg 2004). The EDS has excellent psychometric properties, and has been translated into several languages, including Spanish (Sicilia & González-Cutre 2011).

The Exercise Addiction Inventory (EAI)

As noted earlier, loss of control and negative consequences are prime classifying components of the morbidity. Griffiths (2005) proposed a modified *components model* for behavioral addictions based on the earlier work of Brown (1993). This model comprises six symptoms that are common to all addictions: 1) salience, 2) mood modification, 3) tolerance, 4) withdrawal, 5) conflict, and 6) relapse. Griffiths conceptualizes that addictions are part of a biopsychosocial process and there is growing evidence that all addictions appear to share these symptoms.

These six components of addictions served the theoretical basis for the Exercise Addiction Inventory (EAI) (Terry, Szabo, & Griffiths 2004). The EAI is a short, psychometrically validated questionnaire that comprises only six statements, each corresponding to one of the six symptoms in the components model of addictions (Griffiths 2005). Each statement is rated on a five-point Likert scale, ranging from 1 (strongly disagree) to 5 (strongly agree). The suggested EAI cutoff score for individuals considered at-risk of exercise addiction was originally defined as 24 (i.e. most answers agree or strongly agree with the presence of the six symptoms). However, this cut-off point was never tested psychometrically. The EAI was developed on the basis of an opportunistic sample of 200 habitual exercisers in the United Kingdom. The internal reliability of the original scale was excellent ($\alpha = 0.84$) and concurrent validity was also high ($r = 0.80$). A recent study by Griffiths et al. (2015) collated data gathered from EAI use in five nations (N=6,031). Confirmatory factor analyses using the combined dataset supported the configural invariance and metric invariance, but not scalar invariance, showing a vulnerability in cultural and gender-related interpretation. The internal reliability was also lower than that of the original scale, ranging between $\alpha = .58$ (USA) to $\alpha = .80$ (UK). It should also be noted that the U.S. sample consisted of elite triathletes, and as discussed later, may have interpreted EAI items differently than non-elite athletes.

Other Less Frequently Used Tools in the Assessment of Exercise Addiction

Before the development of reliable and psychometrically validated tools for gauging the risk for exercise addiction, exercise morbidity was investigated by using in-depth interviews (Sachs & Pargman 1979) and the *Commitment to Running Scale* (CRS) (Carmack & Martens 1979). However, as its name implies, the CRS measures *commitment*, rather than addiction. Therefore, its use in exercise addiction research has been criticized (Szabo 2010). While addiction is clearly a morbidity, commitment to exercise implies a dedicated involvement in the activity for mastery and enjoyment (as already discussed).

The Negative Addiction Scale (NAS) (Hailey & Bailey 1982) has been used primarily with runners. Its items measure the psychological rather than physiological aspects of compulsive running. No psychometric properties were reported for the original scale, but a recent translation in Portuguese reported that the questionnaire had good internal consistency (Cronbach α = .79; Modoio et al. 2011).

The Exercise Beliefs Questionnaire (EBQ) (Loumidis & Wells 1998) assesses personal assumptions about exercise behavior on the bases of four factors: 1) social desirability, 2) physical appearance, 3) mental and emotional functioning, and 4) vulnerability to disease and aging. The instrument's internal reliability is relatively good, ranging between α = .67 and α = .89 and its concurrent validity ranges between r = .67 and r = .77.

Another instrument, the *Bodybuilding Dependency Scale* (BDS) (Smith, Hale, & Collins 1998), was specifically developed to assess excessive exercise in bodybuilders. The BDS contains three subscales: 1) social dependence (i.e. individual's need to be in the weightlifting environment), 2) training dependence (i.e. individual's compulsion to lift weights) and 3) mastery dependence (i.e individual's need to exert control over his/her training schedule). Each subscale shows satisfactory internal consistency (α = 0.78, 0.76, and 0.75 respectively; Hurst, Hale, Smith, & Collins 2000). Because of its sports (or workout) specificity, the BDS has restricted range of applicability in sport and exercise psychology.

Despite the name, the *Compulsive Exercise Test* (CET) (Taranis, Touyz, & Meyer 2011) may not be directly related to exercise addiction. Primarily, it is claimed to be a tool that assesses morbid exercise pattern specifically developed to aid studies examining eating disorders. The CET is based on a cognitive-behavioral framework and has 24 items and five subscales that assess avoidance and rule-driven behavior, weight control form of exercise, mood improvement, lack of exercise enjoyment, and exercise rigidity. The psychometric validation of the CET revealed good internal consistency, content validity, and concurrent validity.

The *Exercise Dependence and Elite Athletes Scale* (EDEAS) was developed by McNamara and McCabe (2013) with the aim to study morbid exercise pattern in competitive athletes. The 24-item EDEAS has six factors: 1) unhealthy eating behavior, 2) conflict and dissatisfaction, 3) more training, 4) withdrawal, 5) emotional difficulties, and 6) continuance behavior. The internal reliability of the scale was reported to be acceptable (i.e. α > .60). Based on the developers' report, the EDEAS appears to have good concurrent validity as well. To date, this instrument has only been adopted in a few studies.

A Note of Caution on the Interpretation of Scalar Tools

As noted earlier, screening tools primarily assess the *risk* for exercise addiction. However, few of those classified at risk – based on scalar measures – may turn into addicts. As discussed later, the more recent models forwarded for exercise addiction recognize a *triggering factor* (a switch) in the onset of the morbidity that contradicts the continuum view or the slowly growing and accumulating evolutionary perspective (Figure 11.2).

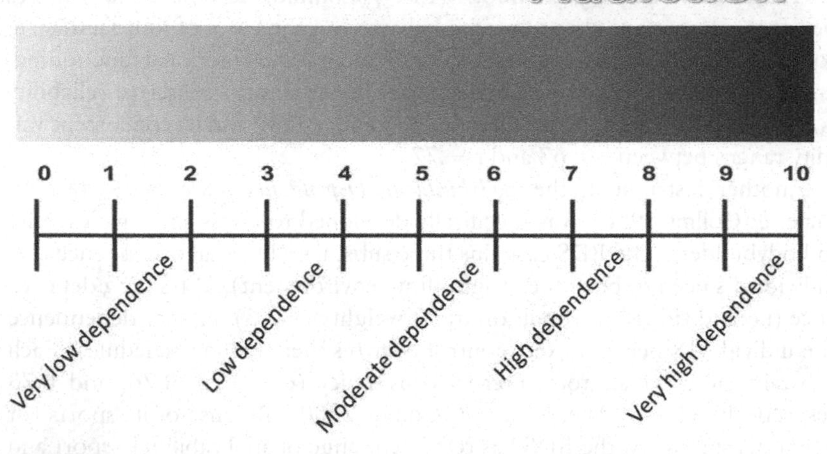

Figure 11.2 A hierarchical and evolutionary conceptualization of exercise addiction

Theoretical Explanations for Exercise Addiction

The Sympathetic Arousal Hypothesis

This physiological model claims that sympathetic adaptation to regular exercise lowers the overall level of arousal (Thompson & Blanton 1987). Lower levels of arousal may be experienced as a lethargic or low-energy state. This uncomfortable feeling urges the person to increase the level of arousal. For the regular exerciser, their exercise activity is an obvious way to increase arousal. However, since the effects of exercise in increasing arousal are relatively short-lasting, increased bouts of exercise are necessary to generate an optimal state of arousal, and eventually leads to tolerance. An issue with this model is that sympathetic adaptation to exercise occurs in everyone, while not all exercisers become addicted.

The Cognitive Appraisal Hypothesis

Another model, proposed by Szabo (1995), suggests that exercise addiction occurs when exercise becomes a means of coping with cognitive stress. Here, repeated and heavy sessions of exercise provide an escape from stress. Once exercise is used for coping with stress, the individual also depends on it to function. When exercise is not possible for any reason, psychological hardship occurs (i.e. withdrawal symptoms). In fact, the loss of exercise may mean the loss – or the inability – of coping. Therefore, the exerciser loses control that generates further vulnerability to stress by amplifying the negative feelings associated with the lack of exercise. The problem can only be resolved via the resumption of the previous pattern of exercise, often at the expense of the other obligations in the individual's daily life. While this model perceives exercise addiction as a means of coping or escape, it only accounts for the maintenance of addiction, but not for its onset.

The Four Phase Model

The four phase hierarchical model was proposed by Freimuth, Moniz, and Kim (2011). In the first phase, they argue that exercise is pleasurable and it is under control. There are no negative experiences, apart from minor strains or muscle soreness. In phase two, the psychological benefits of exercise are more specifically used and the mood-modifying effects can be adopted for coping with stress. Addiction is most likely to occur when exercise becomes the primary or the sole means of coping with stress. Thus, the second phase may pinpoint the onset of exercise addiction, but it does not specify two key issues: 1) a major life stress must exist (whether that is a progressively mounting passive stress or a suddenly appearing uncontrollable form of stress), and 2) it is not known under what conditions or influences exercise will be adopted for coping with the stress. In the third phase, the daily activities are rigidly organized around exercise and the negative consequences are evident. Exercise is performed individually, rather than with friends, in a team, or during scheduled fitness classes. In the fourth stage, the typical symptoms of addiction – such as tolerance, conflict, need for mood modification, withdrawal symptoms and relapse – become evident and the exercise controls the individual rather than vice versa.

The Biopsychosocial Model

The biopsychosocial model contrasts with several conceptualizations of exercise addiction. It was put forward for the explanation of exercise addiction in elite athletes (McNamara & McCabe 2012). In the present authors' view this may be very unlikely. The model states that exercise addiction has a biological factor (e.g. BMI) as its route of origin in the elite athletes. Social and psychological processes may interact to determine whether exercise addiction occurs or not. However, Freimuth et al. (2011) asserted that hard training for long hours, and ambitious strivings in becoming the best – that which characterizes successful elite athletes – should not be confused with the symptoms of addiction.

The Interleukin Six (IL-6) Model

Another theoretical model has highlighted the possible role of interleukin six (IL-6) in exercise addiction (Hamer & Karageorghis 2007). The IL-6 is a pro-inflammatory and anti-inflammatory cytokine secreted by T cells and macrophages to increase the immune response to trauma, such as burning or other types of tissue damage leading to inflammation. The blood concentration level of IL-6 increases during exercise (Aguiló et al. 2014), and higher levels of IL-6 are associated with increased cardiovascular mortality, depression, and negative affect (Puterman et al. 2013). Hamer and Karageorghis (2007) suggest that IL-6 acts as a link from the periphery to the brain. This link may mediate the components of exercise addiction. In people prone to the morbidity, exercise results in a momentary reduction in negative affect. However, at the same time it raises the synthesis of IL-6 and activates the neuroendocrine pathways, which contribute to the negative feelings manifested through the experiencing of withdrawal symptoms. Therefore, exercise acts as a vicious circle by lowering and increasing negative affect. This is a psychoneuroimmunological model that deserves further research attention.

The Monoamine Model

The monoamine model is derived from the early observation that exercise triggers an increase in the levels of catecholamines in the peripheral blood circulation (Cousineau et al. 1977). Later, Szabo, Billett, and Turner (2001) showed that a 30-minute episode of medium to high-intensity aerobic exercise increased uric phenylacetic acid levels – reflecting the phenylethylamine concentration – in healthy males who were habituated to exercise. While catecholamines, among other functions, are involved in the stress response, phenylethylamine is more closely linked to changes in mood. In light of the monoamine hypothesis, it is thought that in addition to an increase in monoamines in the peripheral circulation, the central aminergic activity may also rise in response to exercise. Since brain monoamines are involved in the regulation of mood and affect, their alteration by exercise seems to be an attractive explanation for the role of exercise in the stress response. This is a psychophysiological model that is probably more closely linked to the positive mood-enhancing effects of exercise than exercise addiction per se. Nevertheless, in light of this model, exercise may act as a buffer in the addiction process, in that the negative emotional experiences resulting from life stress are soaked up by the positive effects of exercise.

The Endorphin Model

This endorphin model is popular in the literature, and posits that exercise leads to increased levels of beta-endorphins in the brain, that it turn act as internal psychoactive agents by generating feelings of euphoria. In fact, this hypothesis may be analogous to substance or recreational drug addiction (e.g. heroin, morphine, etc.) with the exception that the psychoactive agent (beta-endorphin) is generated *internally* during exercise instead of being administered externally. Endogenous opioids are involved in modulating several of the sensory, motivational, emotional, and mental functions (McNally & Akil 2002). A novel investigation, using positron emission tomography

(PET), found that acute exercise – performed between aerobic and anaerobic threshold for 60 minutes – resulted in an increase in the availability of μ-opioid receptors in anterior cingulate cortex, prefrontal-, and temporal cortex of young healthy recreational exercising men (Saanijoki et al. 2014). While further research in this area is needed, the opioid response to exercise is likely to be workload- or dose-dependent in addition to individual variability, and could be one of the several explanations for exercise addiction in connection to stress management.

The Pragmatics, Attraction, Communication, Expectation (PACE) Model

The PACE model is not specific to exercise addiction per se, because it was proposed for behavioral addictions in general (Sussman et al. 2011). According to this model, when a situation gets out of control, the individual will *gravitate* towards the means of available coping, reflected by the "Pragmatics" phase in the model. The selection of the coping mechanism is determined by conscious and subconscious analysis via interactions between individual characteristics, situational factors, and (in case of exercise addiction) earlier history of exercise behavior (that is reflected in the "Attraction" part of the model). This attentional focus in the decision forms the "Communication" part of the PACE model, in that experience, interpersonal and intra-personal thoughts, beliefs, and convictions will influence the decision about the means of coping selected by the person. In the final part of the model, the choice is determined by the "Expectation" (i.e. exercise yielding a solution to the problem). The interactional model presented in the next subsection, developed specifically for exercise addiction, is in full accord with the PACE model.

The Interactional Model of Exercise Addiction

A shortcoming of the above listed models for exercise addiction is that they do not provide an explanation for the choice of exercise as a means of coping. Egorov and Szabo (2013) stress that there is an interaction (see Figure 11.3) between several factors, including (but not restricted to) personal/social values, social image, past exercise experience, actual life situations, etc. that determine whether an individual will use exercise for coping or resort to other means of dealing with stress. The number of interactions between situational and personal factors is so large that each case is idiographic in a mindset akin to a "black box." The box can only be opened during diagnosis with the help of mental health professionals. Indeed, exercise addiction, unlike other chemical or behavioral addictions, has a unique characteristic not present in other addictions, which is the physical effort. Earlier it was proposed, based on preliminary laboratory evidence, that exercise acts as cathartic-buffer for stress (Szabo & Tsang 2003). When faced with a stressor, regular exercisers – knowing the mood improving effects of exercise from past experience (Freimuth et al. 2011) – may resort to exercise to cope with the challenge.

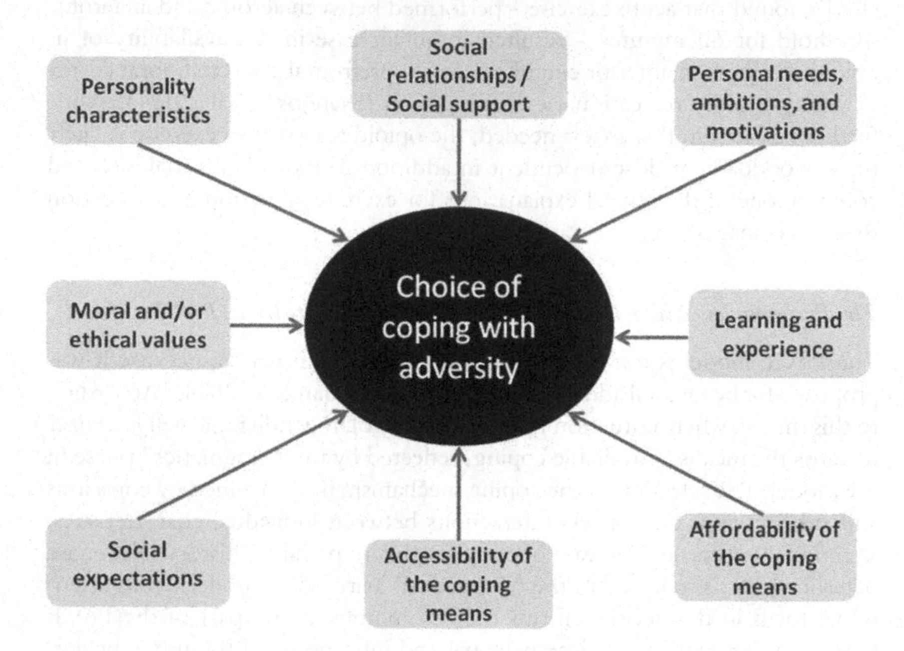

Figure 11.3 The complex interaction through which a person selects a coping strategy with stress

However, not all exercisers will try to reduce the pain of a novel emotional hardship with exercise, but instead may resort to passive forms of escape behaviors or addiction(s). Therefore, the Interactional Model of exercise addiction (Egorov & Szabo 2013) claims that each case is relatively unique and the nomothetic approach in predicting the etiology of the morbidity may be unproductive. Building from case studies (via an inductive approach) may be the way towards a better understanding of the morbidity. A deeper consideration of the model allows the realization that the research into exercise addiction does not – in fact – deal with the morbidity, but simply the presence and severity of the symptoms that occur in it and, therefore, may be perceived as *risk factors.*

Research Trends in Exercise Addiction

A three-year analysis of published research using *PubMed* and *Google Scholar* databases found 128 papers dealing with exercise addiction (Szabo, Griffiths, de la Vega, Mervó, & Demetrovics 2015). This number reveals that there have been approximately 40 publications per year in the area of exercise addiction over recent years. The 128 papers appeared in 89 different journals, indicating that exercise addiction appears to be studied from a multidisciplinary perspective. The majority of the research in the field (>50%) came from the U.S.A.,

U.K., and Australia, but researchers from 25 other nations also contributed to the field (Szabo et al. 2015).

Based on this bibliometric analysis, research examining exercise addiction can be broadly grouped into two categories: (1) survey research, that measures the prevalence and risk for exercise addiction in various exercise- and social contexts, and (2) methodological research aimed at developing, validating, and testing assessment tools, mainly questionnaires (see Table 11.1). Research into exercise addiction significantly rose after two psychometrically valid and reliable tools were developed, namely the EDS and the EAI. In fact, as also illustrated in Table 11.1, most (over 80%) of the studies have adopted the EDS or EAI, with or without additional tools, in their investigations. It is also worth noting that from the past 20 years of research examining exercise addiction, there is an extremely wide range of prevalence estimates reported for its morbidity (see Table 11.1), ranging from 0.3% (Mónok et al. 2012) up to 42% (Lejoyeux, Avril, Richoux, Embouazza, & Nivoli 2008). This is most likely explained via the use of sampling techniques (mostly convenience samples), different samples (different kinds of exercisers, student populations, the general public), and different screening measures employed.

The Prevalence of Risk for Exercise Addiction

Diagnosed cases of exercise addiction have not been reported in the literature, because the morbidity is not listed as a mental disorder in any version of the DSM, and therefore clinical diagnosis is not possible (Szabo et al. 2015). To the best of these authors' knowledge, only three case studies have been published in journals or book chapters (i.e. Griffiths 1997; Kotbagi, Muller, Romo, & Kern 2014; Veale 1995) along with case accounts by Schreiber and Hausenblas (2015) in a recent book that presents several cases in support for the existence and severity of the morbidity. These in-depth case studies clearly reveal an exercise-related morbidity that is consistent with the diagnostic criteria for other behavioral addictions (i.e. disordered gambling), involves the common symptoms of addictions, and shows loss of control over the behavior accompanied by negative consequences to the individual. However, these cases seem to be too few for justifying the research effort reflected in the number of publications in this area. However, the fact is that until now, most studies into exercise addiction have assessed a level of *risk* for the emerging morbidity that may never actually turn into problematic behavior.

The bulk of research on exercise addiction has been carried out using British and American participants. Griffiths, Szabo, and Terry (2005), using the EAI, found that 3.0% of a British sample of university students were identified as *at risk* of exercise addiction. In a series of five studies, examining American university students during the development of the EDS, Hausenblas and Symons Downs (2002) found a risk prevalence that ranged between 3.4% and 13.4%. Among those with deeper involvement in sport, the rate may be higher. Indeed, Szabo and Griffiths (2006) found that 6.9% of British sport

Table 11.1 Summary table of research on exercise addiction over the past 20 years

Year	*Author(s)*	*Type of Study*	*Term*	*n*	*Age (yr)*	*Type of S port*	*QTR*	*Interview*	*Data Source*	*Prevalence (%)*
1995	Thornton and Scott	Survey	EA	40	?	Runners	Other	No	Field	22,5
1998	Slay, Hayaki, Napolitano, and Brownell	Survey	OE	324	?	Runners	Other	No	?	26
2000	Bamber et al.	Survey	ED	194	28,8	Mixed	Other	No	Field	13,9–22,2
2002	Hausenblas and Symons Downs (S1)	MT	ED	266	21,7	Mixed	EDS+Other	Yes	Field	3,4
2002	Hausenblas and Symons Downs (S2)	MT	ED	553	22	Mixed	EDS+ Other	No	Field	13,4
2002	Hausenblas and Symons Downs (S3)	MT	ED	862	21,2	Mixed	EDS+ Other	No	?	3,1
2002	Hausenblas and Symons Downs (S4)	MT	ED	366	20,6	Mixed	EDS+Other	No	?	9,6
2002	Hausenblas and Symons Downs (S5)	MT	ED	46, 373	22,9 20,3	Mixed	EDS+ Other	No	?	9,8
2002	Hausenblas and Fallon	Survey	ED	474	20,3	?	EDS+ Other	No	Field	?
2002	Ackard et al.	Survey	EE	586	20,6	Mixed	Other	No	?	?
2002	Blaydon and Lindner	Survey	ED	203	29,8	Triathlon	Other	No	?	21,6–30,4
2003	Hagan and Hausenblas	Survey	ED	79	21,8	Mixed	EDS+ Other	No	Field	?
2004	Terry et al.	MT	EA, ED, OE	200	21,2	Mixed	EAI, EDS+ Other	No	Field	2,5–3
2004	Hausenblas and Giacobbi Jr.	Survey	ED	390	21,8	?	EDS+ Other	No	Field	?
2004	Symons Downs et al. (S1)	MT	ED	408	20,2	?	EDS+ Other	No	Field	3,6
2004	Symons Downs et al. (S2)	MT	ED	885	20,2	?	EDS+ Other	No	Field	5
2005	Griffiths et al.	MT	EA	279	21,2	Mixed	EAI+ EDS	No	Field	3
2006	Edmunds, Ntoumanis, & Duda	Survey	ED	339	32,1	Mixed	EDS+ Other	No	Field	3,4
2006	Warner and Griffiths	Survey	EA	100	37.6	Gym Users	EAI	No	Field	8
2007	Youngman	Survey	EA	1273	37,9	Triathlon	EAI+ Other	No	Internet	19,9
2006	Szabo and Griffiths	Survey	EA	455	?	Mixed	EAI	No	Field	3.6 – 6.9

2007	Allegre et al.	Survey	ED	95	43,5	Runners	EDS+ Other	No	?	3,2
2008	Lejoyeux et al.	Survey	ED	300	28,6	Mixed	Other	Yes	Field	42
2008	Cook and Hausenblas	Survey	ED	330	19,9	Mixed	EDS+ Other	No	Internet	?
2008	Hausenblas, Gauvin, Symons Downs, & Duley	Survey	ED	40	20,5	Mixed	EDS+ Other	No	Field	?
2009	Chittester and Hausenblas	Survey	ED	113	20,3	Weights	EDS+ Other	No	Lab	?
2009	Lindwall and Palmeira	MT	ED	162	22,6	Mixed	EDS+ Other	No	Field	9,2
2009	Lindwall and Palmeira	MT	ED	269	26,1	Mixed	EDS+ Other	No	Field	5,2
2010	Bratland-Sanda et al.	Survey	ED, EE	38	30,9	Mixed	EDS+ Other	Yes	Field	29
2010	Kern	Survey	ED	484	25,7	Mixed	EDS+ Other	No	Field	?
2010	Smith, Wright, and Winrow	Survey	ED	184	28,1	Runners	EAI+ EDS	No	Field	<10 leisure >30 if also competition
2011	Villella et al.	Survey	EA	2853	16,7	Mixed	EAI+ Other	No	Field	8,7
2011	Pugh and Hadjistavro-poulos	Survey	ED	144	20,7	Mixed	EDS+ Other	No	Internet	?
2011	Sicilia and González-Cutre	Survey	ED	527	29,6	Mixed	EDS+ Other	No	Field	?
2011	Cook, Hausenblas, Tuccitto, & Giacobbi	Survey	ED	539	19,8	Mixed	EDS+ Other	No	Field	?
2011	Bratland-Sanda et al.	Survey	ED	112	30,7	Mixed	EDS+ Other	Yes	Field	?
2011	Modoio et al.	Survey	ED	300	21,2	Mixed	Other	No	?	33
2011	Cook and Hausenblas	Survey	ED	387	20,1	Mixed	EDS+ Other	No	Field	?
2011	Grandi et al.	Survey	ED	79	30	Mixed	Other	No	Field	29,9
2012	Lejoyeux, Guillot, Chalvin, Petit, and Lequen	Survey	ED	500	29	Mixed	EDS+ Other	Yes	Field	29,6
2012	Costa, Cuzzocrea, Hausenblas, Larcan, and Oliva	MT	ED	519	37,1	Mixed	EDS+ Other	No	Field	6,6
2012	Mónok et al.	Survey +MT	EA, ED	474	33,2	Popl. S	EAI+ EDS	Yes	Internet	EAI = 0,5- EDS = 0,3
2014	Lichtenstein, Larsen, Christiansen, Støving, & Bredahl	Survey + MT	EA, EE	590	28,4	Fitness+ Football	EAI+ Other	No	?	5,8

(*continued*)

Table 11.1 (cont.)

Year	*Author(s)*	*Type of Study*	*Term*	*n*	*Age (yr)*	*Type of S port*	*QTR*	*Interview*	*Data Source*	*Prevalence (%)*
2012	McNamara and McCabe	Survey +MT	ED	234	22,5	Mixed	Other	No	Internet	34,8
2013	Cook et al.	Survey	ED	513	19,9	Mixed	EDS+ Other	No	Field	?
2013	Cook et al.	Survey	ED	2660	38,8	Runners	EDS+ Other	No	Internet	1,4
2013	Costa, Hausenblas,Oliva, Cuzzocrea, and Larcan	Survey	ED	409	18–64	Mixed	EDS+Other	No	Field	4.4
2013	Menczel et al.	Survey	ED	1732	31,7	Fitness	EDS+Other	?	Field	?
2013	Szabo, De La Vega, Ruiz-Barquín, and Rivera	Survey	EA	242	27,5	Mixed, Runners	EAI+ Other	No	Field	7–17
2013	Trana	Survey	ED	1546	43	Popl. S	EDS	No	By post	0.4
2014	Cook et al.	Survey	ED	387	20,1	Mixed	EDS+ Other	No	Field	?
2014	Costa et al.	Survey	ED	262	20,9	Mixed	EDS+ Other	No	Field	18.3
2014	Li, Nie, and Ren	Survey	ED	617	19–22	Mixed	EAI	No	Field	8.8
2014	Lichtenstein et al.	Survey	EA	274	16–39	Football, Fitness	EAI	No	Field	7.1, 9.7
2014	Müller et al.	MT	ED	134	22,2	Mixed	EDS+ Other	Yes	Field	10,4
2015	Babusa, Czeglédi, Túry, Mayville, and Urbán	Survey	EA, ED	304	27.8	Body-builders	EAI+ Other	No	Internet	12–23
2015	Maraz, Urbán, Griffiths, and Demetrovic	Survey	EA	447	32.8	Dancers	EAI+ Other	No	Internet	11
2015	Venturella et al.	Survey	EA	686	18–65	Mixed	EAI+ Other	No	Field	15.8

Abbreviations:? = not clear or not known; EA = Exercise Addiction; ED = Exercise Dependence; EAI = Exercise Addiction Inventory; EDS = Exercise Dependence Scale; EE = Excessive Exercise; MT = Test development, validation; n = number of observations; OE = Obligatory Exercise; Other = Other QTRs than the EDS or EAI; Popl. = Population; QTR = Questionnaire; (S) = Study; yr = years science students were considered at risk for the morbidity as based on the EAI. Furthermore, in more intense exercise, the rate could be even higher. As shown in Table 11.1, triathletes, for example, show a risk for the morbidity that ranges between 19.9% (Youngman 2007) to 30.6% (Blaydon & Lindner 2002). However, volume of exercise may not be the sole factor in the prevalence of the risk for exercise addiction because in a study of 95 ultra-marathon runners, Allegre, Therme, and Griffiths (2007) only identified 3.2% to be at risk of exercise addiction. To date, the only population-wide study examining exercise addiction (i.e. Mónok et al. 2012), was conducted on the Hungarian population aged 18–64 years (N = 2,710). The study assessed exercise addiction using both the EDS and the EAI and reported that 0.3% (EDS) and 0.5% (EAI) of the population were at risk for exercise addiction. As noted above, the wide range of the reported risk-prevalence estimates for the morbidity raises serious methodological and conceptual issues in the research area of exercise addiction (and are discussed in more depth in a recent review by Szabo et al. (2015)).

Co-Morbidities and Predisposing Factors in Exercise Addiction

Exercise addiction may be a composite disorder in which other dysfunctions may emerge as a co-morbidity. The eating disorders are the most closely related to exercise addiction, because they often involve exaggerated exercise as a means of weight loss but other comorbid behaviors with exercise addiction have been identified, such as muscle dysmorphia (Foster, Shorter, & Griffiths 2015). Research also suggests that some personality factors may predispose an individual to exercise addiction.

Eating Disorders

It was mentioned earlier that exercise addiction has been classified as both primary and secondary. The latter is the manifestation of excessive exercise in eating disorders as an additional means (to dieting and purging) of weight loss. Excessive involvement in exercise is a common characteristic of eating disorders such as Anorexia Nervosa and Bulimia Nervosa (De Coverley Veale 1987). Involvement in exercise occurs in different "doses" in individuals affected by eating disorders. Over 35 years ago, it was estimated that approximately one-third of anorexic patients may also show abnormally high levels ("doses") of exercise (Crisp, Hsu, Harding, & Hartshorn 1980). Yates, Leehey, and Shisslak (1983) observed a striking resemblance between the psychology of anorexic patients and the very committed runners. They labelled this group of runners as obligatory runners. In the course of their research, the authors interviewed 60 marathoners and examined the traits of a subgroup of male athletes who corresponded to the *obligatory* category. They reported that the male obligatory runners resembled anorexic women in some personality traits, such as feelings of anger, high self-expectation, tolerance of pain, and depression. Yates et al. (1983) related these observations in a unique and hazardous way of establishment of self-identity. This study marked the foundation of research into the relationship between exercise and eating disorders.

Since Yates et al. (1983) published their study, a number of studies have examined the relationship between exercise and eating disorders. A close examination of these studies reveals some controversial findings compared to the original study. For example, three studies comparing anorexic patients with high level, or obligatory, exercisers (Blumenthal et al. 1984; Davis et al. 1995; Knight, Schocken, Powers, Feld, & Smith 1987) failed to demonstrate a relationship between anorexia and excessive exercising. However, Zmijewski and Howard (2003) reported an association between exercise addiction and eating disorders in a student population. A large proportion of the same partipants exhibited symptoms of exercise addiction without any sign of eating disorder. Differences in the adopted research methodology between these inquiries are significant. They all investigated the relationship between exaggerated exercise and anorexia, but from a different perspective. Blumenthal et al. (1984) and Knight et al. (1987) examined mixed gender samples' scores on a popular personality test (the Minnesota Multiphasic Personality Inventory – MMPI). Davis et al. (1995) tested an all-female sample using very specific instruments aimed at assessing compulsiveness, commitment to exercise, and eating disorders. Yates et al. (1983) examined demographic and personality parallels between obligatory runners and anorexic patients. Finally, Zmijewski and Howard (2003) examined a group of healthy students. Furthermore, the classification of the exercise behavior may have differed in these studies. Therefore, the results of all these studies are not directly comparable.

Controversy among the above studies may be reconciled at least in part by considering the results of a study by Wolf and Akamatsu (1994). These authors examined female athletes who exhibited strong tendencies for eating disorders. However, these women did not manifest the personality characteristics associated with eating disorders. Therefore, in agreement with Blumenthal et al.'s (1984) and Knight et al.'s (1987) explanation, differences between people addicted to exercise and the anorexic patients may outweigh the similarities reported by Yates et al. (1983). In a theoretical paper, Yates, Shisslak, Crago, and Allender (1994) also conceded that the comparison of excessive exercisers with eating disordered patients was incorrect because the two populations may be significantly different.

A number of studies conducted by Davis (1990a, 1990b; Davis, Brewer, & Ratusny 1993) examined exercising and non-exercising individuals and their tendency for eating disorders. None of these investigations showed a relationship between exercise behavior and eating disorders. However, a number of studies have reported such a relationship (e.g. French, Perry, Leon, & Fulkerson 1994; Pasman & Thompson 1988; Richert & Hummers 1986; Szymanski & Chrisler 1990; Wolf & Akamatsu 1994). Because, in general, similar measurements were used, the discrepancy between the two sets of studies may be most closely related to the definition of exercise. In the latter set of studies, either excessive exercisers or athletes were tested in contrast to those tested in the first set. However, the definition of *excessive exercise* needs to be standardized in research. Four key factors, including mode, frequency, intensity, and duration, should be reported. Otherwise, it remains unclear what is meant by *excessive exercise* or what is the definition of an *athlete*. Reporting only one or two exercise parameters is not enough, especially in studies dealing with eating disorders, because the latter appears to occur only in a very limited segment of the physically active population.

Several studies suggest that high levels of exercise or athleticism is associated with symptoms of eating disorders. However, the determinants of this relationship are not well understood. Williamson et al. (1995) proposed a psychosocial model for the development of eating disorder symptoms in female athletes (see Figure 11.4). They asserted that overconcern with body size, that is mediated in part by social influence for thinness, anxiety about athletic performance, and negative appraisal of athletic achievement, is a primary and a strong determinant of the etiology of eating disorder symptoms. This model should be given serious consideration in testing several segments of the athletic and exercising population.

Although women appear to be at higher risk for developing eating disorders (Yates et al. 1994), male athletes may be at risk too. For example, Thiel and Hesse (1993) reported a high frequency of eating disorder symptoms and even sub-clinical incidences of diagnosed eating disorders in low weight male

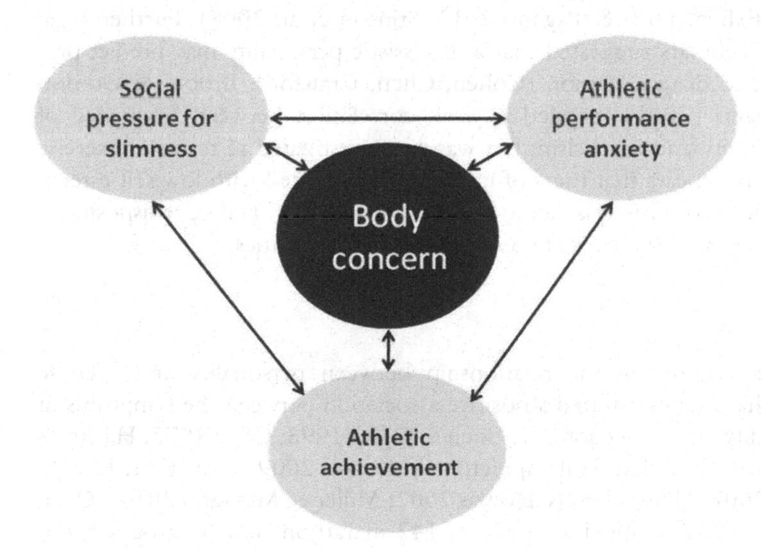

Figure 11.4 A psychosocial model for the development of eating disorder symptoms in female athletes proposed by Williamson et al. (1995); Figure fully redrawn on the basis of Szabo's (2000) work

wrestlers and rowers. This study brings attention to the fact that in some sports (i.e. gymnastics, boxing, wrestling), in which weight maintenance is critical, athletes may be at high risk for developing eating disorders. Athletes in these sports may turn to often *unhealthy* weight control methods (i.e. Enns, Drewnowski, & Grinker 1987). However, this high-risk population has received little attention in the extant literature. In future, more research should be targeted at this segment of the athletic population.

Narcissism

The manifestation of exercise addiction symptoms has been associated with narcissistic personality traits (Flynn 1987; Jibaja-Rusth 1991). However, contrasting findings (at least in women) were also disclosed (Davis & Fox 1993). Recently, Bruno et al. (2014) studied a sample of 150 male and female gym attendees who exhibited higher or lower exercise addiction scores as based on the EAI. While the mean EAI score did not reach the "at risk" category (i.e. mean EAI < 24), gym attendees who scored higher on the EAI also exhibited higher scores of narcissism and lower self-esteem than the other exercisers who scored at the lower end of the EAI. Similar findings were also reported by Miller and Mesagno (2014). These researchers studied 90 male and female gym, fitness centre, and sport contest attendees and found that exercise addiction was positively related to narcissism and perfectionism. Their hierarchical regression analysis revealed that narcissism and self-oriented perfectionism combined predict the risk for exercise addiction. Increased scores of narcissism have also been reported in substance and alcohol addiction (Carter, Johnson, Exline, Post, & Pagano 2012; Stinson et al. 2008). Furthermore, a group of scholars suggested that a narcissistic personality may predict predisposition to drug addiction (Cohen, Chen, Crawford, Brook, & Gordon 2007). Spano (2001) revealed a weak correlation between frequency of physical activity and narcissism, but was not investigated in terms of exercise addiction. It appears that traits of narcissism, combined with low self-esteem and high perfectionism, play a moderating role in individual's predispositions in using exercise as a means of coping with life's adversities.

Perfectionism

Researchers examining the relationship between personality and exercise addiction have demonstrated a positive association between the symptoms of the morbidity and perfectionism (Coen & Ogles 1993; Cook 1997; Hagan & Hausenblas 2003; Hall, Hill, Appleton, & Kozub 2009; Hall, Kerr, Kozub, & Finnie 2007; Hausenblas & Downs 2002; Miller & Mesagno 2014). Coen and Ogles (1993) studied a sample of 142 marathon runners grouped into a high-risk (obligatory) and a low-risk exercise group. The perfectionism scores were higher in the high-risk group in contrast to the low-risk group. However, multidimensional perfectionism only accounted for 11.6% of the variance in exercise addiction. In spite of the low shared variance, the study provided good evidence for the association between exercise addiction and perfectionism. Later, Cook (1997) also reported a positive association between exercise addiction scores and perfectionism, in addition to compulsiveness and body dissatisfaction. Cook also reported a negative relationship between self-esteem and scores of exercise addiction. In another study of 262 exercisers classified as at-risk, symptomatic, and asymptomatic for exercise addiction, the at-risk group scored 15% higher on perfectionism than the asymptomatic subjects (Hausenblas & Downs 2002). In a follow-up study, Symons Downs et al. (2004) replicated these earlier findings with a multidimensional perfectionism tool, on which the group classified as at at-risk for exercise addiction demonstrated higher scores than the asymptomatic and the symptomatic group. The latter group also reported higher perfectionism scores than the asymptomatic group on four of the six perfectionism indices. Similar findings emerged in a study of 246 British middle-distance runners (Hall et al. 2007) in which high ability and perfectionism together explained 29% of the variance in the exercise addiction scores of women. High task and ego orientation, together with perfectionism, explained 27% of the variance in exercise addiction in men. Hall et al. (2009) then replicated the study with 307 British middle-distance runners and found that self-oriented perfectionism and socially prescribed perfectionism were associated with increased scores of exercise addiction.

More recently, Miller and Mesagno (2014) also reported that exercise addiction scores are positively associated with self-oriented perfectionism as well as socially prescribed perfectionism. The positive relationship between perfectionism and indices of exercise addiction may reflect an individual's motivational striving to solve problems according to personal and social expectations, and/or to avoid failure, criticism, and shame. This orientation may also explain why a habituated exerciser with specific personality and behavioral predisposition may choose to rely on a physically effortful coping instead of escape into a passive and instantly gratifying – but marked by a social stigma – form of ill-coping behavior, like drug or alcohol abuse.

Self-Esteem

People at risk for exercise addiction appear to demonstrate lower levels of self-esteem than controls or asymptomatic individuals (Ackard, Brehm, & Steffen 2002; Chittester & Hausenblas 2009; Cook 1997; Grandi, Clementi, Guidi, Benassi, & Tossani 2011; Hall et al. 2009). A study conducted with 155 female aerobic instructors reported a negative correlation between the risk for exercise addiction and self-esteem (Cook 1997). However, the author adopted the Commitment to Exercise Scale to determine the risk for exercise addiction. Therefore, the actual tendency for addiction or a morbid exercise pattern cannot be established in this study. In another survey of 586 college women, Ackard et al. (2002) found that compared to well-adjusted exercisers, women at risk for exercise addiction reported significantly lower self-esteem. Furthermore, a similar study of 113 men also demonstrated a negative correlation between the risk for exercise addiction and self-esteem. However, the shared variance (r^2) was only 4% between the two variables (Chittester & Hausenblas 2009).

A study of 307 British middle-distance runners revealed that labile self-esteem (defined as a fluctuating or unstable self-esteem) mediated the relationship between unconditional self-acceptance and the risk for exercise addiction (Hall et al. 2009). A positive correlation between addiction and labile self-esteem indicated that the two shared 15.2% of the variance. In another study of 107 volunteers, Grandi et al. (2011) also found lower self-esteem in exercisers at risk for exercise addiction in comparison to an asymptomatic control group. In contrast with these findings, Bamber et al. (2000) reported that individuals exhibiting eating disorder symptoms, with or without risk for exercise addiction, had significantly lower levels of self-esteem than a group of control participants and individuals at risk for exercise addiction without symptoms of eating disorders. It appears that self-esteem may play a role in the etiology of exercise addiction, but it may be connected with other personality traits such as perfectionism (Hall et al. 2009; Miller & Mesagno 2014) or other psychological dysfunction like labile self-esteem (Hall et al. 2009). Given that exercise is a socially accepted and rewarded behavior, its adoption for coping with life adversities may preserve – or at least generate the illusion of preservation – one's self-esteem in contrast to destructive and scornful forms of coping like drug or alcohol abuse.

Neuroticism and Extraversion

Neuroticism and extraversion are two relatively widely studied personality traits that have also been associated with the risk for exercise addiction. A study of 246 male and female runners found that among several psychological variables, only neuroticism predicted the risk for exercise addiction (Jibaja-Rusth 1991). Another study (Yates et al. 1992) reported that compulsive runners were more neurotic than asymptomatic runners. Similarly, Kirkby and Adams (1996) found that among a sample of 306 aerobics participants, instructors, and competitors the risk for exercise addiction could be predicted by elevated scores of neuroticism. Bamber et al. (2000) found higher neuroticism scores in people at risk for exercise addiction, who also exhibited eating disorders, in contrast to the exercisers who did not. In another investigation conducted with 390 university students, results demonstrated that extraversion, neuroticism, as well as agreeableness, predicted the risk scores for exercise addiction (Hausenblas & Giacobbi Jr. 2004). These findings were later replicated by Costa and Oliva (2012) who tested 423 volunteers who exercised for at least one full year before the study. Their results suggested that extraversion, neuroticism, and agreeableness are associated with the etiology of exercise addiction. Furthermore, in a study of 806 students and sport association exercisers, Kern (2010) demonstrated a significant inverse relationship between the risk for exercise addiction and emotional stability. Therefore, participants with higher scores of exercise addiction exhibited greater neuroticism. A more recent study conducted with 218 psychology students in Norway has also confirmed that neuroticism may be related to the risk for exercise addiction (Andreassen et al. 2013).

With regard to extraversion, Mathers and Walker (1999) examined 12 exercising students who were at risk for exercise addiction in contrast to 12 asymptomatic and another 12 non-exercising students recruited from the same student population. The "at risk" group did not differ from the asymptomatic group on extraversion, although the two exercise groups were more extraverted than the non-exercise group. The results were interpreted as evidence against the claim that extraversion is a component of the addictive exercise behavior. Similar negative findings were presented by Davis (1990b) in an investigation of 96 exercising women. The author failed to report an association between the risk for exercise addiction and neuroticism and extraversion. Bamber et al. (2000) showed that participants affected by the risk of exercise addiction, and also having symptoms of eating disorders, showed the lowest levels of extraversion. Recently, a group of Danish researchers (Lichtenstein, Christiansen, Bilenberg, & Støving 2014) failed to find a difference in neuroticism between a group of exercisers at risk for exercise addiction and an asymptomatic exercise control group. However, the two groups differed in extraversion, but the effect size was only small to moderate. It appears that findings in the context of an association between the risk for exercise addiction and personality traits are inconsistent.

Neuroticism has also been associated with other behavioral addictions, including internet addiction (Lei & Liu 2006; Tamanaeifar, Arfeei, & Gandomi 2014), pathological gambling (Bagby et al. 2007), compulsive shopping mainly for clothes (Johnson & Attmann 2009), and addiction to pornography and sex (Egan & Parmar 2013; Pinto, Carvalho, & Nobre 2013). Consequently, some personality traits may act as predisposing factors for a wide range of behaviors, including addictive behaviors, that may depend strongly on an interaction with the social and physical environment. For example, the motivation of the individual is determined by past experience and learning. Therefore, non-exercisers who have no affinity for exercise will be unlikely to escape in sports and exercise at times of coping with adversity. However, the exerciser having some personality predisposition, interacting with personal and social motives and/or values, may use exercise to escape from stress in the same way as the social drinker who initially enjoyed alcohol in moderation, but upon experiencing a life trauma starts to drink for pain relief.

It can be posited that some personality traits may increase the risk of escape into excessive exercise. Exercise is not only a part of the everyday life of the habitually physically active individual, but also a conditioned routine behavior. This conditioned aspect of exercise behavior is illustrated by the fact that all exercisers report feelings of deprivation when the exercise routine must be stopped for whatever reason (Szabo 2010). Conditioned behaviors are reinforced with a reward, either positively or negatively. A gray area between the two may depict the transition from the healthy and positively reinforced behavior, that is carried out in moderation, and the morbid pattern of exaggerated exercise behavior, fuelled by negative reinforcement. Not all exercisers, or even the therapeutic exercisers, as illustrated in the interactional model (Egorov & Szabo 2013) will choose exercise as their means of escape from stress and hardship. Personality and learned factors, in combination with environmental factors, will all contribute to the cognitive decision-making process (see Figure 11.3). This complex set of interactions is most likely the moderating factor in the transition from mastery to a *therapeutic* exercise (Egorov & Szabo 2013; Freimuth et al. 2011) as well as in the appraisal of the means of coping when a major or uncontrollable life event strikes. It should be noted that all determinants (gray boxes in Figure 11.3) also interact with each other and all of them bears a different weight at any given time in every person's case. This is why the pathological reliance on exercise behavior can (and should) only be studied from an idiographic perspective as suggested by the interactional model (Egorov & Szabo 2013).

Addiction or Psychological Escape?

To understand what fuels exercise addiction, the motivations for exercise that may trigger an irresistible urge for the behavior need to be considered. People exercise for a reward and the reward may come in different forms. Some common incentives of exercise may include being in good physical shape, looking good, making friends, staying healthy, building muscle, losing weight, etc. The personal experience of the anticipated reward strengthens the exercise behavior. Behaviorists maintain that human action may be understood and explained through reinforcement and punishment. Operant conditioning theory suggests that there are three basic principles of behavior: 1) positive-, and 2) negative reinforcement, and 3) punishment (Bozarth 1994). Positive reinforcement is an incentive for doing something to *gain a reward* that is pleasant or desirable (e.g. increased muscle tone). The reward increases the likelihood of that behavior to reoccur. In contrast, negative reinforcement is an incentive for engaging in a behavior to *escape* a noxious or unpleasant event (e.g. gaining weight). The successful escape is then the reward, which strengthens the behavior. It should be noted that both positive and negative reinforcers increase the likelihood of a particular behavior to reoccur (Bozarth 1994), but their mechanism is different, because in positive reinforcement there is a *gain* that follows the action (e.g. feeling good), whereas in behaviors motivated by negative reinforcement one attempts to *escape* something bad or unpleasant *before happening* that otherwise would occur (e.g. feeling guilty or fat if a planned exercise session is missed).

Punishment refers to situations in which the imposition of some noxious or unpleasant stimulus or the removal of a pleasant or a desired stimulus reduces the probability of a given behavior to reoccur. In contrast to reinforcement, punishment suppresses the behavior and, therefore, exercise should never be used as a means of punishment. Paradoxically, exercise addiction may be perceived as self-punishing behavior. It is a very rare form of addiction that requires substantial physical effort, often to the point of exhaustion.

Individuals addicted to exercise may be motivated by both positive and negative reinforcement. In the former case, high achievers may have an unrealistic self-expectation concerning their goal in exercise that is beyond their physical capacity. Hall et al. (2007) found that ego orientation, perceived ability, task orientation, concern about mistakes, and personal standards accounted for 31% of the variance in exercise addiction. But what are the other factors that account for the rest of the variance? The answer is speculative, but a number of personality factors and also escape behavior(s), associated with the expectancy that exercise can provide a solution for a problem, could be a part of it. Indeed, Duncan (1974), in relation to drug addiction, suggests that addiction is almost identical with, and semantically is just another name for, avoidance or *escape behavior* when the unpleasant feeling is being negatively reinforced by drug or alcohol intake. People addicted to exercise, in this view, reach for a means – with which they had past relief-inducing experience – that provides them with *temporary escape* from an ongoing state of emotional distress and pain. In Duncan's view, all addictions represent similar negatively reinforced behaviors. The term *obligatory exercise* (i.e. Hall et al. 2007) implies a negatively reinforced exercise behavior through which one can avoid something noxious or unwanted. In this view, morbid exercise behavior serves as an escape from stress and adversity.

Conclusions

Exaggerated and morbid patterns of exercise behavior may represent a means for achieving something that is beyond the capacity of the person, or escape from the emotional pain of a reality, that usually appears in the form of chronic and passive stress. The sudden appearance of the latter could be the triggering factor in exercise abuse. Therefore, morbid exercising may be rather "revolutionary" (suddenly appearing) than evolutionary (slowly progressing). High volumes of exercise are not necessarily problematic, considering athletic training at advanced levels. Exercise becomes morbid when it results in harm or exerts negative consequences for the individual. This form of morbidity is known as exercise addiction, and it is different from exaggerated volumes of exercise that accompany various eating disorders with the objective of complementary means to diet for weight loss. The scholastic work on exercise addiction is relatively widespread and it may be out of proportion when considering how rare the disorder appears to be. Indeed, most studies in the area only assess the presumed risk for exercise addiction by using two popular and well validated instruments translated into several languages (i.e. the EAI and EDS). Very few studies have followed up questionnaire data with interviews. The latter is needed, because it becomes increasingly evident that the triggering causes of exercise addiction are very personal and subjective. From a scientific perspective, the subjective aspect of the morbidity makes it difficult to study the disorder with the nomothetic approach. Only a level of risk, that is often misinterpreted as a predisposition for the disorder, can be addressed via the scientific method. However, most of those showing high risk scores on the various exercise addiction screening instruments may in fact never become addicted to exercise. Therefore, as several newer models may suggest, the exploration of individual cases could yield a more profound understanding of the morbidity surrounding exercise addiction.

Note

* The writing of this Chapter was supported by the Hungarian National Research, Development and Innovation Office (Grant numbers: K111938, KKP126835).

References

Ackard, D. M., B. J. Brehm, & J. J. Steffen. 2002. "Exercise and eating disorders in college-aged women: Profiling excessive exercisers." *Eating Disorders 10* (1): 31–47. doi:10.1080/10640260275357354O

Aguiló, A., M. Monjo, C. Moreno, P. Martinez, S. Martínez, & P. Tauler. 2014. "Vitamin C supplementation does not influence plasma and blood mononuclear cell IL-6 and IL-10 levels after exercise." *Journal of Sports Sciences 32* (17): 1659–69. doi:10.1080/02640414.2014.912759

Allegre, B., P. Therme, & M. Griffiths. 2007. "Individual factors and the context of physical activity in exercise dependence: A prospective study of 'ultra-marathoners.'" *International Journal of Mental Health and Addiction 5* (3): 233–43. doi:10.1007/s11469-007-9081-9

Andreassen, C. S., M. D. Griffiths, S. R. Gjertsen, E. Krossbakken, S. Kvam, & S. Pallesen. 2013. "The relationships between behavioral addictions and the five-factor model of personality." *Journal of Behavioral Addictions 2* (2): 90–9. doi:10.1556/JBA.2.2013.003

APA. 1994. "Diagnostic and Statistical Manual of Mental Disorders (DSM-IV)." *Diagnostic and Statistical Manual of Mental Disorders (DSM-IV)*. American Psychiatric Association.

APA. 2013. *Diagnostic and Statistical Manual of Mental Disorders (DSM-5®)*. https://books.google.hu/books/about/Diagnostic_and_Statistical_Manual_of_Men.html?hl=hu&id=-JivBAAAQBAJ.

Babusa, B., E. Czeglédi, F. Túry, S. B. Mayville, & R. Urbán. 2015. "Differentiating the levels of risk for muscle dysmorphia among Hungarian male weightlifters: A factor mixture modeling approach." *Body Image 12* (January): 14–21. doi:10.1016/j.bodyim.2014.09.001

Bagby, R. Michael, D. D. Vachon, E. L. Bulmash, T. Toneatto, L. C. Quilty, & P. T. Costa. 2007. "Pathological gambling and the five-factor model of personality." *Personality and Individual Differences 43* (4): 873–80. doi:10.1016/j.paid.2007.02.011

Bamber, D., I. M. Cockerill, & D. Carroll. 2000. "The pathological status of exercise dependence." *British Journal of Sports Medicine 34* (2): 125–32. doi:10.1136/bjsm.34.2.125

Blaydon, M. J., & K. J. Lindner. 2002. "Eating disorders and exercise dependence in triathletes." *Eating Disorders 10* (1): 49–60. doi:10.1080/106402602753573559

Blumenthal, J. A., O'Toole, L. C., & J. L. Chang. 1984. "Is running an analogue of anorexia nervosa? An empirical study of obligatory running and anorexia nervosa." *JAMA 252* (4): 520–3. doi:10.1001/jama.1984.03350040050022

Bozarth, M. A. 1994. "Physical dependence produced by central morphine infusions: An anatomical mapping study." *Neuroscience and Biobehavioral Reviews 18* (3): 373–83. doi:10.1016/0149-7634(94)90050–7

Bratland-Sanda, S., E. W. Martinsen, J. H. Rosenvinge, Ø. Rø, A. Hoffart, & J. Sundgot-Borgen. 2011. "Exercise dependence score in patients with longstanding eating disorders and controls: The importance of affect regulation and physical activity intensity." *European Eating Disorders Review 19* (3): 249–55. doi:10.1002/erv.971

Bratland-Sanda, S., J. Sundgot-Borgen, Ø. Rø, J. H. Rosenvinge, A. Hoffart, & E. W. Martinsen. 2010. "Physical activity and exercise dependence during inpatient treatment of longstanding eating disorders: An exploratory study of excessive and non-excessive exercisers." *International Journal of Eating Disorders 43* (3): 266–73. doi:10.1002/eat.20769

Brown, R. I. F. 1993. "Some contributions of the study of gambling to the study of other addictions." In W. R. Eadington and J. A. Cornelius (Eds.), *Gambling behavior and problem gambling* (pp. 241–72). Reno, NV: University of Nevada.

Bruno, A., D. Quattrone, G. Scimeca, C. Cicciarelli, V. M. Romeo, G. Pandolfo,…A. Muscatello. 2014. "Unraveling exercise addiction: The role of narcissism and self-esteem." *Journal of Addiction*, 2014 Oct; vol. 2014 e987841. doi:10.1155/2014/987841. Epub 2014 Oct 28.

Carmack, M. A., & R. Martens. 1979. "Measuring commitment to running: A survey of runners' attitudes and mental states." *Journal of Sport Psychology 1* (1): 25–42.

Carter, R. R., S. M. Johnson, J. J. Exline, S. G. Post, & M. E. Pagano. 2012. "Addiction and 'Generation Me': Narcissistic and prosocial behaviors of adolescents with substance dependency disorder in comparison to normative adolescents." *Alcoholism Treatment Quarterly 30* (2): 163–78. doi:10.1080/07347324.2012.663286

Chittester, N. I., & H. A. Hausenblas. 2009. "Correlates of drive for muscularity the role of anthropometric measures and psychological factors." *Journal of Health Psychology 14* (7): 872–7. doi:10.1177/1359105309340986

Cockerill, I. M., & M. E. Riddington. 1996. "Exercise dependence and associated disorders: A review." *Counselling Psychology Quarterly 9* (2): 119–29. doi:10.1080/09515079608256358

Coen, S. P., & B. M. Ogles. 1993. "Psychological characteristics of the obligatory runner: A critical examination of the anorexia analogue hypothesis." *Journal of Sport and Exercise Psychology 15*: 338–54.

Cohen, P., H. Chen, T. N. Crawford, J. S. Brook, & K. Gordon. 2007. "Personality disorders in early adolescence and the development of later substance use disorders in the general population." *Drug and Alcohol Dependence 88* Supplement 1: S71–84. doi:10.1016/j.drugalcdep.2006.12.012

Cook, B., S. Engel, R. Crosby, H. Hausenblas, S. Wonderlich, & J. Mitchell. 2014. "Pathological motivations for exercise and eating disorder specific health-related quality of life." *International Journal of Eating Disorders 47* (3): 268–72. doi:10.1002/eat.22198

Cook, B., H. Hausenblas, D. Tuccitto, & P. R. Giacobbi. 2011. "Eating disorders and exercise: A structural equation modelling analysis of a conceptual model." *European Eating Disorders Review 19* (3): 216–25. doi:10.1002/erv.1111

Cook, B. J., & H. A. Hausenblas. 2008. "The role of exercise dependence for the relationship between exercise behavior and eating pathology mediator or moderator?" *Journal of Health Psychology 13* (4): 495–502. doi:10.1177/1359105308088520

Cook, B. J., & H. A. Hausenblas. 2011. "Eating disorder-specific health-related quality of life and exercise in college females." *Quality of Life Research 20* (9): 1385–90. doi:10.1007/s11136-011-9879-6

Cook, B., T. M. Karr, C. Zunker, J. E. Mitchell, R. Thompson, R. Sherman,…R. D. Crosby. 2013. "Primary and secondary exercise dependence in a community-based sample of road race runners." *J. Sport Exerc. Psychol 35*: 464–9.

Cook, C. A. 1997. *The psychological correlates of exercise dependence in aerobics instructors.* National Library of Canada.

Costa, S., F. Cuzzocrea, H. A. Hausenblas, R. Larcan, & P. Oliva. 2012. "Psychometric examination and factorial validity of the exercise dependence scale-revised in Italian exercisers." *Journal of Behavioral Addictions 1* (4): 186–90. doi:10.1556/JBA.1.2012.009

Costa, S., H. A. Hausenblas, P. Oliva, F. Cuzzocrea, & R. Larcan. 2013. "The role of age, gender, mood states and exercise frequency on exercise dependence." *Journal of Behavioral Addictions 2* (4): 216–23. doi:10.1556/JBA.2.2013.014

Costa, S., H. A. Hausenblas, P. Oliva, F. Cuzzocrea, & R. Larcan. 2014. "Perceived parental psychological control and exercise dependence symptoms in competitive athletes." *International Journal of Mental Health and Addiction 13* (1): 59–72. doi:10.1007/s11469-014-9512-3

Costa, S., & P. Oliva. 2012. "Examining relationship between personality characteristics and exercise dependence." *Review of Psychology 19* (1): 5–11.

Cousineau, D., R. J. Ferguson, J. de Champlain, P. Gauthier, P. Cote, & M. Bourassa. 1977. "Catecholamines in coronary sinus during exercise in man before and after training." *Journal of Applied Physiology 43* (5): 801–6.

Crisp, A. H., L. K. G. Hsu, B. Harding, & J. Hartshorn. 1980. "Clinical features of anorexia nervosa: A study of a consecutive series of 102 female patients." *Journal of Psychosomatic Research 24* (3–4): 179–91. doi:10.1016/0022-3999(80)90040–9

Davis, C. 1990a. "Weight and diet preoccupation and addictiveness: The role of exercise." *Personality and Individual Differences 11* (8): 823–7. doi:10.1016/0191-8869(90)90191-S

Davis, C. 1990b. "Body image and weight preoccupation: A comparison between exercising and non-exercising women." *Appetite 15* (1): 13–21. doi:10.1016/0195-6663(90)90096-Q

Davis, C. 2000. "Exercise abuse." *International Journal of Sport Psychology 31* (2): 278–89.

Davis, C., H. Brewer, & D. Ratusny. 1993. "Behavioral frequency and psychological commitment: Necessary concepts in the study of excessive exercising." *Journal of Behavioral Medicine 16* (6): 611–28. doi:10.1007/BF00844722

Davis, C., & J. Fox. 1993. "Excessive exercise and weight preoccupation in women." *Addictive Behaviors 18* (2): 201–11. doi:10.1016/0306-4603(93)90050-J

Davis, C., S. H. Kennedy, E. Ralevski, M. Dionne, H. Brewer, C. Neitzert, & D. Ratusny. 1995. "Obsessive compulsiveness and physical activity in anorexia nervosa and high-level exercising." *Journal of Psychosomatic Research 39* (8): 967–76. doi:10.1016/0022-3999(95)00064-X

De Coverley Veale, D. M. W. 1987. "Exercise dependence." *British Journal of Addiction 82* (7): 735–40. doi:10.1111/j.1360-0443.1987.tb01539.x

Duncan, D. F. 1974. "Drug abuse as a coping mechanism." *American Journal of Psychiatry 131* (6): 724. doi:10.1176/ajp.131.6.724

Eberle, S. 2004. "Compulsive exercise: Too much of a good thing." *National Eating Disorders Association.*

Edmunds, J., N. Ntoumanis, & J. L. Duda. 2006. "Examining exercise dependence symptomatology from a self-determination perspective." *Journal of Health Psychology 11* (6): 887–903. doi:10.1177/1359105306069091

Egan, V., & R. Parmar. 2013. "Dirty habits? Online pornography use, personality, obsessionality, and compulsivity." *Journal of Sex and Marital Therapy 39* (5): 394–409. doi:10.1080/0092623X.2012.710182

Egorov, A. Y., & A. Szabo. 2013. "The exercise paradox: An interactional model for a clearer conceptualization of exercise addiction." *Journal of Behavioral Addictions 2* (4): 199–208. doi:10.1556/JBA.2.2013.4.2

Enns, M. P., A. Drewnowski, & J. A. Grinker. 1987. "Body composition, body size estimation, and attitudes toward... : Psychosomatic medicine." *LWW.* February. http://journals.lww.com/psychosomaticmedicine/Fulltext/1987/01000/Body_composition,_body_size_estimation,_and.5.aspx.

Flynn, B. R. 1987. *Negative addiction: Psychopathology and changes in mood states in runners.* Unpublished doctoral dissertation.

Foster, A., G. Shorter, & M. Griffiths. 2015. "Muscle dysmorphia: Could it be classified as an addiction to body image?" *Journal of Behavioral Addictions 4* (1): 1–5. doi:10.1556/JBA.3.2014.001

Freimuth, M., S. Moniz, & S. R. Kim. 2011. "Clarifying exercise addiction: Differential diagnosis, co-occurring disorders, and phases of addiction." *International Journal of Environmental Research and Public Health 8* (10): 4069–81. doi:10.3390/ijerph8104069

French, S. A., C. L. Perry, G. R. Leon, & J. A. Fulkerson. 1994. "Food preferences, eating patterns, and physical activity among adolescents: Correlates of eating disorders symptoms." *Journal of Adolescent Health 15* (4): 286–94. doi:10.1016/1054-139X(94)90601-7

Goodman, A. 1990. "Addiction: Definition and implications." *British Journal of Addiction 85* (11): 1403–8. doi:10.1111/j.1360-0443.1990.tb01620.x

Grandi, S., C. Clementi, J. Guidi, M. Benassi, & E. Tossani. 2011. "Personality characteristics and psychological distress associated with primary exercise dependence: An exploratory study." *Psychiatry Research 189* (2): 270–5. doi:10.1016/j.psychres.2011.02.025

Griffiths, M. 1997. "Exercise addiction: A case study." *Addiction Research 5* (2): 161–8. doi:10.3109/16066359709005257

Griffiths, M. 2005. "A 'components' model of addiction within a biopsychosocial framework." *Journal of Substance Use 10* (4): 191–7. doi:10.1080/14659890500114359

Griffiths, M. D., R. Urbán, Z. Demetrovics, M. B. Lichtenstein, R. de la Vega, B. Kun,...& A. Szabo. 2015. "A cross-cultural re-evaluation of the Exercise Addiction Inventory (EAI) in five countries." *Sports Medicine – Open 1* (1): 5. doi:10.1186/s40798-014-0005-5

Griffiths, M. D., A. Szabo, & A. Terry. 2005. "The exercise addiction inventory: A quick and easy screening tool for health practitioners." *British Journal of Sports Medicine 39* (6): e30. doi:10.1136/bjsm.2004.017020

Hagan, A. L., & H. A. Hausenblas. 2003. "The relationship between exercise dependence symptoms and perfectionism – ProQuest." *American Journal of Health Studies 18* (2/3): 133–7.

Hailey, B. J., & L. A. Bailey. 1982. "Negative addiction in runners: A quantitative approach." *Journal of Sport Behavior 5* (3): 150–4.

Hall, H. K., A. P. Hill, P. R. Appleton, & S. A. Kozub. 2009. "The mediating influence of unconditional self-acceptance and labile self-esteem on the relationship between multidimensional perfectionism and exercise dependence." *Psychology of Sport and Exercise 10* (1): 35–44. doi:10.1016/j.psychsport.2008.05.003

Hall, H. K., A. W. Kerr, S. A. Kozub, & S. B. Finnie. 2007. "Motivational antecedents of obligatory exercise: The influence of achievement goals and multidimensional perfectionism." *Psychology of Sport and Exercise 8* (3): 297–316. doi:10.1016/j.psychsport.2006.04.007

Hamer, M., & C. I. Karageorghis. 2007. "Psychobiological mechanisms of exercise dependence." *Sports Medicine 37* (6): 477–84. doi:10.2165/00007256-200737060-00002

Hausenblas, H. A., & D. Symons Downs. 2002. "How much is too much? The development and validation of the exercise dependence scale." *Psychology and Health 17* (4): 387–404. doi:10.1080/0887044022000004894

Hausenblas, H. A., & E. A. Fallon. 2002. "Relationship among body image, exercise behavior, and exercise dependence symptoms." *International Journal of Eating Disorders 32* (2): 179–85. doi:10.1002/eat.10071

Hausenblas, H. A., L. Gauvin, D. S. Downs, & A. R. Duley. 2008. "Effects of abstinence from habitual involvement in regular exercise on feeling states: An ecological momentary assessment study." *British Journal of Health Psychology 13* (2): 237–55. doi:10.1348/135910707X180378

Hausenblas, H. A., & P. R. Giacobbi Jr. 2004. "Relationship between exercise dependence symptoms and personality." *Personality and Individual Differences 36* (6): 1265–73. doi:10.1016/S0191-8869(03)00214-9

Hurst, R., B. Hale, D. Smith, & D. Collins. 2000. "Exercise dependence, social physique anxiety, and social support in experienced and inexperienced bodybuilders and weightlifters." *British Journal of Sports Medicine 34* (6): 431–5. doi:10.1136/bjsm.34.6.431

Jibaja-Rusth, M. L. 1991. *The development of a psycho-social risk profile for becoming an obligatory runner*. Microform Publications, College of Human Development and Performance, University of Oregon.

Johnson, T., & J. Attmann. 2009. "Compulsive buying in a product specific context: Clothing." *Journal of Fashion Marketing and Management: An International Journal 13* (3): 394–405. doi:10.1108/13612020910974519

Kern, L. 2010. "Relationship between exercise dependence and big five personality." *L'Encephale 36* (3): 212–18. doi:10.1016/j.encep.2009.06.007

Kirkby, R., & J. Adams. 1996. "Exercise dependence: The relationship between two measures." *Perceptual and Motor Skills 82* (2): 366.

Knight, P. O., D. D. Schocken, F. S. Powers, J. Feld, & J. T. Smith. 1987. "396: Gender comparison in anorexia nervosa and oblicate running." *Medicine and Science in Sports & Exercise 19* (2): S66.

Kotbagi, G., I. Muller, L. Romo, & L. Kern. 2014. "Pratique problématique d'exercice physique: Un cas clinique." *Annales Médico-Psychologiques, Revue Psychiatrique 172* (10): 883–7. doi:10.1016/j.amp.2014.10.011

Lei, L., Y. Yang, & M. Liu. 2006. "The relationship between adolescents' neuroticism, internet service preference, and internet addiction. " *Acta Psychologica Sinica 38* (3): 375–81.

http://en.cnki.com.cn/Article_en/CJFDTOTAL-XLXB200603008.htm.

Lejoyeux, M., M. Avril, C. Richoux, H. Embouazza, & F. Nivoli. 2008. "Prevalence of exercise dependence and other behavioral addictions among clients of a Parisian fitness room." *Comprehensive Psychiatry 49* (4): 353–8. doi:10.1016/j.comppsych.2007.12.005

Lejoyeux, M., C. Guillot, F. Chalvin, A. Petit, & V. Lequen. 2012. "Exercise dependence among customers from a Parisian sport shop." *Journal of Behavioral Addictions 1* (1): 28–34. doi:10.1556/JBA.1.2012.1.3

Lichtenstein, M. B., E. Christiansen, N. Bilenberg, & R. K. Støving. 2014. "Validation of the exercise addiction inventory in a Danish sport context." *Scandinavian Journal of Medicine and Science in Sports 24* (2): 447–53. doi:10.1111/j.1600-0838.2012.01515.x

Lichtenstein, M. B., K. S. Larsen, E. Christiansen, R. K. Støving, & T. V. G. Bredahl. 2014. "Exercise addiction in team sport and individual sport: Prevalences and validation of the Exercise Addiction Inventory." *Addiction Research and Theory 22* (5): 431–7. doi:10.3109/16066359.2013.875537

Li, M., J. Nie, & Y. Ren. 2014. "The correlation analysis between exercise dependence and exercise motivation." *Computer Modelling and New Technologies 18* (12A): 513–16.

Lindwall, M., & A. Palmeira. 2009. "Factorial validity and invariance testing of the exercise dependence scale-revised in Swedish and Portuguese exercisers." *Measurement in Physical Education and Exercise Science 13* (3): 166–79. doi:10.1080/10913670903050313

Loumidis, K. S., & A. Wells. 1998. "Assessment of beliefs in exercise dependence: The development and preliminary validation of the exercise beliefs questionnaire." *Personality and Individual Differences 25* (3): 553–67. doi:10.1016/S0191-8869(98)00103-2

McNally, G. P., & H. Akil. 2002. "Role of corticotropin-releasing hormone in the amygdala and bed nucleus of the stria terminalis in the behavioral, pain modulatory, and endocrine consequences of opiate withdrawal." *Neuroscience 112* (3): 605–17. doi:10.1016/S0306-4522(02)00105-7

McNamara, J., & M. P. McCabe. 2012. "Striving for success or addiction? Exercise dependence among elite Australian athletes." *Journal of Sports Sciences 30* (8): 755–66. doi:10.1080/02640414.2012.667879

McNamara, J., & M. P. McCabe. 2013. "Development and validation of the exercise dependence and elite athletes scale." *Performance Enhancement and Health 2* (1): 30–6. doi:10.1016/j.peh.2012.11.001

Maraz, A., R. Urbán, M. D. Griffiths, & Z. Demetrovics. 2015. "An empirical investigation of dance addiction." *PLOS ONE 10* (5): e0125988. doi:10.1371/journal.pone.0125988

Mathers, S., & M. B. Walker. 1999. "Extraversion and exercise addiction." *Journal of Psychology 133*: 125–8.

Menczel, Z., E. Kovacs, J. Farkas, A. Magi, A. Eisinger, B. Kun, & Z. Demetrovics. 2013. "Prevalence of exercise dependence and eating disorders among clients of fitness centers in Budapest." *Journal of Behavioral Addictions 2*: 23–4.

Miller, K. J., & C. Mesagno. 2014. "Personality traits and exercise dependence: Exploring the role of narcissism and perfectionism." *International Journal of Sport and Exercise Psychology 12* (4): 368–81. doi:10.1080/1612197X.2014.932821

Modoio, V. B., H. K. M. Antunes, P. R. B. de Gimenez, M. L. De Mello Santiago, S. Tufik, & M. T. de Mello. 2011. "Negative addiction to exercise: Are there differences between genders?" *Clinics 66* (2): 255–60. doi:10.1590/S1807-59322011000200013

Mónok, K., K. Berczik, R. Urbán, A. Szabo, M. D. Griffiths, J. Farkas,...A. Magi. 2012. "Psychometric properties and concurrent validity of two exercise addiction measures: A population wide study." *Psychology of Sport and Exercise 13* (6): 739–46. doi:10.1016/j.psychsport.2012.06.003

Müller, A., B. Cook, H. Zander, A. Herberg, V. Müller, & M. de Zwaan. 2014. "Does the German version of the exercise dependence scale measure exercise dependence?" *Psychology of Sport and Exercise 15* (3): 288–92. doi:10.1016/j.psychsport.2013.12.003

O'Brien, C. P., N. Volkow, & T.-K. Li. 2006. "What's in a word? Addiction versus dependence in DSM-V." *American Journal of Psychiatry 163* (5): 764–5. doi:10.1176/appi.ajp.163.5.764

Ogden, J., D. Veale, & Z. Summers. 1997. "The development and validation of the exercise dependence questionnaire." *Addiction Research 5* (4): 343–55. doi:10.3109/16066359709004348

Pasman, L., & J. K. Thompson. 1988. "Body image and eating disturbance in obligatory runners, obligatory weightlifters, and sedentary individuals." *International Journal of Eating Disorders 7* (6): 759–69. doi:10.1002/1098-108X(198811)7:6<759::AID-EAT2260070605>3.0.CO;2-G

Pinto, J., J. Carvalho, & P. J. Nobre. 2013. "The relationship between the FFM personality traits, state psychopathology, and sexual compulsivity in a sample of male college students." *Journal of Sexual Medicine 10* (7): 1773–82. doi:10.1111/jsm.12185

Pugh, N. E., & H. D. Hadjistavropoulos. 2011. "Is anxiety about health associated with desire to exercise, physical activity, and exercise dependence?" *Personality and Individual Differences 51* (8): 1059–62. doi:10.1016/j.paid.2011.08.025

Puterman, E., E. S. Epel, A. O'Donovan, A. A. Prather, K. Aschbacher, & F. S. Dhabhar. 2013. "Anger is associated with increased il-6 stress reactivity in women, but only among those low in social support." *International Journal of Behavioral Medicine 21* (6): 936–45. doi:10.1007/s12529-013-9368-0

Richert, A. J., & J. A. Hummers. 1986. "Patterns of physical activity in college students at possible risk for eating disorder." *International Journal of Eating Disorders 5* (4): 757–63. doi:10.1002/1098-108X(198605)5:4<757::AID-EAT2260050414>3.0.CO;2-0

Saanijoki, T., L. Tuominen, L. Nummenmaa, E. Arponen, K. Kalliokoski, & J. Hirvonen. 2014. "Physical exercise activates the {mu}-opioid system in human brain." *Society of Nuclear Medicine Annual Meeting Abstracts 55* (Supplement 1): 1909.

Sachs, M. L., & D. Pargman. 1979. "Running addiction: A depth interview examination." *Journal of Sport Behavior 2* (3): 143–55.

Schreiber, K., & H. A. Hausenblas. 2015. *The truth about exercise addiction: Understanding the dark side of thinspiration*. Lanham, MD: Rowman and Littlefield.

Sicilia, Á., & D. González-Cutre. 2011. "Dependence and physical exercise: Spanish validation of the Exercise Dependence Scale-Revised (EDS-R)." *Spanish Journal of Psychology 14* (01): 421–31. doi:10.5209/rev_SJOP.2011.v14.n1.38

Slay, H. A., J. Hayaki, M. A. Napolitano, & K. D. Brownell. 1998. "Motivations for running and eating attitudes in obligatory versus nonobligatory runners." *International Journal of Eating Disorders 23* (3): 267–75. doi:10.1002/(SICI)1098-108X(199804)23:3<267::AID-EAT4>3.0.CO;2-H

Smith, D., B. Hale, & D. Collins. 1998. "Measurement of exercise dependence in bodybuilders." *Journal of Sports Medicine and Physical Fitness 38* (1): 66–74.

Smith, D., C. Wright, & D. Winrow. 2010. "Exercise dependence and social physique anxiety in competitive and non-competitive runners." *International Journal of Sport and Exercise Psychology 8* (1): 61–9.

Spano, L. 2001. "The relationship between exercise and anxiety, obsessive-compulsiveness, and narcissism." *Personality and Individual Differences 30* (1): 87–93. doi:10.1016/S0191-8869(00)00012-X

Stinson, F. S., D. A. Dawson, R. B. Goldstein, S. P. Chou, B. Huang, S. M. Smith,...W. J. Ruan. 2008. "Prevalence, correlates, disability, and comorbidity of DSM-IV narcissistic personality disorder: Results from the Wave 2 national epidemiologic survey on alcohol and related conditions." *Journal of Clinical Psychiatry 69* (7): 1033–45.

Sussman, S., A. Leventhal, R. N. Bluthenthal, M. Freimuth, M. Forster, & S. L. Ames. 2011. "A framework for the specificity of addictions." *International Journal of Environmental Research and Public Health 8* (8): 3399–415. doi:10.3390/ijerph8083399

Symons Downs, D., H. A. Hausenblas, & C. R. Nigg. 2004. "Factorial validity and psychometric examination of the exercise dependence scale-revised." *Measurement in Physical Education and Exercise Science 8* (4): 183–201. doi:10.1207/s15327841mpee0804_1

Szabo, A. 1995. "The impact of exercise deprivation on well-being of habitual exercises." *Australian Journal of Science and Medicine in Sport 27* (3): 68–75.

Szabo, A. 2010. *Addiction to Exercise: A Symptom or a Disorder?* New York: Nova Science.

Szabo, A., E. Billett, & J. Turner. 2001. "Phenylethylamine, a possible link to the antidepressant effects of exercise?" *British Journal of Sports Medicine 35* (5): 342–3. doi:10.1136/bjsm.35.5.342

Szabo, A., M. D. Griffiths, R. de la Vega, B. Mervó, & Z. Demetrovics. 2015 "Methodological and conceptual limitations in exercise addiction research." *Yale Journal of Biology and Medicine 88* (3): 303–8.

Szabo, A., & E. C. Tsang. 2003. "Motivation for increased self-selected exercise intensity following psychological distress: Laboratory-based evidence for catharsis?" *Journal of Psychosomatic Research 55* (2): 133.

Szabo, A. 2000. "Physical activity as a source of psychological dysfunction." *Physical Activity and Psychological Well-Being* 130–53.

Szabo, A., R. De La Vega, R. Ruiz-Barquín, & O. Rivera. 2013. "Exercise addiction in Spanish athletes: Investigation of the roles of gender, social context and level of involvement." *Journal of Behavioral Addictions 2* (4): 249–52. doi:10.1556/JBA.2.2013.4.9

Szabo, A., & M. D. Griffiths. 2006. "Exercise addiction in British sport science students." *International Journal of Mental Health and Addiction 5* (1): 25–8. doi:10.1007/s11469-006-9050-8

Szymanski, L. A., & J. C. Chrisler. 1990. "Eating disorders, gender-role and athletic activity." *Psychology: A Journal of Human Behavior 27* (4): 20–9.

Tamanaeifar, M. R., F. S. Arfeei, & Z. Gandomi. 2014. "Relationship between internet addiction with neuroticism in High School students, Kashan, Iran." *Bimonthly Journal of Hormozgan University of Medical Sciences 17* (1): 69–75.

Taranis, L., S. Touyz, & C. Meyer. 2011. "Disordered eating and exercise: Development and preliminary validation of the Compulsive Exercise Test (CET)." *European Eating Disorders Review 19* (3): 256–68. doi:10.1002/erv.1108

Terry, A., A. Szabo, & M. Griffiths. 2004. "The Exercise Addiction Inventory: A new brief screening tool." *Addiction Research and Theory 12* (5): 489–99.

Thaxton, L. 1982. "Physiological and psychological effects of short-term exercise addiction on habitual runners." *Journal of Sport Psychology 4*: 73–80.

Thiel, A., H. Gottfried, & F. W. Hesse. 1993. "Subclinical eating disorders in male athletes." *Acta Psychiatrica Scandinavica 88* (4): 259–65. doi:10.1111/j.1600-0447.1993.tb03454.x

Thompson, J., & L. Pasman. 1991. "The obligatory exercise questionnaire." *Behavior Therapist* (January) *137*.

Thompson, K. J., & P. Blanton. 1987. "Energy conservation and exercise dependence: A sympathetic arousal hypothesis." *Medicine and Science in Sports and Exercise 19* (2): 91–9. doi:10.1249/00005768-198704000-00005

Thornton, E. W., & S. E. Scott. 1995. "Motivation in the committed runner: Correlations between self-report scales and behaviour." *Health Promotion International 10* (3): 177–84. doi:10.1093/heapro/10.3.177

Trana, I. 2013. "Prevalence and characteristics of exercise dependence among Norwegian men." *24*. http://brage.bibsys.no/xmlui/handle/11250/271210.

Veale, David. 1995. "Does primary exercise dependence really exist." *Exercise Addiction: Motivation for Participation in Sport and Exercise* 1–5.

Venturella, F., G. Uccello, F. Aiello, C. Sanfilippo, G. Passavanti, & M. Mandala. 2015. "Rilevazione statistica della diffusione e della conoscen-za di sostanze dopanti, integratori e dell'exercise addiction." *Italian Journal on Addiction 4* (3). https://iris.unipa.it/bitstream/10447/124921/1/Rilevazione%20Statistica%20Doping.pdf.

Villella, C., G. Martinotti, M. Di Nicola, M. Cassano, G. La Torre, M. D. Gliubizzi,… I. Messeri. 2010. "Behavioural addictions in adolescents and young adults: Results from a prevalence study." *Journal of Gambling Studies 27* (2): 203–14. doi:10.1007/s10899-010-9206-0

Warner, R., & M. D. Griffiths. 2006. "A qualitative thematic analysis of exercise addiction: An exploratory study." *International Journal of Mental Health and Addiction 4* (1): 13–26. doi:10.1007/s11469-006-9000-5

Williamson, D. A., R. G. Netemeyer, L. P. Jackman, D. A. Anderson, C. L. Funsch, & J. Y. Rabalais. 1995. "Structural equation modeling of risk factors for the development of eating disorder symptoms in female athletes." *International Journal of Eating Disorders 17* (4): 387–93. doi:10.1002/1098-108X(199505)17:4<387::AID-EAT2260170411>3.0.CO;2-M

Wolf, E. M., & T. J. Akamatsu. 1994. "Exercise involvement and eating-disordered characteristics in college students." *Eating Disorders 2* (4): 308–18. doi:10.1080/10640269408249129

Yates, A., K. Leehey, & C. M. Shisslak. 1983. "Running — an analogue of anorexia?" *New England Journal of Medicine 308* (5): 251–5. doi:10.1056/NEJM198302033080504

Yates, A., C. M. Shisslak, J. Allender, M. Crago, & K. Leehey. 1992. "Comparing obligatory to nonobligatory runners." *Psychosomatics 33* (2): 180–9. doi:10.1016/S0033-3182(92)71994-X

Yates, A., C. M. Shisslak, M. Crago, & J. Allender. 1994. "Overcommitment to sport: is there a relationship to the eating disorders? " *Clinical Journal of Sport Medicine.* http://journals.lww.com/cjsportsmed/Fulltext/1994/01000/Overcommitment_to_Sport__Is_There_a_Relationship.6.aspx.

Youngman, J. 2007. "Risk for exercise addiction: A comparison of triathletes training for sprint-, Olympic-, half-Ironman-, and Ironman-distance triathlons." *Open Access Dissertations,* December. http://scholarlyrepository.miami.edu/oa_dissertations/12.

Zmijewski, C. F., & M. O. Howard. 2003. "Exercise dependence and attitudes toward eating among young adults." *Eating Behaviors 4* (2): 181–95. doi:10.1016/S1471-0153(03)00022-9

12 Aerobic Exercise in People with Schizophrenia

From Efficacy to Effectiveness

Davy Vancampfort, Simon Rosenbaum, Michel Probst, and Brendon Stubbs

Introduction

Schizophrenia is one of the most debilitating psychiatric disorders (Rössler, Salize, van Os, & Riecher-Rössler 2005). The Diagnostic and Statistical Manual of Mental Disorders *criteria* for schizophrenia include positive and negative symptomatology severe enough to cause social and occupational dysfunction (American Psychiatric Association 2013). Positive symptoms reflect an excess or distortion of normal functioning and include delusions, hallucinations or disorganized speech and behavior (American Psychiatric Association 2013). Negative symptoms reflect a reduction or loss of normal functioning and include symptoms such as affective flattening, apathy, avolition, and social withdrawal (American Psychiatric Association 2013). The lifetime prevalence and incidence of schizophrenia range from 0.30% to 0.66% and from 10.2 to 22.0 per 100,000 person-years, respectively (McGrath, Saha, Chant, & Welham 2008).

People with schizophrenia have consistently higher levels of mortality and morbidity than the general population. Mortality rates remain persistently high at around two to three times those of the general population (Walker, McGee, & Druss 2015). The life expectancy of people with schizophrenia is shortened by 10 to 20 years (Laursen, Munk-Olsen, & Vestergaard 2012) and the *mortality gap* with the general population continues to grow (Saha, Chant, & McGrath 2007). This widening differential gap in premature mortality suggests that people with schizophrenia have not yet fully benefited from the health care improvements observed in the general population. The underlying causes for the increased risk for premature mortality of this population is complex and multi-factorial. It is well established that people with schizophrenia are at an increased risk for cardiovascular disease (Li, Fan, Tang, & Cheng 2014), metabolic syndrome (Vancampfort et al. 2015c), diabetes (Vancampfort et al. 2016a), and respiratory diseases (De Hert et al. 2011). Although genetic factors (Andreassen et al. 2013) and shared pathophysiological mechanisms (Manu et al. 2014) contribute significantly to this high risk profile, unhealthy lifestyle habits such as smoking (Stubbs et al. 2015), poor diet (Heald et al. 2015), and low levels of physical activity (Stubbs et al. 2016) play the most prominent role. In addition, treatment-related factors contribute to the increased risk for somatic co-morbidity (Correll, Detraux, De Lepeleire, & De Hert 2015).

For example, *psychotropic medication* can lead to significant metabolic problems such as weight gain, lipid abnormalities, and changes in glucose regulation. Although in previous years there has been a greater focus on improving the health status of people with schizophrenia (De Hert et al. 2011), there are already clear health inequalities present at the early stages of the disease (Curtis et al. In press), with patients having less access to, and poorer quality of medical care (Mitchell & De Hert 2015). Concurrently, the nature of the mental illness, including the persistence of negative, cognitive and positive symptoms, may affect adherence to treatment as many individuals place unique challenges for the adoption and maintenance of healthy lifestyles (Vancampfort et al. 2015a; Vancampfort et al. 2015b; Vancampfort, Stubbs, Venigalla, & Probst 2015d).

Although psychotropic medication is effective in reducing positive, psychotic symptoms, usually within the early stages of treatment (Leucht et al. 2013), it is less effective for treating the negative symptoms (Fusar-Poli et al. 2014) and cognitive deficits (Nielsen et al. 2015). Unfortunately, it is these symptoms that cause most long-term disability (Vancampfort et al. 2012b). *Cognitive-behavioral therapy* has limited beneficial effects on negative and cognitive symptoms (Jauhar et al. 2014), yet these interventions have considerable societal costs (van der Gaag, Stant, Wolters, Buskens, & Wiersma 2011). Thus, other low-cost treatments that decrease negative symptoms, reduce cognitive deficits, and promote physical health and functional recovery are warranted. Recent meta-analyses (Rosenbaum, Tiedemann, Sherrington, Curtis, & Ward 2014; Firth, Cotter, Elliott, French, & Yung 2015) clearly demonstrated that *exercise* is one possible strategy that could meet this need. The UK National Institute for Health and Care Excellence recommends using exercise and dietary advice to improve the physical health of people with schizophrenia (Kuipers, Yesufu-Udechuku, Taylor, & Kendall 2014). However, current treatment guidelines do not present clear evidence-based recommendations on how to implement exercise within multidisciplinary treatment programs.

The first aim of this chapter is to provide a systematic overview of intervention characteristics, exercise outcomes, and motivational skills used in recent (i.e. last decade) randomized controlled trials investigating the effect of aerobic exercise in schizophrenia. In order to obtain a comprehensive assessment of the impact of aerobic exercise, we conducted a meta-analysis of all studies assessing the impact of aerobic exercise on mental and physical health parameters (if reported in at least three studies) in people with schizophrenia. The second aim is to provide evidence-based clinical and research recommendations regarding the prescription of exercise in schizophrenia.

Methods

Search Strategy

We searched PubMed from 2005 until May 15 2015. We used the key words: "exercise" AND "schizophrenia" in the title, abstract or index term fields. Manual searches were also conducted using the reference lists from recovered articles. After the removal of duplicates, the titles and abstracts of all potentially eligible articles were screened.

Eligibility Criteria

Studies were only included if: (a) randomized controlled trials; (b) published since 2005; (c) based on first-episode or chronic schizophrenia (spectrum) diagnosed according to ICD-10 or DSM-IV (V) criteria, and (d) focused on physical and/or mental health outcomes of aerobic exercise interventions. Aerobic exercise interventions were defined as physical activity that is planned, structured, repetitive, and purposive in the sense that improvement or maintenance of physical fitness or mental or physical health is the primary objective (Caspersen, Powell, & Christenson 1985). Interventions that only used yoga, tai-chi, mindfulness exercises, muscular relaxation techniques or adventure activities were excluded, since their effects are often theoretically derived from factors distinct from the aerobic exercise itself. Multi-modal programs that incorporated aerobic exercise within a broader lifestyle or psychosocial intervention were also excluded, as the effects of aerobic exercise alone cannot be determined. Eligible control conditions included (a) treatment as usual, (b) wait-list, or (c) non-aerobic exercise interventions. Studies were also excluded if: (a) they did not report the effect of exercise on at least one quantitative measure of physical or mental health, (b) they were not published in a peer-reviewed, English language journal, and (c) conference abstracts.

Data Extraction

The following data was extracted: (a) number and characteristics of participants (age and gender), (b) intervention characteristics (frequency, intensity, type, and time of exercise intervention), (c) medical clearance guidelines and serious adverse events, (d) physical and mental health outcomes (cardio-metabolic parameters, physical fitness, psychiatric symptoms and quality of life), (e) dropout rates, and (f) motivational strategies used by clinicians or patients.

Data Analysis

Random-effects meta-analyses were conducted using Comprehensive Meta-Analysis software (Version 3, Biostat, Englewood, New Jersey). Intervention effect sizes (differences between intervention and control groups) were calculated using Hedges' g statistic, and 95% confidence intervals (CIs). Statistical heterogeneity was quantified using the I^2 statistic. An I^2 of more than 75% was considered to indicate considerable heterogeneity, I^2 of 50%–75% was considered to indicate substantial heterogeneity, and an I^2 of less than 40% was considered to indicate limited heterogeneity.

Results

Search Results

An overview of the search results is presented in Figure 12.1. In total 16 RCTs (Beebe et al. 2005; Acil, Dogan, & Dogan 2008; Marzolini, Jensen, & Melville 2009; Pajonk et al. 2010; Behere et al. 2011; Methapatara & Srisurapanont 2011; Scheewe et al. 2012; Varambally et al. 2012; Falkai et al. 2013; Scheewe et al. 2013a; Scheewe et al. 2013b; Battaglia et al. 2013, Kaltsatou et al. 2014; Oertel-Knochel et al. 2014; Kimhy et al. 2015; Svatkova et al. 2015) involving 373 unique persons with ICD or DSM schizophrenia spectrum disorders met the inclusion criteria. The sample size of the included studies ranged from 10 (Beebe et al. 2005) to 73 (Varambally et al. 2012). The majority of the included participants were males.

Characteristics of the Aerobic Exercise Interventions

There was considerable heterogeneity among the identified aerobic exercise interventions. The frequency of sessions ranged from two times per week to daily sessions. Intensity ranged from light exercises to 85% of the maximal heart rate, the time of sessions ranged from 30mins to 120mins, while the duration of interventions ranged from four weeks to eight months. The types

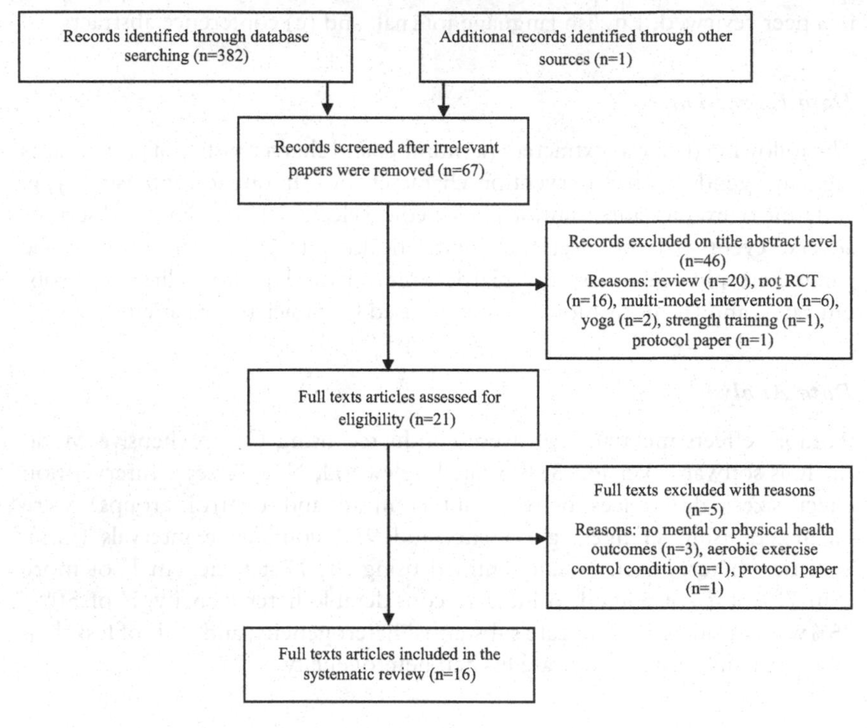

Figure 12.1 Flow diagram for the search results

of aerobic interventions researched were treadmill walking, stationary cycling, various cardiovascular exercise modalities (including aerobic exercises and/or videogames), pedometer walking, soccer, and traditional dancing.

Physical and Mental Health Outcomes of Aerobic Exercise

Cardio-Metabolic Risks

The body mass index was the most common cardio-metabolic outcome examined. Data from five studies (see Figure 12.2) found that exercise did not significantly reduce the body mass index (hedges g =–0.32, 95%CI=–0.79 to 0.16, p=0.16; Figure 12.2). There was some heterogeneity (I^2=57%).

Cardiorespiratory Fitness

It was possible to pool data from three RCTs to compare the influence of aerobic exercise on maximal (VO_2max), predicted or peak (VO_2peak) oxygen uptake with other interventions. This analysis established that exercise significantly improved predicted VO_2max or VO_2peak compared to a control group

Study name	Statistics for each study			
	Hedges's g	Lower limit	Upper limit	p-Value
Marzolini 2009	0.543	−0.458	1.544	0.288
Scheewe 2012	−0.099	−0.626	0.429	0.714
Methapatara 2011	−0.200	−0.685	0.285	0.420
Battaglia 2013	−0.535	−1.438	0.367	0.245
Kalsatou 2014	−1.209	−1.958	−0.460	0.002
	−0.316	−0.789	0.156	0.190

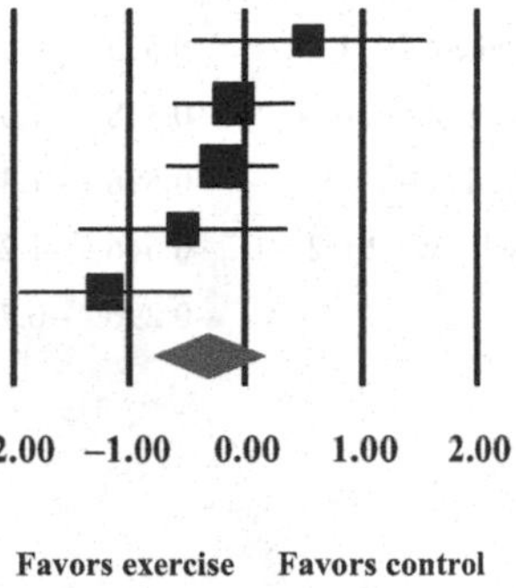

Hedges's g and 95% CI

−2.00 −1.00 0.00 1.00 2.00

Favors exercise Favors control

Figure 12.2 Meta-analysis investigating the effect of aerobic exercise on body mass index

Study name	Statistics for each study			
	Hedges's g	Lower limit	Upper limit	p-Value
Kimhy 2015	0.842	0.173	1.511	0.014
Scheewe 2012	0.257	−0.412	0.925	0.452
Pajonk 2010	0.225	−0.704	1.155	0.635
	0.483	0.061	0.904	0.025

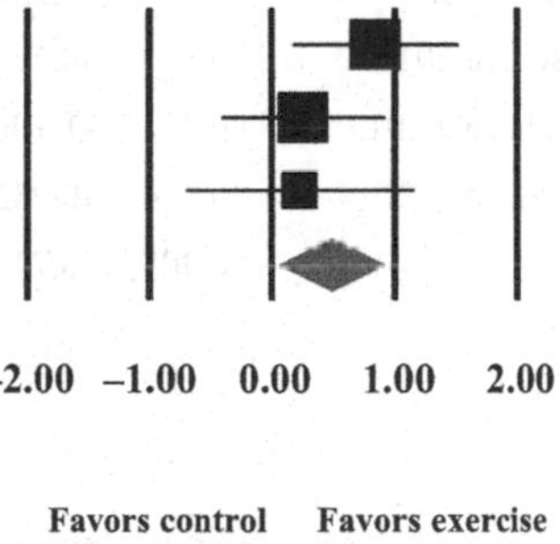

Hedges's g and 95% CI

−2.00 −1.00 0.00 1.00 2.00

Favors control Favors exercise

Figure 12.3 Meta-analysis investigating the effect of aerobic exercise on cardiorespiratory fitness

(hedges g=0.48, 95% CI=0.06 to 0.90, p=0.02) (see Figure 12.3). There was no heterogeneity (I^2=0%)

Psychiatric Symptoms

Data from four studies found that exercise did not significantly reduce positive symptoms (hedges g =−0.30, 95%CI=−0.78 to 0.18, p=0.23; Figure 12.4). There was some heterogeneity (I^2=52%).

Study name	Hedges's g	Lower limit	Upper limit	p-Value
Behere 2011	0.373	−0.252	0.999	0.242
Kaltsatou 2014	−0.358	−1.050	0.334	0.310
Acil 2008	−0.596	−1.308	0.117	0.101
Scheewe 2012	−0.646	−1.278	−0.014	0.045
	−0.296	−0.776	0.185	0.228

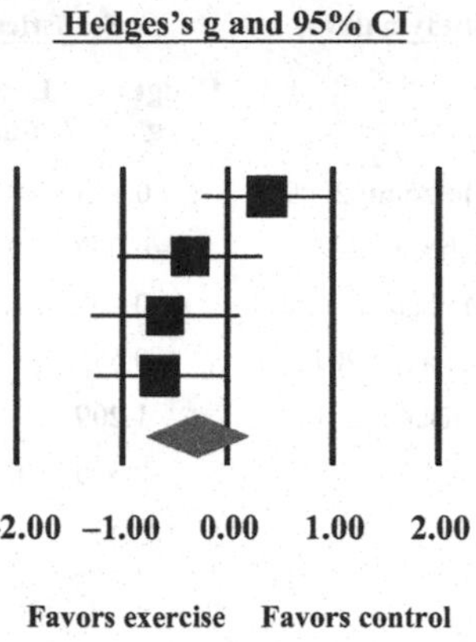

Hedges's g and 95% CI

−2.00 −1.00 0.00 1.00 2.00

Favors exercise Favors control

Figure 12.4 Meta-analysis investigating the effect of aerobic exercise on positive symptoms

Study name	Hedges's g	Lower limit	Upper limit	p-Value
Acil 2008	−0.810	−1.536	−0.085	0.029
Behere 2011	−0.173	−0.794	0.448	0.585
Scheewe 2012	−0.475	−1.100	0.149	0.135
Kaltsatou 2014	−0.254	−0.943	0.435	0.471
	−0.408	−0.739	−0.078	0.015

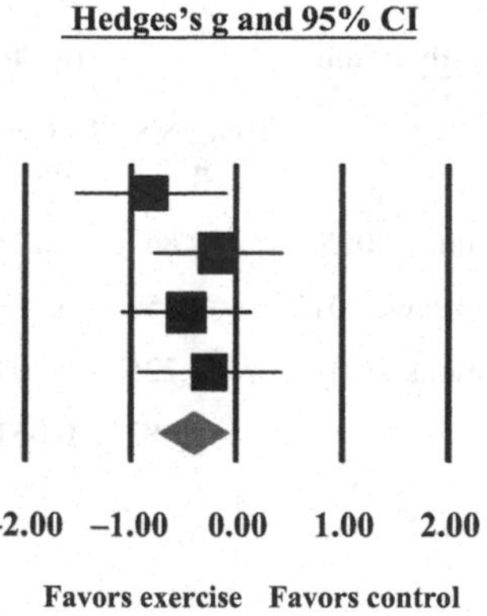

Hedges's g and 95% CI

−2.00 −1.00 0.00 1.00 2.00

Favors exercise Favors control

Figure 12.5 Meta-analysis investigating the effect of aerobic exercise on negative symptoms

In contrast, negative symptoms decreased significantly following exercise (hedges g =−0.41, 95% CI=−0.74 to −0.08, p=0.015; Figure 12.5). There was no heterogeneity (I^2=0%).

Brain Health and Neurocognition

Three RCTS examined the effects of exercise on brain volume. Only Pajonk et al. (2010) observed a significant main effect, as exercise increased hippocampal volume by 12% (significantly more than the table-football control). Both Pajonk et al. (2010) and Scheewe et al. (2013b) found that increased cardiorespiratory fitness was significantly correlated with increases in brain volume. Pajonk et al. (2010) also examined cognition, and found that exercise improved verbal short-term memory by 34% ($p < 0.05$). Kimhy et al. (2015) showed that improvement in cardiorespiratory fitness predicted 15% of the neurocognitive improvement variance following aerobic exercise.

Medical Clearance Guidelines

None of the RCTs explicitly stated the (inter)national guidelines applied for medical clearance of the participants (see Table 12.1). No serious adverse events were reported.

Drop-Out Rates Exercisers

The overall drop-out rate in exercisers was 24.4% (53/217) (see Table 12.1).

Motivational Strategies Used

The most often reported strategies used were (see Table 12.1): (a) gradual progression of dose (frequency, intensity, time), (b) discussion of barriers and how to cope with them, (c) maintaining a log book, (d) stimulating social interaction and social support, (e) financial compensation for transport. Other reported strategies were: (a) applying motivational interviewing skills, (b) setting specific, measurable, achievable, realistic and time-bound goals, (c) a structured approach, and (d) creating a positive climate.

Discussion

General Findings

More than a decade after a 2005 call (Richardson et al. 2005) for more rigorous research investigating the efficacy of aerobic exercise as a treatment strategy for people with schizophrenia, the current systematic review and meta-analysis clearly demonstrates that this call has been heeded. In particular in the last five years, the number of RCTs has increased significantly (13/16). There is currently sufficient scientific evidence to conclude that aerobic exercise reduces negative symptomatology and improves cardiorespiratory fitness. Next to this, there are some indications for beneficial neurocognitive effects. The current scientific evidence also demonstrates that the beneficial effects of aerobic exercise are observed irrespective of the frequency, intensity, duration or type of the aerobic exercise intervention. Nevertheless, there seems to be a tendency towards a dose-response relationship. Due to inconsistency in reporting we were however not able to explore this in more detail.

Table 12.1 Summary of identified studies

Study	*Participants*	*Medical clearance*	*Intervention characteristics (versus controls)*	*Physical and mental health outcomes (versus controls)*	*Drop-out rates in exercisers*	*Motivational strategies used*
Beebe et al. 2005	10 (8♂) outpatients with schizophrenia (DSM-IV), (40-63years).	Obtained from primary care provider, but no guidelines reported.	Treadmill walking, 16 weeks, 3/week: 10 min warm-up session, treadmill walking, then 10 min cool-down exercises. Treadmill walking session gradually increased from 5 to 30 min versus usual care.	Lower body fat percentage (-3.7% vs. -0,02%, p= 0,03), lower BMI (-1.3% vs. -0,02, p > 0,05), 6MWT (+10% vs. +4%, p > 0,05), less positive and negative symptoms (-13.5% vs. +5%, p>0,05).	33% (2/6) dropped-out; 75% attended half of sessions, 50% attended 2/3.	Stepwise increase of intensity based on initial physical fitness assessment
Acil et al. 2008	30 (18♂) inpatients with schizophrenia (DSM-IV) (21-45years).	No guidelines reported.	10 weeks, 3/week, 40min: 10min warming up, 25min aerobic exercise, 5min cool-down, max HR 220-age	total SAPS score -7 (p<0.05) (vs.-1), total SANS score -10.4 (p<0.05) (vs. +2.86), BSI -0.34 (p<0.05) (vs. +0.20), improved physical +1.8 (p<0.05) (vs. +0.67) and mental +2.07 (p<0.05) (vs.-0.8) WHO QoL scores	0% (0/15).	None reported.
Marzolini et al. 2009	13 (8♂) in- and out-patients with schizophrenia or schizoaffective disorder (DSM-IV), (mean age=44.6years).	No guidelines reported.	Aerobic and resistance training 12 weeks, 2/week: from 60% HR to 80% HR and 11–14 on Borg, 60 RM at local recreation center + once per week additional aerobic exercise session individually or during home visit from mental health professional versus usual care.	6MWT score +5.1% (vs. -5.5%) (difference: p=0.1), muscle strength +28,3% (p<0.001) (vs. +12.5%, p=0.2) (difference: p = 0,01), MHI total score +9.2 (=0.03) (vs. +2.8) (difference: p=0.33), no sign reductions in resting blood pressure or BMI.	0% dropped out (0/7) mean attendance was 72% (range 54–87.5%).	Stepwise increase of intensity based on initial physical fitness assessment, barriers were discussed.

Pajonk et al. 2010 Falkai et al. 2013	16 ♂ outpatients with schizophrenia (DSM-IV) (32.9±10.6years)	No guidelines reported.	Ergometer cycling: 12 weeks, 3/week, 30 min at heart rate (±10 beats/min) corresponding to a blood lactate concentration of about 1.5 to 2 mmol/L versus table top football with comparable levels of stimulation to that provided for aerobic exercise.	Hippocampal volume + 12% (vs. -1%, p=0.002). No exercise-related changes in cortical regions. The severity of total symptoms reduced (-9%) versus (+13% higher). (F1,14=6.76; P=0.02).	38% (5/13), average of sessions attended=85%.	Stepwise increase of intensity based on initial physical fitness assessment
Behere et al. 2011 [40]	39(29♂) outpatients with schizophrenia (DSM-IV) (30.2±8.0years)	No guidelines reported.	3 months (first month supervised), 5/week brisk walking, jogging, and exercises in standing and sitting postures and relaxation versus waitlist.	No significant changes in PANSS, SOFS and TRACS scores following exercise or vs. wait list controls.	45% (14/31).	Maintain a logbook
Methapatara et al. 2011	64 (41♂) patients with schizophrenia (DSM-IV) (18–65 years)	No guidelines reported.	Daily pedometer walking, 12 weeks, from baseline up to 5000–8000 steps/day after 1week 5 hours education + leaflets versus care as usual.	Reductions in body weight in exercisers (mean difference 2.2kg, p<0.03).	0% (0/32).	Motivational interviewing, SMART goals, stepwise increase of volume based on initial assessment, discussion of barriers.

(continued)

Table 12.1 (cont.)

Study	*Participants*	*Medical clearance*	*Intervention characteristics (versus controls)*	*Physical and mental health outcomes (versus controls)*	*Drop-out rates in exercisers*	*Motivational strategies used*
Scheewe et al. 2012, 2013a, 2013b Svatkova et al. 2015	39 outpatients with schizophrenia spectrum (DSM-IV) (18–48 years) (per protocol analyses)	ACSM guidelines were followed (although this was not specified for the medical clearance).	6 months, 2/week, 60min aerobic (week 1–3: 45%, week 4–12:65%, week 13–26:75% HRR) and muscle exercises (6 exercises weekly; 3x 10–15 repetitions maximum for biceps, triceps, abdominal, quadriceps, pectoral, deltoid muscles) versus same amount of occupation therapy (creative and recreational activities).	Significant deterioration in VO_2 peak in controls -9.2%) not in exercisers -0.3%); only exercise significantly decreased triglycerides (- 13.5%) (vs. - 2.4%), no effect on BMI, HDL or WC, no significant changes in brain volume, PANSS total -20.7% (vs. +3.3%) (difference: $p<0.001$), MADRS -36.6% (vs. -4.4%) (difference: $p=0.012$), CAN -22% (vs. -4%) (difference: $p=0.05$), improved white matter integrity in fiber tracts, in particular those implicated in motor functioning.	42% (13/31), exercise adherence was 79%.	Stepwise increase of intensity based on initial physical fitness assessment
Varambally et al. 2012	73 (55♂)) in- and outpatients with schizophrenia (DSM-IV) (mean age=30.6years)	No guidelines reported.	Aerobic exercises 4 weeks, 25 sessions 45min (lower intensity) versus wait list.	-17.3% in SOFS score (vs.-7.5%), no changes in PANSS scores 4 months post-intervention.	37.8% (14/37).	social support, transport tickets.

Battaglia et al. 2013	23♂ patients with schizophrenia (DSM-IV) (mean age=35–36 years)	Medical approval certificate needed but no guidelines reported.	Soccer, 12 weeks 2/week, 100–120 min: (20 min week 1 to 5; 25 min week 5 to 8; and 30 min from week 9 to 10); mean HR=50–85% of estimated max HR versus care as usual.	-4.6% ($p<0.001$) in bodyweight and BMI (vs. +5.4% and +1.8%), physical and mental SF-12 +10.5% and 10.8% ($p<0.001$) and improved performances on the 30 meter sprint test and slalom test running with a ball.	16.7% (2/12).	Using the preferred activity, daily log, standard sequence, positive feedback, stimulating social interactions, using behavioral skills to cope with social anxiety, peer modeling, a positive climate, gradual increase in volume; autonomy in stopping the activity.
Kaltsatou et al. 2014	31 (25♂) patients with schizophrenia (DSM-IV) (mean age=59–60 years)	No guidelines reported.	Traditional Greek dancing: 8 months, 3/week: 10min warm-up and cool-down with stretching + 40 min single group dance in a hemi-cycle, HR=60–70% of estimated max HR versus care as usual.	PANSS total -7%, GAF +12%, QLESQ +17.1%, 6MWT +44%, sit-to-stand test time -21.7%, Berg Balance score +17%, maximal isometric force of lower limbs +74.6% (all $p<0.05$) (vs. no sign. changes), and lower BMI, no changes in hand grip strength.	(0% (0/16).	Gradual increase in complexity, overall simple and structured dance steps, stimulating social interactions, reinforcement by the health care staff
Oertel-Knöchel et al. 2014	29 (12♂) patients with schizophrenia (DSM-IV) (39.4±12.3years)	No guidelines reported	Cardio + resistance training: 4 weeks, 3/week, 45min (10min warming-up, 25min aerobic exercise, 10min cool-down) HR=60–70% of estimated max HR + 30min cognitive training versus 45min relaxation + 30min cognitive training versus waiting control group.	An increase in cognitive performance in the domains visual learning, working memory and speed of processing, a decrease in state anxiety and an increase in subjective quality of life, PANSS positive changed not, PANSS decreased significantly.	0% (0/8); overall 32%.	Stepwise increase of intensity based on initial physical fitness assessment, structured (always same setting, time of the day, standard session, number of participants).

(*continued*)

Table 12.1 (cont.)

Study	*Participants*	*Medical clearance*	*Intervention characteristics (versus controls)*	*Physical and mental health outcomes (versus controls)*	*Drop-out rates in exercisers*	*Motivational strategies used*
Kimhy et al. 2015	33(12♂) outpatients with schizophrenia (DSM-IV) (18-55years)	ACSM guidelines were followed (although this was not specified for the medical clearance).	Aerobic training, 12 weeks, 3/week, 60min aerobic (week 1: 60%, week 2: 70%, week 4–12:75% HR max using also video games versus care as usual.	VO_2 peak +18.0% (vs –0.5%) (p=0.002), neurocognition +15.1% (vs –2.0%) (p=0.031).	21% (3/16).	Stepwise increase of intensity based on initial physical fitness assessment, video games, $5 reimbursement for each session attended (paid weekly) to defray the costs of round-trip public transportation.

1RM= one-repetition maximum, 6MWT= 6 minute walk test, ACSM= American College of Sports Medicine, BSI=Brief Symptom Inventory, BMI= body mass index, CDSS= Calgary Depression Scale for Schizophrenia, CAN=Camberwell Assessment of Needs, GAF= Global Assessment of Functioning; HDL= high density lipoproteins, HR=heart rate, HRR=heart rate reserve, MADRS, Montgomery and Åsberg Depression Scale, MHI= Mental Health Inventory, PANSS=Positive and Negative Syndrome Scale, QLESQ=Quality of Life Enjoyment and Satisfaction Questionnaire, SF-12= health related quality of life questionnaire short form 12, SOFS, Socio-Occupational Functioning Scale, SAPS= Scale of Assessment of Positive Symptoms, SANS=Scale of Assessment of Negative Symptoms, SMART= specific, measurable, achievable, realistic and time-bound, TRACS=Tool for Recognition of Emotions in Neuropsychiatric Disorders Accuracy Score, VO_2=oxygen uptake, WC= waist circumference, WHO QoL=World Health Organization Quality of Life scores.

Interestingly, our meta-analysis shows that aerobic exercise results in significant improvements in *cardiorespiratory fitness* but not in reductions in body mass index. Recent debate over the importance of "fitness vs. fatness" has led to increasing consensus regarding the superiority of poor fitness as a better predictor of morbidity and mortality as opposed to fatness (Church, LaMonte, Barlow, & Blair 2005). This is of particular relevance in schizophrenia, given the growing interest in lifestyle interventions aiming to reduce cardiovascular disease risk. Reducing weight and body mass index is notoriously challenging in people with schizophrenia (Chwastiak 2015). However, even in the absence of a reduction in body mass index, improvements in cardiorespiratory fitness following exercise can significantly improve health and reduce mortality (Blair et al. 1995). Considering these new findings we advocate that a shift in clinical focus away from "fatness" and towards "fitness" is justified. Improving cardiorespiratory fitness may be a more feasible, realistic, and clinically meaningful outcome for people with schizophrenia. Overall the trials identified in this review were of short to moderate length and demonstrate that improvements can occur over a short period (most up to 12 weeks). Cardiorespiratory fitness is relatively straightforward to measure in clinical practice and clinicians should consider monitoring this as a "vital sign," given its significant relationship with all-cause morbidity and mortality. In addition, achieving success in changes in cardiorespiratory fitness will provide valuable feedback to patients and clinicians, many of whom experience seemingly inevitable weight gain following treatment. Thus, we propose a shift to start to consider cardiorespiratory fitness in the routine health monitoring of people with schizophrenia.

Another important finding with clinical implications is that the current aerobic exercise-based evidence in people with schizophrenia shows a mean drop-out rate of around 25%, which compares favorably with antipsychotic medication trials, where drop-out rates of around 50% are observed (Martin et al. 2006). However, the drop-out rates ranged considerably from 0% to 45%, clearly demonstrating that one of the challenges for health care professionals and researchers lies in the adherence towards the aerobic exercise interventions provided. We therefore advocate that the focus in clinical practice and future research should not only be on the most ideal dose-response, that is efficacy, but also on how people with schizophrenia may include aerobic exercise as part of their daily routine, or rather the effectiveness of aerobic exercise. Based on the current literature, it is reasonable to assume that individualized and group-based structured, supervised, and progressive aerobic exercise may yield superior results compared to non-structured and unsupervised aerobic exercise (Vancampfort et al. 2016b). Aerobic exercise is also not a one-size-fits-all intervention. Symptoms, previous aerobic exercise history, motivation, and access to services will impact on whether people with schizophrenia participate in aerobic exercise (Vancampfort, Stubbs, Ward, Teasdale, & Rosenbaum 2015e; Vancampfort, Stubbs, Ward, Teasdale, & Rosenbaum 2015f). Inexperience with intense physical effort, associated fatigue and discomfort, increased risk of physical injuries, limited availability of aerobic exercise facilities and specialized equipment, and the cost associated with accessing facilities or training can all act as barriers to moderate-vigorous aerobic exercise (Vancampfort et al. 2012a). It might be stated based on the current evidence that those people with schizophrenia who are unable or unwilling to exercise at moderate to vigorous intensity can still benefit from engaging in some aerobic exercise. However, daily clinical practice and future research in people with schizophrenia requires: (a) clear recommendations for a standardized risk stratification, and (b) greater knowledge on strategies to improve motivation and adherence towards all types of aerobic exercise.

A Standardized Risk Stratification

The safe prescription of aerobic exercise requires adequate medical clearance based on standardized criteria. Surprisingly, the current scientific literature does not report which (inter)national standards are being used for this purpose, reflecting ad hoc clinical practice in this area. Given the greatly increased somatic comorbidity and physical pain among people with schizophrenia (Stubbs et al. 2014), physical therapists or exercise physiologists should identify high-risk persons via standardized screening procedures prior to prescribing an aerobic exercise program. It has been recently recommended that clinicians and researchers should apply the American College of Sports Medicine (ACSM) 2005 *risk stratification* guidelines when prescribing exercise to people with schizophrenia. Stratification involves determining the presence of absolute and relative contraindications to exercise including previously diagnosed somatic co-morbidities and physical pain, evaluation of the total number of risk factors, and consideration of signs and/or symptoms suggestive of possible underlying somatic co-morbidities. Within the ACSM 2005 guidelines people with schizophrenia should be stratified as at low, moderate, or high risk. The stratification of a person with schizophrenia into the low-risk category requires absence of diagnosed cardiovascular or pulmonary disease and no more than one relevant sign and/or symptoms suggestive of underlying cardiovascular, pulmonary, or metabolic disease, whereas the moderate risk designation is appropriate when presenting with two or more risk factors (e.g. pain or discomfort that may be due to ischemia, shortness of breath at rest or with mild exertion, orthopnea, ankle oedema, palpitations or tachycardia). One should be stratified at the high risk level when there is a diagnosis of cardiovascular or pulmonary disease. The ACSM implies that the guidelines provide only a general guide, and well-trained healthcare professionals should use their discretion to a certain degree when applying the criteria. Physician involvement is not necessary for low-risk persons with schizophrenia performing a submaximal test or commencing an exercise prescription of moderate or vigorous exercise. For moderate-risk individuals, guidelines recommend a medical examination before prescribing aerobic exercise at a vigorous intensity. For submaximal cardiorespiratory fitness testing and moderate exercise prescription for those at the moderate risk level, physician involvement is not deemed necessary, although it is not discouraged.

Evidence-Based Strategies for Improving Motivation and Adherence to Aerobic Exercise in People with Schizophrenia

Outlining an appropriate aerobic exercise program for people with schizophrenia requires that health care professionals assure not only safe participation, but also enable participants to feel some success from the program, thus allowing progression. The available evidence demonstrates that prescribing aerobic exercise that is realistic, achievable and agreeable seems to be a good way of increasing self-confidence, in turn enhancing adherence. There is also evidence that people with schizophrenia can manage barriers to aerobic exercise with support from three psychological constructs: (a) the need for *autonomy* (i.e. experiencing a sense of psychological freedom when engaging in exercise), (b) the need for *competence* (i.e. ability to attain desired outcomes following the exercise program), and (c) the need for *social relatedness* (i.e. being socially connected when being physically active). Therefore, as a first step, exploring the mental and physical health benefits of regular aerobic exercise and determining which benefits are most salient to each participant is essential. For inventorying perceived barriers and benefits, health care professions may use a decision balance that facilitates people with schizophrenia in weighing up the pros and cons of aerobic exercise participation. In Table 12.2, a brief overview of strategies used in the current literature is presented.

Points to Consider

In this chapter we focused solely on studies that administered aerobic exercise as the sole component of the intervention (allowing for resistance training as a co-intervention or other co-interventions if similar in the control condition). Therefore, studies utilizing broader lifestyle interventions were excluded. This however prevents us from evaluating how aerobic exercise can work in synergy with other lifestyle changes, such as changes in eating habits and substance use. Second, data on drug-naïve and first episode patients are currently lacking. All participants in the studies reviewed were receiving concomitant psychotropic medications, which could impact the aerobic exercise outcomes reported. If participants are more physically active at the same time as a change in medication, it is not possible to ascribe the source of any outcomes to either or both of these potential interventions. Third, RCTs are not ideal to make any conclusions on the safety issues. Almost all RCTs to date have been selective in the recruitment of participants, excluding high-risk patients with somatic co-morbidities while comprehensive adverse event reporting has been limited within the published literature. The biased populations and limited information on adverse events pose a challenge to documenting the safety of aerobic exercise for people with schizophrenia, demonstrating the importance

Table 12.2 Overview of motivational strategies to improve adherence to aerobic exercise interventions in people with schizophrenia

1. Develop aerobic exercise interventions based on the person's current preferences and expectations and the initial cardiorespiratory fitness and physical activity assessment.
2. Assist in developing an individual action plan taking into account barriers people with schizophrenia are confronted with.
3. Assist the person in setting specific, measurable, achievable, realistic, time-bound, and rewarding goals which lead to successful experiences.
4. Adapt the progression taking into account the individual's health status and physical abilities, age, current fitness status, physical activity history, expectations and goals, side-effects of psychotropic medication, exercise tolerance, and perceived exertion.
5. Use cognitive-behavioral strategies such as self-monitoring, stimulus cuing, goal-setting, and contracting. For example, provide exercise cards and a logbook and use regular progress feedback.
6. Avoid between-peers comparisons but stimulate enjoyable social interactions.
7. Emphasize the short-term benefits after single aerobic exercise sessions: improvements in mood and energy level and reductions in state anxiety, stress levels, distraction of negative thoughts, and the ability to concentrate and focus: aerobic exercise can give people with schizophrenia a sense of power over their recovery.
8. Facilitate autonomous reasons for participation by focusing on the positive experiences of aerobic exercise itself, as well as helping to develop an identity of a physical active person.
9. Focus after several weeks on perceived cardiorespiratory fitness gains, achievement of personal goals and mastery experiences.
10. Discuss problem-solving around barriers, reinforce all progress toward change (even if initially very small progress has been made), and encourage modification of goals if needed.
11. Seek support of others such as family and friends.
12. Use relapse behaviors/strategies: it is important to explain to persons with schizophrenia that relapses are part of the process of change, and that responding with guilt, frustration, and self-criticism may decrease their ability to maintain physical activity. Relapse prevention strategies such as realistic goals setting, planned activity, realistic expectations, identifying and modifying negative thinking, and focusing on benefits of single aerobic exercise sessions seem to be effective.

of thorough medical clearing. Longitudinal and observational studies should also systematically report adverse events and target higher-risk patients with schizophrenia rather than excluding them. The most appropriate safety screening and medical clearance approach for aerobic exercise in clinical and community settings for people with schizophrenia remains a topic for future research.

Conclusions

The evidence provided in this chapter demonstrates that aerobic exercise is a therapeutic intervention in which simultaneous improvements in both psychiatric symptoms and physical health can be achieved. Inclusion of aerobic exercise programs and facilitating easy access to aerobic exercise facilities is warranted within all mental health care services worldwide. Our review findings also show that the future of aerobic exercise as a viable and accepted component of usual care within mental health settings will be enhanced by more methodologically rigorous clinical research focusing not only on the efficacy (i.e. dose-response relationship), but also on its effectiveness (i.e. how people with schizophrenia may include aerobic exercise in their daily lives) and ultimately its cost effectiveness. Although there seems to be a tendency towards a dose-response relationship, our review shows that beneficial effects are observed irrespective of the frequency, intensity, duration or type of the aerobic exercise intervention. We strongly recommend that people with schizophrenia should initially not be advised on which frequency, intensity, and time is needed to achieve benefits, nor should they be asked in the first place to comply with general recommendations. These recommendations comprise at least 150 minutes a week of moderate intensity physical activity, five days a week, or 75 minutes of vigorous intensity physical activity spread across the week [64]. In addition, physical activity to improve strength should be performed on at least two days each week [64]. Rather, we advocate that people with schizophrenia should be informed that such recommendations are aspirational goals and that small changes in daily life will have beneficial effects for health. For example, we encourage health care professionals to take immediate action and promote reducing sedentary behaviors by introducing light aerobic exercise throughout the day. Advice on how to accumulate time spent in light aerobic exercise could include adding five-minute walks throughout the day, for example walking short distances rather than using motorized transport. Adopting small, but incremental lifestyle changes may better position sedentary people with schizophrenia to transition to brief bouts of moderate intensity aerobic exercise. Such an approach will be less constrained by socio-economic, environmental, and organizational barriers, such as lower income, lower education, lack of access to aerobic exercise facilities or time available for leisure. At a minimum, mental health professionals should briefly assess current aerobic exercise behaviors at every consultation, and discuss specific, measurable, achievable, realistic, and time-bound goals that could be adopted, with support and follow-up. Changes in physical and mental health parameters, such as cardiorespiratory fitness, can then be monitored.

References

Acil, A. A., S. Dogan, & O. Dogan. 2008. "The effects of physical exercises to mental state and quality of life in patients with schizophrenia." *Journal of Psychiatric and Mental Health Nursing 15* (10): 808–15.

American Psychiatric Association. 2013. *Diagnostic and Statistical Manual of Mental Disorders (DSM-5)*. Washington, DC: American Psychiatric Association.

Andreassen, O. A., S. Djurovic, W. K. Thompson, A. J. Schork, K. S. Kendler, M. C. O'Donovan,...A. P. Morris. 2013. "Improved detection of common variants associated with schizophrenia by leveraging pleiotropy with cardiovascular-disease risk factors." *American Journal of Human Genetics 92* (2): 197–209.

Battaglia, G., M. Alesi, M. Inguglia, M. Roccella, G. Caramazza, M. Bellafiore, & A. Palma. 2013. "Soccer practice as an add-on treatment in the management of individuals with a diagnosis of schizophrenia." *Neuropsychiatric Disease and Treatment 9*: 595–603.

Beebe, L. H., L. Tian, N. Morris, A. Goodwin, S. S. Allen, & J. Kuldau. 2005. "Effects of exercise on mental and physical health parameters of persons with schizophrenia." *Issues in Mental Health Nursing 26* (6): 661–76.

Behere, R., R. Arasappa, A. Jagannathan, S. Varambally, G. Venkatasubramanian, J. Thirthalli,…B. Gangadhar. 2011. "Effect of yoga therapy on facial emotion recognition deficits, symptoms and functioning in patients with schizophrenia." *Acta Psychiatrica Scandinavica 123* (2): 147–53.

Blair, S. N., H. W. Kohl, C. E. Barlow, R. S. Paffenbarger, L. W. Gibbons, & C. A. Macera. 1995. "Changes in physical fitness and all-cause mortality: A prospective study of healthy and unhealthy men." *JAMA 273* (14): 1093–8.

Caspersen, C. J., K. E. Powell, & G. M. Christenson. 1985. "Physical activity, exercise, and physical fitness: Definitions and distinctions for health-related research." *Public Health Reports 100* (2): 126.

Church, T. S., M. J. LaMonte, C. E. Barlow, & S. N. Blair. 2005. "Cardiorespiratory fitness and body mass index as predictors of cardiovascular disease mortality among men with diabetes." *Archives of Internal Medicine 165* (18): 2114–20.

Chwastiak, L. 2015. "Making evidence-based lifestyle modification programs available in community mental health centers: Why so slow?" *Journal of Clinical Psychiatry 76* (4): e519–20.

Correll, C. U., J. Detraux, J. De Lepeleire, & M. De Hert. 2015. "Effects of antipsychotics, antidepressants and mood stabilizers on risk for physical diseases in people with schizophrenia, depression and bipolar disorder." *World Psychiatry 14* (2): 119–136.

Curtis, J., A. Watkins, S. Rosenbaum, S. Teasdale, M. Kalucy, K. Samaras, & P. Ward. In press. "Keeping the Body in Mind: An individualised lifestyle and life-skills intervention to prevent antipsychotic-induced weight gain in first episode psychosis." *Early Intervention in Psychiatry.*

De Hert, M., C. U. Correll, J. Bobes, M. Cetkovich-Bakmas, D. A. N. Cohen, I. Asai,…D. M. Ndetei. 2011. "Physical illness in patients with severe mental disorders. I. Prevalence, impact of medications and disparities in health care." *World Psychiatry 10* (1): 52–77.

Falkai, P., B. Malchow, T. Wobrock, O. Gruber, A. Schmitt, W. G. Honer,…T. D. Cannon. 2013. "The effect of aerobic exercise on cortical architecture in patients with chronic schizophrenia: A randomized controlled MRI study." *European Archives of Psychiatry and Clinical Neuroscience 263* (6): 469–73.

Firth, J., J. Cotter, R. Elliott, P. French, & A. R. Yung. 2015. "A systematic review and meta-analysis of exercise interventions in schizophrenia patients." *Psychological Medicine 45* (7): 1343–61.

Fusar-Poli, P., E. Papanastasiou, D. Stahl, M. Rocchetti, W. Carpenter, S. Shergill, & P. McGuire. 2014. "Treatments of negative symptoms in schizophrenia: Meta-analysis of 168 randomized placebo-controlled trials." *Schizophrenia Bulletin*: sbu170.

Heald, A., K. Sein, S. Anderson, J. Pendlebury, M. Guy, V. Narayan,…M. Livingston. 2015. "Diet, exercise and the metabolic syndrome in schizophrenia: A cross-sectional study." *Schizophrenia Research 169* (1): 494–5.

Jauhar, S., P. J. McKenna, J. Radua, E. Fung, R. Salvador, & K. R. Laws. 2014. "Cognitive-behavioural therapy for the symptoms of schizophrenia: Systematic review and meta-analysis with examination of potential bias." *British Journal of Psychiatry 204* (1): 20–9.

Kaltsatou, A., E. Kouidi, K. Fountoulakis, C. Sipka, V. Theochari, D. Kandylis, & A. Deligiannis. 2014. "Effects of exercise training with traditional dancing on functional capacity and quality of life in patients with schizophrenia: A randomized controlled study." *Clinical Rehabilitation*: 0269215514564085.

Kimhy, D., J. Vakhrusheva, M. N. Bartels, H. F. Armstrong, J. S. Ballon, S. Khan,...A. Lister. 2015. "The impact of aerobic exercise on Brain-Derived Neurotrophic Factor and neurocognition in individuals with schizophrenia: A single-blind, randomized clinical trial." *Schizophrenia Bulletin*: sbv022.

Kuipers, E., A. Yesufu-Udechuku, C. Taylor, & T. Kendall. 2014. "Management of psychosis and schizophrenia in adults: Summary of updated NICE guidance." *British Medical Journal 348*.

Laursen, T. M., T. Munk-Olsen, & M. Vestergaard. 2012. "Life expectancy and cardiovascular mortality in persons with schizophrenia." *Current Opinion in Psychiatry 25* (2): 83–8.

Leucht, S., A. Cipriani, L. Spineli, D. Mavridis, D. Orey, F. Richter,...J. M. Davis. 2013. "Comparative efficacy and tolerability of 15 antipsychotic drugs in schizophrenia: A multiple-treatments meta-analysis." *Lancet 382* (9896): 951–62.

Li, M., F. Ying-Li, T. Zhen-Yu, & C. Xiao-Shu. 2014. "Schizophrenia and risk of stroke: A meta-analysis of cohort studies." *International Journal of Cardiology 173* (3): 588–90.

McGrath, J., S. Sukanta, D. Chant, & J. Welham. 2008. "Schizophrenia: A concise overview of incidence, prevalence, and mortality." *Epidemiologic Reviews 30* (1): 67–76.

Manu, P., C. U. Correll, M. Wampers, A. J. Mitchell, M. Probst, D. Vancampfort, & M. Hert. 2014. "Markers of inflammation in schizophrenia: Association vs. causation." *World Psychiatry 13* (2): 189–92.

Martin, J. L. R., V. Pérez, M. Sacristán, F. Rodríguez-Artalejo, C. Martínez, & E. Álvarez. 2006. "Meta-analysis of drop-out rates in randomised clinical trials, comparing typical and atypical antipsychotics in the treatment of schizophrenia." *European Psychiatry 21* (1): 11–20.

Marzolini, S., B. Jensen, & P. Melville. 2009. "Feasibility and effects of a group-based resistance and aerobic exercise program for individuals with severe schizophrenia: A multidisciplinary approach." *Mental Health and Physical Activity 2* (1): 29–36. doi:10.1016/j.mhpa.2008.11.001

Methapatara, W., & M. Srisurapanont. 2011. "Pedometer walking plus motivational interviewing program for Thai schizophrenic patients with obesity or overweight: A 12-week, randomized, controlled trial." *Psychiatry and Clinical Neurosciences 65* (4): 374–80.

Mitchell, A. J., & M. De Hert. 2015. "Promotion of physical health in persons with schizophrenia: Can we prevent cardiometabolic problems before they begin?" *Acta Psychiatrica Scandinavica 132* (2): 83–5.

Nielsen, R. E., S. Levander, G. K. Telleus, S. Jensen, T. Östergaard Christensen, & S. Leucht. 2015. "Second-generation antipsychotic effect on cognition in patients with schizophrenia – a meta-analysis of randomized clinical trials." *Acta Psychiatrica Scandinavica 131* (3): 185–96.

Oertel-Knochel, V., P. Mehler, C. Thiel, K. Steinbrecher, B. Malchow, V. Tesky,...F. Hansel. 2014. "Effects of aerobic exercise on cognitive performance and individual psychopathology in depressive and schizophrenia patients." *European Archives of Psychiatry and Clinical Neurosciences 264* (7): 589–604.

Pajonk, F.-G., T. Wobrock, O. Gruber, H. Scherk, D. Berner, I. Kaizl,...& P. Falkai. 2010. "Hippocampal plasticity in response to exercise in schizophrenia." *Archives of General Psychiatry 67* (2): 133–43.

Richardson, C. R., G. Faulkner, J. McDevitt, G. S. Skrinar, D. S. Hutchinson, & J. D. Piette. 2005. "Integrating physical activity into mental health services for persons with serious mental illness." *Psychiatric Services 56* (3): 324–31.

Rosenbaum, S., A. Tiedemann, C. Sherrington, J. Curtis, & P. B. Ward. 2014. "Physical activity interventions for people with mental illness: A systematic review and meta-analysis." *Journal of Clinical Psychiatry 75* (9): 964–74.

Rössler, W., H. J. Salize, J. van Os, & A. Riecher-Rössler. 2005. "Size of burden of schizophrenia and psychotic disorders." *European Neuropsychopharmacology 15* (4): 399–409.

Saha, S., D. Chant, & J. McGrath. 2007. "A systematic review of mortality in schizophrenia: Is the differential mortality gap worsening over time?" *Archives of General Psychiatry 64* (10): 1123–31.

Scheewe, T. W., F. J. G. Backx, T. Takken, F. Jörg, A. C. P. Strater, A. G. Kroes,...W. Cahn. 2013a. "Exercise therapy improves mental and physical health in schizophrenia: A randomised controlled trial." *Acta Psychiatrica Scandinavica 127*: 464–73.

Scheewe, T. W., T. Takken, R. S. Kahn, W. Cahn, & F. J. Backx. 2012. "Effects of exercise therapy on cardiorespiratory fitness in patients with schizophrenia." *Med Sci Sports Exerc 44* (10): 1834–42.

Scheewe, T. W., N. E. M. van Haren, G. Sarkisyan, H. G. Schnack, R. M. Brouwer, M. de Glint,...C. Wiepke. 2013b. "Exercise therapy, cardiorespiratory fitness and their effect on brain volumes: A randomised controlled trial in patients with schizophrenia and healthy controls." *European Neuropsychopharmacology 23* (7): 675–85.

Stubbs B., J. Firth, A. Berry, F. B. Schuch, S. Rosenbaum, F. Gaughran,...D. Vancampfort. 2016. "How much physical activity do people with schizophrenia engage in? A systematic review, comparative meta-analysis and meta-regression." *Schizophrenia Research.*

Stubbs, B., A. J. Mitchell, M. De Hert, C. U. Correll, A. Soundy, M. Stroobants, & D. Vancampfort. 2014. "The prevalence and moderators of clinical pain in people with schizophrenia: A systematic review and large scale meta-analysis." *Schizophrenia Research 160* (1): 1–8.

Stubbs, B., D. Vancampfort, J. Bobes, M. De Hert, & A. J. Mitchell. 2015. "How can we promote smoking cessation in people with schizophrenia in practice? A clinical overview." *Acta Psychiatrica Scandinavica.*

Svatkova, A., R. C. W. Mandl, T. W. Scheewe, W. Cahn, R. S. Kahn, & H. E. Hulshoff Pol. 2015. "Physical exercise keeps the brain connected: Biking increases white matter integrity in patients with schizophrenia and healthy controls." *Schizophrenia Bulletin 41* (4): 869–78.

van der Gaag, M., A. D. Stant, J. K. Wolters, E. Buskens, & D. Wiersma. 2011. "Cognitive–behavioural therapy for persistent and recurrent psychosis in people with schizophrenia-spectrum disorder: Cost-effectiveness analysis." *British Journal of Psychiatry 198* (1): 59–65.

Vancampfort, D., C. U. Correll, B. Galling, M. Probst, M. De Hert, P. B. Ward,... B. Stubbs. 2016a. "Diabetes mellitus in people with schizophrenia, bipolar disorder and major depressive disorder: A systematic review and large scale meta-analysis." *World Psychiatry.*

Vancampfort, D., M. De Hert, B. Stubbs, P. B. Ward, S. Rosenbaum, A. Soundy, & M. Probst. 2015a. "Negative symptoms are associated with lower autonomous motivation towards physical activity in people with schizophrenia." *Comprehensive Psychiatry 56*: 128–32.

Vancampfort, D., J. Knapen, M. Probst, T. Scheewe, S. Remans, & M. De Hert. 2012a. "A systematic review of correlates of physical activity in patients with schizophrenia." *Acta Psychiatrica Scandinavica 125* (5): 352–62.

Vancampfort, D., T. Madou, H. Moens, T. De Backer, P. Vanhalst, C. Helon,...M. Probst. 2015b. "Could autonomous motivation hold the key to successfully implementing lifestyle changes in affective disorders? A multicentre cross sectional study." *Psychiatry Research 228* (1): 100–6.

Vancampfort, D., M. Probst, T. Scheewe, J. Knapen, A. De Herdt, & M. De Hert. 2012b. "The functional exercise capacity is correlated with global functioning in patients with schizophrenia." *Acta Psychiatrica Scandinavica 125* (5): 382–7.

Vancampfort, D., S. Rosenbaum, F. B. Schuch, P. B. Ward, M. Probst, & B. Stubbs. 2016b. "Prevalence and predictors of treatment dropout from physical activity interventions in schizophrenia: A meta-analysis" *General Hospital Psychiatry 39*: 15–23.

Vancampfort, D., B. Stubbs, A. J. Mitchell, M. De Hert, M. Wampers, P. B. Ward,...C. U. Correll. 2015c. "Risk of metabolic syndrome and its components in people with schizophrenia and related psychotic disorders, bipolar disorder and major depressive disorder: A systematic review and meta-analysis." *World Psychiatry 14* (3): 339–47.

Vancampfort, D., B. Stubbs, S. K. Venigalla, & M. Probst. 2015d. "Adopting and maintaining physical activity behaviours in people with severe mental illness: The importance of autonomous motivation." *Preventive Medicine.*

Vancampfort, D., B. Stubbs, P. B. Ward, S. B. Teasdale, & S. Rosenbaum. 2015e. "Integrating physical activity as medicine in the care of people with severe mental illness." *Australian and New Zealand Journal of Psychiatry 49* (8): 681–2.

Vancampfort, D., B. Stubbs, P. B. Ward, S. B. Teasdale, & S. Rosenbaum. 2015f. "Why moving more should be promoted for severe mental illness." *Lancet Psychiatry 2* (4): 295.

Varambally, S., B. N. Gangadhar, J. Thirthalli, A. Jagannathan, S. Kumar, G. Venkatasubramanian,...H. R. Nagendra. 2012. "Therapeutic efficacy of add-on yogasana intervention in stabilized outpatient schizophrenia: Randomized controlled comparison with exercise and waitlist." *Indian Journal of Psychiatry 54* (3): 227.

Walker, E. R., R. E. McGee, & B. G Druss. 2015. "Mortality in mental disorders and global disease burden implications: A systematic review and meta-analysis." *JAMA Psychiatry 72* (4): 334–41.

13 Exercise and Anxiety Disorders

Jennifer Mumm, Sophie Bischoff, and Andreas Ströhle

What Is Anxiety?

Anxiety is a very important emotion that helps us to react appropriately in dangerous situations. This reaction is not under conscious control. It is a fast and automatic response to danger initiated by the limbic system (a brain region) in order to prevent damage. Thereby the body is getting prepared to fight or to flight. When this happens without an objective danger, when the reaction is too strong in relation to the danger or when the expectation of a potential anxiety reaction is very high, there may be an anxiety disorder.

The World Health Organization (Dilling & Freyberger 2014) distinguishes different anxiety disorders in the ICD-10. For this chapter focus is set on panic disorder, agoraphobia, social anxiety disorder, generalized anxiety disorder, obsessive-compulsive disorder and post-traumatic stress disorder.

Panic disorder is characterized by unforeseeable and recurrent panic attacks. During panic attacks patients suffer from intense fear and different body symptoms like tremor, sweating, nausea, palpitations, dryness of the mouth or breathing difficulties. The 12-month prevalence of panic disorder is about 0.7–2.2%. The first onset of the disease is in adolescence or early adulthood and the women-to-men ratio of patients with panic disorder is 2:1 (Goodwin et al. 2005).

Patients with *agoraphobia* show fear and/or avoidance behavior concerning at least two different situations (e.g. crowds or public transport). These situations provoke a lot of fear when patients are confronted with them. They often try to avoid these situations or use safety behavior to be able to stand the confrontation. In severe cases of avoidance patients are very impaired and cannot leave their apartments any more. Safety behavior might be an accompanying person, a cell phone, a bottle of water or special cognitions (e.g. counting down from 100 to 0). The 12-month prevalence of agoraphobia without panic disorder is about 0.7–2.0% and women are affected twice as frequently as men. The first manifestation of the disorder is normally between the ages of 20 and 30 (Goodwin et al. 2005; Michael, Zetsche, & Margraf 2007).

Patients with *social anxiety disorder* fear being in the center of attention or doing something embarrassing. They often try to avoid social situations. Patients display symptoms like blushing, tremor, fear of vomiting and fear of micturition in anxiety-provoking social situations. The lifetime prevalence is 1.6–13.7% and women are affected two times more frequently than men. Normally the onset of the disorder is before the age of 20 (Wittchen et al. 2011; Michael et al. 2007).

Worries are the main symptom of *generalized anxiety disorder*. Patients often fear something terrible could happen to them or their relatives (e.g. an accident) and this apprehension leads to fear and intense bodily symptoms (like palpitation, sweating, nausea, tremor, breathing difficulties, weakness). The lifetime prevalence is about 0.4–5.7% and women are affected more frequently with a ratio of 2.1:1. The average age of onset is either in late adolescence or in elderly (Michael et al. 2007).

Patients with *obsessive-compulsive disorder* (OCD) have either obsessions or compulsions or both. Patients often try to resist those obsessions or compulsions without success. Men and women are nearly affected equally (1.6:1) and the onset of OCD is between 14–39 years. Lifetime prevalence is 0.5–1.6% (Wittchen et al. 2011; Michael et al. 2007).

Post-traumatic stress disorder (PTSD) can follow a traumatic event. Reactivation, flashbacks or nightmares of the trauma, indifference concerning other people or the surrounding environment and avoidance of situations or activities reminding the patient of the trauma are symptoms of PTSD

(ICD-10). Lifetime prevalence is 1.3–11.7% and women are more affected than men with a ratio of 3.4:1 (Wittchen et al. 2011; Michael et al. 2007).

Treatment

First line treatments of anxiety disorders are psychological and pharmacological interventions, based on psychoeducation. Cognitive behavioral therapy and selective serotonin (and norepinephrine) reuptake inhibitors have the best empirical evidence according to randomized controlled clinical trials and are recommended by guidelines (Bandelow et al. 2014). In addition, relaxation techniques but also physical activity and exercise may be used. The currently available evidence for the effectiveness of sports therapy in the treatment of anxiety disorders will be described. As you will see, we are just at the beginning to understand how exercise may be used in the treatment of anxiety disorders and much has to be further studied.

Incidence of Anxiety Disorders in Physically Active Subjects

The relationship between physical activity and the incidence of anxiety disorders has been investigated in several studies. Looking at the data of the National Comorbidity Survey (Goodwin 2003), the likelihood of having an anxiety disorder was reduced for subjects reporting to be regularly physically active. In this study 5877 U.S. adults were interviewed and 60.3% indicated they were regularly active. The likelihood of those subjects to have agoraphobia, panic attacks, generalized anxiety disorder, specific phobia or social anxiety disorder was significantly decreased compared to subjects being only occasionally, rarely or never physically active. Furthermore there was a dose-response relation showing that the higher the frequency of self-reported physical activity, the less the likelihood of having a generalized anxiety disorder, panic attacks, agoraphobia, specific phobia, and social anxiety disorder.

The Early Developmental Stages of Psychopathology study (Ströhle et al. 2007) is investigating a cohort of adolescents and young adults aged 14–24 at baseline assessment. A group of 2548 responders participated at baseline and the second follow-up and were included in the study. At baseline 50.2% of participants indicated being regularly physically active, 14.6% non-regularly, and 35.2% not physically active. At follow-up 44.5% reported regular physical activity, 17% non-regular and 38.5% no physical activity. Cross sectional analysis revealed that at baseline subjects reporting regular physical activity showed a significant lower 12-month prevalence rate of any anxiety disorder than individuals not being physically active. Individuals with regular physical activity at baseline showed a significantly reduced likelihood to have any anxiety disorder 4 years later compared to those not being active.

Cross-sectional data was collected in the Canadian Community Health Survey of Mental Health and Well-Being (Meng & D'Arcy 2013). In total

36,984 subjects were investigated regarding mental diseases in the past year and leisure time physical activity in the past 3 months. Physical inactivity was a risk factor for the prevalence of any anxiety disorder.

Those three studies report data of cross-sectional and longitudinal surveys investigating the incidence of anxiety disorders and a possible relation to physical activity. The results show that physically active individuals show a reduced likelihood to have an anxiety disorder compared to inactive subjects. Furthermore individuals reporting physical activity had a reduced risk of developing an anxiety disorder 4 years later.

Exercise and Anxiety: Current State of Research

In this section studies investigating the effects of exercise on different anxiety disorders are summarized. For an overview of the included studies see Table 13.1.

Panic Disorder with/without Agoraphobia

Several research groups evaluated physical exercise as an intervention for patients with panic disorder with or without agoraphobia. In the following randomized controlled trials are presented exclusively.

Broocks et al. (1998) were the first conducting a placebo-controlled, randomized study to examine the effects of exercise in patients with panic disorder with or without agoraphobia. In total 46 patients were randomly assigned to three 10-week treatment protocols: the exercise group was instructed to run a 4-mile route with increasing workload; in the other conditions patients received either 112.5 mg clomipramine a day or placebo pills. Both exercise and clomipramine were more effective than placebo for all outcome measures. Still clomipramine improved anxiety symptoms significantly earlier and more effectively.

Anxiolytic effects of acute exercise were investigated by Esquivel et al. (2008). Eighteen patients performed either moderate/hard or very light bicycle-ergometer training for up to 15 minutes. Afterwards they were asked to inhale 35% CO2. Reactivity to CO2 was assessed by the number of experienced panic symptoms and a Visual Analogue Anxiety Scale. Panic reactions were significantly smaller in patients that performed moderate/hard exercise. Similar results were reported by Ströhle et al. (2009). They could show that acute bouts of exercise reduce the severity of CCK-4-induced panic and anxiety. These studies may help to understand the underlying mechanisms of the beneficial effects of exercise on panic.

Wedekind et al. (2010) combined exercise and drug treatment. Seventy-five patients either received paroxetine or placebo and additionally had to perform either an exercise or relaxation training. Paroxetine-treated patients improved significantly better than placebo-treated patients. Response and remission rates were higher in the paroxetine compared to pill placebo groups.

Table 13.1 Overview of studies

Authors	*Diagnosis*	*N*	*Intervention*	*Extent*	*Result*
Broocks et al. (1998)	PD w/wo AG	46 1: n=16 2: n=15 3: n=15	1: running a 4-mile route 2: 112.5 mg/day clomipramine (applied step wisely) 3: placebo pill	Exercise: 10 weeks, at least three times a week, increasing workload	Exercise was significantly more effective than placebo in all outcome measures. Clomipramine improved anxiety symptoms significantly earlier and more effectively than exercise.
Esquivel et al. (2008)	PD w/wo AG	18 Exercise: n=10 Control: n=8	Exercise: moderate/ hard exercise on bicycle ergometer before a 35% CO2 panic provocation Control: very-light exercise on bicycle ergometer before a 35% CO2 panic provocation	Exercise: One bout of exercise, moderate/ hard (80–90% heart rate reserve) vs. light intensity	Panic reactions to CO2 were significantly smaller in patients who performed moderate/hard exercise
Wedekind et al. (2010)	PD w/wo AG	75 1: n=21 2: n=20 3: n=17 4: n=17	Paroxetine vs. placebo with Exercise or Relaxation: 1: Paroxetine + Exercise 2: Placebo + Exercise 3: Paroxetine + Relaxation 4: Placebo + Relaxation Exercise: running constantly without resting over 45 minutes. Paroxetine: 40 mg/day applied step wisely Relaxation: one session similar to Autogenic training daily	Exercise: 10 weeks, 3–4 times a week	Paroxetine-treated patients improved significantly better than placebo-treated patients. While paroxetine was superior to placebo, aerobic exercise did not differ from relaxation.

Hovland et al. (2013)	PD w/wo AG	36 Exercise: n=17 CBT: n=19	Exercise: aerobic fitness, long-distance-walking/running and muscular strength exercises alternately CBT: group sessions	Exercise: 12 weeks, three weekly group sessions CBT: once a week	Patients of both groups improved significantly, although CBT leads to a better treatment outcome
Gaudlitz et al. (2015)	PD w/wo AG	47 Exercise: n=24 Control: n=23	Exercise: Group CBT + moderate/high intensity exercise (treadmill) Control: Group CBT + low intensity exercise (e.g. stretching)	Exercise: 8 weeks, 3 times a week, 30 minutes, 70% VO2max CBT: 1 month	moderate/high intensity exercise group yielded a better improvement on the primary outcome measure than the control group
Manger and Motta (2005)	PTSD	26	Walking or jogging on a treadmill	10 weeks, 3 times a week, 30 minutes, 60–80% heart rate reserve	Significant reductions in symptoms of PTSD, depression and anxiety after ten weeks and 1 month follow-up
Newman and Motta (2007)	PTSD	15, all female, age: 14–17	group sessions including warm-up and aerobic exercise	8 weeks, 3 times a week, 40 minutes, 60–80% heart rate reserve	Significant reductions in symptoms of PTSD, depression and anxiety
Diaz and Motta (2008)	PTSD	12 institutionalized female adolescents	Outdoor group sessions	5 weeks, 3 times a week, 25 minutes, 60–90% heart rate reserve	Strong effects for PTSD and trauma symptom severity, no effects for anxiety and depression
Fetzner and Asmundson (2014)	PTSD (full or subsyndromal)	33 1: n=11 2: n=11 3: n=11	Exercise on stationary bicycle 1: cognitive distraction 2: interoceptive prompts 3: no attention task	2 weeks, 6 times in two weeks, 20 minutes, 60–80% heart rate reserve	Patients of all groups showed an improvement in PTSD symptoms, depression and anxiety sensitivity to the same degree

(*continued*)

Table 13.1 (cont.)

Authors	*Diagnosis*	*N*	*Intervention*	*Extent*	*Result*
Rosenbaum, Sherrington, and Tiedemann (2015)	PTSD	81 Exercise: n=39 Usual care: n=42	1: Usual care + exercise 2: Usual care alone(psychotherapy, pharmacotherapy)	Exercise: 12 weeks, 3 sessions per week + walking program, individualized intensity, increasing workload Usual care: 12 weeks	Usual care + exercise improved significantly more considering psychological outcomes (PTSD, depression, stress, sleep) and physical outcomes (waist circumference, body fat and sitting time) in comparison to usual care alone
Powers et al. (2015)	PTSD	9	1: prolonged exposure + exercise (treadmill) prior to each exposure session 2: prolonged exposure alone	12 sessions, 30 minutes, 70% heart rate reserve	Exercise yielded a significant increase of BDNF and a stronger reduction of PTSD symptoms
(Brown et al. 2007)	OCD	15 patients already receiving CBT and/or pharmacological treatment for OCD for at least 3 month at the beginning of the intervention.	Moderate-intensity activity: treadmill, recombinant bicycles and elliptical machines CBT component of training for motivation to exercise	12 weeks, once a week guided, 2–3 times alone, 20–40 minutes of exercise CBT component 20-30min. once a week	Significant reduction of OCD symptoms from baseline to end of treatment, baseline to 3-week follow-up, baseline to 6-week follow-up, baseline to 6-month follow-up, quality of life score significantly increased from baseline to end of intervention
(Lancer, Motta, and Lancer 2007)	OCD	16	Walking at fitness centers	6 weeks, 3 times per week, 30 minutes	Significant reduction of OCD symptoms and depression from the end of baseline to post-intervention and follow-up

(Abrantes et al. 2009)	OCD	15 patients already receiving CBT and/or pharmacological treatment for OCD for at least 3 month at the beginning of the intervention.	Moderate-intensit y activity: treadmill, recombinant bicycles and elliptical machines CBT component of training for motivation to exercise	12 weeks, once a week guided, 2–3 times alone, 20–40 minutes of exercise CBT component 20-30min. once a week	Reductions in negative mood, OCD-symptoms and anxiety at the end of each session relative to beginning of the session. Reduction of obsessions and compulsions during the 12 weeks
(Rector et al. 2015)	OCD	11	CBT + exercise	12 weeks CBT (1 session per week) 12 weeks exercise in the gym (3 times per week). Duration: 15–30 minutes from week 1–4 and 30–45 minutes from week 5–12). Intensity was increased in sessions 5–12.	Significant improvements of OCD-symptoms and depression
(Jazaieri et al. 2012)	SAD	56 MBSR: n=31 Aerobic exercise n =25 Control group: n=29	MBSR and aerobic exercise	MBSR: 8 weeks, daily training, once a week guided Aerobic exercise: 8 weeks, two times per week individual training, once a week group aerobic exercise session	In both intervention groups social anxiety and depression decreased, well-being increased (at post treatment and 3-month follow-up); compared to untreated SAD group MBSR and aerobic exercise group showed decreased symptoms of social anxiety after intervention

(*continued*)

Table 13.1 (cont.)

Authors	*Diagnosis*	*N*	*Intervention*	*Extent*	*Result*
(Jazaieri et al. 2015)	SAD	47 Aerobic exercise: n=20 MBSR: n=27	MBSR (public course) or aerobic exercise (2 month gym membership)	8 weeks, 3 times a week	In both groups anxiety symptoms decreased every week.
(Herring et al. 2012)	GAD	30 women	Resistance training, aerobic exercise or waitlist	2 weekly sessions for 6 weeks for resistance and aerobic exercise training with a duration of 46 minutes	Remission rate for resistance exercise training was 60%, for aerobic exercise training 40% and for waitlist 30%. Worry symptoms were significantly reduced in combined exercise conditions versus waitlist
(Martinsen, Hoffart, and Solberg 1989)	PD w AG SAD GAD	79 PD w AG: n=56 SAD: n=13 GAD: n=10	Aerobic vs non-aerobic exercise	8 weeks, 1h, 3 times per week	Both groups showed significantly decreased anxiety scores, no significant difference between groups
(Merom et al. 2008)	PD GAD SAD	CBT + exercise =38 CBT+ education = 36	CBT group and exercise CBT group and educational sessions	CBT: 90-minute sessions, once a week (8 weeks for PD and GAD, 10 weeks for SAD), Exercise: 5 times or more per week for a total of at least 150 minutes	Significant higher reduction in depression, anxiety and stress scores for CBT+ exercise group compared to CBT + education

PD = panic disorder, w/wo = with/without, AG = agoraphobia, PTSD = post-traumatic stress disorder, OCD = obsessive-compulsive disorder, SAD = social anxiety disorder, GAD = generalized anxiety disorder, CBT = cognitive behavioral therapy, MBSR = mindfulness-based stress reduction.

While paroxetine was superior to placebo, aerobic exercise did not differ from relaxation.

Hovland et al. (2013) allocated 36 patients with panic disorder randomly to either an exercise program or a group cognitive behavioral therapy, both running for 12 weeks. The exercise group performed long-distance walking, muscular strength exercises, and sport games on three days a week in a group setting. The cognitive behavioral therapy (CBT) sessions were conducted once a week. Patients of both groups improved significantly. The within effect sizes in the exercise group were large for the improvement in fear of bodily sensations and fear of fear and small to medium for the change in avoidance behavior. For CBT all corresponding effect sizes were large. The authors conclude that physical exercise accounts for improvements in panic symptoms, although CBT leads to a better treatment result.

Gaudlitz, Plag, Dimeo, and Ströhle (2015) investigated the augmenting effect of exercise on the effectiveness of a group cognitive behavior therapy. Forty-seven patients with panic disorder with or without agoraphobia received a group therapy lasting for one month. In addition they were instructed to perform physical exercise over eight weeks with either low intensity (movements with very little strain-like stretching) or moderate/high intensity (treadmill). The experimental group (moderate/high intensity) yielded a better improvement on the primary outcome measure than the control group. As a result aerobic exercise adds a benefit to CBT.

Summed up, there is evidence that physical activity has a therapeutic effect in the treatment of panic disorder with or without agoraphobia. The results of the described studies were in favor for exercise in comparison to placebo conditions. Still, pharmacotherapy and psychotherapy yielded better treatment outcomes than exercise. Therefore exercise should be recommended as additional treatment or whenever usual care is not practicable. Furthermore, one single bout of exercise directly has an anxiolytic effect.

Post-Traumatic Stress Disorder

Single arm pre/post studies indicate a beneficial effect of exercise for patients with post-traumatic stress disorder (PTSD). Three studies found that an exercise program lead to an improvement in PTSD symptoms (Diaz & Motta 2008; Manger & Motta 2005; Newman & Motta 2007). Due to small sample sizes (n=9 to 15) and missing control groups these results can be regarded as preliminary only.

Fetzner and Asmundson (2014) recruited 33 participants with full or subsyndromal PTSD not receiving a concurrent psychotherapy. They were randomly allocated to three exercise groups. Participants of all groups performed a 20-minute session on a stationary bicycle at 60–80% of heart rate reserve six times in two weeks. The attention of group 1 was drawn away from bodily sensations by watching a distracting video. Participants of group 2 were watching a real-time video of themselves and as a result increased their attention to somatic sensations during exercise. Group 3 exercised without an additional attention task.

Patients of all groups showed an improvement in PTSD symptoms, depression and anxiety sensitivity to the same degree indicating therapeutic effects of exercise like the studies described above. Furthermore the results show that this effect is not due to the attentional focus on somatic sensations serving as an interoceptive exposure. A limitation of this study is the lack of a non-active control group.

Finally, Rosenbaum, Sherrington, and Tiedemann (2015) investigated with a randomized controlled study the augmenting effect of exercise on usual care for PTSD patients. Eighty-one in-patients received either usual care (control group) or usual care plus exercise (experimental group). Usual care involved psychotherapy and pharmacotherapy. The protocol consisted of three exercise sessions a week and a walking programme. After 12 weeks the experimental group improved significantly more considering psychological outcomes (PTSD, depression, stress, and sleep) and physical outcomes (waist circumference, body fat, and sitting time) with medium to large effect sizes in comparison to usual care alone.

This result was confirmed by a pilot study with nine patients with PTSD (Powers et al. 2015). They all received a 12-session treatment with prolonged exposure therapy. While the control group received no further intervention the experimental group completed a 30-minute treadmill task at 70% of heart rate reserve prior to each exposure session. Physical exercise was associated with significant increase of brain-derived neurotrophic factor (BDNF) in serum and a stronger reduction of PTSD symptoms. BDNF is known to play an important role in cognitive processes like fear extinction. It is hypothesized that changes in BDNF due to exercise account for the better improvement in PTSD symptoms in the experimental group compared to the control group.

Due to a lack of high-quality randomized controlled trials, no final conclusion can be drawn. There are indications that physical activity reduces PTSD symptoms. Exercise as an augmenting method increases the effects of usual care.

Obsessive-Compulsive Disorder

For the effect of exercise on obsessive-compulsive disorder (OCD) no randomized controlled trials could be found. There are four studies examining exercise as an addition to CBT or exercise alone.

Brown et al. (2007) investigated 15 patients diagnosed with OCD and conducted a moderate-intensity training for 12 weeks. Patients were training for 20 minutes at the beginning and increased to 40 minutes in session 12. Training was guided once a week and patients were advised to conduct 2–3 training sessions a week alone. To guarantee a moderate-intensity, heart rate was monitored during the sessions and was between 55–69% of age-predicted maximal heart rate. Furthermore patients received CBT sessions for motivation to exercise. In this study exercise was used as an adjunct to an already for at least three-month ongoing CBT and/or pharmacological treatment of OCD. Results showed significant reductions of OCD symptoms from baseline to the other assessment times (end of treatment, 3-week follow-up, 6-week follow-up, 6-month follow-up). The quality of life score significantly increased from baseline to the end of intervention.

Abrantes et al. (2009) used the same sample described above to look at acute effects of exercise. Therefore they analyzed a weekly assessment conducted directly before and after the guided training. They found acute reductions of OCD symptom severity, anxiety and negative mood at the end of each training session, compared to the beginning. Furthermore patients showed less pre-exercise obsessions and compulsions during the 12-week program.

Rector, Richter, Lerman, and Regev (2015) used exercise as an additive component to CBT. They investigated 11 OCD patients taking part at 12 weeks CBT program (1 session per week) combined with exercise in the gym (three times per week). Exercise duration was 15–30 minutes from week 1–4 and 30–45 minutes from week 5 to week 12 with an increase in intensity from sessions 5–12. Pre- post OCD-symptoms and depression scores improved significantly and patients adhered well to the exercise protocol (80%). Comparing effect sizes of this study to effect sizes of monotherapies (a meta-analysis of CBT for OCD and two studies investigating exercise for OCD) the present results indicate a benefit of a combination of exercise and CBT instead of a monotherapy.

Lancer, Motta, and Lancer (2007) investigated 16 OCD patients. Patients had already been treated with standard treatment for OCD (medication and behavior therapy) and reported no more symptom improvement with standard treatment. Exercise was used additionally to standard treatment. Patients were walking at fitness centers for 30 minutes, three times per week for a period of six weeks. There was a significant reduction of OCD symptoms and depression from the end of baseline to post intervention and follow-up.

Summarizing exercise appears to have a beneficial effect on OCD symptoms (even when only used once) and can be used additive to CBT or alone. A limiting factor is that all four studies included small samples (N = 11–16 patients) and had no control group. Because of that results can only be seen as preliminary.

Social Anxiety Disorder

For social anxiety disorder (SAD) there are two studies comparing aerobic exercise and mindfulness-based stress reduction (MBSR). In the first randomized controlled trial Jazaieri, Lee, Goldin, and Gross (2012) randomly assigned 56 social anxiety disorder patients to an 8-week MBSR (N=31) or aerobic exercise (N=25) program. Twenty-nine patients with a diagnosis of SAD served as a control group. During the 8 weeks MBSR was guided once a week and patients were advised to practice daily. For aerobic exercise group patients received a gym membership for 2 months and had to join a weekly aerobic exercise session and conduct individual trainings two times per week. In both intervention groups social anxiety severity and depression decreased and well-being increased (pre-treatment compared to post-treatment and 3-month follow-up). Compared to untreated SAD group MBSR and aerobic exercise groups showed decreased symptoms of social anxiety.

In the second study Jazaieri et al. (2015) treated 47 patients with social anxiety disorder with either MBSR or aerobic exercise for 8 weeks. In both groups social anxiety symptoms decreased every week and there was no difference between the two groups. Taking pre-treatment anxiety into account, patients with low-pre-treatment social anxiety severity in the MBSR group showed less weekly anxiety ratings than in the aerobic exercise group. Contrary, patients with high pre-treatment severity had higher anxiety symptoms during MBSR versus aerobic exercise. The authors conclude that it might be beneficial to take the pre-treatment anxiety severity into account for the decision of the kind of intervention.

Altogether aerobic exercise seems to have a positive effect in social anxiety disorder. Social anxiety symptoms decreased and well-being increased after an exercise intervention. Compared to an untreated group of social anxiety disorder patients, the aerobic exercise group showed less social anxiety symptoms.

Generalized Anxiety Disorder

In their randomized controlled trial Herring, Jacob, Suveg, Dishman, and O'Connor (2012) conducted two different exercise trainings and compared them to a waitlist condition. The exercise trainings were resistance training and an aerobic exercise. In the resistance training predominantly leg press, leg curl and leg extension exercises were performed, while leg cycling was used as aerobic exercise. Patients with generalized anxiety disorder were randomly assigned to one of the three groups and exercised for 6 weeks, two times per week or waited for treatment. After the treatment remission rates were 60% for resistance training, 40% for aerobic exercise training and 30% for waitlist. Worry symptoms were significantly reduced in combined exercise conditions versus waitlist.

This randomized controlled trial investigating the effects of exercise on generalized anxiety disorder symptoms has promising findings. Patients of the exercise groups had beneficial outcomes compared to waitlist. A limiting factor is that there was no interventional control group.

Mixed Samples

In their randomized controlled trial Martinsen, Hoffart, and Solberg (1989) compared a non-aerobic and aerobic exercise training for patients with anxiety disorders (panic disorder with agoraphobia, social anxiety disorder and generalized anxiety disorder). Patients exercised three times per week for one hour. The treatment duration was 8 weeks. In both groups anxiety symptoms significantly decreased after the treatment. There was no difference between the two groups.

Merom et al. (2008) treated three different anxiety diagnoses (panic disorder, generalized anxiety disorder, and social anxiety disorder) in their randomized controlled trial. Patients were randomly allocated to a group CBT + education or group CBT + exercise group. Group CBT sessions were for one specific diagnosis (social anxiety disorder, generalized anxiety disorder or panic disorder), and no mixed diagnosis was in the same CBT group. The exercise group met directly after the weekly CBT sessions with a trainer for 15–30 minutes to talk about pedometer use, benefits of exercise, collecting logbook etc., while the education group met three times with this trainer and talked about healthy eating. Exercise was carried out by the patients alone and should have a moderate intensity. The CBT + exercise group showed a significantly higher reduction in depression, anxiety and stress scores than the CBT + education group.

Meta-Analysis

Bartley, Hay, and Bloch (2013) included seven studies with 407 anxiety patients in their meta-analysis to find out if aerobic exercise is effective for the treatment of anxiety disorders. The authors conclude that CBT is beneficial compared to exercise and that pharmacotherapy has a tendency to be superior to exercise. But when comparing exercise to waitlist or placebo controls, exercise showed a great benefit.

In their meta-analysis Wipfli, Rethorst, and Landers (2008) investigated 49 randomized, controlled trials with 3,566 participants. Individuals exercising showed higher reductions in anxiety than no treatment control groups. The effect sizes of reductions of anxiety for participants exercising 3–4 times per week instead of 1–2 times per week or 5+ times per week were significantly higher. Comparing effect sizes of exercise group to other treatment groups (CBT, pharmacotherapy, group therapy, light exercise, relaxation/meditation, stress management education, music therapy), exercise was superior in reducing anxiety, only pharmacotherapy was better and CBT equal. Looking at the dose-response relationship Wipfli et al. (2008) could not find a significant correlation.

Kind of Exercise

In the studies described above the effect of different kinds of exercise on anxiety symptoms was investigated. Some used walking, running or bicycle training, some moderate versus light intensity training, resistance training or non-aerobic exercise. Furthermore duration, extent of training sessions and the presence of control groups varied. In this chapter the studies mentioned above are examined by looking at the kind of intervention used.

Different Kinds of Exercise

Authors stated three different kinds of exercise: 1) Walking, running, bicycle and aerobic training; 2) Non-aerobic training, and 3) Resistance training.

1. Looking at walking, running, bicycle and aerobic training there are eight studies comparing this kind of exercise to a control group. Four of them used exercise as a monotherapy, while the other four used exercise as an additional treatment.

   The monotherapy studies (Broocks et al. 1998; Hovland et al. 2013; Jazaieri, Goldin, Werner, Ziv, & Gross 2012; Jazaieri et al. 2015) compared aerobic exercise to clomipramine, placebo pill, CBT, waitlist and MBSR. Clomipramine and CBT were superior to exercise regarding the magnitude of reduction of anxiety symptoms. MBSR was as effective as exercise. But exercise was superior to placebo pill and waitlist and significantly reduced anxiety symptoms in all of the four studies.

   For the four studies using exercise as an additional treatment (Rosenbaum et al. 2015; Powers et al. 2015; Merom et al. 2008; Wedekind et al. 2010) exercise was used as an adjunct to CBT, paroxetine, placebo, CBT and/or pharmacotherapy and exercise prior to prolonged exposures. Control groups were CBT plus educational sessions, paroxetine plus relaxation, placebo plus relaxation CBT and/or pharmacotherapy and prolonged exposures without exercise respectively. In three of the four studies additional exercise was significantly better in reducing anxiety scores than the no exercise control groups. Only paroxetine and placebo plus exercise versus relaxation yielded no significant difference between exercise and relaxation.

   One study without control group (Rector et al. 2015) used exercise as an addition to CBT and anxiety symptoms improved significantly after the exercise training and CBT.

   Six studies conducted an exercise treatment as a monotherapy without control groups (Abrantes et al. 2009; Lancer, Motta, & Lancer 2007, Brown et al. 2007; Fetzner & Asmundson 2014; Manger & Motta 2005; Newman & Motta 2007). The patients participating in those studies improved significantly regarding their anxiety symptoms after the intervention. Furthermore anxiety symptoms decreased from the beginning to the end of each session. A limiting factor for those studies is the small sample.
2. Regarding non-aerobic training there is only one study comparing non-aerobic training to aerobic training (Martinsen et al. 1989). In both groups anxiety scores decreased significantly and there was no difference between aerobic and non-aerobic treatment.
3. Finally one study investigated resistance training (Herring et al. 2012). Resistance training was compared to aerobic exercise and waitlist. Remission rate was 60% for patients conducting resistance training, 40%

for aerobic exercise and 30% for waitlist. Anxiety scores after treatment of resistance training and aerobic exercise groups were significantly lower than scores of the waitlist group.

Taken together all kinds of exercise (aerobic, non-aerobic, and resistance training) seem to have a beneficial effect on anxiety. Results for non-aerobic and resistance training should be regarded carefully, because there were only two studies investigating those treatments.

Additional Exercise vs. Monotherapy

In all of the studies using exercise as a monotherapy, anxiety scores decreased after exercise treatment. Comparing exercise as a monotherapy to other treatments, CBT and Clomipramine were superior to exercise regarding the magnitude of reduction of anxiety. MBSR was as effective as exercise and exercise was significantly better in reducing anxiety than placebo and waitlist.

For those studies using exercise as an addition to other treatments (CBT, pharmacotherapy, placebo, exercise prior to exposure), results indicated a benefit of additional exercise than not implementing it. Only one study using paroxetine and placebo with exercise or relaxation yielded no significant difference between relaxation and exercise (Wedekind et al. 2010). Only one study (Rector et al. 2015) investigated exercise plus CBT without a control group and patients improved significantly regarding anxiety scores. The authors compared the effect sizes of their study to effect sizes of monotherapy studies (CBT or exercise only) and results seem to be beneficial for the additional treatment compared to monotherapy.

Extent of Exercise

Looking at the extent of exercise studies varied a lot. Extent was from 5 to 12 weeks with two to five exercise sessions per week and a duration of 15 to 60 minutes. Most studies required a period of 8 to 12 weeks and implemented three training sessions per week lasting 30 to 45 minutes.

In their meta-analysis Wipfli et al. (2008) found out that effect sizes for participants exercising 3–4 times per week instead of 1–2 times per week or 5+ times per week were significantly higher in reducing anxiety.

Intensity of Exercise

Two studies compared different intensities of exercise.

Gaudlitz et al. (2015) compared a moderate/high-intensity exercise to a low-intensity exercise additional to CBT. Patients of the moderate/high-intensity group improved better than the low-intensity group.

Another study investigated the effects of exercise on panic symptoms and anxiety after a 35% CO_2 inhalation (Esquivel et al. 2008). Patients were advised to exercise moderate/hard or very light directly before inhalation.

Panic symptoms and anxiety ratings after inhaling CO2 were significantly smaller in the moderate/hard training group.

Taken together those two studies indicate a benefit of moderate/hard intensity exercise compared to low intensity. In contrast Wipfli et al. (2008) could not find a significant dose-response relationship.

Control Groups

Looking at the different kinds of control groups used in the above-listed studies (Table 13.1) MBSR and relaxation were as effective as exercise in reducing anxiety symptoms. Pharmacotherapy and CBT were superior to exercise. But exercise yielded significantly higher reductions in anxiety scores than placebo pill, waitlist or educational sessions. Unfortunately many studies had no control groups.

Bartley et al. (2013) reported in their meta-analysis that exercise led to large effects in reducing anxiety when there was no control group, waitlist or placebo. But compared to other control groups exercise had no superior effect sizes. Therefore it is important to conduct more studies with adequate control groups.

Mechanisms of Action

Psychological Mechanisms

As already mentioned one underlying mechanism for the beneficial effects of exercise is its function as interoceptive exposure. There is empirical evidence that repeated exposure to running reduces cognitive and affective responses to bodily sensations over time in patients with high anxiety sensitivity due to altered attribution and emotional valence (Sabourin, Stewart, Watt, & Krigolson 2015).

In contrast, Fetzner and Asmundson (2014) found that attention to somatic sensations does not explain the positive effects of exercise. PTSD patients benefited from exercise to the same degree regardless of their attention being drawn away from or being directed to somatic sensations.

Further psychological explanations for the effects of exercise are an increase in self-efficacy and self-esteem. Patients feeling stuck due to their anxiety disorder may experience a new feeling of mastery being able to successfully complete a training task (Imayama et al. 2011; Imayama et al. 2013; Katula, Blissmer, & McAuley 1999). Change in body scheme and experienced internal locus of control may additionally contribute to an improvement of mental well-being through exercise.

Biological Mechanisms

As underlying biological mechanisms especially the neurotransmitter system, stress system and neurotrophic factors are discussed.

There is evidence that patients with panic disorder and/or agoraphobia show increased sensitivity to the selective 5-HT(1A) receptor agonist ipsapirone in comparison to healthy control subjects. Ipsapirone influences serotonergic mechanisms in the brain through binding to 5-HT(1A) receptor subtypes, resulting in decreased serotonergic transmission. The administration of ipsapirone leads to psychological and somatic symptoms and full panic attacks in patients with panic disorder (Broocks et al. 2000).

Broocks et al. (2003) investigated whether an effective treatment of the anxiety disorder is attended by a modification of this responsivity. Therefore, they performed ipsapirone challenges before and after a 10-week treatment with exercise, clomipramine or placebo. All psychological responses to ipsapirone were significantly reduced in the exercise and clomipramine group other than in the placebo group. After all, an anxiety treatment had divergent effects on physiologic responses (body temperature, cortisol) to ipsapirone.

It is likely that exercise leads to a normalization of the serotonin metabolism as endurance training was followed by an increased serotonin synthesis and turnover in rat brains (Dey 1994).

It is known that physical activity decreases cortisol levels in the body or inhibits an increase (Rimmele et al. 2009; Broocks et al. 1999; Kraemer et al. 1999). A few research projects explored whether an exercise treatment for anxiety is associated with alterations in the stress system. Wedekind et al. (2008) allocated 29 patients with panic disorder to four treatment conditions: paroxetine vs. placebo combined with either relaxation or exercise. Patients were asked to collect their night-time urine for measurement of cortisol excretion. Although all subjects improved significantly, no differences in cortisol excretion at pre/post measurements or between groups could be detected.

Plag et al. (2014) reported that no differences between CBT+ low level exercise and CBT+ high level endurance training for panic patients could be found considering cortisol levels during 1-month treatment. However, at the 7-month follow-up patients in the high-level group showed lower cortisol levels than patients in the low-level group, indicating a decelerated effect of aerobic exercise on the stress system (HPA system).

Furthermore, the positive effect of physical anxiety might be attributable to an increase of neurotrophic factors which protect existing neurons and enhance the formation of new ones in the human brain. Panic patients show decreased levels of brain-derived neurotrophic factor (BDNF) compared to healthy controls (Ströhle et al. 2010). Lower BDNF concentrations are associated with impaired extinction learning (Singewald, Schmuckermair, Whittle, Holmes, & Ressler 2015). Therefore it is an important finding that physical activity increases BDNF levels in patients with panic disorder (Ströhle et al. 2010; Szuhany, Bugatti, & Otto 2015). Pre-clinical studies found improved extinction learning in rats through exercise (Van Kummer & Cohen 2015; Andero & Ressler 2012). The influence of physical activity on the extinction of fear in humans was studied by Powers et al. (2015) in a pilot study in patients with post-traumatic stress disorder (n=9). The results show an increase in BDNF levels and a decrease of PTSD symptoms due to physical activity prior to each exposure session indicating an improved extinction.

Is Exercise Safe for Panic Patients?

Cross-sectional studies have proven that subjects with a diagnosed anxiety disorder are less likely to engage in physical exercise than persons without mental disorder (Goodwin 2003; Broocks et al. 1997; Abu-Omar, Rutten, & Robine 2004). This result can be interpreted in two different ways: on the one hand physical activity prevents the incidence of mental disorders and as a result more active people will develop a new disorder more unlikely. This is what longitudinal studies confirm (Ströhle et al. 2007; Jonsdottir, Rodjer, Hadzibajramovic, Borjesson, & Ahlborg 2010). On the other hand the existence of an anxiety disorder may present a barrier to exercise regularly. One of the key symptoms for panic disorder is the fear of somatic changes like an increasing heart rate. These somatic sensations may be misinterpreted as a sign for a serious disease (e.g. heart attack). Bodily sensations during exercise are similar to those experienced during panic attacks which explains why panic patients avoid engaging in physical activity. Also, panic disorder patients seem to experience more somatic symptoms after exercise than healthy control subjects (Ströhle et al. 2009). This is where the vicious circle begins: physical activity like climbing stairs is avoided wherever possible. As a result physical fitness decreases, which in turn leads to a cumulative appearance of unpleasant sensations. Consistently avoidance behavior further increases.

From a psychotherapeutic view this fear-inducing effect of exercise is even desired and may be used for interoceptive exposure. Interoceptive exposures give panic patients the chance to adapt to bodily sensations, to experience them as harmless and to reattribute them to physiological mechanisms.

Cameron and Hudson (1986) reported that exercise acutely increased anxiety in 20% of panic patients. But does exercise actually cause panic attacks?

A few studies have reported the amount of experienced panic attacks during physical exercise. Taylor et al. (1987) found that one subject out of 40 panic patients experienced a panic attack while exercising on a treadmill but was able to continue the training. Gaffney, Fenton, Lane, and Lake (1988) explored ten panic patients undergoing a stationary leg cycling program of whom none experienced panic. Another study investigated 16 patients with panic disorder and found that leg cycling led to no panic attacks except in one subject (Stein et al. 1992).

Martinsen, Raglin, Hoffart, and Friis (1998) explored the association of an increase of lactate level and experienced panic. In research lactate infusions are used to provoke panic attacks. Interestingly, infusions cause a lower concentration of blood lactate level (5 to 6 mmol/L) than maximal exercise (>6 up to 14 mmol/L) (Liebowitz et al. 1985). However, the authors hypothesize that exercise does not produce panic attacks despite increased lactate levels. Panic patients were instructed to exercise on a bicycle ergometer with increasing workload. Thirty-five completed the submaximal task and 24 patients completed both the submaximal and supramaximal task. Although lactate levels were higher than caused by infusion only 4% (n=1) of the subjects reported panic during exercise. In contrast, 67% of patients with panic disorder experience a panic attack after lactate infusion. The authors conclude that panic patients can safely engage in vigorous exercise with the chances of experiencing a panic attack being small.

Ramos, Sardinha, Nardi, and de Araujo (2014) draw the same conclusion: 52 panic patients and 52 healthy controls underwent a cardiopulmonary exercise task (cycling) while their physiological functions were observed. Panic patients were found to have a lower maximal oxygen uptake and experienced a higher heart rate during exercise. None of the panic patients experienced abnormal blood pressure, heart rate or respiratory responses. No panic attack was reported.

It has to be noted that the described studies have some limitations: a significant number of patients refused to exercise on a maximal intensity level. It is possible that especially these subjects have a higher chance to experience panic attacks. Furthermore presence of study staff could be interpreted as safety signal and thereby inhibit panic attacks.

Summed up there are no objective limitations for panic patients in performing physical exercise. The chance of experiencing a full panic attack is remarkably small. However, even if physical activity causes anxiety it should not be avoided but encouraged as part of therapy.

Conclusion

Altogether, studies indicate a beneficial effect of exercise in the treatment of anxiety disorders. Especially for the treatment of panic disorder with/or without agoraphobia randomized controlled trials revealed a positive effect of exercise. For post-traumatic stress disorder, obsessive-compulsive disorder, social anxiety disorder, and generalized anxiety disorder results are promising but more randomized controlled trials are needed to draw further conclusions.

In general it seems to be the best to use exercise as an additional treatment to cognitive behavioral therapy or pharmacotherapy. If usual care is not applicable, exercise still has an anxiety-reducing effect when used as a monotherapy.

Most studies used aerobic exercise as exercise intervention, but resistance training and non-aerobic exercise seem to be helpful as well. Further research is necessary to find out what kind of exercise with which intensity and frequency is most suitable.

References

Abrantes, A. M., D. R. Strong, A. Cohn, Y. Cameron, B. D. Greenberg, M. C. Mancebo, & R. A. Brown. 2009. "Acute changes in obsessions and compulsions following moderate-intensity aerobic exercise among patients with obsessive-compulsive disorder." *J Anxiety Disord 23* (7): 923–7. doi:10.1016/j.janxdis.2009.06.008

Abu-Omar, K., A. Rutten, & J.-M. Robine. 2004. "Self-rated health and physical activity in the European Union." *Soz Praventivmed 49* (4): 235–42.

Andero, R., & K. J. Ressler. 2012. "Fear extinction and BDNF: Translating animal models of PTSD to the clinic." *Genes Brain Behav 11* (5): 503–12. doi:10.1111/j.1601-183X.2012.00801.x

Bandelow, B., J. Wiltink, G. W. Alpers, C. Benecke, J. Deckert, A. Eckhardt-Henn,... M. E. Beutel. 2014. *Deutsche S3-Leitlinie Behandlung von Anststörungen*. www.awmf.org/leitlinien.html.

Bartley, C. A., M. Hay, & M. H. Bloch. 2013. "Meta-analysis: Aerobic exercise for the treatment of anxiety disorders." *Prog Neuropsychopharmacol Biol Psychiatry* 45: 34–9. doi:10.1016/j.pnpbp.2013.04.016

Broocks, A., B. Bandelow, A. George, C. Jestrabeck, M. Opitz, U. Bartmann,...G. Hajak. 2000. "Increased psychological responses and divergent neuroendocrine responses to m-CPP and ipsapirone in patients with panic disorder." *Int Clin Psychopharmacol 15* (3): 153–61.

Broocks, A., B. Bandelow, G. Pekrun, A. George, T. Meyer, U. Bartmann,...E. Ruther. 1998. "Comparison of aerobic exercise, clomipramine, and placebo in the treatment of panic disorder." *Am J Psychiatry 155* (5): 603–9.

Broocks, A., T. F. Meyer, B. Bandelow, A. George, U. Bartmann, E. Ruther, & U. Hillmer-Vogel. 1997. "Exercise avoidance and impaired endurance capacity in patients with panic disorder." *Neuropsychobiology 36* (4): 182–7.

Broocks, A., T. Meyer, A. George, U. Hillmer-Vogel, D. Meyer, B. Bandelow,...E. Ruther. 1999. "Decreased neuroendocrine responses to meta-chlorophenylpiperazine (m-CPP) but normal responses to ipsapirone in marathon runners." *Neuropsychopharmacology 20* (2): 150–61. doi:10.1016/S0893-133X(98)00056-6

Broocks, A., T. Meyer, M. Opitz, U. Bartmann, U. Hillmer-Vogel, A. George,...B. Bandelow. 2003. "5-HT1A responsivity in patients with panic disorder before and after treatment with aerobic exercise, clomipramine or placebo." *Eur Neuropsychopharmacol 13* (3): 153–64.

Brown, R. A., A. M. Abrantes, D. R. Strong, M. C. Mancebo, J. Menard, S. A. Rasmussen, & B. D. Greenberg. 2007. "A pilot study of moderate-intensity aerobic exercise for obsessive compulsive disorder." *J Nerv Ment Dis 195* (6): 514–20. doi:10.1097/01.nmd.0000253730.31610.6c

Cameron, O. G., & C. J. Hudson. 1986. "Influence of exercise on anxiety level in patients with anxiety disorders." *Psychosomatics 27* (10): 720–3. doi:10.1016/s0033-3182(86)72622-4

Dey, S. 1994. "Physical exercise as a novel antidepressant agent: Possible role of serotonin receptor subtypes." *Physiol Behav 55* (2): 323–9.

Diaz, A. B., & R. Motta. 2008. "The effects of an aerobic exercise program on post-traumatic stress disorder symptom severity in adolescents." *Int J Emerg Ment Health 10* (1): 49–59.

Dilling, H., & H. J. Freyberger. 2014. *Taschenführer zur ICD-10-Klassifikation psychischer Störungen*. 7 ed: World Health Organization und Verlag Hans Huber, Hogrefe AG, Bern.

Esquivel, G., J. Diaz-Galvis, K. Schruers, C. Berlanga, C. Lara-Munoz, & E. Griez. 2008. "Acute exercise reduces the effects of a 35% CO2 challenge in patients with panic disorder." *J Affect Disord 107* (1–3): 217–20. doi:10.1016/j.jad.2007.07.022

Fetzner, M. G., & G. J. Asmundson. 2014. "Aerobic exercise reduces symptoms of post-traumatic stress disorder: A randomized controlled trial." *Cogn Behav Ther*: 1–13. doi:10.1080/16506073.2014.916745

Gaffney, F. A., B. J. Fenton, L. D. Lane, & C. R. Lake. 1988. "Hemodynamic, ventilatory, and biochemical responses of panic patients and normal controls with sodium lactate infusion and spontaneous panic attacks." *Arch Gen Psychiatry 45* (1): 53–60.

Gaudlitz, K., J. Plag, F. Dimeo, & A. Ströhle. 2015. "Aerobic exercise training facilitates the effectiveness of cognitive behavioral therapy in panic disorder." *Depress Anxiety 32* (3): 221–8. doi:10.1002/da.22337

Goodwin, R. D. 2003. "Association between physical activity and mental disorders among adults in the United States." *Prev Med 36* (6): 698–703.

Goodwin, R. D., C. Faravelli, S. Rosi, F. Cosci, E. Truglia, R. de Graaf, & H.-U. Wittchen. 2005. "The epidemiology of panic disorder and agoraphobia in Europe." *Eur Neuropsychopharmacol 15* (4): 435–43. doi:10.1016/j.euroneuro.2005.04.006

Herring, M. P., M. L. Jacob, C. Suveg, R. K. Dishman, & P. J. O'Connor. 2012. "Feasibility of exercise training for the short-term treatment of generalized anxiety disorder: A randomized controlled trial." *Psychother Psychosom 81* (1): 21–8. doi:10.1159/000327898

Hovland, A., I. H. Nordhus, T. Sjobo, B. A. Gjestad, B. Birknes, E. W. Martinsen,...S. Pallesen. 2013. "Comparing physical exercise in groups to group cognitive behaviour therapy for the treatment of panic disorder in a randomized controlled trial." *Behav Cogn Psychother 41* (4): 408–32. doi:10.1017/S1352465812000446

Imayama, I., C. M. Alfano, L. A. C. Bertram, C. Wang, L. Xiao, C. Duggan,...A. McTiernan. 2011. "Effects of 12-month exercise on health-related quality of life: A randomized controlled trial." *Prev Med 52* (5): 344–51. doi:10.1016/j.ypmed.2011.02.016

Imayama, I., C. M. Alfano, C. E. Mason, C. Wang, L. Xiao, C. Duggan,...A. McTiernan. 2013. "Exercise adherence, cardiopulmonary fitness and anthropometric changes improve exercise self-efficacy and health-related quality of life." *J Phys Act Health 10* (5): 676–89.

Jazaieri, H., P. R. Goldin, K. Werner, M. Ziv, & J. J. Gross. 2012. "A randomized trial of MBSR versus aerobic exercise for social anxiety disorder." *J Clin Psychol 68* (7): 715–31. doi:10.1002/jclp.21863

Jazaieri, H., I. A. Lee, P. R. Goldin, & J. J. Gross. 2015. "Pre-treatment social anxiety severity moderates the impact of mindfulness-based stress reduction and aerobic exercise." *Psychol Psychother.* doi:10.1111/papt.12060

Jonsdottir, I. H., L. Rodjer, E. Hadzibajramovic, M. Borjesson, & G. Ahlborg, Jr. 2010. "A prospective study of leisure-time physical activity and mental health in Swedish health care workers and social insurance officers." *Prev Med 51* (5): 373–7. doi:10.1016/j.ypmed.2010.07.019

Katula, J. A., B. J. Blissmer, & E. McAuley. 1999. "Exercise intensity and self-efficacy effects on anxiety reduction in healthy, older adults." *J Behav Med 22* (3): 233–47.

Kraemer, W. J., K. Hakkinen, R. U. Newton, B. C. Nindl, J. S. Volek, M. McCormick,... W. J. Evans. 1999. "Effects of heavy-resistance training on hormonal response patterns in younger vs. older men." *J Appl Physiol (1985) 87* (3): 982–92.

Lancer, R., R. Motta, & D. Lancer. 2007. "The effect of aerobic exercise on obsessive-compulsive disorder, anxiety, and depression: A preliminary investigation." *Behavior Therapist 30* (3): 53–62.

Liebowitz, M. R., J. M. Gorman, A. J. Fyer, M. Levitt, D. Dillon, G. Levy,...D. F. Klein. 1985. "Lactate provocation of panic attacks. II. Biochemical and physiological findings." *Arch Gen Psychiatry 42* (7): 709–19.

Manger, T. A., & R. W. Motta. 2005. "The impact of an exercise program on post-traumatic stress disorder, anxiety, and depression." *Int J Emerg Ment Health 7* (1): 49–57.

Martinsen, E. W., A. Hoffart, & O. Solberg. 1989. "Aerobic and non-aerobic forms of exercise in the treatment of anxiety disorders." *Stress Medicine 5*: 115–120.

Martinsen, E. W., J. S. Raglin, A. Hoffart, & S. Friis. 1998. "Tolerance to intensive exercise and high levels of lactate in panic disorder." *J Anxiety Disord 12* (4): 333–42.

Meng, X., & C. D'Arcy. 2013. "The projected effect of increasing physical activity on reducing the prevalence of common mental disorders among Canadian men and women: A national population-based community study." *Prev Med 56* (1): 59–63. doi:10.1016/j.ypmed.2012.11.014

Merom, D., P. Phongsavan, R. Wagner, T. Chey, C. Marnane, Z. Steel,...A. Bauman. 2008. "Promoting walking as an adjunct intervention to group cognitive behavioral therapy for anxiety disorders – a pilot group randomized trial." *J Anxiety Disord 22* (6): 959–68. doi:10.1016/j.janxdis.2007.09.010

Michael, T., U. Zetsche, & J. Margraf. 2007. "Epidemiology of anxiety disorders." *Psychiatry 6* (4): 136–42.

Newman, C. L., & R. W. Motta. 2007. "The effects of aerobic exercise on childhood PTSD, anxiety, and depression." *Int J Emerg Ment Health 9* (2): 133–58.

Plag, J., K. Gaudlitz, S. Schumacher, F. Dimeo, T. Bobbert, C. Kirschbaum, & A. Ströhle. 2014. "Effect of combined cognitive-behavioural therapy and endurance training on cortisol and salivary alpha-amylase in panic disorder." *J Psychiatr Res 58*: 12–19. doi:10.1016/j.jpsychires.2014.07.008

Powers, M. B., J. L. Medina, S. Burns, B. Y. Kauffman, M. Monfils, G. J. Asmundson,...J. A. Smits. 2015. "Exercise augmentation of exposure therapy for PTSD: Rationale and pilot efficacy data." *Cogn Behav Ther*: 1–14. doi:10.1080/16506073.2015.1012740

Ramos, P. S., A. Sardinha, A. E. Nardi, & C. G. de Araujo. 2014. "Cardiorespiratory optimal point: A submaximal exercise variable to assess panic disorder patients." *PLoS One 9* (8): e104932. doi:10.1371/journal.pone.0104932

Rector, N. A., M. A. Richter, B. Lerman, & R. Regev. 2015. "A pilot test of the additive benefits of physical exercise to CBT for OCD." *Cogn Behav Ther*: 1–13. doi:10.1080/16506073.2015.1016448

Rimmele, U., R. Seiler, B. Marti, P. H. Wirtz, U. Ehlert, & M. Heinrichs. 2009. "The level of physical activity affects adrenal and cardiovascular reactivity to psychosocial stress." *Psychoneuroendocrinology 34* (2): 190–8. doi:10.1016/j.psyneuen.2008.08.023

Rosenbaum, S., C. Sherrington, & A. Tiedemann. 2015. "Exercise augmentation compared with usual care for post-traumatic stress disorder: A randomized controlled trial." *Acta Psychiatr Scand 131* (5): 350–9. doi:10.1111/acps.12371

Sabourin, B. C., S. H. Stewart, M. C. Watt, & O. E. Krigolson. 2015. "Running as interoceptive exposure for decreasing anxiety sensitivity: Replication and extension." *Cogn Behav Ther*: 1–11. doi:10.1080/16506073.2015.1015163

Singewald, N., C. Schmuckermair, N. Whittle, A. Holmes, & K. J. Ressler. 2015. "Pharmacology of cognitive enhancers for exposure-based therapy of fear, anxiety and trauma-related disorders." *Pharmacol Ther 149*: 150–90. doi:10.1016/j.pharmthera.2014.12.004

Stein, J. M., L. A. Papp, D. F. Klein, S. Cohen, J. Simon, D. Ross,...& J. M. Gorman. 1992. "Exercise tolerance in panic disorder patients." *Biol Psychiatry 32* (3): 281–7.

Ströhle, A., B. Graetz, M. Scheel, A. Wittmann, C. Feller, A. Heinz, & F. Dimeo. 2009. "The acute antipanic and anxiolytic activity of aerobic exercise in patients with panic disorder and healthy control subjects." *J Psychiatr Res 43* (12): 1013–7. doi:10.1016/j.jpsychires.2009.02.004

Ströhle, A., M. Hofler, H. Pfister, A. G. Muller, J. Hoyer, H.-U. Wittchen, & R. Lieb. 2007. "Physical activity and prevalence and incidence of mental disorders in adolescents and young adults." *Psychol Med 37* (11): 1657–66. doi:10.1017/S003329170700089X

Ströhle, A., M. Stoy, B. Graetz, M. Scheel, A. Wittmann, J. Gallinat,...R. Hellweg. 2010. "Acute exercise ameliorates reduced brain-derived neurotrophic factor in patients with panic disorder." *Psychoneuroendocrinology 35* (3): 364–8. doi:10.1016/j.psyneuen.2009.07.013

Szuhany, K. L., M. Bugatti, & M. W. Otto. 2015. "A meta-analytic review of the effects of exercise on brain-derived neurotrophic factor." *J Psychiatr Res 60*: 56–64. doi:10.1016/j.jpsychires.2014.10.003

Taylor, C. B., R. King, A. Ehlers, J. Margraf, D. Clark, C. Hayward,...S. Agras. 1987. "Treadmill exercise test and ambulatory measures in panic attacks." *Am J Cardiol 60* (18): 48J–52J.

Van Kummer, B. H., & R. W. Cohen. 2015. "Exercise-induced neuroprotection in the spastic Han Wistar rat: The possible role of brain-derived neurotrophic factor." *Biomed Res Int*: 834543. doi:10.1155/2015/834543

Wedekind, D., A. Broocks, N. Weiss, K. Engel, K. Neubert, & B. Bandelow. 2010. "A randomized, controlled trial of aerobic exercise in combination with paroxetine in the treatment of panic disorder." *World J Biol Psychiatry 11* (7): 904–13. doi:10.3109/15622975.2010.489620

Wedekind, D., A. Sprute, A. Broocks, G. Huther, K. Engel, P. Falkai, & B. Bandelow. 2008. "Nocturnal urinary cortisol excretion over a randomized controlled trial with paroxetine vs. placebo combined with relaxation training or aerobic exercise in panic disorder." *Curr Pharm Des 14* (33): 3518–24.

Wipfli, B. M., C. D. Rethorst, & D. M. Landers. 2008. "The anxiolytic effects of exercise: A meta-analysis of randomized trials and dose-response analysis." *J Sport Exerc Psychol 30* (4): 392–410.

Wittchen, H.-U., F. Jacobi, J. Rehm, A. Gustavsson, M. Svensson, B. Jönsson,...H. C. Steinhausen. 2011. "The size and burden of mental disorders and other disorders of the brain in Europe 2010." *Eur Neuropsychopharmacol 21* (9): 655–79. doi:10.1016/j.euroneuro.2011.07.018

14 Exercise and ADHD

Implications for Treatment

Lorna McWilliams

Introduction

Attention deficit hyperactivity disorder (ADHD) is one of the most common neurodevelopmental disorders in Western society. With a worldwide prevalence rate of around 3–5% and characterized with core symptoms of impulsivity, inattention, and hyperactivity, there is currently no single underlying cause or remedial treatment (Polanczyk, de Lima, Horta, Biederman, & Rohde 2007). At present, the prevailing treatment is medication, which inhibits the re-uptake of norepinephrine (e.g. atomoxetine) and dopamine (e.g. methylphenidate; MPH), primarily reducing cognitive function deficits commonly associated with the disorder whilst improving some of the behavioral indicators. However, these types of medication often have adverse side effects and symptoms of ADHD can longitudinally persist into adulthood in up to 50% of cases diagnosed in childhood (Lara et al. 2009).

Aside from behavioral difficulties, the disorder also manifests itself with functioning deficits, particularly impairments in executive functions. Executive function (EF) is often described as a set of higher-order cognitive processes based within the prefrontal or frontal-striatal brain regions, underlying the ability to successfully plan and carry out goal-directed behavior (Seidman 2006). The development of EF takes place throughout childhood and is linked with neurological and brain structure changes, thus factors that positively influence and contribute to EF development are important areas of research (Lamm, Zelazo, & Lewis 2006).

Gaining recent interest is the idea that exercise may have an ameliorative effect on symptoms of ADHD (Halperin, Berwid, & O'Neill 2014). Further, exercise is known to enhance EF processes evidenced by positive neurobiological changes that have been documented as a result of physical activity, mainly in cateholaminergic pathway regulation, improved brain-derived neurotropic factor (BDNF) levels, as well as increased neural activity in brain structures and networks involved in successful EF processes, all of which have also been implicated in ADHD (Dishman et al. 2006; Rommel, Halperin, Mill, Asherson, & Kuntsi 2013). Despite the promising findings supporting the effectiveness of physical activity on EF processes, there is a dearth of literature involving randomized controlled trial (RCT) and longitudinal designs in children, particularly in ADHD populations (Sibley & Etnier 2003; Tomporowski, Davis, Miller, & Naglieri 2008). This chapter aims to provide a comprehensive overview of the implications that exercise may have on the long-term beneficial treatment to improve the behavioral and functional symptoms of ADHD.

What Is ADHD?

This pervasive neurodevelopmental disorder is primarily defined as a "*persistent pattern of inattention and/or hyperactivity-impulsivity that interferes with functioning or development*" using the Diagnostic and Statistical Manual of Mental Disorders, 5th Edition, (DSM-V; American Psychiatric Association 2013). With three sub-types – predominantly inattentive, predominantly hyperactive-impulsive, and combined type based on the three core features of ADHD – there is longitudinal evidence that the sub-types of the disorder exist, yet the way that the disorder is expressed over time is unstable largely due to the heterogeneity of the disorder (Faraone, Biederman, Weber, & Russell 1998). The second system is similar to the DSM-V manual and terms ADHD as "*hyperkinetic disorders*" using the International Statistical Classification of Diseases (ICD-10; World Health Organization 1992) and tends to identify more severe manifestations of the disorder.

Children diagnosed with ADHD often also have co-morbid diagnoses such as externalizing disorders (oppositional defiant disorder, ODD, and conduct disorder), internalizing mood disorders (anxiety, depression, and bipolar disorder), autism spectrum disorder (ASD), and obsessive compulsive disorder (OCD), highlighting the complexity of the disorder and its treatment (Spencer, Biederman, & Mick 2007; Wilens et al. 2002). Children with ADHD are also more likely to have learning difficulties (Simonoff, Pickles, Wood, Gringras, & Chadwick 2007) and problems with motor development such as those found in children with Developmental Co-ordination Disorder (DCD; Baerg et al. 2011), which certainly influences the effects that exercise may have on the physical and behavioral manifestations of ADHD.

ADHD has been found to have a high heritability level of around 0.76 (Coghill & Hogg 2012) and several genes have been implicated in the disorder, mainly within the catecholaminergic system such as neurotransmitter genes DAT1, a dopamine transporter and DRD4, a dopamine receptor gene as well as noradrenergic genes (see reviews by Gizer, Ficks, & Waldman 2009; Li, Sham, Owen, & He 2006). The BDNF gene, thought to be involved in neurogenesis and brain plasticity, has also been associated with the disorder, although with inconsistent findings (Archer, Oscar-Berman, & Blum 2011). Several studies provide comprehensive reviews of neuroimaging and neuro-connectivity studies to demonstrate differences between ADHD samples and typically developing control groups on specific brain regions and neural circuits including the prefrontal cortex and structures responsible for the release of dopamine (Bush, Valera, & Seidman 2005; Konrad & Eickhoff 2010; Sonuga-Barke & Castellanos 2007). It is worthwhile to note that the brain regions, neurotransmitter systems and neural networks identified here are also predominantly responsible for carrying out cognitive processes, all of which exercise has been found to influence.

In contrast, the idea surfaces that gene-environment interactions are important in understanding the causes of the disorder, particularly in the treatment of ADHD (Biederman & Faraone 2005). Although it is not clear whether the role that the environment plays in the development of ADHD is causal, moderate effect sizes have been found between environmental risk factors and the disorder, such as pre-natal risk factors, social deprivation including diet, adverse family outcomes, parental stress and parental symptoms of the disorder (Aarnoudse-Moens, Weisgas-Kuperus, van Goudoever, & Oosterlaan 2009; Graziano, McNamara, Geffken, & Reid 2011; Counts, Nigg, Stawicki, Rappley, & Von Eye 2005; Stevenson et al. 2010). Such environmental factors also contribute to the successful treatment of the disorder, given that parents are primarily responsible for treatment decisions for their children as well as influencing the effectiveness of non-pharmacological treatment strategies involving parenting training and education. Given that parents are primarily responsible for their children's lifestyle, including diet and physical activity participation, parenting factors could be a particular issue in children with symptoms of ADHD. For example, it has been shown that parents influence peer relationships in children with ADHD, important for positive experiences during physical activity participation (Mikami, Emeh, & Stephens 2010).

As already mentioned, regardless of underlying cause, ADHD is associated with impairment on EF tasks such as working memory (WM), response inhibition, attention and planning, as well as non-executive tasks such as processing speed, and for the most part are the primary outcome measures in studies assessing the effectiveness of exercise as a treatment for the disorder. Such processes are perhaps functional impairments as a result of the underlying neurobiological origins of the disorder (Bush et al. 2005). It is worthy to note at this point that this construct (EF) still lacks a concrete definition and the cognitive processes included under this umbrella term vary from study to study (Barkley 2012). The "*unity and diversity of EF*" theory proposed that EF is primarily composed of working memory, the ability to shift attention (set shifting) and response inhibition processes (Miyake et al. 2000). Based on confirmatory factor analyses, it was revealed that performance on tasks designed to specifically assess these constructs were correlated but also distinct. Given that the sample used to demonstrate this was a sample of young adults using a cross-sectional design, it is difficult to apply this theory to a neurodevelopmental perspective. Certainly, evidence suggests that these constructs emerge at a young age; however, the processes are likely to be highly changeable (and interchangeable) throughout development, particularly for those with ADHD (Huizinga, Dolan, & van der Molen 2006). Further, it should also be noted that non-executive cognitive processes such as perception and processing speed are likely to contribute to EF task performance and so such measures are often included when assessing any single EF process (Salthouse 2005).

Individuals with ADHD are likely to exhibit impaired performance on neuropsychological tasks, even in adulthood, although the extent to which and on which specific EF measures is highly variable, particularly since the different ADHD subtypes tend to exhibit different EF profiles (e.g. Nigg, Willcutt, Doyle, & Sonuga-Barke 2005). The research evidence is inconclusive given that other studies find that children with ADHD can perform as accurately as typically developing children on EF tasks and that EF deficits can improve using longitudinal designs (Halperin, Trampush, Miller, Marks, & Newcorn 2008).

There are several different theories that attempt to conceptualize these EF deficits as contributing to the disorder, such as the cognitive-energetic model (Sergeant 2000) and dual-pathway model (Sonuga-Barke 2003). The principal idea being that deficits are due to: 1) core impairment in response inhibition, the basis for self-control or self-regulation, and/or 2) disrupted motivation systems that lead to a delay in arousal and produce reinforcement difficulties (Castellanos, Sonuga-Barke, Milham, & Tannock 2006). These concepts are also grounded in neurobiological research, identifying specific brain regions (e.g. prefrontal cortex) and networks thought to be responsible for EF processes (Liddle et al. 2011) and also implicate the dopaminergic systems again, since dopamine has a role in both response inhibition and reward signalling (Blum et al. 2008). Figure 14.1 illustrates the possible ways in which neurobiological, EF and motivation risk factors may underlie the behavioral symptoms of the disorder (Tripp & Wickens 2009).

In contrast, Sonuga-Barke and Halperin (2010) suggest that there may be multiple developmental pathways related to the environmental risk factors for ADHD development, as mentioned earlier. This is perhaps primarily due to the underlying response inhibition or motivation deficits being moderated by environmental factors such as punitive reinforcement from parents (Sonuga-Barke & Halperin 2010). Further evidence supports this

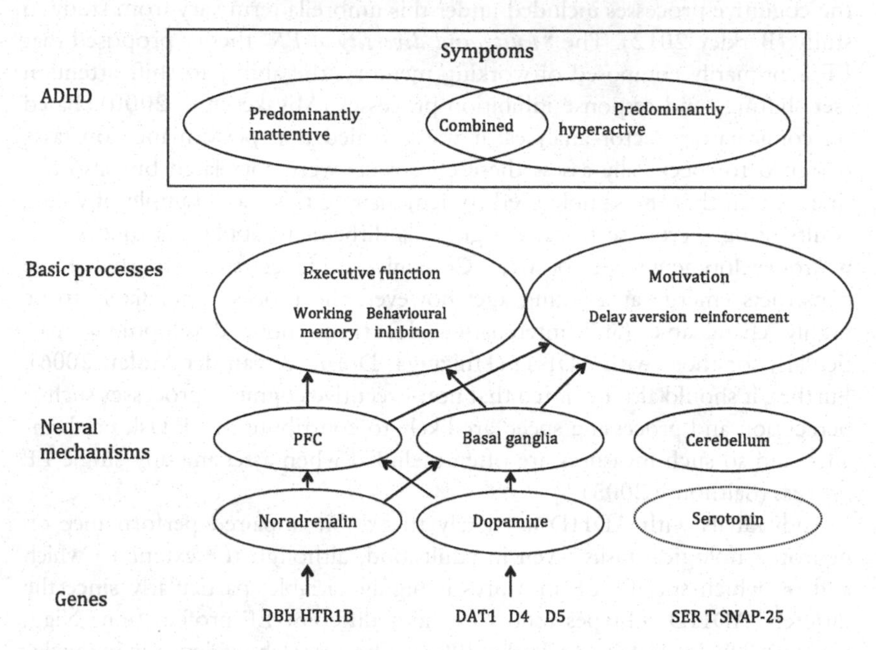

Figure 14.1 Model depicting factors underlying ADHD symptoms (Tripp and Wickens 2009). (Reprinted with permission from Elsevier)

idea whereby a disorganized family life has been linked with reduced longitudinal improvement on EF task performance in typical child development (Hughes & Ensor 2009). In addition, other environmental factors such as levels of physical fitness and obesity in childhood and adolescents have been found to affect EF task performance whereby lower levels of physical fitness and obesity are linked to poorer performance (Hillman, Castelli, & Buck 2005; Ruiz et al. 2010). Given the lack of a unifying theory to explain or treat the disorder, it could be suggested that treatment methods would be effective if developed to improve the EF impairments as well as behavioral symptoms of the disorder.

Current Treatment for ADHD

At present, pharmacological medication tends to be the primary treatment for symptom reduction in ADHD, although often recommended in combination with psychosocial interventions (e.g. National Institute for Clinical Excellence 2009). Primarily, psychostimulant medication is linked to improving altered dopaminergic systems, although up to 30% of children fail to benefit long term from this type of medication (Banaschewski, Roessner, Dittmann, Janardhanan Santosh, & Rothenberger 2004). In general, whilst medication has been found to improve behavioral symptoms, EF processes and classroom productivity (Prasad et al. 2013), side effects include sleep difficulties, loss of appetite, and mood dysregulation (Tobaiqy et al. 2011). Further, when medication is terminated, improvements in the profile of the disorder are not likely to be maintained (Buitelaar & Medori 2010).

Various other treatment strategies aim to improve functioning for those with ADHD, including cognitive training and neurofeedback (Klingberg et al. 2005; Arns, de Ridder, Strehl, Breteler, & Coenen 2009). However, such types of cognitive intervention have yet to show effectiveness in terms of transferring treatment from a laboratory (lab) setting to home and school-based environments (Rapport, Orban, Kofler, & Friedman 2013). Other interventions focus on environmental aspects aimed at improving ADHD outcomes principally based on psychosocial, family or parent-based training programmes and behavior modification techniques as well as dietary elimination/supplements (e.g. exclusion of certain food types/inclusion of fatty acids) (Fabiano et al. 2009; Hodgson, Hutchinson, & Denson 2012).

A meta-analysis comparing the effectiveness of 26 randomized controlled trials (RCT) that either used methylphenidate treatment, psychosocial interventions or both combined, in children aged 6–12 years, revealed that both types of treatment can be effective in improving teacher and parent-rated ADHD symptoms (and ODD) as well as social behaviors (van der Oord, Prins, Oosterlaan, & Emmelkamp 2008). However, both types of treatment, even when combined, were found to be limited in improving academic outcomes for the child (effect size range 0.19 to 0.35). Although a review by Prasad and colleagues (2013) suggests that medication does improve academic outcomes in children, according to Langberg and Becker (2012) long-term medication has minimal benefits for academic achievement in studies that followed participants for at least 3 years. Given that it has already been mentioned that EF deficits are implicated in academic achievement, treatments for ADHD should aim to target this aspect of associated impairment.

An additional meta-analysis of non-pharmacological treatment was conducted recently (Sonuga-Barke et al. 2013). The authors compared non-pharmacological interventions with treatment as usual (medication) using strict inclusion criteria (e.g. randomized controlled trials, RCTs). "Probably blind" analyses whereby one could expect that parents or teachers known to the intervention may bias outcome measures if they report child symptoms used to assess intervention effectiveness. Although initial effect sizes were promising (0.21–0.64), the "probably blind" analyses revealed that only dietary supplements and artificial food coloring elimination interventions remained significant (0.16 and 0.42, respectively). The authors did not include any studies that use exercise as a treatment for ADHD, mostly due to lack of sound methodologies, illustrating how novel this idea exists within the literature. Further, treatment outcomes only relied on either observational methods or parent and teacher-rated symptoms. Studies should aim to include neuropsychological measures in addition to assessing behavioral symptoms. Several authors have begun to postulate the efficacy of exercise in improving outcomes in ADHD (e.g. Berwid & Halperin 2012; Gapin, Labban, & Etnier 2011; Halperin, Bédard, & Curchack-Lichtin 2012).

Exercise as Treatment for ADHD

The investigation of exercise used in the treatment of ADHD is based on speculative evidence from animal studies suggesting biological underpinnings and supports a neurobiological basis for the disorder (Nithianantharajah & Hannan 2009). Studies report improved behavioral symptoms (Lufi & Parish-Plass 2011) and, importantly, enhanced performance on measures of EF in samples of children with a clinical diagnosis of ADHD (Medina et al. 2010).

The term physical activity is "*any bodily movement produced by skeletal muscles that results in energy expenditure*"; used interchangeably with exercise, which is "*physical activity that is planned, structured, repetitive, and purposive in the sense that improvement or maintenance of one of more components of physical fitness is an objective*" and both result in physical fitness, "*a set of attributes that people have or achieve*" (Caspersen, Powell, & Christenson 1985). For the purposes of this chapter, exercise and physical activity will be used interchangeably. There are two main methods of participating in exercise; acute "*concerns itself with a single bout of exercise*" and chronic "*concerns itself with the repetition of bouts of exercise over time*" (Audiffren 2009). Both types may have differential effects on ADHD.

In typically developing populations, studies have reported significant improvements in performance on tasks assessing non-EF tasks such as processing speed (Ellemberg and St. Louis-Deschênes 2010); as well as EF tasks of inhibitory control (Hillman et al. 2009); attention (Cereatti, Casella, Manganelli, & Pesce 2009) and memory (Pesce, Crova, Cereatti, Casella, & Bellucci 2009) following acute periods of physical activity in children and adolescents. Research has also demonstrated that typically developing children with higher levels of physical fitness perform better on a task of response inhibition and a memory task in addition to increased brain volume in the areas implicated in attention and inhibitory control (basal ganglia and hippocampus), compared to those with lower levels of fitness (Chaddock et al. 2010).

Preliminary evidence does suggest that physical activity can improve EF task performance in children with ADHD (Kang, Choi, Kang, & Han 2011; Pontifex, Saliba, Raine, Pichietti, & Hillman 2013; Smith et al. 2013; Verret, Guay, Berthiaume, Gardiner, & Béliveau 2012). Compared to methylphenidate (MPH) treatment, physical activity was found to have similar effects on a task of sustained attention performance in a sample of 25 boys diagnosed with ADHD (Medina et al. 2010). The design consisted of the completion of the Continuous Performance Test (CPT) at baseline, immediately after 30 minutes of exercise on a treadmill (moderate intensity heart rate; HR) and after one minute of stretching. Of the 25 boys, 16 were being treated for ADHD using MPH whilst 9 were not receiving any medication as treatment.

Within-group analyses revealed significant improvements in reaction time and improved accuracy, indicating that physical activity may help improve ADHD symptoms regardless of whether medication is prescribed as treatment for the disorder. Between-group analyses revealed no significant benefit from MPH treatment albeit those on MPH refrained from taking their medication 24 hours prior to testing. Nevertheless, all children completed the CPT task before exercise for the first time and immediately after the exercise session for the second time and so learning effects and any sustained improvements cannot be investigated.

THE UNDERLYING MECHANISMS OF CHANGE BY PHYSICAL ACTIVITY

It has been suggested that regular physical activity can lead to altered brain function, including those thought to underlie the cognitive improvements found when exercising (Verret et al. 2012). Neurobiological improvements as a result of exercise include increased cerebral flow, increased dopamine levels, capillary growth, increased tissue density, and improved neural connections (e.g. Colcome et al. 2004). The increase in neural connections and catecholaminergic pathways (including dopamine) could potentially mitigate symptoms of ADHD in children (Gapin et al. 2011). It is also suggested by Gapin and colleagues that the effects of physical activity may be more pronounced in individuals who have cognitive impairments such as those found in children with ADHD. For example, Tantillo, Kesick, Hynd, and Dishman (2002) evaluated the effect of acute treadmill walking in children (aged 6–12 years) with and without clinical levels of ADHD by assessing eye blinking, thought to be used as a proxy measure for dopaminergic functioning. The study found that the children with ADHD demonstrated improved regulation of their dopaminergic pathways with an increase in spontaneous eye blinking and slower acoustic startle eye blink response compared to typically developing children as well as a rest (inactive) condition (Tantillo et al. 2002). Although it is speculative whether the test used can accurately reflect changes in dopaminergic functioning, the study links the neurobiological evidence with measures that are related to functioning. Other research links exercise-related improvements with increased release of BDNF, indicating regulation of neural growth, found following an acute period of physical activity in a healthy adult sample (Rasmussen et al. 2009).

A cross-sectional study in a sample of 18 boys (aged 8–12 years) with a clinical ADHD diagnosis assessed the relationships between physical activity levels and EF task performance (Gapin & Etnier 2010). Each child completed EF tasks assessing response inhibition, planning, working memory and a non-EF task (processing speed) and recorded physical activity using an accelerometer and daily activity log over a 1-week period. Regression analyses revealed that higher reported physical activity levels predicted improved performance on the planning task. The other cognitive measures were in the expected direction although did not reach statistical significance. The positive relationship between physical activity and planning was not due to order effects (order of task completion was randomized); however, the sample only consisted of boys and the correlational design prevents the investigation of whether a causal relationship exists between physical activity and enhanced EF processes or whether children with ADHD would benefit more or less than typically developing children.

Further, as with Gapin and Etnier's (2010) findings in an ADHD sample, higher levels of physical fitness have been related to improved EF task performance in children compared to those with lower levels of fitness (Chaddock, Hillman, Buck, & Cohen 2011). Davis and Lambourne (2009) suggest that physical activity (although they argue that it should be high-intensity exercise) improves neurobiological functions, which in turn improve performance on EF tasks (see Figure 14.2). However, many of the research findings using acute exercise designs are reported in relation to lab-based settings and so studies that are naturalistic, and thus representative of real-world participation, would

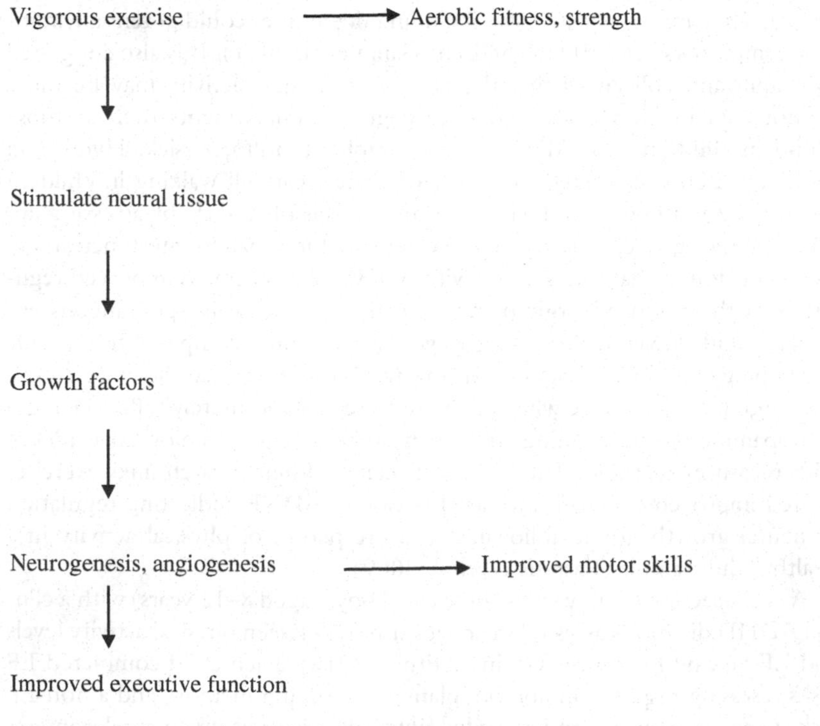

Figure 14.2 Proposed model of effect of physical activity on cognitive function (Davis and Lambourne 2009). (Reprinted with permission from John Wiley and Sons)

improve the ecological validity of any findings illustrating improvements in EF processes, particularly for those with ADHD.

Exercise Effects on Behavioral Symptoms

Alongside physical health improvements that could be associated with ADHD, a meta-analytic review concluded that exercise interventions have a small effect size (–0.38) on improving mental health outcomes in children (Ahn & Fedewa 2011). Specifically for ADHD, RCT trials had an effect size of –0.92 on symptom reduction. The authors also report that physical activity programmes can also decrease symptoms of depression (–0.41), anxiety (–0.35), and improve self-concepts and levels of self-esteem (0.29). The type of interventions successful in improving psychological outcomes included both circuit training and combination approaches indicating positive findings in relation to ADHD and other behavioral difficulties.

A small study (n = 32), found that a physical activity intervention improved parent-rated and self-report behavior in a group of boys diagnosed with ADHD (n = 15) as well as a group of children (n = 17) with other behavioral problems (Lufi & Parish-Plass 2011). The intervention consisted of 20 90-minute periods of individual and team-based physical activity over the course of a year. The study found significant improvements on parent-rated behavior subscales (mainly reduced aggression and anxiety scores) and anxiety in the self-report subscales at 1-year follow-up. This was found in both groups; however, there was no control group to compare findings with those who did not receive the intervention. These findings suggest that physical activity may improve other behavior difficulties in addition to ADHD.

In a sample of adolescents (n = 84; aged 11–16 years) diagnosed with ADHD, a 10-week physical activity intervention resulted in improvements on the attention, motor skills, academic and classroom behavior subscales of the Behavior Rating Scale compared to a control group who did not receive the intervention (Ahmed & Mohamed 2011). The authors do not report which informant completed the behavior questionnaire and it is not clear how the participants were diagnosed with ADHD or if receiving treatment, so the findings are difficult to interpret. The inclusion of an EF task in addition to the behavioral assessments would have further supported the authors' results. Similarly, Mahar and colleagues (2006) assessed the immediate effects of a school-based physical activity intervention in 243 children on observed symptoms of inattention in the classroom. The intervention involved 10 minutes of classroom-based activity per school day for 12 weeks compared to the control condition of no additional physical activity. Compared to the control group, children who received the intervention were rated (by researchers) as more "on task" in classroom-based activities, indicating improved levels of attention (Mahar et al. 2006). However, the observers were not blinded to condition and the study took place in a single school with no long-term follow-up to identify whether the improvements could be sustained over time, with no indication as to whether any children had a clinical diagnosis of ADHD in order to conduct sub-group analyses.

McKune, Pautz, and Lombard (2003) carried out a study with 19 children diagnosed with ADHD to assess the effect of a 5-week physical activity intervention on symptoms of the disorder as rated by parents. Thirteen children received the intervention, which involved 60 minutes of physical activity five times per week compared to 6 children in the control arm (no additional physical activity). Surprisingly, it was found that all children's behavior improved post-intervention, regardless of treatment group, compared to baseline behavior ratings (McKune et al. 2003). The findings may reflect that the children were all from the same school and parents were not blind to the study aims, generating the possibility of halo effects. Furthermore, the intervention and control groups had an uneven sample size, with a particularly small number of children in the control group, limiting any clear conclusions from this study.

Chuang, Tsai, Chang, Huang, and Hung (2015) assessed the effect of a 30-minute period of exercise using a treadmill (intensity of 60% HR reserve) on response inhibition performance using a Go No/Go task compared to a sedentary period (viewing a video) in a sample of 19 children diagnosed with ADHD (age 8–12 years). They found that reaction time on Go trials was significantly quicker following the exercise condition; however, the authors did not tease apart any differing contributions to the effects over and above medication as 9 children were receiving pharmacological treatment for the disorder. Additionally, the lack of control group limits the interpretation of findings.

Exercise Effects on Both Behavioral Symptoms and Cognitive Functions

A recent study included both behavioral and cognitive measures in the assessment of a physical activity intervention in young children (n = 17; aged 4 to 9 years) with symptoms of ADHD indicating clinical risk (Smith et al. 2013). Using a before and after design, the children received 30 minutes of additional physical activity on each school day over a period of 9 weeks. Cognitive measures included set shifting, planning, response inhibition, and working memory tasks designed for younger children (aged 5–8 years). There was a significant improvement on one sub-test of the response inhibition task (The Shape School) compared to baseline performance, although the effect size was small (0.32). Although no other findings were significant, the varied age range could explain this. Despite the small sample size, the findings are promising for future research with young children in this area with the development of between-group designs.

Verret et al. (2012) assessed the impact of a 10-week physical activity intervention on cognitive and behavioral measures of ADHD in 10 children (aged 7–12 years) diagnosed with ADHD on cognitive and behavioral outcome measures. The intervention lasted 45 minutes, three times a week over the study period whilst a control group of 11 children with ADHD received no intervention. The cognitive tasks used in the study assessed attention and response inhibition (Test of Everyday Attention for Children) whilst parents and teachers rated symptoms of ADHD. Although there were no differences on the behavioral assessment or response inhibition task post-intervention, children in the intervention group were significantly better at the sustained attention task compared to the control group (Verret et al. 2012). However, the children were not randomly assigned to the intervention or control group, thus a possibility of bias may have been introduced. It would have been valuable to know whether the improvements on the sustained attention task remained over time; however, the follow-up testing took place within a 1-week post-intervention period.

The studies mentioned so far have used before and after designs to assess the impact of physical activity on improving symptoms of ADHD and/or EF processes. Using randomization procedures for group allocation (treatment vs. control) helps to ensure that no systematic differences between groups affect the outcome measures and therefore improves the robustness of results.

Randomized Controlled Trials and Exercise Effects on EF Processes

A recent meta-analysis regarding whether exercise improves cognitive performance in children, adolescents, and young adults included five RCT studies that used a chronic physical activity design and EF tasks as outcome measures (Verburgh, Königs, Scherder, & Oosterlaan 2013). The authors found a small, non-significant effect size of 0.14 for the effect of chronic physical activity on improving EF outcomes, with a total of only 358 participants across the studies. This is surprising given that cognitive processes such as response inhibition performance in children have been implicated in the same brain structures thought to be positively affected by physical activity (Kramer & Erickson 2007). One study included as one of the chronic physical activity studies assessed the immediate effects (rather than chronic design) of a 30-minute period of stationary cycling on a choice reaction task compared to sedentary television viewing in preadolescent boys (Ellemberg & St. Louis-Deschênes 2010). Despite the lack of a large effect size, some individual studies (with RCT designs) do report positive findings in relation to exercise and cognition (EF) in child and adolescent samples (e.g. Fisher et al. 2011; Davis et al. 2007), albeit often excluding children with ADHD. Interest in this research area has however increased, with several RCT studies recently published, two of which assess acute effects of exercise and a fourth study evaluating the effectiveness of a chronic intervention on outcomes in ADHD samples.

Acute Studies

Using a sample of young adults (18–25 years), Gapin, Labban, Bohall, Wooten, and Chang (2015) investigated the acute effects of exercise on response inhibition (Stroop task), task switching (Trail Making task) and working memory (forward/backward digit span task) between college students with (n=10) and without (n = 10) an ADHD diagnosis. The authors also collected blood samples to assess BDNF levels for both groups before and after the exercise period to investigate as a marker for neurobiological improvements. Each participant completed the three tasks before the 30-minute exercise period completed on a treadmill (50–65 HR reserve intensity) and again after 5 minutes to return HR to resting levels. Following the period of acute exercise, performance improved on all three tasks in the non-ADHD group whilst those with ADHD indicated an improvement only on the Stroop task. Despite encouraging findings, it is impossible to tease apart whether these improvements demonstrate a learning effect. Additionally, there were no post-exercise changes to BDNF levels in either group.

With a similar study design, Piepmeier and colleagues (2015), in a sample of 32 adolescents (mean age 10.7 years), assessed the acute effects of 30 minutes of cycling compared to viewing a 30-minute nature documentary on EF processes whilst including a non-ADHD control group. Measures of EF included the trail-making test to assess set shifting, a Stroop task (response inhibition), and the Tower of London task (planning and problem solving). The ADHD group consisted of 14 adolescents, 10 of whom were being treated with stimulant medication. The authors reported improvements in response time for the Stroop and trail-making tasks following the exercise condition, although there were no effects on planning and problem solving or any effects of group. This could be explained by a lack of power required for this study design and the children with ADHD taking medication continued treatment as usual, which may have influenced their performance on the EF tasks.

Chronic Studies

Kamijo and colleagues (2011), using an RCT design, investigated the effects of a physical activity intervention on working memory and corresponding EEG activity in a sample of children aged 7–9 years including five children who had received a clinical diagnosis of ADHD (Kamijo et al. 2011). The intervention involved circuit-training and motor-skill development games for a period of 2 hours after each school day for 9 months. There was a wait-list control group who received no physical activity intervention. The working memory task was based on the Sternberg Task, responding to whether an encoded stimulus was present in an array of letters (using three set sizes of letters; 1, 3, and 5), completed pre- and post-intervention period. Compared to the control group, children who received the intervention were both more accurate and faster at responding on the working memory task at follow-up supported by positive EEG activity changes underlying working memory processes. However, it is worthy to note that five children with ADHD were being treated with medication and it is not clear which group these children were assigned to and so this may have influenced the study findings.

Using a crossover RCT design, Hill and colleagues found that one school week of an additional 15 minutes of in-school physical activity improved performance on paper and pencil tasks designed to assess EF processes compared to a week of no additional physical activity (Hill et al. 2010). A second study was conducted by this research group with the addition of a parent-reported ADHD symptoms measure (Hill, Williams, Aucott, Thomson, & Mon-Williams 2011). Their original findings were replicated and, although children with no ADHD symptoms performed better on the EF tasks, this did not significantly influence the overall positive effect of physical activity. The effects of physical activity were found for all children independent of symptoms of ADHD. Although promising, the lack of long-term follow-up testing limits the ability to assess whether there was a sustained improvement on cognitive processes as a result of the intervention. Further, there is little information on the reliability and validity of the cognitive measures used in this study, making it difficult to interpret the findings.

Another study assessed the effects of yoga on symptoms of ADHD using an RCT design (Jensen & Kenny 2004). In a sample of boys diagnosed with the disorder, 20 weeks of an hourly yoga session (n = 11) was compared to an active non exercise-based control (n = 8) condition that focussed on improving co-operation skills in the children. Teacher/parent-rated behavior and performance on a sustained attention task (Test of Variables of Attention) were outcome measures and the authors found significant post-intervention improvements in both groups on parent-rated behavior, although not for teacher-rated behavior or sustained attention-task performance. It is not clear when the follow-up data were collected and so the findings are difficult to interpret, particularly as parents were not blind to study aims. It could be that although yoga is a beneficial form of exercise, it is a low-intensity form, suggesting that exercise intensity (mostly moderate-vigorous) is more beneficial in improving EF processes.

A Korean RCT study with children (n = 32; mean age 8.4 years) clinically diagnosed with ADHD assessed the effects of a 12-session physical activity intervention (90 minutes twice a week, 6 weeks in total) on cognitive tests (Trail Making Test B and Digit Symbol Span) of visual attention, memory and task switching (Kang et al. 2011). The control group received 12 sessions of education designed to improve behavior and all of the children started receiving medication treatment during the first week of the intervention. Compared to the control group, children who received the intervention showed larger improvements on both cognitive tasks (faster response time for task switching and improved score for memory) as well as improved attention scores rated by parent and teacher. Cognitive performance may have been influenced by the relatively short intervention timeframe or by immediate follow-up testing (practice effects). Further, possible effects of medication treatment may have improved task performance and behavior. Although positive, it remains to be seen whether the positive EF improvements are sustained over time and can be beneficial as a stand-alone treatment for symptoms of ADHD.

Finally, in an attempt to assess both acute and chronic effects of exercise on both EF and motor performance in children with ADHD, Ziereis and Jansen (2015) conducted an RCT study in Germany. The design involved children (n = 43) aged 7–12 years with an ADHD diagnosis to one of three groups, two of which were experimental. Children either received a 12-week hourly session exercise intervention focusing on specific abilities (group 1, n = 13), sports without a specific focus (group 2, n = 14) or a wait-list control group (group 3, n = 16). No children were receiving stimulant medication as treatment for ADHD. All children completed EF measures assessed using a forward/backward digit span and letter-number sequencing task to measure verbal working memory and the corsi block tapping task to measure visuospatial working memory as well as a motor performance battery assessing manual dexterity, catching, aiming, static and dynamic balance (Movement Assessment Battery for Children; M-ABC) one week prior to the intervention commencing. For group 1, the 12 weekly sessions included catching, throwing, balancing, juggling, and co-ordination exercises whilst group 2 participated in swimming, climbing, wrestling, and athletic events such as hurdles and relay races. Group 3 did not receive any additional exercise treatment. After the first weekly session, each child completed the working memory and motor performance tasks to assess immediate effects of exercise as well as post intervention (week 13). Immediate findings demonstrated a main effect of time for the catching and aiming measures of motor performance although no effect of treatment (in either exercise group) was found following the first week of the interventions. However, after the intervention period, children who had taken part in both exercise conditions exhibited greater performance on both verbal WM tasks as well as total score for the M-ABC test battery compared to the control group. No intervention effects were found for the visuospatial WM task.

Conclusions and Recommendations for Future Research

As the review of the evidence outlined above suggests, it could be assumed that there is a positive effect of physical activity on EF processes and teacher and parent rated behavior symptoms in children with ADHD. It seems that providing additional physical activity improves behavior and, in particular, performance on EF tasks that assess working memory (WM), response inhibition, attention and planning, providing a sound rationale to continue to develop studies within this field in order to ascertain whether these improvements can be sustained over time.

A logical next step in the development of this area of research would be to conduct a full-scale randomized controlled trial with children diagnosed with ADHD compared to treatment as usual (e.g. medication). Another treatment arm, although perhaps ambitious, could use a different type of non-pharmacological intervention (e.g. parenting or psychosocial therapy) to enable the investigation of whether physical activity could have the greatest effect on symptoms of ADHD and functioning deficits. As with the work conducted by Ziereis and Jansen (2015), more studies are required that investigate the possible differential effects of chronic and acute exercise within the one study but with longer-term follow-up periods.

The current studies were unable to disentangle the contribution of different types of ADHD (e.g. DSM-V subtypes), typically due to small sample sizes, in the effectiveness of physical activity on symptoms and further work is required to investigate this. Other studies could focus on the evaluation of implementing an exercise protocol within schools for children with high levels of hyperactivity/inattention, even at sub-threshold levels and compare this to a control group who continue to receive the typical school curriculum. Another avenue to explore, similar to Lufi and Parish-Plass (2011) using an RCT design would be to assess whether exercise only helps "pure" ADHD or also those who have other clinically significant problems like ODD and conduct disorder. Most studies that conduct research in this field use intelligence (IQ) score exclusion criteria, and so it is not yet clear whether exercise has any beneficial effect on children who have ADHD and a co-morbid learning difficulty. This is particularly important given that despite diminishment of behavioral symptoms, those with ADHD still exhibit lower IQ score as young adults (Biederman et al. 2009).

Finally, further work is required to quantify the intensity and duration of exercise required to improve cognitive and behavioral manifestations of ADHD as a viable treatment method.

References

Aarnoudse-Moens, C. S. H., N. Weisgas-Kuperus, B. van Goudoever, & J. Oosterlaan. 2009. "Meta-analysis of neurobehavioural outcomes in very preterm and/or very low birth weight children." *Pediatrics 124*: 717–28.

Ahmed, G. M., & S. Mohamed. 2011. "Effect of regular aerobic exercises on behavioral, cognitive and psychological response in patients with attention deficit-hyperactivity disorder." *Life Science Journal 8* (2): 392–7.

Ahn, S., & A. L. Fedewa. 2011. "A meta-analysis of the relationship between children's physical activity and mental health." *J Pediatr Psychol 36* (4): 385–97.

American Psychiatric Association. 2013. *Diagnostic and statistical manual of mental disorders: DSM-V* (Fifth Edition). Washington, DC: American Psychiatric Association.

Archer, T., M. Oscar-Berman, & K. Blum. 2011. "Epigenetics in developmental disorder: ADHD and endophenotypes." *Journal of Genetic Syndromes and Gene Therapy 2* (104).

Arns, M., S. de Ridder, U. Strehl, M. Breteler, & A. Coenen. 2009. "Efficacy of neurofeedback treatment in ADHD: The effects on inattention, impulsivity and hyperactivity: A meta-analysis." *Clinical EEG and Neuroscience 40* (3): 180–9.

Audiffren, M. 2009. "Acute exercise ad psychological functions: A cognitive-energetic approach." In T. McMorris, P. D. Tomporowski, & M. Audiffren (Eds.), *Exercise and cognitive function* (pp. 3–39). Chichester: John Wiley and Sons.

Baerg, S., J. Cairney, J. Hay, L. Rempel, N. Mahlberg, & B. E. Faught. 2011. "Evaluating physical activity using accelerometry in children at risk of developmental coordination disorder in the presence of attention deficit hyperactivity disorder. [Evaluation Studies Research Support, Non-U.S. Gov't]." *Res Dev Disabil 32* (4): 1343–50.

Banaschewski, T., V. Roessner, R. W. Dittmann, P. Janardhanan Santosh, & A. Rothenberger. 2004. "Non–stimulant medications in the treatment of ADHD." *European Child and Adolescent Psychiatry 13* (1): i102–i116.

Barkley, R. A. 2012. *Executive functions: What they are, how they work, and why they evolved*. New York: Guilford Press.

Berwid, O. G., & J. M. Halperin. 2012. "Emerging support for a role of exercise in attention-deficit/hyperactivity disorder intervention planning." *Curr Psychiatry Rep 14*: 543–51.

Biederman, J., & S. V. Faraone. 2005. "Attention-deficit hyperactivity disorder." *Lancet 366* (9481): 237–48.

Biederman, J., C. R. Petty, S. W. Ball, R. Fried, A. E. Doyle, N. J. Cohen,...S. V. Faraone. 2009. "Are cognitive deficits in attention deficit hyperactivity/disorder related to the course of the disorder? A prospective controlled follow-up study of grown up boys with persistent and remitting course." *Psychiatry Research 170*: 177–82.

Blum, K., A. L. Chen, E. R. Braverman, D. E. Comings, T. J. H. Chen, V. Arcuri,... M. Oscar-Berman. 2008. "Attention-deficit-hyperactivity disorder and reward deficiency syndrome." *Neuropsychiatric Disease and Treatment 4* (5): 893–917.

Buitelaar, J., & R. Medori. 2010. "Treating attention-deficit/hyperactivity disorder beyond symptom control alone in children and adolescents: A review of the potential benefits of long-acting stimulants." *European Child and Adolescent Psychiatry 19* (4): 325–40.

Bush, G., E. M. Valera, & L. J. Seidman. 2005. "Functional neuroimaging of attention-deficit/hyperactivity disorder: A review and suggested future directions." *Biol Psychiatry 57* (11): 1273–84.

Caspersen, C. J., K. E. Powell, & G. M. Christenson. 1985. "Physical activity, exercise, and physical fitness: Definitions and distinctions for health-related research." *100* (2): 126–31.

Castellanos, F., E. J. S. Sonuga-Barke, M. P. Milham, & R. Tannock. 2006. "Characterizing cognition in ADHD: Beyond executive dysfunction." *Trends in Cognitive Sciences 10* (3): 117–23.

Cereatti, L., R. Casella, M. Manganelli, & C. Pesce. 2009. "Visual attention in adolescents: Facilitating effects of sport expertise and acute physical exercise." *Psychology of Sport and Exercise 10* (1): 136–45.

Chaddock, L., K. I. Erickson, R. S. Prakash, J. S. Kim, M. W. Voss, M. VanPatter,...A. F. Kramer. 2010. "A neuroimaging investigation of the association between aerobic fitness, hippocampal volume and memory performance in preadolescent children." *Brain Research 1358* (172–83).

Chaddock, L., C. H. Hillman, S. M. Buck, & N. J. Cohen. 2011. "Aerobic fitness and executive control of relational memory in preadolescent children." *Med Sci Sports Exerc 43* (2): 344–9.

Chuang, L. Y., Y. J. Tsai, Y. K. Chang, C. J. Huang, & T. M. Hung. 2015. "Effects of acute aerobic exercise on response preparation in a Go/No Go task in children with ADHD: An ERP study." *Journal of Sport and Health Science 4*: 82–8.

Coghill, D. R., & K. M. Hogg. 2012. *Molecular Genetics of Attention Deficit–Hyperactivity Disorder (ADHD)*. Online library. John Wiley and Sons.

Colcome, S., A. F. Kramer, K. I. Erickson, P. Scalf, E. McAuley, N. J. Cohen,...S. Elavsky. 2004. "Cardiovascular fitness, cortical plasticity and aging." *PNAS 101* (9): 3316–21.

Counts, C. A., J. T. Nigg, J. A. Stawicki, M. D. Rappley, & A. Von Eye. 2005. "Family adversity in DSM-IV ADHD combined and inattentive subtypes and associated disruptive behavior problems." *Journal of the American Academy of Child and Adolescent Psychiatry 44* (7): 690–8.

Davis, C. L., & K. Lambourne. 2009. "Exercise and cognition in children." In T. McMorris, P. D. Tomporowski, & M. Audiffren (Eds.), *Exercise and cognitive function.* Chichester, UK: John Wiley and Sons.

Davis, C. L., P. D. Tomporowski, C. A. Boyle, J. L. Waller, P. H. Miller, J. Naglieri, & M. Gregowski. 2007. "Effects of aerobic exercise on overweight children's cognitive functioning: A randomized controlled trial." *Research Quarterly for Exercise and Sport 78* (5): 510–19.

Dishman, R. K., H. R. Berthoud, F. W. Booth, C. W. Cotman, V. R. Edgerton, M. R. Fleshner,...M. J. Zigmond. 2006. "Neurobiology of exercise." *Obesity 14* (3): 345–56.

Ellemberg, D., & M. St. Louis-Deschênes. 2010. "The effect of acute physical activity on cognitive function during development." *Psychology of Sport and Exercise 11*: 122–6.

Fabiano, G. A., W. E. Pelham, Jr, E. K. Coles, E. M. Gnagy, A. Chronis-Tuscano, & B. C. O'Connor. 2009. "A meta-analysis of behavioral treatments for attention-deficit/hyperactivity disorder." *Clinical Psychology Review 29* (2): 129–40.

Faraone, S. V., J. Biederman, W. Weber, & R. L. Russell. 1998. "Psychiatric, neuropsychological, and psychosocial features of DSM-IV subtypes of attention-deficit/hyperactivity disorder: Results from a clinically referred sample." *Journal of the American Academy of Child and Adolescent Psychiatry 37* (2): 185–93.

Fisher, A., J. M. E. Boyle, J. Y. Paton, P. Tomporowski, C. Watson, J. H. McColl, & J. J. Reilly. 2011. "Effects of a physical education intervention on cognitive function in young children: Randomized controlled pilot study." *BMC Pediatrics 11*: 97–106.

Gapin, J., & J. L. Etnier. 2010. "The relationship between physical activity and executive function performance in children with attention-deficit hyperactivity disorder." *Journal of Sport and Exercise Psychology 32*: 753–63.

Gapin, J. I., J. D. Labban, S. C. Bohall, J. S. Wooten, & Y. K. Chang. 2015. "Acute exercise is associated with specific executive functions in college students with ADHD: A preliminary study." *Journal of Sport and Health Science 4*: 89–96.

Gapin, J. I., J. D. Labban, & J. L. Etnier. 2011. "The effects of physical activity on attention deficit hyperactivity disorder symptoms: The evidence." *Preventitive Medicine 52*: 570–4.

Gizer, I., C. Ficks, & I. Waldman. 2009. "Candidate gene studies of ADHD: A meta-analytic review." *Human Genetics 126* (1): 51–90.

Graziano, P., J. McNamara, G. Geffken, & A. Reid. 2011. "Severity of children's ADHD symptoms and parenting stress: A multiple mediation model of self-regulation." *Journal of Abnormal Child Psychology 39* (7): 1073–83.

Halperin, J. M., A. C. Bédard, & J. T. Curchack-Lichtin. 2012. "Preventive interventions for ADHD: A neurodevelopmental perspective." *Neurotherapeutics 9*: 531–41.

Halperin, J. M., O. Berwid, & S. O'Neill. 2014. "Healthy body, healthy mind? The effectiveness of physical activity to treat ADHD in children." *Child Adolesc Psychiatric Clin N Am 23*: 899–936.

Halperin, J. M., J. W. Trampush, C. J. Miller, D. J. Marks, & J. H. Newcorn. 2008. "Neuropsychological outcome in adolescents/young adults with childhood ADHD: Profiles of persisters, remitters and controls." *Journal of Child Psychology and Psychiatry 49* (9): 958–66.

Hill, L., J. H. G. Williams, L. Aucott, J. Milne, J. Thomson, J. Greig,...M. Mon-Williams. 2010. "Exercising attention within the classroom." *Developmental Medicine and Child Neurology 52*: 929–34.

Hill, L., J. H. G. Williams, L. Aucott, J. Thomson, & M. Mon-Williams. 2011. "How does exercise benefit performance on cognitive tests in primary-school pupils?" *Developmental Medicine and Child Neurology 53*: 630–5.

Hillman, C. H., D. M. Castelli, & S. M. Buck. 2005. "Aerobic fitness and neurocognitive function in healthy preadolescent children." *Med Sci Sports Exerc 37* (11): 1967–74.

Hillman, C. H., M. B. Pontifex, L. B. Raine, D. M. Castelli, E. E. Hall, & A. Kramer. 2009. "The effect of acute treadmill walking on cognitive control and academic achievement in preadolescent children." *Neuroscience 159*: 1044–54.

Hodgson, K., A. D. Hutchinson, & L. Denson. 2012. "Nonpharmacological treatments for ADHD: A meta-analytic review." *Journal of Attention Disorders.* doi: 10.1177/1087054712444732

Hughes, C. H., & R. A. Ensor. 2009. "How do families help or hinder the emergence of early executive function?" *New Directions for Child and Adolescent Development 2009* (123): 35–50. doi:10.1002/cd.234

Huizinga, M., C. V. Dolan, & M. W. van der Molen. 2006. "Age-related change in executive function: Developmental trends and a latent variable analysis." *Neuropsychologia 44*: 2017–36.

Jensen, P. S., & P. T. Kenny. 2004. "The effects of yoga on the attention and behavior of boys with attention-deficit/hyperactivity disorder (ADHD)." *Journal of Attention Disorders 7*: 205–16.

Kamijo, K., M. B. Pontifex, K. C. O'Leary, M. R. Scudder, C. T. Wu, D. M. Castelli, & C. H. Hillman. 2011. "The effects of an afterschool physical activity program on working memory in preadolescent children." *Developmental Science 14* (5): 1046–58.

Kang, K. D., J. W. Choi, S. G. Kang, & D. H. Han. 2011. "Sports therapy for attention, cognitions and sociality." *Int J Sports Med 32*: 953–9.

Klingberg, T., E. Fernell, P. J. Olesen, M. Johnson, P. Gustafsson, K. Dahlström,... H. Westerberg. 2005. "Computerized training of working memory in children with ADHD – a randomized, controlled trial." *Journal of the American Academy of Child and Adolescent Psychiatry 44* (2): 177–86.

Konrad, K., & S. B. Eickhoff. 2010. "Is the ADHD brain wired differently? A review on structural and functional connectivity in attention deficit hyperactivity disorder." *Human Brain Mapping 31* (6): 904–16.

Kramer, A. F., & K. I. Erickson. 2007. "Effects of physical activity on cognition, well-being, and brain: Human interventions." *Alzheimer's and Dementia 3* (2, Supplement): S45–S51.

Lamm, C., P. D. Zelazo, & M. D. Lewis. 2006. "Neural correlates of cognitive control in childhood and adolescence: Disentangling the contributions of age and executive function." *Neuropsychologia 44* (11): 2139–48.

Langberg, J., & S. Becker. 2012. "Does long-term medication use improve the academic outcomes of youth with attention-deficit/hyperactivity disorder?" *Clinical Child and Family Psychology Review 15* (3): 215–33.

Lara, C., J. Fayyad, R. de Graff, R. C. Kessler, S. Aguilar-Gaxiola, M. Angermeyer,...N. Sampson. 2009. "Childhood predictors of adult ADHD: Results from the WHO world mental health (WMH) survey initiative." *Biological Psychiatry 1*: 46–54.

Li, D., P. C. Sham, M. J. Owen, & L. He. 2006. "Meta-analysis shows significant association between dopamine system genes and attention deficit hyperactivity disorder (ADHD)." *Human Molecular Genetics 15* (14): 2276–84.

Liddle, E. B., C. Hollis, M. J. Batty, M. J. Groom, J. J. Totman, M. Liotti,...P. F. Liddle. 2011. "Task-related default mode network modulation and inhibitory control in ADHD: Effects of motivation and methylphenidate." *Journal of Child Psychology and Psychiatry 52*(7): 761–71.

Lufi, D., & J. Parish-Plass. 2011. "Sports-based group therapy program for boys with ADHD or with other behavioral disorders." *Child and Family Behavior Therapy 33* (3): 217–30.

McKune, A. J., J. Pautz, & J. Lombard. 2003. "Behavioural response to exercise in children with attention-deficit/hyperactivity disorder." *S. Afr. J Sports Med 15*: 17–21.

Mahar, M. T., S. K. Murphy, D. A. Rowe, J. Golden, A. T. Shields, & T. D. Raedeke. 2006. "Effects of a classroom-based program on physical activity and on-task behavior." *Medicine and Science in Sports and Exercise 38* (12): 2086–94.

Medina, J., T. Netto, M. Muszkat, A. Medina, D. Botter, R. Orbetelli, ...M. Miranda. 2010. "Exercise impact on sustained attention of ADHD children, methylphenidate effects." *ADHD Attention Deficit and Hyperactivity Disorders 2* (1): 49–58.

Mikami, A., A. Jack, C. Emeh, & H. Stephens. 2010. "Parental influence on children with attention-deficit/hyperactivity disorder: I. Relationships between parent behaviors and child peer status." *Journal of Abnormal Child Psychology 38* (6): 721–36.

Miyake, A., N. P. Friedman, M. J. Emerson, A. H. Witzki, A. Howerter, & T. Wager. 2000. "The unity and diversity of executive functions and their contributions to complex 'frontal lobe' tasks: A latent variable analysis." *Cognitive Psychology 41*: 49–100.

National Institute for Clinical Excellence. 2009. *Attention deficit hyperactivity disorder: The NICE guideline on diagnosis and management of ADHD in children, young people and adults* Clinical Guideline 72. London: The British Psychological Society and The Royal College of Psychiatrists.

Nigg, J. T., E. G. Willcutt, A. E. Doyle, & E. J. S. Sonuga-Barke. 2005. "Causal heterogeneity in attention-deficit/hyperactivity disorder: Do we need neuropsychologically impaired subtypes?" *Biological Psychiatry 57* (11): 1224–30.

Nithianantharajah, J., & A. J. Hannan. 2009. "The neurobiology of brain and cognitive reserve: Mental and physical activity as modulators of brain disorders." *Progress in Neurobiology 89*: 369–82.

Pesce, C., C. Crova, L. Cereatti, R. Casella, & M. Bellucci. 2009. "Physical activity and mental performance in preadolescents: Effects of acute exercise on free-recall memory." *Mental Health and Physical Activity 2*: 16–22.

Piepmeier, A. T., C. H. Shih, M. Whedon, L. M. Williams, M. E. Davis, D.A Henning,... J. L. Etnier. 2015. "The effects of acute exercise on cognitive performance in children with and without ADHD." *Journal of Sport and Health Science 4*: 97–104.

Polanczyk, G., M. S. de Lima, B. L. Horta, J. Biederman, & L. A. Rohde. 2007. "The worldwide prevalance of ADHD: A systematic review and metaregression analysis." *American Journal of Psychiatry 164*: 942–8.

Pontifex, M. B., B. J. Saliba, L. B. Raine, D. L. Pichietti, & C. H. Hillman. 2013. "Exercise improves behavioural, cognitive, and scholastic performance in children with attention-deficit/hyperactivity disorder." *J Pediatr 162*: 543–51.

Prasad, V., E. Brogan, C. Mulvaney, M. Grainge, W. Stanton, & K. Sayal. 2013. "How effective are drug treatments for children with ADHD at improving on-task behaviour and academic achievement in the school classroom? A systematic review and meta-analysis." *European Child and Adolescent Psychiatry 22* (4): 203–16.

Rapport, M. D., S. A. Orban, M. J. Kofler, & L. M. Friedman. 2013. "Do programs designed to train working memory, other executive functions, and attention benefit children with ADHD? A meta-analytic review of cognitive, academic, and behavioral outcomes." *Clinical Psychology Review*. doi: 10.1016/j.cpr.2013.08.005

Rasmussen, P., P. Brassard, H. Adser, M. V. Pedersen, L. Leick, E. Hart,…H. Pilegaard. 2009. "Evidence for a release of brain-derived neurotrophic factor from the brain during exercise." *Experimental Physiology 94* (10): 1062–9.

Rommel, A. S., J. M. Halperin, J. Mill, P. Asherson, & J. Kuntsi. 2013. "Protection from genetic diathesis in attention-deficit/hyperactivity disorder: Possible complementary roles of exercise." *J Am Acad Child Adolesc Psychiatry 52* (9): 900–10.

Ruiz, J. R., F. B. Ortega, R. Castillo, M. Martin-Matillas, L. Kwak, G. Vicente-Rodriguez … L. A. Moreno. 2010. "Physical activity, fitness, weight status, and cognitive performance in adolescents." *Journal of Pediatrics 157*: 917–22.

Salthouse, T. 2005. "Relations between cognitive abilities and measures of executive functioning." *Neuropsychology 19* (4): 532–45.

Seidman, L. J. 2006. "Neuropsychological functioning in people with ADHD across the lifespan." *Clinical Psychology Review 26*: 466–85.

Sergeant, J. A. 2000. "The cognitive-energetic model: An empirical approach to attention-deficit hyperactivity disorder." *Neuroscience and Biobehavioral Reviews 24* (1): 7–12.

Sibley, B. A., & J. L. Etnier. 2003. "The relationship between physical activity and cognition in children: A meta-analysis." *Pediatric Exercise Science 15*: 243–56.

Simonoff, E., A. Pickles, N. Wood, P. Gringras, & O. Chadwick. 2007. "ADHD symptoms in children with mild intellectual disability." *Journal of the American Academy of Child and Adolescent Psychiatry 46* (5): 591–600.

Smith, A. L., B. Hoza, K. Linnea, J. D. McQuade, M. Tomb, A. J. Vaughn,…H. Hook. 2013. "Pilot physical activity intervention reduces severity of ADHD symptoms in young children." *Journal of Attention Disorders 17* (1): 70–82.

Sonuga-Barke, E. J. S. 2003. "The dual pathway model of AD/HD: An elaboration of neuro-developmental characteristics." *Neuroscience and Biobehavioral Reviews 27* (7): 593–604.

Sonuga-Barke, E. J. S., D. Brandeis, S. Cortese, D. Daley, M. Ferrin, M. Holtmann,…J. Sergeant. 2013. "Nonpharmacological interventions for ADHD: Systematic review and meta-analyses of randomized controlled trials of dietary and psychological treatments." *Am J Psychiatry 170*: 275–89.

Sonuga-Barke, E. J. S., & F. X. Castellanos. 2007. "Spontaneous attentional fluctuations in impaired states and pathological conditions: A neurobiological hypothesis." *Neuroscience and Biobehavioral Reviews 31* (7): 977–86.

Sonuga-Barke, E. J. S., & J. M. Halperin. 2010. "Developmental phenotypes and causal pathways in attention deficit/hyperactivity disorder: Potential targets for early intervention?" *Journal of Child Psychology and Psychiatry 51* (4): 368–89.

Spencer, T. J., J. Biederman, & E. Mick. 2007. "Attention-deficit/hyperactivity disorder: Diagnosis, lifespan, comorbidities, and neurobiology." *Journal of Pediatric Psychology 32* (6): 631–42.

Stevenson, J., E. Sonuga-Barke, K. Grimshaw, K. M. Parker, M. J. Rose-Zerilli, J. W. Holloway, & J. D. Warner. 2010. "The role of histamine degradation gene polymorphisms in moderating the effects of food additives in children's ADHD symptoms." *Am J Psychiatry 167*: 1108–15.

Tantillo, M., C. M. Kesick, G. W. Hynd, & R. K. Dishman. 2002. "The effects of exercise on children with attention-deficit hyperacitvity disorder." *Medicine and Science in Sports and Exercise 34* (2): 203–12.

Tobaiqy, M., D. Stewart, P. J. Helms, J. Williams, J. Crum, C. Steer, & J. McLay. 2011. "Parental reporting of adverse drug reactions associated with attention-deficit hyperactivity disorder (ADHD) medications in children attending specialist clinics in the UK." *Drug Saf 34* (3): 211–19.

Tomporowski, P. D., C. L. Davis, P. H. Miller, & J. A. Naglieri. 2008. "Exercise and children's intelligence, cognition, and academic achievement." *Educational Psychology Review 20* (2): 111–31.

Tripp, G., & J. R. Wickens. 2009. "Neurobiology of ADHD." *Neuropharmacology 57* (7–8): 579–89.

van der Oord, S., P. J. M. Prins, J. Oosterlaan, & P. M. G. Emmelkamp. 2008. "Efficacy of methylphenidate, psychosocial treatments and their combination in school-aged children with ADHD: A meta-analysis." *Clinical Psychology Review 28* (5): 783–800.

Verburgh, L., M. Königs, E. J. A. Scherder, & J. Oosterlaan. 2013. "Physical exercise and executive functions in preadolescent children, adolescents and young adults: A meta-analysis." *British Journal of Sports Medicine*. doi:10.1136/bjsports-2012-091441

Verret, C., M. C. Guay, C. Berthiaume, P. Gardiner, & L. Béliveau. 2012. "A physical activity program improves behaviour and cognitive functions in children with ADHD: An exlporatory study. *Journal of Attention Disorders 16* (1): 71–80.

Wilens, T. E., J. Biederman, S. Brown, S. Tanguay, M. C. Monuteaux, C. Blake, & T. J. Spencer. 2002. "Psychiatric comorbidity and functioning in clinically referred preschool children and school-age youths with ADHD." *Journal of the American Academy of Child and Adolescent Psychiatry 41* (3): 262–8.

World Health Organization. 1992. *International classificiation of mental and behavioural disorders (ICD-10)*. Geneva: World Health Organization.

Ziereis, S. & P. Jansen. 2015. "Efffects of physical activity on executive function and motor performance in children with ADHD." *Research in Developmental Disabilities 38*: 181–91.

15 Can Physical Activity Prevent or Treat Clinical Depression?[1]

Nanette Mutrie, Katie Richards, Stephen Lawrie, and Gillian Mead

Chapter Aims

In this chapter we aim to:

- Define clinical depression and discuss its prevalence and link to physical health;
- describe prospective epidemiological studies;

- consider systematic reviews of physical activity and exercise as treatment for clinical depression;
- outline neurobiological mechanisms that could provide plausible explanation of the antidepressant effect of exercise; and
- note the existence of guidelines about the role of physical activity/exercise in the treatment of depression.

Introduction

People who take regular exercise know that this helps them "feel good"; and it is commonly understood that being active has a role to play in positive mental health. In this chapter we explore whether or not regular activity could prevent or treat depression. To begin this exploration we need to define what we mean by physical activity and exercise and depression.

Physical Activity and Exercise

Physical activity is any bodily movement that requires skeletal muscle activity, and exercise is a planned structured regime of physical activity that is designed to improve one or more components of physical fitness. The components of fitness that relate to health include aerobic (or cardiorespiratory fitness) muscle strength and flexibility. Applying this definition in practice, particularly when scrutinising papers to determine whether an intervention fulfils the definition of exercise or not, can be difficult and sometimes an element of judgment is required. In most of the studies that focus on treatment it is correct to use the term exercise but in other areas the differentiation of physical activity and exercise is less clear.

Defining "Clinical" Depression

The term "depression" is sometimes loosely used to describe episodes of unhappiness that affect most people from time to time. If however depression persists over time and interferes with the ability to function in work or relationships, it may be classified using standard diagnostic criteria. Clinically defined depression is known as major depressive disorder in the Diagnostic and Statistical Manual of Mental Disorders-5 (DSM-5) and as a depressive episode in International Classification of Diseases- 10 (ICD-10) (American Psychiatric Association 2013; World Health Organization 2010). Depression can be secondary to other conditions, such as alcohol and drug misuse, and is often associated with chronic diseases, such as type 2 diabetes, HIV, and cardiac disease.

The DSM-5 and ICD-10 are commonly referenced when classifying mental or psychiatric disorders. Table 15.1 shows the various codes used in ICD-10. These classification systems allow both clinicians and researchers to have a common language and standard criteria for diagnosis. The DSM-5

Table 15.1 ICD-10 codes for mental and behavioral disorders

Numerical code	*Description*
F00-F09	Organic, including symptomatic, mental disorders
F10-F19	Mental and behavioral disorders due to psychoactive substance use
F20-F29	Schizophrenia, schizotypal and delusion disorders (e.g. paranoid schizophrenia)
F30-F39	Mood (affective) disorders (e.g. depression)
F40-F48	Neurotic, stress-related and somatoform disorders (e.g. phobias)
F50-F59	Behavioral syndromes associated with physiological disturbances and physical factors (e.g. eating disorders)
F60-F69	Disorders of adult personality and behavior (e.g. kleptomania)
F70-F79	Mental retardation (e.g. mild mental retardation)
F80-F89	Disorders of psychological development (e.g. developmental disorders of speech and language)
F90-F98	Behavioral and emotional disorders with onset usually occurring in childhood and adolescence (e.g. hyperkinetic disorders)
F99	Unspecified mental disorder

Source: ICD10Data.com

criteria for major depressive disorders are summarized in Table 15.2 and the symptoms noted by ICD-10 for depression are summarized in Table 15.3. Clinical depression is diagnosed using these standard instruments or clinical interviews. To formally diagnose depression, a clinical judgment is made using criteria listed in the DSM-5 or the ICD-10. A fuller exploration of the epidemiology of common mental disorders, including depression, is provided in Chapter 1 of this book by Robert Kohn.

KEY POINT: THE DIAGNOSTIC AND STATISTICAL MANUAL OF MENTAL DISORDERS (DSM-5) AND THE INTERNATIONAL CLASSIFICATION OF DISEASES – 10 (ICD-10) (World Health Organization, 2010) ARE THE TWO CLASSIFICATION SYSTEMS WHICH PROVIDE CLEAR DEFINITIONS OF CLINICAL DEPRESSION

Until recently, a lack of consistency amongst researchers concerning the criteria for defining depression plagued our understanding of the relationship between physical activity and depression. Many older studies and reviews have included cases of "depression" that would not reach clinically defined criteria, and may be better described as transitory negative affect or minor depression. Clinically defined depression will be the main focus of the discussion in this chapter.

Table 15.2 Summary of DSM-5 criteria for major depressive episode

Category	*Criteria*
A	At least five of the following symptoms have been present during the same 2-week period, nearly every day, and represent a change from previous functioning. At least one of the symptoms must be either (1) depressed mood or (2) loss of interest or pleasure
A(1)	Depressed mood (or alternatively can be irritable mood in children and adolescents)
A(2)	Markedly diminished interest or pleasure in all, or almost all, activities
A(3)	Significant weight loss or weight gain when not dieting
A(4)	Insomnia or hypersomnia
A(5)	Psychomotor agitation or retardation
A(6)	Fatigue or loss of energy
A(7)	Feelings of worthlessness or excessive or inappropriate guilt
A(8)	Diminished ability to think or concentrate
A(9)	Recurrent thoughts of death, recurrent suicidal ideation without a specific plan, or a suicidal attempt or a specific plan for committing suicide
B	Symptoms are not better accounted for by a Mood Disorder Due to a General Medical Condition, a Substance-Induced Mood Disorder, or Bereavement (normal reaction to death of a loved one)
C	Symptoms are not better accounted for by a Psychotic Disorder (e.g. Schizo-affective Disorder)

Source: American Psychiatric Association. 2013. *Diagnostic and Statistical Manual of Mental Disorders (Dsm-5)*. Fifth ed.

Table 15.3 Summary of ICD-10 symptoms for depression

General criteria	In mild, moderate and severe episodes the individual usually suffers from depressed mood, loss of interest and enjoyment, and reduced energy leading to increased fatiguability and diminished activity. Marked tiredness after only slight effort is common.
Additional symptoms include	reduced concentration and attention reduced self-esteem and self-confidence ideas of guilt and unworthiness (even in a mild type of episode) bleak and pessimistic views of the future ideas or acts of self-harm or suicide disturbed sleep diminished appetite

Source: ICD10Data.com

KEY POINT: THERE IS A RANGE OF MOOD STATES THAT MIGHT BE CLASSED AS NEGATIVE BUT NOT ALL OF THEM WOULD MEET DIAGNOSTIC CRITERIA FOR "CLINICAL" DEPRESSION

The most common self-completed questionnaire used to measure depression severity in research is the Beck Depression Inventory (BDI) (Beck, Ward, Mendelsohn, Mock, & Erbaugh 1961). This was updated to include the DSM-IV criteria and is called the BDI-II (Beck, Steer, Ball, & Ranieri 1996). It is important to note the version of the BDI when interpreting intervention effects as the newer version uses different cut-off scores to classify depression severity. In all versions of the questionnaire a higher score signifies more severe depression. Several exercise studies have included people who at baseline had a score lower than the moderate level of depression as measured by the BDI. Such a score would be considered as mild or perhaps transitory depression. Recent systematic reviews will be alert to this definition but this was probably not the case in the early narrative reviews in this area.

While the failure to differentiate between clinical and minor depression has led to confusion, or even an overestimation of the effect of physical activity on depression, it would be wrong to ignore less severe depression. In the UK, NICE guidelines for the diagnosis and treatment of depression highlighted the need to identify what has been termed as "sub-threshold" depression. In this case a clinical diagnosis of depression may not be clear, and instead could be considered as minor depression (National Institute for Health and Clinical Excellence 2009). There is also a potential role for physical activity in helping this sub-threshold level of depression.

KEY POINT: EXERCISE MAY IN THEORY IMPROVE BOTH MAJOR AND MINOR DEPRESSION

Prevalence of Depression

In the UK mental health illnesses account for 230 per 1000 referrals to primary care services. Data from social trends analysis in the year 2000 showed that 1 in 6 adults living in the UK experience a mental health illness at any given time, and 1 in 4 will experience at least one illness over the course of a year (Office of National Statistics 2001). Depression is one of the most common mental health illnesses. It is estimated that between 3–8% of the population worldwide might suffer from clinical depression, with a higher proportion of females affected by the condition than males (Ferrari et al. 2013).

An analysis of the public health impact of chronic disease in 60 countries concluded that depression produces the greatest reduction in self-reported health, ahead of angina, arthritis, asthma, and diabetes (Moussavi et al. 2007). In 1997 it was estimated that by the year 2020 major depression will be the second leading cause of disease burden in terms of disability-adjusted life years after ischaemic heart disease (Murray & Lopez 1997). In 2006 that estimate was updated and it was projected that by the year 2030 depression will be the leading cause of disease burden (Mathers & Loncar 2006). These statistics clearly show the impact of depression, making it a major public health issue.

KEY POINT: IT HAS BEEN ESTIMATED THAT BY THE YEAR 2030 DEPRESSION WILL BE THE LEADING CAUSE OF DISEASE BURDEN

Depression Is Linked to Physical Health Problems

In addition to the importance of mental health in its own right, it has now been recognized that negative emotions, particularly depression, may have a detrimental impact upon self-care behaviors and physical health. The manifestation of depression has been highlighted as a risk factor for the development of chronic diseases, such as coronary heart disease and diabetes (Anderson, Freedland, Clouse, & Lustman 2001).

Another new and timely area of interest is the idea that childhood depression may increase the risk of adult weight gain and obesity (Hasler et al. 2005). Given that childhood depression is treatable and given the worldwide concern about obesity levels, this association needs further study. These connections between negative emotion and various diseases suggest an increased role for activity since it may provide a means of improving positive emotions in those who are at risk of disease because of poor mental health.

KEY POINT: DEPRESSION IS ASSOCIATED WITH A WIDE RANGE OF PHYSICAL DISEASES

Prevention of Depression by Physical Activity and Exercise

The strongest epidemiological evidence for a causal relationship between inactivity and developing depression comes from prospective cohort studies. Cross-sectional studies that have sought associations between mood and activity at a single time point cannot provide causal information because it is just as likely that inactivity might cause depression as it is that depression has caused inactivity. One longitudinal study has clearly shown this bidirectional relationship using the data from the British Whitehall Cohort study. Those who were not depressed at baseline but remained regularly active were less likely to become depressed at the 8-year follow-up than those who were not regularly active. Conversely, those who were depressed at baseline were less likely to be regularly active at the 8-year follow-up than those who were not depressed at baseline (Azevedo Da Silva et al. 2012).

Studies on depression where the outcome was clinically defined by questionnaire or interview are listed in Tables 15.4 and 15.5. In all of these studies, statistical adjustments for potential confounding variables, such as age and socio-economic background, were made. Studies that have measured depressive symptoms but do not define a clinical cut-off or category have been excluded.

Table 15.4 shows seven prospective studies in middle to older aged adults, with follow-up ranging from 3 years to 27 years; these studies all showed that physical activity protected against later depression. A lower level of physical activity was associated with an increased risk of clinically defined depression at a later date (relative risk was 1.6 – 2, although the 95% CI is wider). It is clear that the inactivity preceded the depression and thus the possibility that the association is simply a result of those who are depressed being inactive can be refuted. Since people may well be inactive because they are disabled or have other illnesses, it is important to account for physical health status. One study showed that including or excluding those who were unable to walk did not attenuate the relationship (Strawbridge, Deleger, Roberts, & Kaplan 2002). However, there are other reasons, such as lack of social skills or socio-economic status, which could also predict both inactivity and depression that may not have been fully accounted for.

In comparison to Table 15.4, Table 15.5 shows three prospective studies that did not show that activity was associated with a lower risk of later depression. The reasons for this lack of association are unclear, but the measures of activity are not detailed, relying on self-report of defined categories of activity. Such measurements are likely to underestimate the physical activity and depression relationship. It is also possible that publication bias which favors positive significant results over non-significant outcomes has operated and this is why we have more positive than non-significant results. It should also be noted that no published studies have reported statistically significant negative effects in which depression scores have increased with increasing activity levels.

In 2013, Mammen and Faulkner systematically reviewed studies with a prospective-based, longitudinal design examining relationships between physical activity and depression over at least two time intervals. A total of 25 of the 30 studies found a significant, inverse relationship between baseline physical activity and follow-up depression, suggesting that physical activity can prevent the onset of depression. Given the heterogeneity in physical measurement in the reviewed studies, a clear dose-response relationship between physical activity and reduced depression was not readily apparent. However, this systematic review offers promising evidence that any level of physical activity, including low levels, can prevent future depression (Mammen & Faulkner 2013).

Table 15.4 Prospective studies with measures of clinically defined depression showing a protective effect from physical activity

Authors and country	*Participants*	*Design*	*Measures of depression and physical activity*	*Results*
Farmer et al. (1988) U.S.A	1497 respondents to the National Health and Nutrition Examination Survey follow up study (NHANES)	Prospective longitudinal 8 year follow-up	Center for Epidemiological Studies Depression Scale (CES-D) using a cut-off point of 16 to indicate depression. Physical activity categorised as did "little or none" or "much or moderate" recreational and non-recreational activity.	Women who had engaged in little or no recreational activity were twice as likely to develop depression at follow-up as those who had engaged in "much" or "moderate" activity (95% CI = 1.1–3.2). There was no significant association over the same time period for men or for non-recreational activity in a usual day for either women or men. For men who were depressed at baseline, inactivity was a strong predictor of continued depression at the 8-year follow-up.
Camacho, Roberts, Lazarus, Kaplan, & Cohen (1991) U.S.A.	8,023 in systematic sample from community	9 and 18 year follow-up after baseline in 1965	Depression measured by standard instrument used in Human Population Laboratory studies. Cut-off of 5 used to indicate depression Frequency of commonly reported physical activity resulting in a physical activity index ranging from 0 (no activity) to 14 high active with strenuous activity.	In the first wave of follow-up (1974), the relative risk (RR) of developing depression was significantly greater for both men and women who were low active in 1965 (RR 1.8 for men, 1.7 for women) compared to those who were high active (although there was a protective effect for moderate levels of activity for women but not men). Some evidence of dose response. The second wave of follow-up (1983) suggests that decreasing activity levels over time increases the risk of subsequent depression (OR = 2.02) although the odds ratio was not significant once the model was fully adjusted (OR = 1.61).

(*continued*)

Table 15.4 (cont.)

Authors and country	*Participants*	*Design*	*Measures of depression and physical activity*	*Results*
Paffenbarger, Lee, & Leung (1994) U.S.A	10, 201 Harvard Alumni (men only)	23–27 year follow-up	Physician diagnosed depression. Rich data on all types and intensities of physical activity – self-reported.	Men who engaged in 3 or more hours of sport activity per week at baseline had a 27% reduction in the risk of developing depression at follow-up compared to those who played for less than 1 hour per week. Statistical evidence of dose response.
(Strawbridge et al. 2002) U.S.A.	1,947 community dwelling adults (age 50–94)	Baseline 1994 and follow-up 1999	Diagnostic and Statistical Manual (DSM) criteria used to define depression. Physical activity measured on an 8 point scale.	Physical activity was protective of incident depression at 5-year follow-up (OR .83). Excluding disabled participants, who may be depressed but unable to be active, did not attenuate results.
van Gool et al. (2003) Netherlands	1,280 community dwelling late middle aged and older people	6-year follow-up	Dutch version of Center for Epidemiologic Studies Depression scale (CES-D) scale with a cut-off point of 16 or above (using 20 items) defining depression. Minutes of physical activity per week were self-reported from a variety of possible activities.	155 people became depressed from baseline to follow-up and this was associated with changing from an active to a sedentary lifestyle (RR 1.62) and significantly associated to a decrease in minutes of physical activity.

(Bernaards et al. 2006) Netherlands	1747 workers from 34 companies	Prospective longitudinal with 3-year follow-up	Dutch version of CES-D used to measure depression. Those scoring 6 and above (using 11 items from this scale) defined as depressed. Physical activity measured by response to "How often within the past four months did you participate in strenuous sports activities or strenuous physical activities that lasted long enough to become sweaty?"	For workers with a sedentary job, strenuous leisure time physical activity (1–2 times per week) was significantly associated with a reduced risk of future depression and emotional exhaustion. Higher frequencies (≥3 times per week) did not show this relationship
(Gallegos-Carrillo et al. 2013) Mexico	1047 participants in Health Worker Cohort Study, Mexico who were free of depressive symptoms at baseline	Prospective longitudinal with 6-year follow-up	Center for Epidemiological Studies Depression Scale (CES-D) using a cut-off point of 16 to indicate depression. Physical activity measured by self-report questionnaire of recreational activities over the last year and converted into average MET/hours per week	Three patterns of physical activity were created: inactive, moderately active, and highly active. Those in the inactive category (just over one third of the sample) had a higher incidence of depression after 6 years (16.5%) than among those in the other two categories (10.6%).

Table 15.5 Prospective studies with measures of clinically defined depression that do not show a protective effect from physical activity

Authors and country	*Participants*	*Design*	*Measures*	*Results*
Weyerer (1992) Germany	1536 community dwelling adults	Cross-sectional and prospective (5-year follow –up)	All people in this study were interviewed by a research psychiatrist and 8.3% were identified as depressed using a clinical scale. Physical activity was self-reported as regular, occasional or none.	Cross-sectional data showed that those reporting only occasional physical activity were 1.55 times more likely to have depression than those who were regularly physically active, although this was not statistically significant. Low physical activity was not a predictor of depression at a 5-year follow-up
Cooper-Patrick, Ford, Mead, Chang, & Klag (1997) U.S.A.	973 physicians (men only)	Prospective study. Baseline taken during medical school and follow-up happens every 5 years. In this study baseline was 1978 and follow-up 1993.	Self-reported clinical depression Self-reported physical activity (frequency of exercising "to a sweat")	There was no evidence for increased risk of depression for those reported exercising "to a sweat" compared to those who did not or for those who became inactive over the follow-up period (15 years).
Kritz-Silverstein, Barrett-Connor, & Corbeau (2001) U.S.A	Adults aged 50–89 in 1984–7 (932 men 1097 women) and followed up 1992–5 (404 men and 540 women)	Cross-sectional and prospective (12 year follow up)	Modified Beck Depression Inventory (BDI) with scores below 13 being considered as not depressed. Physical activity graded low medium high by yes/no responses to 2 questions about strenuous physical activity at leisure or work.	Cross-sectional data showed that physical activity significantly related to lower depression scores. No evidence of predictive effect of low activity on follow-up depression scores.

KEY POINT: THE AVAILABLE EVIDENCE SUGGESTS THAT PHYSICAL ACTIVITY HAS A PREVENTATIVE ROLE FOR DEPRESSION

Reviewing the Evidence Base for Physical Activity as a Treatment for Depression

Settings for treatment of depression include general practice, hospitals, specialist clinics (or resource centres), private therapy, and informal settings. When people do seek out treatment the first port of call may be the GP, in which case some form of medication may be offered. Using data from EU member states from 2000 to 2010, it has been shown that the use of antidepressants had increased by over 80% in that decade (OECD 2012). This trend suggests that medication is the first line of treatment and the most frequent.

Some people may be offered counselling style therapies. Trained professionals who might undertake this therapy include psychiatrists, clinical psychologists, counselling psychologists, hypnotherapists, and social workers. Within this area there are many approaches ranging from psychoanalysis (which perhaps is the lay person's impression of all therapy), to client-centred and cognitive-behavioral approaches. A particular style that has become popular is cognitive behavioral therapy (CBT). Sometimes the whole family may be involved in the therapeutic process. In extreme cases of depression, Electro Convulsive Therapy (ECT) or psychosurgery (such as prefrontal lobotomy) could be performed. A good source of reference for understanding these treatments is the Royal College of Psychiatrists in the UK (www.rcpsych.ac.uk/default.aspx). Of course, of key interest to us in this chapter is the role of physical activity and exercise in the treatment of clinical depression.

There is an extensive literature on the role of physical activity and exercise as a treatment for depression. Systematic reviews have become the gold standard for the synthesis of results in an area where several quality studies are available and are often accompanied by meta-analysis if studies have used the same outcome variables. In this chapter we have focused on systematic reviews and meta-analyses as the most reliable form of evidence.

Systematic Reviews

The systematic reviews and meta-analyses published prior to 2001 had limitations, such as trials with no clinical definition of depression and non-randomised designs. In 2001 the first robust systematic review in the area was published (Lawlor & Hopker 2001). In Lawlor and Hopker's (2001) systematic review and meta-analysis, studies had to have defined depression in the clinical range by a recognized method, and studies had to be randomized controlled trials. From the 14 studies included, the mean effect size (ES) for exercise compared to no treatment was −1.1 (95% CI −1.5 to −0.6) which is a large effect. Cognitive therapy had a similar ES to exercise. There were not sufficient studies to compare exercise to medication. The mean difference between exercise and control groups in BDI score was –7.3 (95%CI –10 to –4.6). Effectiveness has been shown by the ES in this meta-analysis but the clinical significance of this level of change in the BDI was questioned by Lawlor and Hopker who concluded that: "The effectiveness of exercise in reducing symptoms of depression cannot be determined because of lack of good quality research on clinical populations with adequate follow up."

That rather negative conclusion has been criticized because the restriction to clinical levels of depression and the inclusion of only RCT designs made this a stringent review of the best quality evidence at the time. In fact, the effect sizes were similar to other therapies for depression and so arguably the authors could have concluded that exercise could be used in the management of depression (Biddle & Mutrie 2008).

The science of systematic reviewing has improved in the last decade and most systematic reviews now follow the guidance of Cochrane. This is an organization of 37,000 contributors from over 130 countries (www.cochrane.org). The goal of Cochrane is to make information about the effects of various aspects of health care readily available. The reviews held in the library help practitioners and policy makers, and the public, make decisions about what the evidence suggests about any health intervention. The methods adopted involve systematic reviewing of all available literature on the topic in question.

The Cochrane review of exercise for depression is highly cited and downloaded. The 2012 update identified 30 studies that met their inclusion criteria, which is substantially more than the 14 studies in the Lawlor and Hopker (2001) review completed just 11 years previously. The 2012 meta-analysis by Rimer et al. showed a moderate effect size (–0.67, 95% confidence interval (CI) –0.90 to –0.43) for exercise versus no treatment control conditions.

When studies that had potential bias due to lack of allocation concealment or lack of blinding to outcomes, a further meta-analysis of the remaining four trials showed a reduced effect but it was still significant. Finally, the authors compared the exercise effects to those of cognitive behavioral therapy for the six trials that had these comparisons and found no significant difference. This confirmed the findings of the Lawlor and Hopker review that exercise had a similar effect size to other recognised therapies for depression. However, there were still insufficient studies to compare exercise to drug therapies.

There are, however, some methodological limitations of the trials included in the review. The main risks of bias are:

- Allocation concealment where there is a risk that the person making the allocation to intervention or control knows the sequence and it is therefore not random;

- blinding the outcome assessor where there is a risk that the person assessing physical activity levels or depression levels knows which group the person was in; and
- incomplete outcome data where there is a risk that bias occurs because only those completing the intervention have data.

Despite these concerns about bias, the authors of this review concluded that it is reasonable to recommend exercise as a treatment for depression.

A further update of this review has now been completed (Cooney et al. 2013). In this update, a more stringent definition of exercise was used: the intervention had to fulfil the American College of Sports Medicine definition of "exercise." This meant that trials of activity that did not fulfil the definition of exercise were not included. Thus a large trial of advice to add activity to usual care for depression, published in 2012, was not included (Chalder et al. 2012). This trial showed no effect on depression, possibly because the participants were already receiving other treatments for their depression.

In the most recent Cochrane review, the summary effect size for exercise compared with no treatment was –0.62 (Cooney et al. 2013). A subsequent analysis suggested that this was equivalent to a difference in the Beck Depression Inventory of 5 points (Cooney, Dwan, & Mead 2014). For the six trials considered to be at low risk of bias (adequate allocation concealment, intention-to-treat analyses, and blinded outcome assessment), a further analysis showed a small clinical effect in favor of exercise which did not reach statistical significance. Finally, seven trials compared the exercise effects to those of cognitive behavioral therapy and found no significant difference. Four trials compared exercise with antidepressant medication and no significant difference was found.

A consistent finding from all the systematic reviews that we have discussed is that exercise is as effective as other psychological therapies. In the most recent review there were sufficient studies that compared exercise to antidepressant medication and again no significant difference was found. This lack of difference between exercise and other standard therapies is a very important finding. It means that exercise works as well as other approaches to the treatment of depression. However, authors have also noted that there are many unanswered questions such as the exact dosage and mode of activity that might work best.

One team of systematic reviewers have taken action on this advice and examined the physical activity mode of walking (Robertson, Robertson, Jepson, & Maxwell 2012). They found eight trials that met their inclusion criteria and showed a large effect size of 0.86 (CI 1.12, 0.61). The authors concluded that "Walking has a statistically significant, large effect on the symptoms of depression in some populations but the current evidence base from randomised, controlled trials is limited."

KEY POINT: A COCHRANE SYSTEMATIC REVIEW OF THE STUDIES THAT HAVE USED EXERCISE AS A TREATMENT FOR DEPRESSION CONCLUDED THAT IT IS REASONABLE TO RECOMMEND EXERCISE TO PEOPLE WITH DEPRESSIVE SYMPTOMS

Challenges for Researchers Conducting Trials with Physical Activity or Exercise as a Treatment for Depression

There are several challenges for researches wanting to conduct trials of exercise for depression:

- Recruiting a large enough sample to ensure statistical power;
- ensuring that participants in the intervention and control arms have the same amount of contact with the person delivering the intervention – otherwise it is difficult to conclude whether any benefits are due to social interaction;
- conducting "double-blind" studies. In exercise studies this is not possible to achieve because the patient will certainly know that they are exercising, in a way that is different from receiving an unmarked tablet. While every effort should be made to keep studies at least single blind (that is the researcher taking outcome measures should not know the group assignment), it remains a challenge to achieve this. Clients and patients may inadvertently reveal their assignment to researchers, who are otherwise blind to assignment, in conversation, such as "I really enjoyed that weight training";
- potential participants may not wish to be randomized to "routine" or placebo condition, and those who do agree to take part but are randomized to no treatment may feel demoralized or resentful. One possible solution is to name the control condition as a waiting list for the intervention so that people assigned to that condition know that they have to wait 3 or 6 months or sometimes 12 months before receiving the intervention;
- participants randomized to the control intervention may find their own motivation or take up local opportunities to become more active. This may "dilute" any treatment effects;
- controlling for the effects of the positive characteristics of an exercise leader or the social effect of a group. The social aspect of group exercise may add further benefits;
- conducting long term follow-up. Very few studies have managed to fund adequate follow-up of 1 or 2 years;
- finding adequate measures of the variables of interest. For physical activity measures, objective monitoring should ideally be undertaken because

over-estimation of exercise occurs with self-report measures. For measuring depression a clinical interview might be considered the best option rather than a self-complete questionnaire.

The CONSORT group have considered these inadequacies in trials of a diverse range of medical interventions and have provided excellent guidance on how to design and report trials that avoid many of these challenges. See the website for further details (www.consort-statement.org). For trials of complex interventions, there is a new set of guidelines (TIDieR) which aims to enhance the reporting of trials of complex interventions by providing sufficient details about the intervention components to allow replication (Hoffmann et al. 2014).

There are also challenges performing systematic reviews and meta-analyses of exercise for depression. These include 1) ensuring that all available evidence is identified, including small trials that may be published only in abstract form, 2) making decisions about whether an intervention fulfils the definition of exercise, 3) deciding which "arm" of a trial to include when there may be several "arms" (e.g. different doses of exercise), 4) decisions about whether to perform subgroup analyses (e.g. according to how depression was diagnosed), and 5) assessing risk of bias when important aspects of methodology may not be clearly reported in trials. Having a clear protocol for a review avoids some of these difficulties, ensures transparency in the review process, and avoids needing to make post hoc decisions about, for example, which trials to include.

KEY POINT: EXERCISE IS A COMPLEX INTERVENTION AND DESIGNING ROBUST TRIALS OF EXERCISE HAS INHERENT METHODOLOGICAL DIFFICULTIES. THE DIFFERENT DESIGNS OF TRIALS CAN MAKE COMBINING THEM INTO A META-ANALYSIS CHALLENGING

Further Systematic Reviews in Relation to Exercise and Depression

In one of the first overviews of this area it was noted that there were no studies on young people (Biddle & Mutrie 1991). A Cochrane review has now focussed on young people (Larun, Nordheim, Ekeland, Hagen, & Heian 2006). The review also included anxiety. The review found 16 studies and although some promising results for exercise were noted for depression, the reviewers concluded that the effect of exercise on anxiety and depression for young people was not known at this time because of the lack of studies (Larun et al. 2006). A recent prospective population study also found no evidence of a protective effect from physical activity on the onset of depression in adolescence (Stavrakakis et al. 2013).

There is clearly more work to be done and there are two further registered systematic reviews in the Cochrane database in relation to physical activity and depression. The first is related to older adults and the second relates to the use of dance therapy as a mode of exercise for people with depression.

Physical Activity and Depression: What Are the Potential Neurobiological Mechanisms?

The evidence suggests that physical activity and exercise can reduce the risk of depression, and that exercise can be used as a treatment for people who have already developed the condition. Many plausible mechanisms have been suggested to explain these findings (Biddle, Mutrie, & Gorely 2015). These explanations range from psychological mechanisms, such as increased self-esteem or distraction from other aspects of life, to neurobiological mechanisms, such as the release of endorphins. It has also been noted that all potential mechanisms may operate at the same time and in a synergistic way, but a detailed analysis of the neurobiological mechanisms is rarely provided. In this section, we explore in detail the neurobiological mechanisms by which exercise might have an antidepressant effect. These mechanisms are also explored in detail, and with discussion of evidence supporting the mechanisms, in Chapter 2 by Terry McMorris and Jo Corbett.

Neurotransmitters and Neuropeptides

Neurotransmitters are biochemical substances that send signals from one nerve cell to another. Monoamines are a group of neurotransmitters thought to play a critical role in the regulation of emotions and in the treatment of depression (Dunn & Dishman 1991). Long-term voluntary exercise in rats leads to an increased expression of monoamines in the locus coeruleus (LC) and dorsal raphe nucleus (DRN), and these increases are accompanied by a reduction in depressive behaviors (Dishman et al. 1997; Kim, Lim, Baek, Ryu, & Seo 2015). In humans, brain monoamine activity is indirectly estimated from blood and urine, and evidence is mixed as to whether exercise leads to an increase in monoamine metabolites or not (Dunn & Dishman 1991).

Exercise also activates the opioid system, and this may account for the "runner's high" phenomenon (Boecker et al. 2008). Much of the attention has focused on the opioid β-endorphin. Several studies have shown increased β-endorphin levels in blood plasma following high-intensity exercise (Farrell, Gates, Maksud, & Morgan 1982), and these increases correlate with mood improvements (Wildmann, Kruger, Schmole, Niemann, & Matthaei 1986). However, contradictory associations with depressive symptoms question its role in the therapeutic effect of exercise (Darko, Risch, Gillin, & Golshan 1992). As with monoamines, peripheral estimates of β-endorphin may not provide an accurate valuation of central opioidergic activity, but a recent neuroimaging study provided the first evidence that exercise increases the release of opioids in the human brain (Boecker et al. 2008).

KEY POINT: THERE IS MIXED EVIDENCE FOR THE INFLUENCE OF EXERCISE ON THE NEUROTRANSMITTER AND NEUROPEPTIDE SYSTEMS ASSOCIATED WITH DEPRESSION

Stress and the Hypothalamic-Pituitary-Adrenal Axis (HPA)

The hypothalamic-pituitary-adrenal (HPA) axis, the body's neuroendocrine stress response system, is strongly implicated in the pathophysiology of depression. Clinical depression is frequently precipitated by stress and this association is supported by extensive evidence of higher peripheral cortisol in depressed patients (Gibbons & McHugh 1962). Approximately one half of individuals with depression display abnormally high HPA axis activity (Arana, Wilens, & Baldessarini 1985).

Exercise, as with other stressors, activates the HPA axis resulting in the production of glucocorticoids (Stranahan, Lee, & Mattson 2008). Longer-term exercise training can lead to beneficial adaptations to the HPA axis in animals and humans. For example, exercise attenuates the glucocorticoid and "emotionality" responses in rats to novel environments (Droste, Chandramohan, Hill, Linthorst, & Reul 2007), and trained individuals generally have lower tissue sensitivity to glucocorticoids at rest, with transiently higher sensitivity following exercise (Duclos, Gouarne, & Bonnemaison 2003).

Pro-Inflammatory Cytokines and Exercise

Cytokines are proteins that regulate the immune system. In response to infection pro-inflammatory cytokines can induce "sickness behaviors," including disturbances in sleep, mood, and appetite (Hart 1988). A meta-analysis of observational studies established that pro-inflammatory cytokines, interleukin (IL) -6 and tumor necrosis factor α (TNF-α), were significantly higher in patients with depression (Dowlati et al. 2010). The beneficial effects of exercise on depression could be mediated by anti-inflammatory mechanisms. Regular exercise increases anti-inflammatory activity by influencing the quantity of IL-1 β, IL-6, and TNF-α (Goldhammer et al. 2005).

Rethorst and colleagues (2013) measured cytokines over a 12-week exercise programme for previously unremitted depression. A significant positive correlation was found between changes in IL-1 β and depression scores. Additionally, TNF-α was a predictor of treatment outcome, with higher baseline TNF-α associated with enhanced reductions in depression (Rethorst et al. 2013). Further exploration of the possible immune function of exercise and relationship to depression can be found in Chapter 4 of this book by Aderbal Silva Aguiar, Jr., and Alexandra Latini.

KEY POINT: PRELIMINARY EVIDENCE SUGGESTS THAT THE POSITIVE EFFECT OF EXERCISE MAY BE MEDIATED BY CHANGES IN CYTOKINES

Metabolism, Blood Flow, and Angiogenesis

Metabolic, vascular, and cerebral blood flow (CBF) abnormalities are present in people with depression. During symptom remission some of these underlying abnormalities normalize (Drevets 2000). Exercise interventions may, therefore, alleviate depression by directly rectifying the brain's underlying vasculature and metabolism. There are clear alterations in blood flow and metabolism as a consequence of engaging in exercise (Ide & Secher 2000).

CBF and metabolism remain fairly stable in response to moderate exercise, but increase in response to higher intensities (Ide & Secher 2000). Longer-term exercise, for at least four months, enhances cerebral activity and vasculature in human brain regions implicated in depression, including the hippocampus and prefrontal cortex (Burdette et al. 2010) and these increases are often accompanied by better performance in cognitive and memory tasks (Colcombe et al. 2004)

Exercise-induced changes in cerebral vasculature occur in parallel with angiogenesis (Swain et al. 2003). Angiogenesis is broadly defined as the formation of new blood vessels from existing vasculature and is tightly regulated by angiogenic growth factors. Exercise elevates the expression of the angiogenic growth factors, insulin growth factor-1 (IGF-1) and vascular endothelial growth factor (VEGF). VEGF and IGF-1 appear to be important in mediating exercise-induced changes in vasculature and mood (Duman et al. 2009). It is possible therefore that the effect of exercise on depression is facilitated by vascular and metabolic alterations in the brain, but it is noteworthy that exercise has been shown to reduce depression irrespective of changes in cerebrovascular fitness (Craft, Freund, Culpepper, & Perna 2007).

KEY POINT: EXERCISE MAY INFLUENCE DEPRESSION AND CARDIOVASCULAR SYSTEMS BY INCREASING ANGIOGENESIS

Neurogenesis and Neurotrophic Factors

Exercise treatments activate molecular and cellular cascades that support neurogenic processes (Voss, Vivar, Kramer, & van Praag 2013). Neurogenesis refers to the process of generating new neurons from precursor cells. Depression is associated with regional reductions in hippocampal volume (Videbech & Ravnkilde 2004). These morphological abnormalities in depressed patients have been attributed to deficits in adult neurogenesis, and antidepressants may work by inhibiting or reversing neurogenic impairments (Duman 2005).

Greater hippocampal neurogenesis in rodents is one of the most reliably reported observations in exercise research (e.g. Neeper, Gomez-Pinilla, Choi, & Cotman 1995). Voluntary wheel running can increase neuronal spinal density, synaptic plasticity, and the expression of neurotrophic factors (van Praag 2008). Neurotrophic factors facilitate the maturation, proliferation, and survival of neurons, and have been proposed to mediate the neurogenic effects of exercise (Voss et al. 2013). Most neurotrophic investigations have fixated on brain-derived neurotrophic factor (BDNF). A positive correlation has been shown between voluntary exercise and BDNF expression in the hippocampus and cerebral cortex of rats (Neeper et al. 1995). In rodent models of depression exercise-induced increases in BDNF were associated with reductions in depressive behaviors (Duman, Schlesinger, Russell, & Duman 2008).

Central BDNF levels in humans are indirectly inferred from blood plasma and serum. A systematic review and meta-analysis of 29 studies established that a single session of exercise has a moderate effect on circulating BDNF (Szuhany, Bugatti, & Otto 2015). Longer-term programmes of regular exercise led to small increases in resting BDNF and intensify the moderate increases following a single bout of exercise, although the effects may be smaller in women. There were small to moderate effect sizes in studies examining BDNF changes in depressed patients (Szuhany et al. 2015).

Alongside angiogenesis, growth factors IGF-1 and VEGF increase hippocampal cell proliferation and neurogenesis in exercising rodents (Fabel et al. 2003). Studies generally indicate a close coupling of exercise-induced angiogenic and neurogenic processes (Pereira et al. 2007). Independently altering the activity of BDNF, IGF-1, VEGF, β-endorphins, HPA axis activity, cytokines, or monoamines appears to influence the neurogenic and antidepressant effects of exercise (Duman et al. 2009; Kiuchi, Lee, & Mikami 2012; Zunszain, Anacker, Cattaneo, Carvalho, & Pariante 2011). This suggests that greater neurogenic abilities may be a key mechanism mediating the effects of exercise in depression.

KEY POINT: SEVERAL OF THE NEUROBIOLOGICAL MECHANISMS ASSOCIATED WITH EXERCISE AND DEPRESSION COULD BE MEDIATED BY ADULT NEUROGENESIS

Neurobiological Mechanisms: Conclusions

No single neurobiological mechanism can adequately account for antidepressant effect of exercise. As with other antidepressant treatments, the effect of exercise on the brain could possibly be mediated by increases in monoamine transmission, attenuated HPA axis activity, improvements in cerebral vasculature and adult neurogenesis. To some extent, these possibilities represent the currently available technologies and changing conceptualizations of depression in recent decades. Given the rather limited amount of research to date, it is not possible to either confirm or refute any of these neurobiological explanations.

The bulk of the evidence on the putative neurobiological mechanisms of the effects of exercise in depression is drawn from animal studies. Animal studies provide an invaluable source of information as they allow experimental approaches and give insights into the molecular and cellular mechanisms that cannot be investigated in humans, but translating this evidence to humans requires care due to the physiological and behavioral differences between humans and other species. Further, integrated investigations are required to identify and disentangle the neurobiological mechanisms mediating the beneficial effects of exercise in depression.

Published Guidelines

Several policy guidelines groups in the UK approached the topic of exercise and depression. Such groups have also reviewed the evidence about the effectiveness of physical activity in the treatment of depression and they have made more positive conclusions than the Lawlor and Hopker (2001) review. The National Institute for Health and Clinical Excellence reviewed the evidence in 2009 and recommended structured, supervised exercise programmes, three times a week (of 45 minutes to 1 hour) over 10 to 14 weeks, as an intervention for mild to moderate depression (National Institute for Health and Clinical Excellence 2009). A more recent guideline published by the Scottish Intercollegiate Guidelines Network (SIGN) in 2010, on the topic of non-pharmaceutical management of depression in adults, also recommended that structured exercise could be considered as a treatment option for patients with depression. The evidence on which this recommendation was made was graded "B" within the SIGN system (Scottish Intercollegiate Guidelines Network (SIGN) 2010). SIGN guidance uses an A B C D grading system. A set of evidence given a Grade A is awarded to a body of evidence with almost all studies or reviews or meta-analysis being judged to be well conducted with very little risk of bias. Grade B is given to evidence judged to be from high-quality systematic reviews of case control or cohort or studies or high-quality studies with a very low risk of confounding or bias and a high probability that the relationship is causal. Thus SIGN is confident about the evidence, although we might want the evidence to be of even higher quality and achieve the Grade A in future.

A further important policy document that has endorsed the use of exercise for the treatment and prevention of depression is the consensus document of the Chief Medical Officers of England, Scotland, Wales and Northen Ireland (Department of Health 2011) which underpinned national physical activity guidelines for the UK.

Increasing Acknowledgment from Mental Health Professionals of the Role of Physical Activity and Exercise

Over the past 20 or so years the literature on physical activity, exercise, and depression has been growing. The evidence, however, has taken some time to filter to professionals and organizations who might be involved with the treatment of depression. For example, in 1997 in the UK an overview of depression and its treatment did not mention the value of exercise at all (Hale, 1997).

On a more positive note, there is some evidence that any reluctance to consider the "body" in the treatment of mental health may be shifting. In the UK, the National Health Service has produced a website to enable understanding of a variety of conditions and to know what treatment choices they might have, including self-help strategies (see www.nhs.uk/Pages/HomePage.aspx). In describing treatment for mild depression, the website suggests that "there is evidence that exercise may help depression and it is one of the main treatments if you have mild depression. Your GP may refer you to a qualified fitness trainer for an exercise scheme, or you can find out more about starting exercise here" (see Table 15.6 and www.nhs.uk/Conditions/Depression/Pages/Treatment.aspx). In addition, recent leaflets about depression from the Royal College of Psychiatrists in the UK suggest that exercise is a good self-help strategy for depression (see www.rcpsych.ac.uk/expertadvice/treatmentswellbeing/physicalactivity.aspx). The Scottish Association for Mental Health (SAMH) has produced a video and website proposing five ways to better mental health (www.samh.org.uk/). These are evidence based and are:

- Staying connected;
- learning;
- giving;
- taking notice; and
- keeping active.

Table 15.6 Advice from the Royal College of Psychiatrists about how to use exercise to help depression

- Physical activity should:
- Be **enjoyable** – if you don't know what you might enjoy, try a few different things.
- Help you to feel more **competent**, or **capable**. Gardening or DIY projects can do this, as well as getting you more active.
- Give you a **sense of control** over your life – that you have choices you can make (so it isn't helpful if you start to feel that you *have* to exercise). The sense that you are looking after yourself can also feel good.
- Help you to **escape** for a while from the pressures of life.
- Be **shared.** The **companionship** involved can be just as important as the physical activity.

(see: www.rcpsych.ac.uk/expertadvice/treatmentswellbeing/physicalactivity.aspx).

The emphasis on the importance of physical activity has been endorsed by Sir Chris Hoy, the Olympic cyclist, who has become an Ambassador for SAMH.

Another aspect of the treatment of depression that suggests that it is worthwhile to pursue exercise is that of patient choice. In the UK, drugs continue to be the most frequently used treatment for depression. Patients often report that they do not want drugs (Scott 1996). Consequently, exercise is a reasonable option with few negative side effects and could be cost effective in comparison to other non-drug options such as psychotherapy. Studies on the cost-effectiveness and cost-benefit of exercise versus drugs or other therapies must be undertaken so that the potential economic advantages of exercise can be measured. Perhaps the economic arguments will be the most powerful in persuading mental health professionals to include exercise as a treatment option.

KEY POINT: INCREASINGLY, THE MENTAL HEALTH BENEFITS OF EXERCISE ARE BEING RECOGNIZED AND THERE ARE NUMEROUS RESOURCES AVAILABLE TO HELP PEOPLE TO START EXERCISING

Chapter Summary

- The weight of evidence shows that prospective studies suggest a protective effect from physical activity on the development of depression, but not all studies show this;
- systematic reviews with meta-analytic findings show moderate effect sizes from studies that have used exercise as a treatment for depression;
- the weight of the evidence suggests that there is a causal connection between physical activity/exercise and reduction of depression;
- there are several neurobiological mechanisms that have potential to provide explanations of the antidepressant effect of exercise, but more research is needed in humans; and
- current guidance from relevant authorities all include advice to increase physical activity levels to help depression.

Note

1 This chapter is an update and expansion of a chapter first published in Biddle, S. J. H., N. Mutrie, and T. Gorely (2015). *Psychology of physical activity. Determinants, well-being and interventions* (3rd ed.). London: Routledge.

References

Anderson, R. J., K. E. Freedland, R. E. Clouse, & P. J. Lustman. 2001. "The prevalence of comorbid depression in adults with diabetes: A meta-analysis." *Diabetes Care 24* (6): 1069–78.

Arana, G. W., T. E. Wilens, & R. J. Baldessarini. 1985. "Plasma corticosterone and cortisol following dexamethasone in psychiatric patients." *Psychoneuroendocrinology 10* (1): 49–60.

Azevedo Da Silva, M., A. Singh-Manoux, E. J. Brunner, S. Kaffashian, M. J. Shipley, M. Kivimäki, & H. Nabi. 2012. "Bidirectional association between physical activity and symptoms of anxiety and depression: The Whitehall II study." *European Journal of Epidemiology 27* (7): 537–46.

Beck, A. T., R. A. Steer, R. Ball, & W. Ranieri. 1996. "Comparison of Beck Depression Inventories -IA and -II in psychiatric outpatients." *J Pers Assess 67* (3): 588–97. doi:10.1207/s15327752jpa6703_13

Beck, A. T., C. H. Ward, M. Mendelsohn, J. Mock, & H. Erbaugh. 1961. "An inventory for measuring depression." *Archives of General Psychiatry 4*: 561–71.

Bernaards, C. M., M. P. Jans, S. G. van den Heuvel, I. J. Hendriksen, I. L. Houtman, & P. M. Bongers. 2006. "Can strenuous leisure time physical activity prevent psychological complaints in a working population?" *Occup Environ Med 63* (1): 10–16.

Biddle, S. J. H., & N. Mutrie. 1991. *Psychology of physical activity and exercise: A health-related perspective.* London: Springer-Verlag.

Biddle, S. J. H., & N. Mutrie. 2008. *Psychology of physical activity: Determinants, well-being and interventions* (2nd ed.). London: Routledge.

Biddle, S. J. H., N. Mutrie, & T. Gorely. 2015. *Psychology of physical activity. Determinants,well being and interventions* (3rd ed.). London: Routledge.

Boecker, H., T. Sprenger, M. E. Spilker, G. Henriksen, M. Koppenhoefer, K. J. Wagner,...T. R. Tolle. 2008. "The runner's high: Opioidergic mechanisms in the human brain." *Cereb Cortex 18* (11): 2523–31. doi:10.1093/cercor/bhn013

Burdette, J. H., P. J. Laurienti, M. A. Espeland, A. Morgan, Q. Telesford, C. D. Vechlekar,...W. J. Rejeski. 2010. "Using network science to evaluate exercise-associated brain changes in older adults." *Front Aging Neurosci 2*: 23. doi:10.3389/fnagi.2010.00023

Camacho, T. C., R. E. Roberts, N. B. Lazarus, G. A. Kaplan, & R. D. Cohen. 1991. "Physical activity and depression: Evidence from the Alameda County Study." *Am J Epidemiol 134* (2): 220–31.

Chalder, M., N. J. Wiles, J. Campbell, S. P. Hollinghurst, A. M. Haase, A. H. Taylor, ...G. Lewis. 2012. "Facilitated physical activity as a treatment for depressed adults: Randomised controlled trial." *BMJ 344*: e2758. doi:10.1136/bmj.e2758

Colcombe, S. J., A. F. Kramer, K. I. Erickson, P. Scalf, E. McAuley, N. J. Cohen,...S. Elavsky. 2004. "Cardiovascular fitness, cortical plasticity, and aging." *Proc Natl Acad Sci U S A 101* (9): 3316–21. doi:10.1073/pnas.0400266101

Cooney, G. M., K. Dwan, C. A. Greig, D. A. Lawlor, J. Rimer, F. R. Waugh,...G. E. Mead. 2013. "Exercise for depression." *Cochrane Database Syst Rev 9*: CD004366. doi:10.1002/14651858.CD004366.pub6

Cooney, G., K. Dwan, & G. Mead. 2014. "Exercise for depression." *JAMA 311* (23): 2432–3. doi:10.1001/jama.2014.4930

Cooper-Patrick, L., D. E. Ford, L. A. Mead, P. P. Chang, & M. J. Klag. 1997. "Exercise and depression in midlife: A prospective study." *Am J Public Health 87* (4): 670–3.

Craft, L. L., K. M. Freund, L. Culpepper, & F. M. Perna. 2007. "Intervention study of exercise for depressive symptoms in women." *J Womens Health (Larchmt) 16* (10): 1499–1509. doi:10.1089/jwh.2007.0483

Darko, D. F., S. C. Risch, J. C. Gillin, & S. Golshan. 1992. "Association of beta-endorphin with specific clinical symptoms of depression." *Am J Psychiatry 149* (9): 1162–7.

Department of Health. 2011. *Start Active, Stay Active. A report on physical activity for health from the four home countries' Chief Medical Officers*. London: Author.

Dishman, R. K., K. J. Renner, S. D. Youngstedt, T. G. Reigle, B. N. Bunnell, K. A. Burke,...J. L. Meyerhoff. 1997. "Activity wheel running reduces escape latency and alters brain monoamine levels after footshock." *Brain Res Bull 42* (5): 399–406.

Dowlati, Y., N. Herrmann, W. Swardfager, H. Liu, L. Sham, E. K. Reim, & K. L. Lanctot. 2010. "A meta-analysis of cytokines in major depression." *Biological Psychiatry 67* (5): 446–57. doi:10.1016/j.biopsych.2009.09.033

Drevets, W. C. 2000. "Neuroimaging studies of mood disorders." *Biological Psychiatry 48* (8): 813–29.

Droste, S. K., Y. Chandramohan, L. E. Hill, A. C. Linthorst, & J. M. Reul. 2007. "Voluntary exercise impacts on the rat hypothalamic-pituitary-adrenocortical axis mainly at the adrenal level." *Neuroendocrinology 86* (1): 26–37. doi:10.1159/000104770

Duclos, M., C. Gouarne, & D. Bonnemaison. 2003. "Acute and chronic effects of exercise on tissue sensitivity to glucocorticoids." *J Appl Physiol (1985) 94* (3): 869–75. doi:10.1152/japplphysiol.00108.2002

Duman, C. H., L. Schlesinger, R. Terwilliger, D. S. Russell, S. S. Newton, & R. S. Duman. 2009. "Peripheral insulin-like growth factor-I produces antidepressant-like behavior and contributes to the effect of exercise." *Behav Brain Res 198* (2): 366–71. doi:10.1016/j.bbr.2008.11.016

Duman, R. S. 2005. "Neurotrophic factors and regulation of mood: Role of exercise, diet and metabolism." *Neurobiol Aging, 26 Suppl 1*: 88–93. doi:10.1016/j.neurobiolaging.2005.08.018

Dunn, A. L., & R. K. Dishman. 1991. "Exercise and the neurobiology of depression." *Exercise and Sport Sciences Reviews 19*: 41–98.

Fabel, K., K. Fabel, B. Tam, D. Kaufer, A. Baiker, N. Simmons,...T. D. Palmer. 2003. "VEGF is necessary for exercise-induced adult hippocampal neurogenesis." *Eur J Neurosci 18* (10): 2803–12.

Farmer, M. E., B. Z. Locke, E. K. Moscicki, A. L. Dannenberg, D. B. Larson, & L. S. Radloff. 1988. "Physical activity and depressive symptoms: The NHANES I Epidemiologic Follow-up Study." *Am J Epidemiol 128* (6): 1340–51.

Farrell, P. A., W. K. Gates, M. G. Maksud, & W. P. Morgan. 1982. "Increases in plasma beta-endorphin/beta-lipotropin immunoreactivity after treadmill running in humans." *J Appl Physiol Respir Environ Exerc Physiol 52* (5): 1245–9.

Ferrari, A. J., A. J. Somerville, A. J. Baxter, R. Norman, S. B. Patten, T. Vos, & H. A. Whiteford. 2013. "Global variation in the prevalence and incidence of major depressive disorder: A systematic review of the epidemiological literature." *Psychological Medicine 43* (3): 471–81. doi:10.1017/S0033291712001511

Gallegos-Carrillo, K., Y. N. Flores, E. Denova-Gutierrez, P. Mendez-Hernandez, L. D. Dosamantes-Carrasco, S. Henao-Moran,...J. Salmeron. 2013. "Physical activity and reduced risk of depression: Results of a longitudinal study of Mexican adults." *Health Psychol 32* (6): 609–15. doi:10.1037/a0029276

Gibbons, J. L., & P. R. McHugh. 1962. "Plasma cortisol in depressive illness." *J Psychiatr Res 1*: 162–71.

Goldhammer, E., A. Tanchilevitch, I. Maor, Y. Beniamini, U. Rosenschein, & M. Sagiv. 2005. "Exercise training modulates cytokines activity in coronary heart disease patients." *Int J Cardiol 100* (1): 93–9. doi:10.1016/j.ijcard.2004.08.073

Hale, A. S. 1997. "ABC of mental disorders: Depression. *British Medical Journal 315*: 43–6.

Hart, B. L. 1988. "Biological basis of the behaviour of sick animals." *Neuroscience and Biobehavioral Reviews 12* (2): 123–37.

Hasler, G., D. S. Pine, D. G. Kleinbaum, A. Gamma, D. Luckenbaugh, V. Ajdacic,... J. Angst. 2005. "Depressive symptoms during childhood and adult obesity: The Zurich Cohort Study." *Molecular Psychiatry 10* (9): 842–50.

Hoffmann, T. C., P. P. Glasziou, I. Boutron, R. Milne, R. Perera, D. Moher,...S. Michie. 2014. "Better reporting of interventions: Template for intervention description and replication (TIDieR) checklist and guide." *BMJ 348*: g1687. doi:10.1136/bmj.g1687

Ide, K., & N. H. Secher. 2000. "Cerebral blood flow and metabolism during exercise." *Prog Neurobiol 61* (4): 397–414.

Kim, T. W., B. V. Lim, D. Baek, D. S. Ryu, & J. H. Seo. 2015. "Stress-induced depression is alleviated by aerobic exercise through up-regulation of 5-Hydroxytryptamine 1A receptors in rats." *Int Neurourol J 19* (1): 27–33. doi:10.5213/inj.2015.19.1.27

Kiuchi, T., H. Lee, & T. Mikami. 2012. "Regular exercise cures depression-like behavior via VEGF-Flk-1 signaling in chronically stressed mice." *Neuroscience 207*: 208–17. doi:10.1016/j.neuroscience.2012.01.023

Kritz-Silverstein, D., E. Barrett-Connor, & C. Corbeau. 2001. "Cross-sectional and prospective study of exercise and depressed mood in the elderly: The Rancho Bernardo study." *Am J Epidemiol 153* (6): 596–603.

Larun, L., L. V. Nordheim, E. Ekeland, K. B. Hagen, & F. Heian. 2006. "Exercise in prevention and treatment of anxiety and depression among children and young people." *Cochrane Database of Systematic Reviews 3*. doi:10.1002/14651858.CD004691.pub2

Lawlor, D. A., & S. W. Hopker. 2001. "The effectiveness of exercise as an intervention in the management of depression: Systematic review and meta-regression analysis of randomised controlled trials." *BMJ 322* (7289): 763–7.

Mammen, G., & G. Faulkner. 2013. "Physical activity and the prevention of depression: A systematic review of prospective studies." *Am J Prev Med 45* (5): 649–57. doi:10.1016/j.amepre.2013.08.001

Mathers, C. D., & D. Loncar. 2006. "Projections of global mortality and burden of disease from 2002 to 2030." *PLoS Med 3*(11): e442. doi:10.1371/journal.pmed.0030442

Moussavi, S., S. Chatterji, E. Verdes, A. Tandon, V. Patel, & B. Ustun. 2007. "Depression, chronic diseases, and decrements in health: Results from the World Health Surveys." *Lancet 370* (9590): 851–8. doi:10.1016/S0140-6736(07)61415-9

Murray, C. J., & A. D. Lopez. 1997. "Alternative projections of mortality and disability by cause 1990–2020: Global Burden of Disease Study." *Lancet 349*: 1498–504.

National Institute for Health and Clinical Excellence. 2009. Depression: The treatment and management of depression in adults (updated edition). British Psychological Society and Royal College of Psychiatrists.

Neeper, S. A., F. Gomez-Pinilla, J. Choi, & C. Cotman. 1995. "Exercise and brain neurotrophins." *Nature 373* (6510): 109. doi:10.1038/373109a0

OECD. 2012. Health at a Glance: Europe 2012.

Office of National Statistics. 2001. Social Trends (Vol. 31). London: Stationary Office.

Paffenbarger, Jr., R. S., I. M. Lee, & R. Leung. 1994. "Physical activity and personal characteristics associated with depression and suicide in American college men." *Acta Psychiatr Scand Suppl 377*: 16–22.

Pereira, A. C., D. E. Huddleston, A. M. Brickman, A. A. Sosunov, R. Hen, G. M. McKhann,...S. A. Small. 2007. "An in vivo correlate of exercise-induced neurogenesis in the adult dentate gyrus." *Proc Natl Acad Sci U S A 104* (13): 5638–43. doi:10.1073/pnas.0611721104

Rethorst, C. D., M. S. Toups, T. L. Greer, P. A. Nakonezny, T. J. Carmody, B. D. Grannemann,...M. H. Trivedi. 2013. "Pro-inflammatory cytokines as predictors of antidepressant effects of exercise in major depressive disorder." *Molecular Psychiatry 18* (10): 1119–24. doi:10.1038/mp.2012.125

Robertson, R., A. Robertson, R. Jepson, & M. Maxwell. 2012. "Walking for depression or depressive symptoms: A systematic review and meta-analysis." *Mental Health and Physical Activity 5*: 66–75.

Scott, J. 1996. "Cognitive therapy of affective disorders: A review." *Journal of Affective Disorders 37*: 1–11.

Scottish Intercollegiate Guidelines Network (SIGN). 2010. *Non-pharmaceutical management of depression in adults.* Edinburgh: Scottish Intercollegiate Guidelines Network.

Stavrakakis, N., A. M. Roest, F. Verhulst, J. Ormel, P. de Jonge, & A. J. Oldehinkel. 2013. "Physical activity and onset of depression in adolescents: A prospective study in the general population cohort TRAILS." *J Psychiatr Res.* doi:10.1016/j.jpsychires.2013.06.005

Stranahan, A. M., K. Lee, & M. P. Mattson. 2008. "Central mechanisms of HPA axis regulation by voluntary exercise." *Neuromolecular Med 10* (2): 118–27. doi:10.1007/s12017-008-8027-0

Strawbridge, W. J., S. Deleger, R. E. Roberts, & G. A. Kaplan. 2002. "Physical activity reduces the risk of subsequent depression for older adults." *American Journal of Epidemiology 156* (4): 328–34.

Swain, R. A., A. B. Harris, E. C. Wiener, M. V. Dutka, H. D. Morris, B. E. Theien,... W. T. Greenough. 2003. "Prolonged exercise induces angiogenesis and increases cerebral blood volume in primary motor cortex of the rat." *Neuroscience 117* (4): 1037–46.

Szuhany, K. L., M. Bugatti & M. W. Otto. 2015. "A meta-analytic review of the effects of exercise on brain-derived neurotrophic factor." *J Psychiatr Res 60*: 56–64. doi:10.1016/j.jpsychires.2014.10.003

van Gool, C. H., G. I. Kempen, B. W. Penninx, D. J. Deeg, A. T. Beekman, & J. T. van Eijk. 2003. "Relationship between changes in depressive symptoms and unhealthy lifestyles in late middle aged and older persons: Results from the Longitudinal Aging Study Amsterdam." *Age Ageing 32* (1): 81–7.

van Praag, H. 2008. "Neurogenesis and exercise: Past and future directions." *Neuromolecular Med 10* (2): 128–40. doi:10.1007/s12017-008-8028-z

Videbech, P., & B. Ravnkilde. 2004. "Hippocampal volume and depression: A meta-analysis of MRI studies." *Am J Psychiatry 161* (11): 1957–66. doi:10.1176/appi.ajp.161.11.1957

Voss, M. W., C. Vivar, A. F. Kramer, & H. van Praag. 2013. "Bridging animal and human models of exercise-induced brain plasticity." *Trends in Cognitive Science 17* (10): 525–44.

Weyerer, S. 1992. "Physical inactivity and depression in the community. Evidence from the Upper Bavarian Field Study." *Int J Sports Med 13* (6): 492–96.

Wildmann, J., A. Kruger, M. Schmole, J. Niemann, & H. Matthaei. 1986. "Increase of circulating beta-endorphin-like immunoreactivity correlates with the change in feeling of pleasantness after running." *Life Sci 38* (11): 997–1003.

Zunszain, P. A., C. Anacker, A. Cattaneo, L. A. Carvalho, & C. M. Pariante. 2011. "Glucocorticoids, cytokines and brain abnormalities in depression." *Prog Neuropsychopharmacol Biol Psychiatry 35* (3): 722–9. doi:10.1016/j.pnpbp.2010.04.011

Section 5

Implications for the Health Sector and School

16 Prescribing Exercise for Mental Health

Mode and Dose-Response Considerations

Brandon L. Alderman and Ryan L. Olson

I think that I cannot preserve my health and spirits unless I spend four hours a day at least – and it is commonly more than that – sauntering through the woods and over the hills and fields absolutely free from all worldly engagements.

Henry David Thoreau

Introduction

Henry David Thoreau, one of the world's greatest thinkers, was an avid exerciser. He wrote about his daily walking practice and the influence it had on his life and his thinking. If we are careful, we can take away several important points from his quote above. First and perhaps most importantly, he was taking many thousands of steps more than the average human living in most developed countries today (Hallal et al. 2012). Current international guidelines recommend approximately 150 minutes per week of moderate-intensity physical activity, often expressed as 30 minutes of brisk walking or an equivalent activity performed five or more days per week (Haskell et al. 2007; Tremblay et al. 2011; World Health Organization 2010). In terms of steps, current recommendations call for approximately 10,000 steps per day for adults (Tudor-Locke, Johnson, & Katzmarzyk 2009). Although walking "through the woods and over the hills and fields" would likely result in fewer total steps than walking on a flat surface without obstacles, clearly spending at least four hours a day engaged in physical activity would result in exceeding our current physical activity recommendations. Second, Thoreau was sauntering *in nature*, free from the cares and worries of the time. In today's world, that means exercising without a cell phone and without access to the internet (i.e. unplugged), since most people spend far too much time completely connected and plugged in with technology (Kakabadse, Porter, & Vance 2007; Small & Vorgan 2008). In addition, most people also live in large, metropolitan areas and spend far less time outdoors than individuals from previous generations. This is an interesting but paradoxical trend given that urbanization may result in a higher risk for anxiety, depression, and other stress-related illnesses (Dohrenwend and Dohrenwend 1974; Firdaus & Ahmad 2014; Hidaka 2012; Peen, Schoevers, Beekman, & Dekker 2010). Recent studies also suggest that spending time in nature or natural environments may actually enhance overall well-being. Notably, these positive effects have been shown to be accompanied by structural and functional brain changes (Bratman, Hamilton, Hahn, Daily, & Gross 2015). Even without being armed with today's scientific understanding, Thoreau recognized the importance of daily physical activity.

Two aspects of Thoreau's statement are particularly relevant to the present chapter and the current state of the literature on the mental health benefits of physical activity. First, Thoreau's choice (or mode) of physical activity was walking. Although more aerobic (i.e. oxygen-demanding) forms of exercise such as walking or jogging have received the bulk of the research attention related to improvements in psychological health, recent efforts have focused on alternative modes of exercise to determine whether or not they confer similar benefits to overall well-being. Resistance exercise or weight lifting has become an increasingly popular form of exercise and given the consistent benefits for musculoskeletal health (Kraemer, Ratamess, & French 2002; Lambert, Nelson, Jovanovic, & Cerda 2015), it has recently been included in updated physical activity guidelines (Garber et al. 2011). Accompanying its rise in popularity, there has also been considerable attention devoted to the cognitive and mental health benefits of resistance exercise.

Outside of traditional aerobic and resistance forms of exercise, humans from all walks of life are now engaging in a wide variety of physical activities that have yet to receive much research attention (e.g. yoga, tai chi, qigong, aerobic dance, sports). There is also an increasing focus of research on the mental health benefits derived from integrative activities that combine physical and cognitive demands (see Chapter 7). Thus, it is important to determine the unique and generalized mode-specific benefits of exercise and/or physical activity. We also know that Thoreau walked at least four hours a day to preserve his health and spirits. Yet, we do not know whether this activity was more of a hike or a leisurely walk, or for that matter his preference for exercise intensity (i.e. the percentage of maximal effort). This information is critical not only for determining the overall energy expenditure of the activity, but recent evidence also suggests there are dose-dependent effects of exercise duration and intensity on a number of physical and mental health outcomes, including all-cause mortality. Dose-response refers to the amount of exercise (i.e. duration, intensity, and frequency) that results in optimal benefits for health. Many topic areas within exercise science are currently focused on establishing dose-response relationships. Although some individuals, such as Thoreau, are able to self-administer an optimal dose of exercise for their own health and sanity, it is vital to elucidate these complex dose-response relationships between exercise and health in order to advance the use of exercise as medicine (Nagamatsu et al. 2014).

In this chapter, we will first present evidence on the various types or modes of exercise that many people engage in, and follow with the current dose-response evidence regarding the mental health benefits derived from participating in these specific forms of exercise. Exercise is effective at enhancing a broad array of mental and cognitive health variables (e.g. ADHD, sleep, schizophrenia, various aspects of brain function); however, in the interest of space we will primarily focus our attention on the current evidence for depression, anxiety, and cognitive functioning. We will conclude the chapter by outlining some general recommendations for prescribing exercise for mental health.

Mode of Exercise

The mode of exercise refers to the type of activity (e.g. walking, bicycling, playing tennis) in which one participates. It is important to know the type(s) of activity one engages in since this information is necessary to determine the total amount of energy or calories expended and to understand the physical and mental health benefits accrued from participating in different physical activities. Indeed, one of the primary reasons that people exercise is to develop and/or maintain one or more components of physical fitness (e.g. enhanced aerobic capacity, muscular strength, or body composition).

Aerobic

Many people specifically engage in *aerobic exercise* to improve their health status, reduce disease risk, and modify body composition (Kravitz & Vella

2002). This preference for exercise is supported by the research literature as the majority of studies on depression, anxiety, and cognitive function have focused on aerobic forms of exercise. However, even within the broad category of aerobic exercise, there is an extremely wide variety of health effects that may accompany specific physical activities. For instance, walking and jogging are typically engaged in at a relatively consistent intensity and are not dependent on a participant's skill level. Activities such as aerobic dance, running, and hiking result in an increase in energy expenditure but are not able to be sustained for the same duration as walking and jogging. Skill-based activities (e.g. basketball, racquet sports, volleyball), on the other hand, are highly variable in terms of energy expenditure, are more cognitively demanding, and have higher performance demands compared to other physical activities. Given the amount of research attention devoted to walking, treadmill running, and stationary cycling, we will primarily focus our attention on these modes of exercise. However, we will also emphasize other modes of exercise that may have clinical and practical utility in enhancing overall mental health and cognitive well-being.

Resistance

Relative to aerobic forms of exercise, there is less research dedicated to the potential benefits of resistance exercise. *Resistance exercise* or *resistance training* refers to a specialized type of exercise that involves the progressive use of resistive loads that cause muscles to contract against an external resistance with the goal of increasing muscular strength, muscular endurance, and/or body composition. Although the terms resistance exercise, resistance training, strength training, and weight training are often used synonymously, the term resistance exercise encompasses the broad range of training modalities and goals to enhance health (Kraemer & Ratamess 2004). For many years, resistance exercise programs were prescribed for the development of athletes due to their effects on muscular strength, power, hypertrophy, and sport-specific fitness. More recently, the public's increasing interest in resistance exercise programs has been driven by its more general health benefits, including improvements in blood pressure, body composition and weight loss, bone mass, lower back pain, glucose tolerance and insulin resistance, osteoporosis, and risk for colon cancer, among others. Additionally, it has a beneficial impact on the quality of life and functional capacity for individuals of all ages. The study of resistance exercise and the advancement of resistance exercise prescription for mental health is challenging due to the large number of program variables that can be manipulated in a quality resistance exercise program: intensity (or loading), volume (the number of sets and repetitions), exercises selected (major muscle groups, upper body, lower body), the order of the exercises, rest intervals between sets, velocity of contraction, and exercise frequency. Despite these challenges, the number of studies aimed at examining the potential benefits of resistance exercise for mental health and cognitive function has increased (Conn 2010). This is important because the benefits obtained from resistance exercise might be different than those that are commonly associated with aerobic exercise.

Integrated Mind and Body

In America and many other countries around the world, walking, jogging, cycling, or lifting weights is the primary choice of exercise. Today, if you walk into any health club in the United States you are likely to see treadmills, cycle ergometers, elliptical machines, and a variety of resistance exercise equipment. Nevertheless, there are millions of people around the world who engage in alternative modes of physical activity. For the purpose of this chapter, we will focus on some of the more common complementary and alternative forms of exercise (Clarke, Black, Stussman, Barnes, & Nahin 2015), including yoga, tai chi, and qigong. Yoga originated from ancient Indian philosophy and is typically characterized as a combination of physical postures, controlled breathing techniques, and meditation or relaxation. Similar to traditional forms of exercise, yoga encompasses multiple styles and forms, with Hatha yoga being the most commonly practiced form in the United States and Europe. Tai chi originated in China as a form of martial arts and is sometimes referred to as "moving meditation" due to the slow, gentle movements combined with deep breathing and awareness techniques. Tai chi is practiced individually or in groups, and is classified as a low-impact, weight-bearing aerobic exercise that includes a mental training component. Similar to tai chi, qigong is a discipline developed from traditional Chinese medicine that combines gentle physical movements, mental focus, and deep breathing. The benefits of these alternative forms of exercise range from increased muscular strength, coordination, and flexibility to improvements in mood and affective states and sleep. Importantly, these forms of exercise involve training of the *mind and the body*, and thus are often referred to as *mind-and-body practices.* A number of these practices are receiving increasing research attention in the United States and elsewhere (e.g. Jahnke, Larkey, Rogers, Etnier, & Lin 2010; Saeed, Antonacci, & Bloch 2010). Indeed, there is even a branch of the National Institutes of Health (NIH), the National Center for Complementary and Integrative Health (NCCIH), devoted to the advancement of scientific research on practices such as yoga, tai chi, and qigong that are not generally considered part of traditional medicine. Although the scientific basis for these modes of physical activity is still in its infancy, emerging evidence supports their beneficial effects for physical, mental, and cognitive health (Jahnke et al. 2010).

Dose-Response Considerations

As we move towards a better understanding of the health benefits of exercise, it is important to consider both the type and amount of physical activity necessary for promoting physical and mental health. This question was addressed in a consensus symposium published in *Medicine and Science in Sports and Exercise* in 2001 (Kesaniemi et al. 2001), where experts from around the world reviewed the evidence to determine whether or not a dose-response relationship existed between physical activity and a number of physical and mental health outcomes, including all-cause mortality. Relative to all-cause mortality, it was concluded that an inverse linear relationship exists such that individuals who engage in greater amounts of physical activity are at a reduced risk of dying prematurely (Lee & Skerrett 2001). However, little evidence was available at the time for the physical activity characteristics of frequency (per week), intensity, and duration. The dose-response evidence for exercise on depression and anxiety was also reviewed (Dunn, Trivedi, & O'Neal 2001) and consisted predominantly of cross-sectional data from observational studies. Reductions in depressive symptoms were observed following both aerobic and resistance modes of exercise. Unfortunately, little evidence was available at the time to support any firm conclusions regarding the efficacy of exercise for reducing anxiety. Overall, Dunn and colleagues regarded the conclusions as "suggestive but not convincing." These initial consensus reviews served as the foundation for additional research aimed at understanding the dose-response relationships of exercise with a variety of health outcomes, including depression, anxiety, and cognitive function.

Next, we review more recent studies that have examined the effects of aerobic, resistance, and mind-and-body exercise on these mental health outcomes. We pay particular attention to those studies that address dose-response relationships. In addition, we will examine meta-analytic reviews to help establish dose-response recommendations. A *meta-analysis* is a statistical technique for combining findings from independent studies and systematic reviews, often with the goal of determining the magnitude of the treatment effect of an independent variable (e.g. exercise) on an outcome variable (e.g. all-cause mortality). Fortunately, a number of meta-analyses have been conducted to summarize the effects of exercise on depression, anxiety, and cognitive functioning (see Chapter 13 and Chapter 15 in this volume). Importantly, these systematic reviews have also included moderating analyses of the primary exercise program variables of frequency, intensity, duration, and mode of exercise, which are helpful in establishing general recommendations for the prescription of exercise for mental health.

Evidence for Exercise and Depression

Depression (or major depressive disorder; MDD) is a serious mood disorder that interferes with a person's ability to function in work and daily life (see Chapter 15 this volume for an expanded definition of clinical depression). It is one of the most prevalent and debilitating of all the psychiatric disorders, and is associated with considerable morbidity and mortality (Kessler 2012). Globally, more than 350 million people suffer from depression, and by 2030 it is predicted to be the leading cause of disease burden (Mathers, Fat, & Boerma

2008). Depression is not only debilitating on its own, but also has a high rate of comorbidity with other mental illnesses. Nearly three-quarters of all people who meet criteria for depression at some point during their lifetime also meet the criteria for another psychiatric disorder: approximately three-fifths will be comorbid for one of the anxiety disorders, one-quarter for substance-use disorders, and one-third for impulse-control disorders (DeRubeis, Siegle, & Hollon 2008). The current standard treatment for depression is antidepressant medication (e.g. selective serotonin reuptake inhibitors; SSRIs) along with various forms of cognitive behavioral therapy (CBT) or psychotherapy. These treatments are often associated with unwanted negative side effects, can be time-consuming and costly, and may not be as efficacious as once thought. Thus, there is a clear societal need for alternative and augmentative treatment strategies for depression. Exercise is one such lifestyle intervention that has few negative side effects, can be self-administered, and has shown similar clinical efficacy as more traditional treatments. However, in order to advance empirically based recommendations for the prescription of exercise, specific evidence for mode and dose-response program variables must be firmly established. Below we provide the current state of the evidence for mode and dose-response characteristics of aerobic and resistance exercise in the treatment of depression.

Aerobic Exercise Evidence

There is a long history of research examining the effects of aerobic exercise on depression (Franz & Hamilton 1905; Morgan 1969). The efficacy of exercise for treating and preventing depression has been examined in a number of cross-sectional, prospective, and randomized controlled trial (RCT) studies. The use of exercise as a treatment for MDD has gained considerable support, as evidenced by the mention of exercise as a potential form of treatment in recent guidelines published by the American Psychiatric Association (Gelenberg et al. 2010; Rethorst & Trivedi 2013). The majority of studies examining the effects of exercise for depression have incorporated aerobic forms of exercise, and recently there have been several studies aimed at examining the dose-response relationship between aerobic exercise and symptom reduction in MDD. In one of the few RCTs that have systematically examined the dose of exercise, Dunn et al. (2005) compared two different doses of aerobic exercise (7 kcal/kg/week, low dose (LD) or 17.5 kcal/kg/week, public health dose (PHD)) performed at two different frequencies (3 or 5 days per week) to an attention-controlled placebo group that performed flexibility training in individuals with mild-to-moderate depression. Intensity of exercise was self-selected by participants assigned to the exercise conditions. After the 12-week intervention, the mean Hamilton Rating Scale for Depression (HAM-D) score was reduced by 47% in the PHD group, which was significantly greater than the reductions in HAM-D scores observed in the LD exercise (30%) and attention control groups (29%). The treatment response for the PHD of exercise (defined as a 50% reduction in HAM-D score) is equivalent to reductions typically observed for other treatments for depression, such as antidepressant medication and CBT (Dunn et al. 2005). To our knowledge, this is the only study to date that has compared the frequency of exercise sessions per week (i.e. between exercise performed 3 days vs. 5 days per week) and given that no differences were found for exercise frequency, it appears that the total amount of exercise completed may be more important than the frequency of exercise sessions per week. However, more research on these important exercise program variables is warranted.

In an earlier RCT, Blumenthal and colleagues (1999) examined the effectiveness of aerobic exercise compared with standard antidepressant treatment (sertraline) and a combination of both treatments in 156 adults aged 50 years and older. For the exercise program, participants attended three supervised sessions per week for 16 consecutive weeks in a group setting and were prescribed walking or jogging at a vigorous intensity, ranging from 70% to 85% of *heart rate reserve* (HRR), which reflects the difference between maximal and resting heart rate. Although patients receiving medication alone exhibited the fastest initial response, there were no significant differences in depressive symptoms between the three groups by the end of the 16-week program. As a follow-up to this study, Babyak and colleagues (2000) revisited the participants from the Blumenthal et al. (1999) RCT to determine the long-term effectiveness of exercise compared to medication or combined treatment. Depression was again assessed in approximately 85% of the initially enrolled study population. The 10-month follow-up (6 months post-intervention) showed that individuals assigned to the exercise-only group displayed significantly lower rates of depression (30%) compared to the medication (50%) or combined treatment (55%) groups, suggesting a lower relapse rate following an aerobic exercise treatment program performed at high intensity relative to antidepressant medication. These studies provide preliminary support for the efficacy of exercise in treating depression relative to antidepressant medication.

Although there are promising results, not all studies have found a dose dependent effect of aerobic exercise on depression. Chu, Buckworth, Kirby, and Emery (2009) randomized sedentary women with high levels of depressive symptoms to a low- or high-intensity aerobic exercise condition or to a stretching program for 10 weeks. Participants in the aerobic exercise groups met for one 30–40 minute supervised session and were asked to complete 3–4 additional unsupervised exercise sessions during the week. During unsupervised sessions, participants in the exercise groups chose their preferred modes of aerobic exercise, such as aerobic dancing, walking, or biking. This approach likely served to enhance the motivation and adherence to exercise. All three groups demonstrated a significant reduction in depressive symptoms and no differences were observed between the groups at the end of the 10-week intervention. However, after controlling for baseline BDI-II depression scores, the high-intensity aerobic exercise group reported significantly fewer depressive symptoms than either the low-intensity or stretching groups at weeks 5 and

10, suggesting a dose-dependent effect of exercise on depressive symptoms. More recently, Trivedi and colleagues (2011) manipulated the energy expenditure of exercise to examine the efficacy of aerobic exercise as an augmentation treatment for MDD patients who had not remitted with previous SSRI treatment. Among 126 sedentary individuals (men and women aged 18–70 years) diagnosed with DSM-IV nonpsychotic MDD, those assigned to a higher dose of exercise (16 kcal/kg/week) showed a trend for better remission rates than individuals assigned to a lower dose of exercise (4 kcal/kg/week). However, participants assigned to the low dose of exercise had better adherence rates than those assigned to the higher dose of exercise. This led the authors to conclude that a higher dose of exercise may be more effective for some, while a low dose of exercise may be more effective for others. Thus, although a general prescription for an antidepressant response of aerobic exercise may be defensible for primary care providers, it is likely that important individual differences would need to be considered when prescribing exercise to treat depression.

Resistance Exercise Evidence

Two early meta-analytic reviews of the literature demonstrated significant benefits for both aerobic and resistance forms of exercise for depression (Craft & Landers 1998; North, McCullagh, & Tran 1990). Although a bulk of the literature has focused on aerobic exercise, more recent studies have begun to examine the potential clinical benefits of resistance exercise for depression. In a RCT examining the effects of resistance exercise on depression, Singh, Clements, and Fiatarone (1997) randomized individuals aged 60 years and older with depression or dysthymia to a 10-week resistance exercise program. Participants were randomized to either a supervised 3 days per week high-intensity resistance training program or to a health education control group. The resistance exercise program consisted of upper and lower body exercises performed for 3 sets at 8 repetitions at 80% of the individual *one-repetition maximum (1RM)*, the maximal amount of weight an individual can lift in a single repetition for a given exercise (Garber et al. 2011). Participants in the resistance exercise group reported significantly reduced depressive symptoms compared to the health education control group by the end of the 10-week intervention, with 59% achieving a clinically meaningful response compared with only 26% in the health education control group. Following the intervention, participants were instructed to continue engaging in twice-weekly, unsupervised resistance exercise sessions for another 10 weeks. Depressive symptoms remained significantly lower following this subsequent 10-week period in the resistance exercise group, with 73% of participants being classified as "non-depressed" (defined as a BDI score < 9) compared with 36% of participants in the control group. In 2005, Singh and colleagues (2005) followed this study by conducting a RCT to examine the dose-response effect of resistance exercise on symptoms of depression in older adults. Sixty older adults (> 60 years) with minor or major depression were assigned to a low (20% of 1RM) or high (80% of 1RM) intensity progressive resistance training program or a treatment-as-usual (TAU) condition for 8 weeks. The high-intensity resistance exercise program was more effective at reducing depression symptoms compared to either the low-intensity or TAU condition. Although the results of this trial support a preliminary dose-response relationship between resistance exercise intensity and symptom reductions in depression, the 20% of 1RM condition is considered to be very low intensity by most standards and is not recommended in progressive resistance exercise programs endorsed by authoritative bodies (Kraemer & Ratamess 2004; Nelson et al. 2007). Future research is warranted to establish the optimal dose of resistance exercise for depression.

Meta-Analytic Evidence

There are now hundreds of studies published that have specifically examined the efficacy of aerobic and resistance exercise for depression. As a result, a number of meta-analyses have been conducted to summarize the findings (e.g. North et al. 1990; Rethorst, Wipfli, & Landers 2009; Wegner et al. 2014). Craft and Landers (1998) published one of the earliest systematic reviews in this area and reported a large beneficial effect of exercise (aerobic and nonaerobic) for individuals with clinical depression, although no significant differences were found between the exercise program variables of frequency, intensity, or duration. Interestingly, there was a nonsignificant trend for greater reductions in depression following more moderate-to-vigorous intensities of exercise relative to lower intensities of exercise. Although higher intensities of exercise might be best for reducing depression, individuals with depression may find exercise to be difficult, particularly when it is vigorous in nature. Therefore, it is important to individualize an exercise program based on overall goals, exercise history, and preferences. There was also a significant difference reported for the average number of weeks required to see reductions in depression, with longer (9–12 weeks) interventions resulting in larger effects ($ES = -1.18$) compared to shorter (8 weeks or less) interventions ($ES = -0.54$). Relative to exercise intensity, few published studies and no systematic reviews have addressed the interaction between intensity and duration of aerobic and resistance exercise. This is an important dose-response consideration given that higher intensities of exercise are not able to be sustained for as long as lower intensities of exercise, an issue of particular relevance for individuals suffering with depression. A more recent meta-analysis of 58 randomized trials (Rethorst et al. 2009) demonstrated that exercise sessions lasting 45–59 minutes in duration, performed 5 days per week, and for at least 10–16 weeks showed the largest effect sizes for clinically depressed individuals. High-intensity exercise (activity classified as >75% of maximal aerobic capacity) resulted in the largest effect sizes; however, this was not significantly different from lower intensities of exercise. Both aerobic and resistance exercise demonstrated similar benefits in this review, and among studies that did not use individuals with a formal diagnosis of depression, combined aerobic and resistance exercise programs were best. Thus, the prescription of exercise might change depending on whether an individual has a clinical diagnosis of depression or based on depressive symptom severity.

To date, two systematic reviews of RCTs have been performed to specifically establish dose-response recommendations for exercise in the treatment of depression. The first study by Perraton, Kumar, and Machotka (2010) analyzed 14 studies that met their strict inclusion criteria. They concluded that 30 minutes of supervised aerobic exercise performed 3 days per week for at least 8 weeks is likely to be most effective for reducing symptoms of depression. Due to limited evidence, the authors concluded that further research is required before making any firm recommendations for resistance exercise. Their recommendations for aerobic exercise are similar to current recommendations for healthy populations established by the American College of Sports Medicine (ACSM) (2013) (Garber et al. 2011). The second study by Stanton and Reaburn (2014) reviewed five RCTs and concluded that exercise is beneficial in the treatment of depression when a program of supervised aerobic exercise is performed 3 to 4 times weekly, at low- to moderate-intensity or at a preferred intensity, and sessions lasting 30–40 minutes in duration. In terms of detecting an antidepressant response, they recommend programs lasting at least 9 weeks in duration. In general, we agree with the majority of these recommendations; however, given that no negative effects have been found for resistance exercise programs and that additional benefits may be accrued from participating in resistance exercise (e.g. enhanced bone health, muscular strength and endurance, and posture and balance), we recommend that exercise prescriptions include both aerobic and resistance exercise.

Evidence for Exercise and Anxiety

Anxiety disorders were described as early as the fourth century B.C. in the writings of Hippocrates (Regier et al. 1988), but their importance was not fully appreciated until recently. *Generalized anxiety disorder* (GAD), which was only first defined in the DSM-III, is characterized by fluctuating levels of uncontrollable worry associated with fatigue, insomnia, muscle tension, poor concentration, and irritability. There are also a number of other anxiety disorders, including social anxiety disorder, panic disorder (with or without agoraphobia), specific phobias, and substance or medication-induced anxiety disorder (see Chapter 13 in this volume for a more detailed description of clinical anxiety disorders). Given the increasingly stressful nature of modern life, anxiety and other stress-related disorders are among the most prevalent of all mental health disorders. Similar to depression, conventional approaches for treating anxiety include antidepressants and other psychoactive drugs (e.g. benzodiazepines) and/or psychotherapy. Given the high comorbidity rates of anxiety with depression combined with global disease burden, various lifestyle behaviors including exercise may be a viable alternative or adjunctive form of treatment. Although the therapeutic effects of exercise for anxiety disorders have received less research attention, considerable evidence supports the use of exercise in the treatment and prevention of anxiety and other stress-related illnesses (Herring, Puetz, O'Connor, & Dishman 2012; Petruzzello 2012).

Aerobic Exercise Evidence

An extensive body of research supports the anxiolytic benefits of aerobic exercise in both clinical and nonclinical populations. Much of the research supporting these benefits has focused on the acute effects of exercise on state anxiety. Indeed, reductions in state anxiety following acute aerobic exercise are one of the most commonly reported psychological benefits within the exercise psychology literature. *State anxiety* refers to transient feelings of tension, apprehension, or worry lasting anywhere from moments to hours in duration. While acute exercise (i.e. one bout of exercise) has consistently been shown to reduce state anxiety (Ensari, Greenlee, Motl, & Petruzzello 2015; Petruzzello, Landers, Hatfield, Kubitz, & Salazar 1991), chronic exercise programs lasting from weeks to months have been shown to result in significant improvements in trait anxiety (Petruzzello 2012). *Trait anxiety* refers to a more general predisposition to appraise an event or stimulus as threatening and respond with apprehension, worry, and nervousness. Although the transient, short-lived effects of exercise on state anxiety are meaningful, we will limit our focus here to the findings from chronic exercise interventions and meta-analytic reviews. Admittedly, one of the difficulties in reviewing the evidence of chronic exercise and anxiety is that many studies focus on one particular anxiety disorder or on a very unique clinical sample (e.g. caregivers, cancer patients, individuals with chronic obstructive pulmonary disease). This makes summarizing the overall effect, as well as attempting to clarify mode and dose-response considerations of exercise, quite challenging. Fortunately there are a number of meta-analyses and systematic reviews that can be used to derive evidence-based recommendations for prescribing exercise for anxiety disorders.

In an early study investigating the effects of exercise on anxiety (Sexton, Maere, & Dahl 1989), 52 inpatients diagnosed with non-psychotic anxiety disorder were randomized to either a moderate-intensity walking program or a vigorous intensity jogging program. The exercise program consisted of 4–5 sessions per week lasting 30 minutes in duration for 8 weeks. Anxiety was assessed following the intervention and again at a 6-month follow-up. Both groups exhibited reductions in anxiety that persisted up to 6 months following the exercise programs, with no significant differences emerging between the groups. Fitness improvements were observed only for the jogging group at 8 weeks post-intervention, but disappeared at the 6-month follow-up, and did not correlate with self-reported anxiety reductions. Of note, participants assigned to the walking group had higher adherence rates compared to the jogging group, leading the authors to conclude that although both groups showed reduced levels of anxiety, adherence to moderate-intensity exercise may be better than vigorous intensity exercise among anxiety patients. These findings are promising given that individuals with anxiety might have difficulties starting and maintaining an exercise program. In order to compare the effects of exercise with another promising behavioral intervention, Jazaieri, Goldin, Werner, Ziv, & Gross (2012) conducted an RCT of mindfulness-based stress reduction (MBSR) versus aerobic exercise in adults with social anxiety disorder. The standard MBSR program was comprised of eight 2.5-hour group classes per week, a one-day meditation retreat, and daily home practice. Participants in the exercise condition were provided with a 2-month gym membership and required to complete at least two individual bouts of moderate-intensity aerobic exercise and one group-based aerobic exercise session per week across the 8-week study. Both MBSR and exercise were associated with significant reductions in social anxiety and increases in subjective well-being following the intervention, which persisted for at least 3 months. Notably, beneficial effects for exercise were observed even in the absence of instruction and monitoring of exercise intensity, two factors that could influence psychological outcomes to exercise. However, individuals were allowed to self-select and monitor their own exercise, which is arguably more effective for enhancing exercise adherence among anxiety patients. Recently, Fetzner and Asmundson (2015) conducted a study to examine the effects of aerobic exercise for post-traumatic stress disorder (PTSD). Thirty-three individuals with PTSD engaged in a 2-week vigorous intensity exercise program totaling six sessions of stationary cycling at 60–80% of HRR. Although the primary aim of this study was to assess the impact of an attentional focus manipulation during exercise, most of the participants (89%) reported clinical reductions in PTSD severity after 2 weeks of aerobic exercise. Therefore, the authors proposed that vigorous intensity exercise may be effective for reducing symptoms of PTSD. Based on the revised DSM-5 guidelines, PTSD is now included in a category separate from the anxiety disorders but it is noteworthy that this is the first RCT of exercise for PTSD.

Despite the difficulty in reviewing this body of evidence, Herring, Lindheimer, and O'Connor (2013) recently conducted a comprehensive, systematic review of the available studies and concluded that exercise training is effective at reducing symptoms of anxiety among healthy adults, chronically ill patients, and patients with panic disorder. Preliminary evidence also supports the use of exercise training as an alternative therapy for patients with social anxiety disorder, GAD, and obsessive-compulsive disorder. Importantly, the observed reductions in anxiety were reported to be comparable in nature to other treatments (medication and CBT) for panic and generalized anxiety disorders. Although there is consistent evidence to support the use of aerobic exercise to treat anxiety, much less information is available from RCTs to guide mode and dose-response recommendations. Therefore, the influence of moderating variables (e.g. mode, frequency, intensity, duration) in meta-analytic reviews has been used to guide current recommendations.

Resistance Exercise Evidence

While the anxiolytic effect of aerobic exercise has been established, less research has been conducted on resistance exercise, yet emerging data are encouraging. Similar to aerobic exercise, early studies examining the effects of resistance training on mood and well-being focused on acute changes in state anxiety (e.g. Focht & Koltyn 1999). These early studies suggested that acute bouts of resistance training result in little to no effect on state anxiety, and might even lead to temporary elevations in anxiety symptoms. For instance, a study comparing the effects of an acute bout of resistance and aerobic exercise found an increase in anxiety immediately following resistance exercise, while there was a significant reduction in anxiety only occurring 60 minutes following aerobic exercise (Raglin, Turner, & Eksten 1993). Related to dose, the resistance exercise consisted of three sets of 6–10 repetitions performed at 70–80% of 1RM (i.e. vigorous intensity) for six to seven exercises while the aerobic exercise was comprised of stationary cycling for 30 minutes at 70–80% of age-predicted maximum heart rate (i.e. moderate-to-vigorous intensity). These bouts of exercise did not represent comparable intensities and may have resulted in higher intensities for the resistance exercise condition. It is also possible that the use of a self-reported state anxiety questionnaire consisting of anxiety-related adjectives was confounded by physiological arousal due to exercise. Fortunately, several chronic exercise studies have been conducted to examine effects on trait anxiety. Notably, a study comparing mode of exercise evaluated the effects of 8 weeks of aerobic (e.g. jogging or brisk walking) and non-aerobic (e.g. strength training, coordination, and flexibility training) exercise on panic disorder, agoraphobia without panic attacks, social phobia, and GAD (Martinsen, Hoffart, & Solberg 1989). Both groups achieved significant reductions in anxiety, while no differences were found between aerobic and non-aerobic groups. This is especially interesting given that significant increases in cardiorespiratory fitness (VO_2 max) only occurred for the aerobic exercise group, suggesting that reductions in anxiety following exercise may not be dependent upon changes in aerobic fitness.

In one of the few RCTs aimed at investigating the mode of exercise, Herring et al. (2012) conducted a RCT to compare the effects of 6 weeks of aerobic versus resistance exercise on remission and worry symptoms among anxiety patients. Thirty sedentary women with a primary DSM-IV diagnosis of GAD were randomly assigned to either an aerobic exercise program, a resistance exercise program, or to a wait-list control condition. The resistance exercise program involved twice-weekly sessions of moderate-intensity lower-body weightlifting while the aerobic exercise program included equivalent sessions of moderate-intensity stationary cycling. Participants showed improved remission rates following resistance exercise relative to aerobic exercise and the wait-list condition. These findings suggest that resistance exercise equivalent to a low dose of aerobic exercise may be a feasible, low-risk treatment for women with GAD. At least four additional RCTs have been conducted to examine the effects of resistance exercise on symptoms of anxiety in healthy adults, with a majority of these trials being performed in older adults (Herring et al. 2013). A small meta-analysis of these RCTs resulted in an overall *ES* of 0.54, suggesting a moderate effect of resistance exercise that is comparable to effects typically observed following aerobic exercise programs. However, due to the small number of studies available, important moderating analyses of mode and dose-response program variables were not conducted. Preliminary evidence from these studies and meta-analyses may aid in the development of future RCTs to elucidate the optimal dose of resistance exercise for anxiety.

Meta-Analytic Evidence

One of the earliest meta-analyses examining the effects of exercise on anxiety was conducted by Petruzzello and colleagues (1991), who reviewed the exercise-anxiety literature for state, trait, and psychophysiological correlates of anxiety. Results for state anxiety revealed that exercise was associated with small reductions in anxiety symptoms (*ES* = 0.24) similar to other known anxiety-reducing treatments (e.g. relaxation). There were no differences observed between acute and chronic exercise (*ES* = 0.23 and 0.25, respectively) or for exercise intensity. There was a significant difference between aerobic (n=173) and non-aerobic (n=13) exercise, indicating that aerobic exercise (*ES* = 0.26) may be more effective at reducing anxiety than resistance exercise (*ES* = −0.05); however, there were too few effect sizes available for resistance exercise programs to make any definitive conclusions. Analyses of trait anxiety revealed an overall mean *ES* of 0.34, indicating that chronic exercise was associated with a reduction in self-reported trait anxiety. Regarding program length, studies had to exceed 10 weeks before significant changes in trait anxiety occurred, with programs lasting greater than 16 weeks resulting in the largest effects. Additionally, there were no differences found for exercise intensity, suggesting that reductions in anxiety were similar for lower and higher intensities of exercise. Furthermore, and similar to the acute exercise literature, effect sizes were larger for aerobic (n = 51, *ES* = 0.36) relative to non-aerobic (n = 2, *ES* = −0.16) exercise. Observations of psychophysiological correlates of anxiety revealed an overall mean *ES* of 0.56, indicating that exercise was associated with meaningful reductions in autonomic and peripheral nervous system measures of anxiety. Importantly, the most consistent variable across all three studies was exercise duration, indicating that exercise sessions of at least 21 minutes are necessary to achieve reductions in state and trait levels of anxiety; however, it is essential for future studies and reviews to consider the duration of exercise sessions in concert with the intensity of activity.

More recently, a meta-analysis by Wipfli, Rethorst, and Landers (2008) examined evidence from RCTs (n=49), and found a moderate effect

(*ES* = -0.48) of exercise on anxiety. Exercise also resulted in greater reductions in anxiety relative to known anxiety-reducing treatments (e.g. CBT, group therapy, relaxation/meditation, stress management education) and larger effects were shown for participants who exercised 3–4 times per week compared to those who exercised 1–2 times or 5 or more times per week. Dose-response evidence suggested larger effects for exercise that approached a dose of 12.5 kcal/kg/week, with reduced efficacy as exercise dose increased beyond this amount. This optimal dose is slightly lower than current public health recommendations by the ACSM, indicating that individuals who suffer from more severe forms of anxiety or are less inclined to exercise may warrant a lower dose of exercise. In contrast, Conn (2010) reported larger *ES*s following moderate- or high-intensity physical activity programs than for studies that incorporated lower intensity exercise, although these findings might only occur for supervised exercise programs. Clearly more work is needed before firm empirically based mode and dose-response recommendations for anxiety can be established.

Evidence for Exercise and Cognitive Functioning

The research literature on the effects of exercise on cognition has increased dramatically over the past several decades. In general, both meta-analytic and narrative reviews support the conclusion that exercise results in small but significant benefits for cognitive and brain health (Etnier et al. 1997). Research examining important exercise program characteristics (e.g. mode, intensity) on cognition is significant on several fronts. First, there is a link between cognitive functioning and various mental health disorders. In fact, one criterion for diagnosing MDD is cognitive dysfunction, and deficits in cognitive functioning may further contribute to disability and poor quality of life among individuals with depression or anxiety. Second, individuals across the globe are living longer. In the United States, the combination of the elevated birth rate during the "baby boom" period and increased life expectancy through medical advances is expected to result in a doubling of the population aged 65 years and older from 2010 to 2040, with the number of persons aged 65 years and older increasing from 40 million in 2010 to 81 million in 2040 (Centers for Disease Control and Prevention 2003). This is especially troubling considering individuals of advancing age are at increased risk of age-related cognitive decline and Alzheimer's disease. However, there is heterogeneity in this age-related decline with some individuals maintaining healthy cognition throughout most of their adult lives. This evidence on individual differences has bolstered research efforts aimed at identifying lifestyle behaviors that might reduce the normal pattern of age-related cognitive decline. Exercise is one promising solution that aligns with the current evidence for depression and anxiety. Despite several decades of research demonstrating that physical activity and exercise are related to improvements in various aspects of cognition, there is a reluctance among health care providers to embrace exercise as a prevention and treatment strategy for cognitive decline (Nagamatsu et al. 2014). Clarifying the role of exercise program variables in the complex exercise-cognition relationship may help to advance its use as an effective behavioral strategy for enhancing brain health.

Aerobic Exercise Evidence

There is now a consistent and considerable body of evidence that includes both human and animal studies that demonstrates the cognitive-enhancing benefits of exercise (Hillman, Erickson, & Kramer 2008). A number of large prospective studies have also shown significant associations between exercise and physical activity participation with reduced risk for cognitive impairment and dementia. Importantly, several recent RCTs have shown beneficial effects of aerobic exercise on cognitive function, which have been accompanied by changes in a number of behavioral, endocrinological, and neurophysiological outcomes. Understanding these outcomes allow for better study designs targeting mechanisms underlying these beneficial effects. In a large-scale RCT to determine whether physical activity improves cognitive function in individuals at risk for Alzheimer's disease, Lautenschlager and colleagues (2008) randomly assigned older adults with memory problems to a 24-week home-based exercise program or to an education group. The physical activity intervention encouraged participants to perform at least 150 minutes of moderate-intensity physical activity split into three 50-minute sessions per week, which is in line with current physical activity recommendations. Most participants chose walking or other forms of aerobic exercise as their preferred mode; however, a small number (n=12) also engaged in resistance exercise in combination with their aerobic activity. Improved scores on the Alzheimer Disease Assessment Scale-Cognitive Subscale (ADAS-Cog) were found following the 6-month exercise program, and these improvements persisted for another 12 months following the intervention. This study suggests that current recommendations for physical activity may be useful for attenuating cognitive decline, even among those at greatest risk. In order to better understand the underlying mechanisms associated with age-related memory loss, Erickson et al. (2011) examined the effects of exercise on hippocampal volume in 120 older adults randomly assigned to either a 1-year moderate-intensity aerobic exercise group or to a control group of stretching and toning exercises. The *hippocampus* is a brain region necessary for various types of learning and has been linked with memory and cognition in both human and animal studies (Shors, Olson, Bates, Selby, & Alderman 2014). Participants in the aerobic exercise group participated in a supervised walking program for 40 minutes 3 days per week at an intensity of 60–75% of HRR. Participants in the stretching control group demonstrated a 1.4% decline in hippocampal volume over the 1-year trial while those assigned to the aerobic exercise intervention showed an increase by 2%, which was also associated with improvements in memory function. Because hippocampal volume shrinks by approximately 1–2% annually in older adults, the authors concluded that a 2% increase in hippocampal volume through exercise is enough to offset age-related loss in hippocampal volume by approximately 1–2 years.

It has been challenging to establish dose-response relationships for exercise and various mental health outcomes, and this is especially true for cognitive function. Studies with high internal validity and appropriate comparison groups (i.e. RCTs) provide the most compelling evidence when trying to establish these relationships. Unfortunately, there is a general lack of conclusive research in this area, making it difficult to advance recommendations for exercise prescription. In the only RCT to date that has examined a dose-dependent effect of exercise on cognitive function, Greer, Grannemann, Chansard, Karim, & Trivedi (2015) examined the role of exercise as an augmentation strategy for depression. In particular, they evaluated the effectiveness of exercise on cognitive function in patients with MDD who reported cognitive impairments following treatment with a SSRI. Participants who enrolled in the TReatment with Exercise Augmentation for Depression (TREAD) trial were randomized to receive either a low (4 kcal/kg/week) or high (16 kcal/kg/week) dose of aerobic exercise for 12 weeks. Using a combined supervised and home-based protocol (to reduce participant burden), participants exercised at a self-selected moderate-intensity on a treadmill or cycle ergometer. The high dose of exercise was selected based on current physical activity recommendations (Haskell et al. 2007) and is equivalent to walking 4.0 mph for 210 minutes per week while the low dose is equivalent to walking approximately 3.0 mph for 75 minutes per week. Participants completed the Cambridge Neuropsychological Test Automated Battery (CANTAB) before and after the intervention to assess attention, visual memory, executive function, and working memory. Improvements in spatial working memory were observed following the high dose of exercise, while both doses of exercise improved psychomotor speed, visuospatial memory, and executive functioning. Both doses of exercise also reduced depressive symptom severity, although these improvements in mood were independent of the changes in cognition. This study suggests a dose-response effect of exercise on specific cognitive tasks among depressed persons who partially responded to SSRI treatment, with some cognitive functions improving regardless of exercise dose. These findings support a beneficial role of exercise when combined with SSRI treatment for reducing cognitive impairments, particularly when conducted at a public health dose (i.e. 30 minutes of moderate-intensity aerobic exercise for ≥5 days per week). Additional rigorously designed RCTs are needed to specifically evaluate the dose-response relationship between exercise and cognitive function.

Resistance Exercise Evidence

Considerable evidence supports the role of aerobic exercise for enhancing cognition and attenuating age-related cognitive decline (Brisswalter, Collardeau, & Rene 2002; Colcombe & Kramer 2003; Etnier et al. 1997), and given the benefits for physical and mental health, recent studies have begun to examine the influence of resistance exercise on cognitive functioning. Colcombe and Kramer (2003) published a meta-analysis of chronic exercise interventions among older adults and found general and selective benefits of exercise on cognition, such that all types of cognition were favorably impacted by exercise but the largest effects were found on executive function tasks. This review also supported the use of combined exercise training programs (e.g. resistance and aerobic exercise) relative to those that only included aerobic exercise. Indeed, a more recent review of the resistance exercise literature supported a facilitating effect of resistance exercise on various aspects of cognitive function in healthy older adults (Chang, Pan, Chen, Tsai, & Huang 2012). Although the evidence supporting various modes of exercise, corresponding dose, and intensity considerations stem from the acute exercise literature (Chang, Labban, Gapin, & Etnier 2012), several recent RCTs examining these important program design variables have been conducted, which may help to advance more precise resistance exercise prescription guidelines.

Ansai and Rebelatto (2014) randomly assigned 69 sedentary adults aged 80 years and older to a high-intensity resistance exercise, multicomponent exercise, or control condition for 16 weeks. The resistance exercise consisted of six exercises (leg press, chest press, calf raises, back extension, abdominals, and rowing) performed for 3 sets of 10–12 repetitions with the aim of reaching fatigue by the third set. The multicomponent group performed aerobic exercise on a cycle ergometer, resistance and balance exercises, and a cool-down stretch period. No significant differences were found following the intervention between the three groups on depressive symptoms or cognition, suggesting that either these training programs were ineffective for older adults or that cognition as assessed in this study was relatively stable and unalterable by exercise. Strassnig and colleagues (2015) recently conducted an innovative RCT to examine the feasibility and efficacy of an 8-week high-velocity circuit resistance training program in patients with schizophrenia and bipolar disorder. A sample of 12 overweight or obese community-dwelling patients with schizophrenia (*n*=9) and bipolar disorder (*n*=3) completed a circuit-based resistance training program twice a week that included the following exercises: leg press, calf raise, leg curl, chest press, overhead press, lat pull-down, hip adduction and abduction, seated row, arm curl, and triceps extension. Participants performed 3 circuits of 10–12 repetitions for each of the eleven exercises, moving from machine to machine with minimal recovery time. Participants manipulated the velocity of movement by performing the concentric phase of each exercise as fast as possible and the eccentric phase for two seconds. In addition to improvements in clinical symptoms of depression and psychotic symptoms that are prevalent in bipolar disorder and schizophrenia, improvements in working memory and processing speed were observed following the 8-week intervention. Although this type of an exercise program would be contraindicated for most beginners, particularly those with a current diagnosis of mental illness, it does provide preliminary evidence for the efficacy of higher intensity resistance exercise programs for cognitive and mental health.

It is also possible that aerobic and resistance exercise programs might result in selective or unique effects on cognition. A recent RCT by Iuliano and colleagues (2015) examining multiple exercise modes on cognitive and attentional processes found that a 12-week program of high-intensity aerobic exercise improved performance on tasks measuring selective and sustained attention (Attentive Matrices) and general intelligence and reasoning (Raven's Progressive Matrices). Resistance training, on the other hand, resulted in improvements in praxis, the ability to synthesize and sequence motor tasks, as assessed by time to complete the Drawing Copy Test. If different modes of exercise result in select benefits for cognitive function, a strong case might be made for prescribing integrative or multimodal exercise programs. This idea was tested in a sample of 109 patients with dementia who were residing in a geriatric nursing home (Bossers et al. 2015). Patients were assigned to either 9 weeks of a combined aerobic and resistance exercise group, an aerobic exercise-only group, or to a social interaction group. Each condition consisted of 36 30-minute sessions, with participants in the combined group completing two strength and two walking sessions per week and those in the aerobic exercise-only group completing four walking sessions. Strength training consisted of 3 sets of 8–12 repetitions for four different lower body exercises while the aerobic exercise entailed walking indoors or outdoors at moderate-to-vigorous intensity. Overall, the combined exercise program was more effective than the aerobic exercise program for enhancing cognitive and motor function. Future studies aimed at combining exercise modalities (e.g. Garcia-Soto, Lopez, & Santibanez 2013) or integrating exercise with mental or cognitive training (e.g. Barnes et al. 2013) are warranted and might be best for enhancing cognitive and neural plasticity.

Meta-Analytic Evidence

Meta-analytic reviews have also been conducted to better understand the effects of exercise on cognitive function and identify potential dose-response relationships. An early meta-analysis found a small but significant effect of exercise on cognitive performance across individuals aged 4–94 years, $ES = 0.25$ (Etnier et al. 1997). Moderating analyses failed to indicate a significant influence of mode, intensity, duration, frequency, or length of exercise program. Thus, little evidence was available at the time to support a dose-response relationship between exercise and cognition. Sibley and Etnier (2003) examined physical activity and cognition in children and found an overall $ES = 0.32$, indicating a facilitating effect of physical activity. Intensity and exercise bout duration were not assessed, but mode of exercise did not have a significant moderating influence on the findings. Conversely, a meta-analysis by Colcombe and Kramer (2003) suggests that exercise bout durations lasting 31–45 minutes ($ES = 0.61$) are more beneficial compared to bouts lasting either 15–30 minutes (*ES* = 0.18) or 46–60 minutes (*ES* = 0.47). As mentioned, they also reported that aerobic-only exercise programs demonstrated smaller effects than combined programs consisting of both aerobic and strength training exercises. Collectively, these findings suggest that a dose-response relationship may exist in older adults while a dose-response relationship has not yet been established in the general population and for children. Findings from experimental studies using human and animal models allow us to identify specific mechanisms that may explain the connection between exercise and cognitive function and allow us to prescribe exercise targeted at enhancing cognition. Likely mechanisms include changes in aerobic fitness, sleep quality, depression and anxiety status, and self-efficacy. Candidate biological mechanisms include changes in brain structure (e.g. grey and white matter increases) and function (e.g. enhanced neural networks) that could be impacted by changes in cerebral blood flow, brain glucose metabolism, neurotransmitter concentrations, and neurotrophic factors (e.g. BDNF, IGF-1), which are all known to be influenced by exercise (Etnier 2009). As with the general recommendations for depression and anxiety, it is important for future studies to systematically manipulate the dose of exercise in order to advance our understanding and guide effective policy decisions and recommendations.

Mind-and-Body Exercise for Health

Recent physical activity guidelines for adults have supported the idea of combining aerobic and resistance exercise for mental health (Garber et al. 2011). Growing evidence also suggests that both physical and mental activity can improve cognitive function and mental health, thus several investigators are exploring the possibility of synergistic or additive effects of exercise and mental training (Alderman, Olson, Brush, & Shors 2016; Barnes et al. 2013; Shors et al. 2014). Indeed, an ongoing investigation examining the combined effects of mental and physical (MAP) training on mental health has yielded promising results. MAP training consists of 8 weeks of focused-attention meditation (mental) and traditional moderate-intensity aerobic exercise (physical). This intervention has shown positive results across three different study populations, including healthy and clinically depressed adults (Alderman et al. 2016) and young women who were recently homeless; suffered poverty, trauma, and abuse; and displayed addictive behaviors (Shors et al. 2014). Although no comparison has yet been made between the MAP training intervention and the individual components of meditation and aerobic exercise, this type of integrative approach cuts across traditional boundaries and represents a new frontier for designing behavioral interventions to maximize cognitive and mental health benefits. Although these results are promising, further large-scale RCTs are needed to establish the benefits of these multimodal programs over those that may be accrued from participating in exercise or mental training alone. Also, as mentioned earlier in the chapter, more people from around the world are engaging in a variety of non-traditional physical activities not classically reported in the scientific literature. Yoga, tai chi, and qigong have been used for thousands of years in Eastern cultures (e.g. India and China) yet have only recently begun to receive scientific attention in the West. We will briefly outline findings from the literature, since they may have implications for the prescription of exercise for treating mental health disorders.

Several recent systematic reviews have been conducted to summarize the benefits of yoga for cognition and mental health (Cramer et al. 2013; Gothe & McAuley 2015; Li & Goldsmith 2012). Gothe and McAuley evaluated 15 RCTs for the effects of acute and chronic yoga on cognition. Yoga was found to have the strongest effects for attention and processing speed, followed by executive function and memory. Despite the limited number of available studies and the variety of doses used, it was concluded that yoga is associated with modest improvements in cognitive function. Recent research has also shown favorable results of yoga practice on depression and anxiety. Li and Goldsmith (2012) suggested that due to its relative safety, low cost, and high compliance rate, yoga may serve as an effective adjunctive therapy in the treatment of anxiety and stress-related illnesses. A recent review of RCTs by Cramer et al. (2013) also supported yoga as a viable treatment option for individuals with depression. To date, a majority of studies examining alternative forms of exercise have focused on yoga, but recently there has been a shift towards understanding other mind-and-body practices. One form of exercise with known meditative properties is tai chi, which has shown promising results for mental and physical health. Chou and colleagues (2004) examined the effects of a 3-month tai chi program among older adults reporting high levels of depressive symptoms. Seven participants were randomly assigned to the tai chi training performed three times per week for 45-minute sessions, while seven subjects were assigned to a wait-list control group. Depressive symptoms were significantly reduced following the tai chi program (~52% reduction) compared to the control condition (~20% increase) and even though a very modest sample size was used, these results are suggestive of the potential for tai chi as a treatment for depression. Similar studies have corroborated these findings. Lee, Lee, and Woo (2009) randomized 139 nursing home residents into a 26-week tai chi program or to a control group, where participants continued their typical daily routine. The tai chi program was conducted by a trained certified instructor in a group setting and took place three times per week with each session lasting approximately one hour. Participants in the tai chi program showed significant improvements in mental and physical health, including reductions in symptoms of depression relative to the control group. Recent meta-analytic evidence also suggests tai chi may provide antidepressant and anxiolytic effects. Wang et al. (2014) assessed numerous RCTs and quasi-experimental studies and reported that tai chi is beneficial for both depression and anxiety outcomes compared to usual treatments (e.g. medication, education and wait-list controls, sham exercise). Despite these promising results, an important limitation within this field is the lack of specific details regarding the dose of tai chi. It is generally believed that most common practices of tai chi are similar in exertion to moderate-intensity aerobic exercise (Lan, Chen, & Lai 2004), yet intensity remains to be determined. Doing so might allow for comparisons with more traditional exercise and potentially yield information that could be used in formulating recommendations. Similar to yoga and tai chi practice, qigong has recently increased in popularity and has been shown to result in favorable health benefits. One of the few studies to assess the effects of qigong on cognitive function found significant improvements in self-reported cognitive function, as measured by the Functional Assessment of Cancer Therapy-Cognitive Function and European Organization for Research and Treatment of Cancer CF questionnaires, following 10 weeks of qigong exercise in cancer patients (Oh et al. 2012). Qigong has also been examined in relation to depression and anxiety disorders. Wang and colleagues (2013) summarized 10 RCTs and reported a beneficial effect of qigong exercise on depressive symptoms when compared to wait-list controls or TAU, as well as evidence that it might even be superior to conventional aerobic exercise. Based on the current evidence, the data did not suggest a beneficial effect of qigong on anxiety symptoms. In contrast, a recent meta-analysis found small-to-moderate benefits of qigong for reducing symptoms of depression and anxiety (Yin & Dishman 2014). Many studies that have examined mind-and-body practices vary widely in the dosage and intensity, making it difficult to compare results across studies. The interventions typically range anywhere from 4 weeks to 4 months in duration with sessions lasting between 30 to 120 minutes. The total number of sessions per week also vary considerably between studies, with some participants meeting as little as once per week and others meeting five times per week. Future research should address these important methodological discrepancies and focus on determining the proper dosage of these mind-and-body exercises for promoting overall health and well-being.

General Recommendations

The goal of this chapter was to outline the current evidence regarding mode and dose-response considerations in the prescription of exercise for mental health. It is difficult to derive specific exercise prescription guidelines by combining the findings across the separate domains of depression, anxiety, and cognitive function. Although it might be argued that the treatment or enhancement for each of these areas is unique, there is considerable comorbidity and overlap between them. Thus, general recommendations for enhancing overall mental health through exercise might be justifiable. Relative to the mode of exercise, it is clear that aerobic exercise is effective at promoting mental and cognitive health. The majority of RCTs and meta-analyses support the beneficial effects of aerobic exercise for depression, anxiety, and cognitive function. There is less available research on the use of resistance exercise, but there is sufficient evidence supporting improvements in depression and emerging data for anxiety and cognitive functioning to support combining aerobic and resistance exercise. Current recommendations by the ACSM and the Centers for Disease Control and Prevention (CDC) for maintaining physical and mental health include aerobic and resistance exercise sessions performed five or more days per week. Given the evidence, it seems practical at the current time to endorse these recommendations. It is also important to realize that many people around the world are engaging in various forms or types of mind-and-body activities that have received less scientific attention and are not typical of what has been emphasized for decades in the United States and elsewhere. Many of these mind-and-body practices include a large physical activity component and may be as effective in the treatment of depression and anxiety and in enhancing cognition as more traditional modes of exercise. Many of these mind-and-body practices also include a meditative or mental training component, which has been shown to be efficacious and beneficial for clinical populations (Lutz et al. 2009). Thus, the combined effects of these mind-and-body practices may prove to be particularly effective for some. Future research should address these complementary and alternative forms of exercise. Relatedly, given that the goal is to promote cognitive and mental health, combining exercise with other empirically supported cognitive and mental training programs warrants further investigation (Chapter 7, this volume). Various cognitive and video game training programs have already been shown to improve cognition (Anguera et al. 2013) and recent studies have suggested that neurobehavioral training might be effective for depression and anxiety (Siegle, Ghinassi, & Thase 2007). These cognitive training approaches may have additive or synergistic effects when combined with exercise, as a neuroscientific basis for their combination has been established (Alderman et al. 2016; Shors et al. 2014).

The available dose-response evidence (i.e. frequency, intensity, and duration) is less clear and is predominantly driven through meta-analytic findings (i.e. moderating analyses). There is a problem in relying too exclusively on this information to formulate our exercise prescription guidelines. First, these moderating analyses are often conducted with very small or an unequal number of effects, making the results tentative relative to the overall *ES*. Second, a recent practice has been to only include RCTs in systematic reviews and this often results in overly strict inclusion/exclusion criteria and a small number of RCTs included in the review. Several recent meta-analyses have included five to seven RCTs in their reviews, which are not sufficient enough to conduct meaningful analyses and draw conclusions about important program design variables. This is unfortunate given that several examples of meta-analyses exist which were later contradicted by large RCTs as well as meta-analyses addressing the same issue and reaching opposite conclusions (LeLorier, Gregoire, Benhaddad, Lapierre, & Derderian 1997). Despite these limitations, several conclusions may be proposed based on the large and growing number of meta-analyses that have synthesized the mental health benefits of exercise. First, regarding intensity and total volume of exercise, a minimum threshold might exist that must be exceeded prior to experiencing mental health benefits. Currently, this threshold is not well understood, especially as it relates to individual differences (e.g. fitness level, age and gender, genetic factors). Second, given current recommendations from the ACSM and the American Heart Association (AHA) for the promotion of health (Haskell et al. 2007), and the fact that evidence from RCTs and meta-analyses do not contradict these recommendations, it seems reasonable to support the current recommendations of combining aerobic and resistance exercise for 20–60 minutes on most or all days of the week. This exercise should be performed at moderate-to-vigorous intensities, although it is possible that lower intensities of exercise might provide benefits for some. In order to observe changes in depression and anxiety, as well as improvements in cognitive function, exercise training programs should be adhered to for at least 8 weeks and perhaps for 6 months or longer. Unless exercise is specifically being used to treat anxiety and depressive disorders, or to enhance cognitive function in the short term, it should be viewed as a lifestyle modification and practiced daily throughout life. Although the body of evidence supporting the use of exercise as a treatment for depression and anxiety and for age-related cognitive decline is growing, the clinical prescription of exercise as an alternative or augmentation therapy to other established forms of treatment is needed. It is also important to conduct large-scale, rigorous RCTs to examine dose-response relationships of exercise on these mental health outcomes in order to determine precise guidelines for the optimal type, frequency, intensity, and duration of exercise for clinical and nonclinical populations. Lastly, in order to advance our understanding of the effects of exercise on psychological health outcomes and to enhance the global acceptance of exercise, there needs to be "informed, balanced, and critical evaluations" of the current exercise and mental health literature (Ekkekakis 2013).

References

Alderman, B. L., R. L. Olson, C. J. Brush, & T. J. Shors. 2016. "Map training: Combining meditation and aerobic exercise reduces depression and rumination while enhancing synchronized brain activity." *Translational Psychiatry 6*: e726.

American College of Sports Medicine. 2013. *ACSM's Guidelines for Exercise Testing and Prescription*. 9 ed. Philadelphia, PA: Wolters Kluwer Health/Lippincott Williams and Wilkins.

Anguera, J. A., J. Boccanfuso, J. L. Rintoul, O. Al-Hashimi, F. Faraji, J. Janowich,...A. Gazzaley. 2013. "Video game training enhances cognitive control in older adults." *Nature 501* (7465): 97–101.

Ansai, J. H., & J. R. Rebelatto. 2014. "Effect of two physical exercise protocols on cognition and depressive symptoms in oldest-old people: A randomized controlled trial." *Geriatrics and Gerontology International.*

Babyak, M., J. A. Blumenthal, S. Herman, P. Khatri, M. Doraiswamy, K. Moore,...K. R. Krishnan. 2000. "Exercise treatment for major depression: Maintenance of therapeutic benefit at 10 months." *Psychosomatic Medicine 62* (5): 633–8.

Barnes, D. E., W. Santos-Modesitt, G. Poelke, A. F. Kramer, C. Castro, L. E. Middleton, & K. Yaffe. 2013. "The mental activity and exercise (max) trial: A randomized controlled trial to enhance cognitive function in older adults." *Journal of the American Medical Association Internal Medicine 173* (9): 797–804.

Blumenthal, J. A., M. A. Babyak, K. A. Moore, W. E. Craighead, S. Herman, P. Khatri,...K. R. Krishnan. 1999. "Effects of exercise training on older patients with major depression." *Archives of Internal Medicine 159* (19): 2349–56.

Bossers, W. J., L. H. van der Woude, F. Boersma, T. Hortobagyi, E. J. Scherder, & M. J. van Heuvelen. 2015. "A 9-week aerobic and strength training program improves cognitive and motor function in patients with dementia: A randomized, controlled trial." *American Journal of Geriatric Psychiatry 23* (11): 1106–16.

Bratman, G. N., J. P. Hamilton, K. S. Hahn, G. C. Daily, & J. J. Gross. 2015. "Nature experience reduces rumination and subgenual prefrontal cortex activation." *Proceedings of the National Academy of Sciences 112* (28): 8567–8572.

Brisswalter, J., M. Collardeau, & A. Rene. 2002. "Effects of acute physical exercise characteristics on cognitive performance." *Sports Medicine 32* (9): 555–66.

Centers for Disease Control and Prevention. 2003. "Trends in aging – United States and worldwide." *Morbidity and Mortality Weekly Report 52* (6): 101–4, 106.

Chang, Y. K., J. D. Labban, J. I. Gapin, & J. L. Etnier. 2012. "The effects of acute exercise on cognitive performance: A meta-analysis." *Brain Research 1453*: 87–101.

Chang, Y. K., C. Y. Pan, F. T. Chen, C. L. Tsai, & C. C. Huang. 2012. "Effect of resistance-exercise training on cognitive function in healthy older adults: A review." *Journal of Aging and Physical Activity 20*: 497–501.

Chou, K. L., P. W. H. Lee, E. C. S. Yu, D. Macfarlane, Y. H. Cheng, S. S. C. Chan, & I. Chi. 2004. "Effect of Tai Chi on depressive symptoms amongst Chinese older patients with depressive disorders: A randomized controlled trial." *International Journal of Geriatric Psychiatry 19* (11): 1105–7.

Chu, I. H., J. Buckworth, T. E. Kirby, & C. F. Emery. 2009. "Effect of exercise intensity on depressive symptoms in women." *Mental Health and Physical Activity 2* (1): 37–43.

Clarke, T. C., L. I. Black, B. J. Stussman, P. M. Barnes, & R. L. Nahin. 2015. "Trends in the use of complementary health approaches among adults: United States, 2002–2012." *National Health Statistics Reports 79*: 1–16.

Colcombe, S., & A. F. Kramer. 2003. "Fitness effects on the cognitive function of older adults: A meta-analytic study." *Psychological Science 14* (2): 125–30.

Conn, V. S. 2010. "Depressive symptom outcomes of physical activity interventions: Meta-analysis findings." *Annals of Behavioral Medicine 39* (2): 128–38.

Craft, L. L., & D. M. Landers. 1998. "The effect of exercise on clinical depression and depression resulting from mental illness: A meta-analysis." *Journal of Sport and Exercise Psychology 20* (4): 339–57.

Cramer, H., R. Lauche, J. Lanhorst, & G. Dobos. 2013. "Yoga for depression: A systematic review and meta-analysis." *Depression and Anxiety 30* (11): 1068–83.

DeRubeis, R. J., G. J. Siegle, & S. D. Hollon. 2008. "Cognitive therapy versus medication for depression: Treatment outcomes and neural mechanisms." *Nature Reviews Neuroscience 9*: 788–96.

Dohrenwend, B. P., & B. S. Dohrenwend. 1974. "Psychiatric disorders in urban settings." In G. Caplan (Ed.), *American handbook of psychiatry* (pp. 424–47). New York: Basic Books.

Dunn, A. L., M. H. Trivedi, J. B. Kampert, C. G. Clark, & H. O. Chambliss. 2005. "Exercise treatment for depression: Efficacy and dose response." *American Journal of Preventive Medicine 28* (1): 1–8.

Dunn, A. L., M. H. Trivedi, & H. A. O'Neal. 2001. "Physical activity dose-response effects on outcomes of depression and anxiety." *Medicine and Science in Sports and Exercise 33* (6 Suppl): S587–597; discussion 609–510.

Ekkekakis, P. 2013. "Physical activity as a mental health intervention in the era of managed care: A rationale." In P. Ekkekakis (Ed.), *Routledge handbook of physical activity and mental health* (pp. 1–32). New York: Routledge.

Ensari, I., T. A. Greenlee, R. W. Motl, & S. J. Petruzzello. 2015. "Meta-analysis of acute exercise effects on state anxiety: An update of randomized controlled trials over the past 25 years." *Depression and Anxiety 32* (8): 624–34.

Erickson, K. I., M. W. Voss, R. S. Prakash, C. Basak, A. Szabo, L. Chaddock,...A. F. Kramer. 2011. "Exercise training increases size of hippocampus and improves memory." *Proceedings of the National Academy of Sciences 108* (7): 3017–22.

Etnier, J. L. 2009. "Physical activity programming to promote cognitive function: Are we ready for prescription?" In W. Chodzko-Zajko, A. F. Kramer, & L. W. Poon (Eds.), *Enhancing cognitive functioning and brain plasticity* (pp. 159–175). Champaign, IL: Human Kinetics.

Etnier, J. L., W. Salazar, D. M. Landers, S. J. Petruzzello, M. Han, & P. Nowell. 1997. "The influence of physical fitness and exercise upon cognitive functioning: A meta-analysis." *Journal of Sport and Exercise Psychology 19* (3): 249–77.

Fetzner, M. G., & G. J. G. Asmundson. 2015. "Aerobic exercise reduces symptoms of post-traumatic stress disorder: A randomized controlled trial." *Cognitive Behaviour Therapy 44* (4): 301–13.

Firdaus, G., & A. Ahmad. 2014. "Temporal variation in risk factors and prevalence rate of depression in urban population: Does the urban environment play a significant role?" *International Journal of Mental Health Promotion 16* (5): 279–88.

Focht, B. C., & K. F. Koltyn. 1999. "Influence of resistance exercise of different intensities on state anxiety and blood pressure." *Medicine and Science in Sports and Exercise 31* (3): 456–63.

Franz, S. I., & G. V. Hamilton. 1905. "The effects of exercise upon the retardation in conditions of depression." *American Journal of Insanity 62* (2): 239–56.

Garber, C. E., B. Blissmer, M. R. Deschenes, B. A. Franklin, M.J. Lamonte, I.M. Lee,...D.P. Swain. 2011. "American College of Sports Medicine Position Stand. Quantity and quality of exercise for developing and maintaining cardiorespiratory, musculoskeletal, and neuromotor fitness in apparently healthy adults: Guidance for prescribing exercise." *Medicine and Science in Sports and Exercise 43* (7): 1334–59.

Garcia-Soto, E., D. M. M. Lopez, & M. Santibanez. 2013. "Effects of combined aerobic and resistance training on cognition following stroke: A systematic review." *Revista de Neurologia 57* (12): 535–41.

Gelenberg, A. J., M. P. Freeman, J. C. Markowitz, J. F. Rosenbaum, M. E. Thase, M. H. Trivedi, & R. S. Van Rhoads. 2010. "American Psychiatric Association practice guidelines for the treatment of patients with major depressive disorder, Third Edition." *American Journal of Psychiatry 167* (Suppl. 10): 9–118.

Gothe, N. P., & E. McAuley. 2015. "Yoga and cognition: A meta-analysis of chronic and acute effects." *Psychosomatic Medicine 77* (7): 784–97.

Greer, T. L., B. D. Grannemann, M. Chansard, A. I. Karim, & M. H. Trivedi. 2015. "Dose-dependent changes in cognitive function with exercise augmentation for major depression: Results from the Tread Study." *European Neuropsychopharmacology 25* (2): 248–56.

Hallal, P. C., L. B. Andersen, F. C. Bull, R. Guthold, W. Haskell, & U. Ekelund. 2012. "Global physical activity levels: Surveillance progress, pitfalls, and prospects." *Lancet 380*: 247–57.

Haskell, W. L., I. M. Lee, R. R. Pate, K. E. Powell, S. N. Blair, B. A. Franklin,...A. Bauman. 2007. "Physical activity and public health: Updated recommendation for adults from the American College of Sports Medicine and the American Heart Association." *Circulation 116* (9): 1081–93.

Herring, M. P., J. B. Lindheimer, & P. J. O'Connor. 2013. "The effect of exercise training on anxiety." *American Journal of Lifestyle Medicine*. doi:10.1177/1559827613508542

Herring, M. P., T. W. Puetz, P. J. O'Connor, & R. K. Dishman. 2012. "Effect of exercise training on depressive symptoms among patients with a chronic illness: A systematic review and meta-analysis of randomized controlled trials." *Archives of Internal Medicine 172* (2): 101–11.

Hidaka, B. H. 2012. "Depression as a disease of modernity: Explanations for increasing prevalence." *Journal of Affective Disorders 140* (3): 205–14.

Hillman, C. H., K. I. Erickson, & A. F. Kramer. 2008. "Be smart, exercise your heart: Exercise effects on brain and cognition." *Nature Reviews Neuroscience 9* (1): 58–65.

Iuliano, E., A. di Cagno, G. Aquino, G. Fiorilli, P. Mignogna, G. Calcagno, & A. Di Costanzo. 2015. "Effects of different types of physical activity on the cognitive functions and attention in older people: A randomized controlled study." *Experimental Gerontology 70*: 105–10.

Jahnke, R., L. Larkey, C. Rogers, J. L. Etnier, & F. Lin. 2010. "A comprehensive review of health benefits of Qigong and Tai Chi." *American Journal of Health Promotion 24* (6): e1–e25.

Jazaieri, H., P. R. Goldin, K. Werner, M. Ziv, & J. J. Gross. 2012. "A randomized trial of MBSR versus aerobic exercise for social anxiety disorder." *Journal of Clinical Psychology. 68* (7): 715–31.

Kakabadse, N., G. Porter, & D. Vance. 2007. "Addicted to technology." *Business Strategy Review 18* (4): 81–5.

Kesaniemi, Y. A., E. Danforth, M. D. Jensen, P. G. Kopelman, P. Lefebvre, & B. A. Reeder. 2001. "Dose-response issues concerning physical activity and health: An evidence-based symposium." *Medicine and Science in Sports and Exercise 33* (6): S351–S358.

Kessler, R. C. 2012. "The costs of depression." *Psychiatric Clinics of North America 35* (1): 1–14.

Kraemer, W. J., & N. A. Ratamess. 2004. "Fundamentals of resistance training: Progression and exercise prescription." *Medicine and Science in Sports and Exercise 36* (4): 674–88.

Kraemer, W. J., N. A. Ratamess, & D. N. French. 2002. "Resistance training for health and performance." *Current Sports Medicine Reports 1* (3): 165–71.

Kravitz, L., & C. A. Vella. 2002. "Energy expenditure in different modes of exercise." *American College of Sports Medicine Current Comment*. www.acsm.org/docs/current-comments/energyexpenindifferentexmodes.pdf.

Lambert, K. G., R. J. Nelson, T. Jovanovic, & M. Cerda. 2015. "Brains in the city: Neurobiological effects of urbanization." *Neuroscience and Biobehavioral Reviews 58*: 107–22.

Lan, C., S. Y. Chen, & J. S. Lai. 2004. "Relative exercise intensity of Tai Chi Chuan is similar in different ages and gender." *American Journal of Chinese Medicine 32* (1): 151–60.

Lautenschlager, N. T., K. L. Cox, L. Flicker, J. K. Foster, F. M. van Bockxmeer, J. Xiao,...O. P. Almeida. 2008. "Effect of physical activity on cognitive function in older adults at risk for Alzheimer's disease: A randomized trial." *Journal of the American Medical Association 300* (9): 1027–37.

Lee, I. M., & P. J. Skerrett. 2001. "Physical activity and all-cause mortality: What is the dose-response relation?" *Medicine and Science in Sports and Exercise 33* (6): S459–S471.

Lee, L. Y. K., D. T. F. Lee, & J. Woo. 2009. "Tai Chi and health-related quality of life in nursing home residents." *Journal of Nursing Scholarship 41* (1): 35–43.

LeLorier, J., G. Gregoire, A. Benhaddad, J. Lapierre, & F. Derderian. 1997. "Discrepancies between meta-analyses and subsequent large randomized, controlled trials." *New England Journal of Medicine 337* (8): 536–42.

Li, A. W., & C. A. Goldsmith. 2012. "The effects of yoga on anxiety and stress." *Alternative Medicine Review 17* (1): 21–35.

Lutz, A., H. A. Slagter, N. B. Rawlings, A. D. Francis, L. L. Greischar, & R. J. Davidson. 2009. "Mental training enhances attentional stability: Neural and behavioral evidence." *Journal of Neuroscience 29* (42): 13418–27.

Martinsen, E. W., A. Hoffart, & O. Solberg. 1989. "Comparing aerobic with nonaerobic forms of exercise in the treatment of clinical depression: A randomized trial." *Comprehensive Psychiatry 30* (4): 324–31.

Mathers, C. D., D. M. Fat, & J. T. Boerma. 2008. *The Global Burden of Disease: 2004 Update*. World Health Organization.

Morgan, W. P. 1969. "Physical fitness and emotional health: A review." *American Corrective Therapy Journal 23*: 124–7.

Nagamatsu, L. S., L. Flicker, A. F. Kramer, M. W. Voss, K. I. Erickson, C. L. Hsu, & T. Liu-Ambrose. 2014. "Exercise is medicine, for the body and the brain." *British Journal of Sports Medicine 48* (12): 943–4.

Nelson, M. E., W. J. Rejeski, S. N. Blair, P. W. Duncan, J. O. Judge, A. C. King,...C. Castaneda-Sceppa. 2007. "Physical activity and public health in older adults: Recommendation from the American College of Sports Medicine and the American Heart Association." *Circulation 116* (9): 1094–1105.

North, T. C., P. McCullagh, & Z. V. Tran. 1990. "Effect of exercise on depression." *Exercise and Sport Sciences Reviews 18*: 379–415.

Oh, B., P. N. Butow, B. A. Mullan, S. J. Clarke, P. J. Beale, N. Pavlakis,...J. Vardy. 2012. "Effect of medical Qigong on cognitive function, quality of life, and a biomarker of inflammation in cancer patients: A randomized controlled trial." *Supportive Care in Cancer 20*: 1235–42.

Peen, J., R. A. Schoevers, A. T. Beekman, & J. Dekker. 2010. "The current status of urban-rural differences in psychiatric disorders." *Acta Psychiatr Scand 121* (2): 84–93.

Perraton, L. G., S. Kumar, & Z. Machotka. 2010. "Exercise parameters in the treatment of clinical depression: A systematic review of randomized controlled trials." *Journal of Evaluation in Clinical Practice 16* (3): 597–604.

Petruzzello, S. J. 2012. "Doing what feels good (and avoiding what feels bad) – a growing recognition of the influence of affect on exercise behavior: A comment on Williams et al." *Annals of Behavioral Medicine 44* (1): 7–9.

Petruzzello, S. J., D. M. Landers, B. D. Hatfield, K. A. Kubitz, & W. Salazar. 1991. "A meta-analysis on the anxiety-reducing effects of acute and chronic exercise: Outcomes and mechanisms." *Sports Medicine 11* (3): 143–82.

Raglin, J. S., P. E. Turner, & F. Eksten. 1993. "State anxiety and blood pressure following 30 min of leg ergometry or weight training." *Medicine and Science in Sports and Exercise 25* (9): 1044–8.

Regier, D. A., J. H. Boyd, J. D. Burke, D. S. Rae, J. K. Myers, M. Kramer,...B. Z. Locke. 1988. "One-month prevalence of mental disorders in the United States: Based on five epidemiological catchment area sites." *Archives of General Psychiatry 45* (11): 977–86.

Rethorst, C. D., & M. H. Trivedi. 2013. "Evidence-based recommendations for the prescription of exercise for major depressive disorder." *J Psychiatr Pract 19* (3): 204–12.

Rethorst, C. D., B. M. Wipfli, & D. M. Landers. 2009. "The antidepressive effects of exercise: A meta-analysis of randomized trials." *Sports Medicine 39* (6): 491–511.

Saeed, S. A., D. J. Antonacci, & R. M. Bloch. 2010. "Exercise and meditation for depressive and anxiety disorders." *American Family Physician 81* (8): 981–6.

Sexton, H., A. Maere, & N. H. Dahl. 1989. "Exercise intensity and reduction in neurotic symptoms: A controlled follow-up study." *Acta Psychiatr Scand 80*: 231–5.

Shors, T. J., R. L. Olson, M. E. Bates, E. A. Selby, & B. L. Alderman. 2014. "Mental and physical (map) training: A neurogenesis-inspired intervention that enhances health in humans." *Neurobiology of Learning and Memory 115*: 3–9.

Sibley, B. A., & J. L. Etnier. 2003. "The relationship between physical activity and cognition in children: A meta-analysis." *Pediatric Exercise Science 15* (3): 243–56.

Siegle, G. J., F. Ghinassi, & M. E. Thase. 2007. "Neurobehavioral therapies in the 21st century: Summary of an emerging field and an extended example of cognitive control training for depression." *Cognitive Therapy and Research 31* (2): 235–62.

Singh, N. A., K. M. Clements, & M. A. Fiatarone. 1997. "A randomized controlled trial of progressive resistance training in depressed elders." *Journals of Gerontology, Series A: Biological Sciences and Medical Sciences 52* (1): M27.

Singh, N. A., T. M. Stavrinos, Y. Scarbek, G. Galambos, C. Liber, & M. A. Fiatarone Singh. 2005. "A randomized controlled trial of high versus low intensity weight training versus general practitioner care for clinical depression in older adults." *Journals of Gerontology, Series A: Biological Sciences and Medical Sciences 60* (6): 768–76.

Small, G., & G. Vorgan. 2008. *Ibrain: Surviving the technological alterations of the modern mind.* London: Harper Collins.

Stanton, R., & P. Reaburn. 2014. "Exercise and the treatment of depression: A review of the exercise program variables." *Journal of Science and Medicine in Sport 17* (2): 177–82.

Strassnig, M. T., J. F. Signorile, M. Potiaumpai, M. A. Romero, C. Gonzalez, S. Czaja, & P. D. Harvey. 2015. "High velocity circuit resistance training improves cognition, psychiatric symptoms and neuromuscular performance in overweight outpatients with severe mental illness." *Psychiatry Research 229* (1–2): 295–301.

Tremblay, M. S., A. G. LeBlanc, M. E. Kho, T. J. Saunders, R. Larouche, R. C. Colley,...S. C. Gorber. 2011. "Systematic review of sedentary behaviour and health indicators in school-aged children and youth." *Internation Journal of Behavioral Nutrition and Physical Activity 8* (1): 98.

Trivedi, M. H., T. L. Greer, T. S. Church, T. J. Carmody, B. D. Grannemann, D. I. Galper,...S. N. Blair. 2011. "Exercise as an augmentation treatment for nonremitted major depressive disorder: A randomized, parallel dose comparison." *Journal of Clinical Psychiatry 72* (5): 677–84.

Tudor-Locke, C., W. D. Johnson, & P. T. Katzmarzyk. 2009. "Accelerometer-determined steps per day in US adults." *Medicine and Science in Sports and Exercise 41* (7): 1384–91.

Wang, C. W., C. L. W. Chan, R. T. H. Ho, H. W. H. Tsang, C. H. Y. Chan, & S. M. Ng. 2013. "The effect of Qigong on depressive and anxiety symptoms: A systematic review and meta-analysis of randomized controlled trials." *Evidence-Based Complementary and Alternative Medicine 2013*: 716094.

Wang, F., E. K. O. Lee, T. Wu, H. Benson, G. Fricchione, W. Wang, & A. S. Yeung. 2014. "The effects of Tai Chi on depression, anxiety, and psychological well-being: A systematic review and meta-analysis." *International Journal of Behavioral Medicine 21* (4): 605–17.

Wegner, M., I. Helmich, S. Machado, A. E. Nardi, O. Arias-Carrion, & H. Budde. 2014. "Effects of exercise on anxiety and depression disorders: Review of meta-analyses and neurobiological mechanisms." *CNS and Neurological Disorders – Drug Targets 13* (1002–14).

Wipfli, B. M., C. D. Rethorst, & D. M. Landers. 2008. "The anxiolytic effects of exercise: A meta-analysis of randomized trials and dose-response analysis." *Journal of Sport and Exercise Psychology 30* (4): 392.

World Health Organization. 2010. Global Recommendations on Physical Activity for Health.

Yin, J., & R. K. Dishman. 2014. "The effect of Tai Chi and Qigong practice on depression and anxiety symptoms: A systematic review and meta-regression analysis of randomized controlled trials." *Mental Health and Physical Activity 7* (3): 135–46.

17 Acute vs. Chronic Effects of Exercise on Mental Health

Steven J. Petruzzello, Daniel R. Greene, Annmarie Chizewski, Kathryn M. Rougeau, and Tina A. Greenlee

Introduction

The U.S. Surgeon General's Report on Mental Health defines *mental health* as "a state of successful performance of mental function, resulting in productive activities, fulfilling relationships with other people, and the ability to adapt to change and to cope with adversity" (U.S. Department of Health and Human Services 1999, p. 4). By contrast, *mental illness* refers collectively to all diagnosable mental disorders (p. 5), for which mental function, productive activities, and relationships are impaired to some extent. This would include the anxiety disorders, mood disorders, and other less common disorders (e.g. schizophrenia). Mental disorders "are health conditions that are characterized by alterations in thinking, mood, or behavior (or some combination thereof) associated with distress and/or impaired functioning" (U.S. Department of Health and Human Services 1999, p. 4).

Mental illnesses are more severe forms of "mental health problems." The criteria for mental health problems are related to the signs and symptoms that meet the criteria for a mental disorder, but at an insufficient intensity or duration. However, the diagnosis or awareness of a mental health problem is sufficient to lead to active efforts of health promotion, prevention, and treatment. The mental health-mental illness relationship is often conceptualized as a continuum, with mental health at one end and mental illness on the other, with mental health problems falling somewhere between the two endpoints, but clearly closer to mental illness (Leith 1994; see also Chapter 1 of this volume).

The aim of the present chapter is to provide an overview of the acute and chronic effects of exercise on important mental health outcomes like stress, anxiety, depression, cognitive function, and psychological well-being.[1] Each of the sections that follow will provide a fairly brief synopsis along with some exemplar studies of what is currently known about acute and chronic effects of exercise on each of these outcomes along with what we feel are important implications in each of these areas for health care practitioners (e.g. physicians, physician's assistants, nurses and nurse practitioners, personal trainers) as well as implications for those working in primary and secondary school settings (e.g. administrators, teachers).

Stress

Our world is a fast-paced world, and the pace seems to be getting faster all the time. Increasing levels of stress are often a result of living in such a world. The ever-increasing number of stressors leads to physical ailments (e.g. "simple colds," chronic disease, and disability) along with mental ailments. Anxiety and anxiety disorders along with depression have become increasingly prevalent in our society and their causes can almost always be traced to stress that has become unmanageable.

Stress can be caused by many different things, but tends to be more likely psychological than anything else (Tsatsoulis & Fountoulakis 2006). It is worth pointing out that many of the psychological stressors are what Lupien, McEwen, Gunnar, and Heim (2009) refer to as "anticipatory," that is, "they are based on expectation as the result of learning and memory" (p. 436). One self-regulatory strategy that many individuals use as a way of coping with or handling stress is exercise. What follows is a brief synopsis of what is known regarding how exercise influences the stress response both acutely and chronically.

Defining Stress

Stress has been defined as "a state of disharmony, or threatened homeostasis" (Chrousos & Gold 1992, p. 1245). Credit for the field of study is often given to Hans Selye, who in the 1930s borrowed the term *stress* from the field of physics and used it in conjunction with the results of his initial studies (it is worth pointing out that Walter Cannon's work and his coining of the term "homeostasis" actually pre-dated that of Selye). Selye found that animals exposed to unpleasant conditions (stressors) developed a variety of physiological changes, including peptic ulcers, enlarged adrenal glands, and atrophied immune tissues. He noted that despite the varied nature of the stressors, the physiological responses were similar (i.e. nonspecific); animals exposed to stressors for an extended period of time got sick (see Chrousos & Gold 1992 for a nice overview of the historical development of the field). Lupien, Maheu, Tu, Fiocco, and Schramek (2007) pointed out that while Selye examined responses to physical stressors, his contemporary John Mason focused on psychological conditions that would induce a stress response. Mason's work showed that the "determinants of the stress response are highly specific, and therefore, potentially predictable and measurable" (Lupien et al. 2007, p. 210). Notably, the novelty or newness of the stressor, its degree of predictability, and how much perceived control over the stressor the individual has are important in determining the stress response.

Over many millennia, the body (including the brain) has evolved ways to deal with the stressors we face. In essence, the *stress response* is initiated when some real, perceived or expected threat (or challenge) is encountered (e.g. threat of injury, embarrassment, potential loss) (Lupien et al. 2009; see also Sapolsky 2003 for a readable discussion). The initial reaction is often referred to as "arousal and alarm," a cascade of immediate physiological adaptations that are marshalled to deal with the crisis (i.e. the "fight-or-flight" response; see also the work of Taylor et al. 2000 which highlights the "tend and befriend" response and differences between males and females). These evolutionarily ancient responses appear ideally suited for dealing with short-term stressors (driven primarily by the sympathetic nervous system (SNS): "fight-or-flight"), eventually returning the body to its more stable resting state (largely a function of the sympathetic nervous system's counterpart, the parasympathetic nervous system, PNS: "rest-and-digest"). This initial reaction is often accompanied by feelings of anxiety, irritability, and vulnerability until the stressor is resolved or strain, worry, cynicism, and difficulty sleeping if unresolved. If prolonged or chronic with no resolution, the strain leads to fatigue and numerous, insidious stress-related disorders, including anxiety and depression.

Numerous studies have examined self-reports of stress or perceived stress and whether exercise influences these perceptions. People generally report feeling less stress following acute exercise bouts and feeling less stressed in general when they are physically active as opposed to being sedentary (American Psychological Association 2014). Again, this is an apparent paradox as exercise itself is a stressor, yet yields "relief" from psychosocial stressors. Relevant to the present discussion, exercise itself is a stressor; it disrupts homeostasis and invokes both SNS-Adrenal-Medullary system activation (SAM) and Hypothalamic-Pituitary-Adrenocortical system activation (HPA). Acute bouts of exercise have similar physiological effects as psychological stressors (i.e. increased cardiovascular activity, release of stress hormones), but are not detrimental to health like mental stress can be (Hamer & Steptoe 2013). Often viewed as a positive stressor, regular bouts of acute exercise lead to adaptation of physiological systems that are invoked during the stress response. It has been suspected for some time that regular exercise and/or fitness might beneficially impact the stress response. Animal research has shown physically active animals to be more stress-resistant than sedentary animals (Fleshner 2005). The cross-stressor adaptation hypothesis states that exposure to a stressor of sufficient intensity and/or duration will induce adaptation of stress response systems and lead to cross-stressor tolerance (Sothmann et al. 1996). What follows is a brief review of the findings related to how acute and chronic exercise impact stress reactivity.

Acute Exercise and Stress

Although the topic of exercise and its impact on the stress response is clearly important, there has been relatively little work examining acute exercise. Several early studies found reductions in self-reported measures of stress (e.g. anxiety-related thoughts, anticipation of threat; Rejeski, Thompson, Brubaker, & Miller 1992). Much of the work examining acute exercise effects has tended to focus on cardiovascular responses to a psychosocial stressor following a bout of exercise. For example, a meta-analytic review by Hamer, Taylor, and Steptoe (2006) revealed blood pressure (BP) reactivity was modified in an adaptive fashion following acute bouts of exercise (an average reduction in diastolic BP of 3.0±2.7 mmHg and systolic BP of 3.7±3.9 mmHg). In order to examine the effects of acute exercise on cardiovascular responses to psychosocial stressors, Roemmich, Lambaise, Salvy, and Horvath (2009) conducted two experiments with children (8–12 yrs old). Using an interval exercise protocol (30 s cycling at 80 rpm alternating with 30 s cycling at 50 rpm for 6 min) in comparison to watching videos, HR and BP responses were recorded before, during, and after preparation for and presentation of an interpersonal speech. Following either the exercise or video treatment, the participants read for 20 minutes and then performed the psychosocial stress task. The exercise manipulation effectively attenuated diastolic BP (effect sizes = 0.81 – 1.28), demonstrating that the cardioprotective effects of exercise can be extended to children.

Chronic Exercise and Stress

Whereas the previous section discussed the effects of acute bouts of exercise on the stress response, repeated acute bouts of exercise over time (i.e. chronic exercise) may also have positive effects on the stress response. Often, this is due to increases in physical fitness (i.e. cardiovascular fitness) resulting from engagement in such regular exercise. This section details what is known regarding such chronic exercise effects.

An early meta-analysis (Crews & Landers 1987) examined the effects of aerobic fitness on the response to psychosocial stressors to address a straightforward question: Do more aerobically fit individuals have a smaller response to stress than less fit individuals? Crews and Landers concluded that fitness imparted a buffering effect, with more fit individuals having a sizably smaller stress response than unfit individuals. More recent reviews of the fitness-stress reactivity literature suggest more caution is warranted. A meta-analysis conducted by Forcier and colleagues (2006) examined the effects of physical fitness on cardiovascular reactivity (HR, SBP, DBP) and recovery from psychosocial stressors. Their results also favored a fitness effect, but the magnitude of the effect was much less than that reported by Crews and Landers (1987). They estimated that fit individuals have about a 15% to 25% reduction in reactivity relative to unfit individuals, which could have important clinical implications. In contrast to both Crews and Landers and Forcier et al., Jackson and Dishman (2006) concluded that the evidence does *not* support the hypothesis of reduced reactivity to stressors in more fit individuals. They did find evidence for a slightly faster recovery following stressor offset in more fit individuals, but noted that such an effect was rather small.

Given the conclusions from reviews of this literature and in spite of criticisms of longitudinal exercise-training studies, it seems reasonable to conclude that exercise training has stress-buffering effects. Heydari, Boutcher, and Boutcher (2013) showed positive effects of a 12-week high-intensity interval exercise training program on reactivity and recovery from stress using more sophisticated measures of the cardiovascular response than changes in HR or BP (e.g. arterial stiffness, baroreflex sensitivity, muscle blood flow). A study highlighting the more ecologically valid, real-world nature of the fitness–stress relationship was conducted by King, Baumann, O'Sullivan, Wilcox, and Castro (2002). They examined the effect of a year-long exercise intervention (compared to a year-long nutrition education program) on cardiovascular stress reactivity in initially sedentary women who were providing care for a relative with dementia. HR and BP reactivity were assessed at the beginning and end of the intervention in response to a 6-minute period during which participants were prompted to discuss those "aspects of her caregiving experience that she found to be most frustrating or disturbing" (p. M29). Compared to those in the nutrition intervention, participants in the exercise intervention showed significantly smaller blood pressure reactivity (reduction of 4–9 mmHg almost twice that reported in the Forcier et al. (2006) meta-analysis). This was interpreted as an improvement in the relationship between chronic stress and cardiovascular pathology as a result of exercise training.

While most exercise–stress research has tended to focus on cardiovascular and catecholamine reactivity, some research has examined the hypothalamic-pituitary-adrenal cortical (HPA) axis response to psychosocial stressors. Traustadottir, Bosch, and Matt (2005) examined whether adaptations known to occur within the HPA axis in response to physical stress (e.g. blunted cortisol response to exercise performed at the same absolute intensity, greater maximal capacity of the adrenal glands) also occur in response to psychosocial stress (i.e. cross-stressor adaptation). They were also interested in whether aerobic fitness influenced that response. Younger (19–36 yrs old) and older (59–81 yrs old) women performed a battery of psychosocial stressors lasting about 30 minutes, referred to as the Matt Stress Reactivity Protocol (MSRP; consisted of the Stroop Color and Word Test, mental arithmetic, an anagram task stressing time urgency, a cold pressor task, and an interpersonal interview, in that order). The younger women were classified as unfit along with about half of the older women; the remainder were classified as fit. HR, BP, and plasma cortisol were assessed before the protocol, at the conclusion of each of the tasks, and during recovery from the protocol. Significant increases in cardiovascular and endocrine responses were seen, but the endocrine response (i.e. cortisol) was modified by fitness. The older fit women had significantly lower cortisol responses to the stressors than their older unfit counterparts. This difference in response to stress occurred in spite of there being no significant differences at baseline between the groups. It is worth noting that even during the recovery time period, cortisol responses were still elevated above baseline values for the older unfit group. Although the cortisol effects were pronounced, no differences existed for adrenocorticotropin releasing hormone response, nor for any of the cardiovascular measures, between the fit and unfit. Thus, aerobic fitness did blunt the stress response in these older women. Traustadottir, Bosch, and Matt (2005) interpreted this as evidence that aerobic fitness can influence sensitivity to stress, ultimately affecting the HPA axis in the direction of a reduced cortisol response to psychosocial stress (and thus, a reduced allostatic load).

The conclusion reached by Traustadottir and colleagues has been confirmed in at least two studies since then (Rimmele et al. 2009; Rimmele et al. 2007). Thus, the overall findings indicate that physical activity and fitness seem to blunt the deleterious effects of the stress response (e.g. blunted cortisol) and enhance the facilitative effects (e.g. faster SNS response).

Implications for the Health Sector and School

For those who use exercise as a way of dealing with the daily stressors of life, it is probably a good idea to arrange to exercise when it will be most helpful. Thus, a morning workout might help some people to get ready to take on the day's challenges, a noontime workout might provide a much needed break in the day to recharge the batteries, or some might find a workout later in the day a useful method for "purging" the tensions and worries of the day. This is obviously something that will need to be determined by each person on an individual basis. Given the current level of understanding, it seems likely that exercise can be useful in reducing the stress response. Whether this is in terms of reduced reactivity to the stressor itself or through a faster recovery from the stressor remains unclear. Continued careful examination is needed to clarify both aspects of the stress response and the potential buffering effects that exercise might have. Regardless of the eventual findings, exercise is clearly beneficial, reducing the allostatic load and, ultimately, reducing risk of disease and disability.

Anxiety

Anxiety is a broad term that covers a wide variety of mental disorders and everyday feelings. It is important to distinguish anxiety disorders from normal anxiety. It is common to experience feelings of apprehension, worry, tension, nervousness, and even fear. What separates these everyday feelings of anxiety from an anxiety disorder is severity and duration of symptoms (see also Chapter 1). Anxiety disorders involve chronic or long-term feelings of apprehension, worry, nervousness, and tension (U.S. Department of Health and Human Services 1999). Often these feelings can manifest in situations in which there is little or no reason to worry. Not only do these feelings persist, but they are much stronger in nature, often resulting in altered cognition (e.g. irrational thoughts) and behavior (e.g. avoidance).

Recent evidence suggests anxiety disorders are on the rise. The lifetime prevalence rate of anxiety disorders was 24.9 and 28.8 in 1994 and 2005, respectively (Kessler et al. 1994; Kessler et al. 2005). Further, anxiety has been reported as the most commonly diagnosed mental disorder, affecting young adults and women most frequently. Lifetime prevalence rates for women (32.4%) are notably higher than for men (22.4%; Gum, King-Kallimanis, & Kohn 2009).

Traditionally, anxiety disorders have been treated with medication and psychotherapy. While both methods have been successful in reducing anxiety symptoms, medication and psychotherapy exert a heavy burden on health care costs. Also, anti-anxiety and antidepressant medications (i.e. most commonly prescribed for anxiety disorders) can have negative side effect, and provide only temporary symptom relief (i.e. stop working if medications are not taken). Specifically, benzodiazepines and other anti-anxiety medications have been associated with drowsiness, depression, dizziness, impaired thinking, nausea, blurred vision, and sexual dysfunction (Longo & Johnson 2000).

Given the vast burden anxiety disorders have (e.g. negative side effects from medication, expense), researchers have examined physical activity as a potential treatment or preventative measure to combat anxiety. Not only has physical activity shown promise in reducing the severity of anxiety, but it may also act to treat specific symptoms of anxiety (please see Chapter 13 for more detailed information about exercise and anxiety disorders).

Acute Exercise and Anxiety

There is a sizable body of literature examining exercise and anxiety. Several reviews of this literature have been done (e.g. Asmundson et al. 2013; Jayakody, Gunadasa, & Hosker 2014; Wipfli, Rethorst, & Landers 2008), but the most comprehensive contribution to this research was a meta-analysis on the acute and chronic effects of exercise on anxiety (Petruzzello, Landers, Hatfield, Kubitz, & Salazar 1991). This meta-analysis resulted in an effect size (*d*) of .26 for aerobic exercise on state anxiety, indicating that an acute bout of aerobic exercise was associated with ¼ standard deviation reduction in anxiety. An updated meta-analysis was recently performed on only randomized controlled trials. Ensari, Greenlee, Motl, and Petruzzello (2014) reported an effect size of d = .16 for all randomized controlled trials involving exercise and anxiety, including seven studies using resistance exercise. Taken together, both of these meta-analyses provide evidence that acute bouts of exercise have a significant impact on anxiety.

Many studies included in the abovementioned meta-analyses were composed of convenient samples; participants volunteered for the study and were not actively recruited for specific criteria. This means that many studies assessing exercise effects on anxiety are completed with healthy college-aged individuals with normal to low levels of anxiety. Despite this, significant effects are still prevalent but are limited due to a floor effect (i.e. anxiety can only be reduced so much). To combat this, Motl and Dishman (2004) used caffeine pills to elevate precondition anxiety levels. Participants completed four different conditions: cycling with caffeine, cycling with placebo, quiet rest with caffeine, and quiet rest with placebo. Anxiety was significantly increased after caffeine consumption, but was significantly reduced 10 minutes post exercise (d = .34) only (Motl & Dishman 2004). There was no change in anxiety following placebo in either the exercise or quiet rest conditions. Similarly, Petruzzello Snook, Gliottoni, and Motl (2009) assessed changes in state anxiety following 20 minutes of moderate intensity cycling in individuals with multiple sclerosis who were split into high and low trait anxious groups. The high trait anxious group showed significant improvements in anxiety at 5 (d = 1.0), 20 (d = 1.0), and 60 (d = 1.17) minutes post exercise while the low trait anxious group showed a small improvement (d = .29) after 60 minutes only (Petruzzello et al. 2009). While current reviews and meta-analyses support using exercise to combat generalized anxiety disorders, the empirical evidence is highly conservative. By using convenient samples and not clinical populations, the effects of acute bouts of exercise on anxiety may be grossly underrepresented.

Anaerobic Exercise (e.g. Resistance Exercise)

Anaerobic exercise has not received the same level of attention as aerobic exercise with regards to psychological effects (Strickland & Smith 2014). Of the available literature, findings are highly variable and plagued with methodological discrepancies that make systematically reviewing this body of literature difficult. Specifically, Petruzzello et al. (1991) located only three nonaerobic acute exercise studies for their meta-analysis and concluded that there was not a statistically significant effect size for this type of activity. However, included under nonaerobic exercise was a massage condition (Weinberg, Jackson, & Kolodny 1987), and the remaining studies were unpublished works. The meta-analysis by Ensari, Greenlee, Motl, and Petruzzello (2015) included seven resistance exercise studies, with effect sizes ranging from –.27 to .30. Ensari and colleagues identified a number of significant moderators including: well-designed studies (i.e. a Physiotherapy Evidence Database (PEDro) score ≥ 6), high exercise intensity, female-only samples, sedentary/low active participants, and age (>25 yrs). Such moderators become important as resistance exercise studies resulting in low effect sizes suffer from numerous methodological mistakes. For example, Arent, Alderman, Short, and Landers (2007) had the only negative effect size (i.e. d = –.27), but this study did not include females, had an active sample of college-aged individuals (i.e. aged 18–23), had a PEDro score below 6, and had no anxiety manipulation (e.g. high baseline anxiety).

With regards to anaerobic exercise, there is a potential to misinterpret anxiety data immediately post exercise. As exercise itself is a stressor, individuals may experience an increase in tension/activation that could easily be mistaken as an increase in anxiety post exercise (Rejeski, Hardy, & Shaw 1991). However, assessing anxiety at various time points after the completion of exercise may aid in the understanding of resistance exercise effects on anxiety. O'Connor, Bryant, Veltri, and Gebhardt (1993) assessed state anxiety after an acute bout of either high, moderate, or low intensity resistance exercise. Anxiety assessments were made pre, immediately post, and every 15 minutes post exercise up to 120 minutes. Results indicate that anxiety was significantly reduced from pre at 90, 105, and 120 minutes post moderate intensity resistance exercise. This gives evidence that an acute bout of resistance exercise reduces anxiety below baseline, but that this process may not occur immediately post exercise. Very few studies assess changes in anxiety in excess of 1 hour post.

There have been numerous studies showing reductions in anxiety shortly after resistance exercise. Greene and Petruzzello (2014) found a significant reduction in anxiety 20 minutes post moderate intensity resistance exercise (d = .54), with no change from pre to immediately post. There was a significant increase in anxiety immediately post high intensity resistance exercise (M difference = 2.73, d = –0.60), with anxiety returning to baseline at 20 minutes post. Similarly, Bartholomew and Linder (1998) reported a significant reduction in state anxiety at both 15 and 30 minutes following low-intensity resistance exercise, but a significant increase in anxiety at 5 and 15 minutes following high intensity. While Ensari and colleagues (2015) found larger reductions in anxiety following high-intensity exercise, closer examination of the resistance exercise literature gives evidence for larger reductions in anxiety following low to moderate intensity. High-intensity resistance exercise appears to increase anxiety immediately post exercise, but anxiety levels return to baseline between 20–30 minutes post exercise. The effects of high-intensity resistance exercise beyond 30 minutes post exercise remains unclear.

Resistance exercise may also protect against anxiogenic events. Bartholomew (1999) had participants perform either 20 minutes of resistance exercise or an equal time viewing photos, after being exposed to 20 minutes of either negative, positive, or neutral mood manipulations. Similar to previous research, during the neutral mood manipulation participants reported increased anxiety following resistance exercise that returned to baseline at 15 minutes post. Interestingly, anxiety fell below baseline at 30 minutes post exercise (d = .35), and continued to decrease at 45 (d = .22) minutes post exercise; the control condition showed no change at any time point (Bartholomew 1999). After exposure to the negative mood manipulation, both the control and resistance exercise conditions showed a significant increase in state anxiety that did return to baseline by 15 minutes post exercise, but anxiety continued to decrease below baseline at 30 (d = 2.48) and 45 (d = 2.79) minutes post resistance exercise only. This gives evidence that not only does resistance exercise show reductions in anxiety 30 minutes post exercise, but that these reductions persist even in the presence of additional stressors.

Chronic Exercise and Anxiety

Aerobic Exercise

The long-term effects of aerobic exercise on anxiety have been extensively researched. Most notably, Petruzzello et al. (1991) showed an overall effect size of .34 for self-reported trait anxiety. It was suggested that 10 weeks of training seemed optimal to elicit the largest improvements in trait anxiety. Numerous studies have been conducted since 1991, mostly showing beneficial effects. Herring, Jacob, Suveg, and O'Connor (2011) showed significant reductions in trait anxiety following week 4 (d = .47) and week 6 (d = .54) of an aerobic exercise (leg cycling) intervention. The important component to the Herring et al. work was that the female participants were all diagnosed with generalized anxiety disorder (based on DSM-IV criteria). This is crucial as research has shown exercise can reduce anxiety even in individuals without high baseline anxiety, but aerobic exercise can have much larger effects on individuals with high baseline anxiety.

Guszkowska and Sionek (2009) conducted a similar study on healthy female participants, assessing trait anxiety changes over the course of a 12-week aerobic exercise intervention. The results revealed a significant reduction in trait anxiety (d = .51), giving evidence that chronic exercise not only reduces anxiety in those living with an anxiety disorder but also within healthy populations. Others have shown that individuals with high anxiety sensitivity show a significant reduction in anxiety sensitivity after only 2 weeks of brief bouts of aerobic exercise (Broman-Fulks & Storey 2008). The aerobic exercise sessions consisted of 20 minutes of treadmill running at 60–90% predicted HR_{max}. Additionally, a measure of anxiety sensitivity showed immediate reductions (i.e. following session one of six) that were maintained throughout the duration of the study. This gives evidence that individuals with high anxiety sensitivity can see significant reductions in such sensitivity with acute bouts of aerobic exercise and this can be maintained with continued exercise participation.

Resistance Exercise

There has been limited research assessing the longitudinal effects of resistance exercise on anxiety. Of the limited research done, results appear promising. Herring et al. (2011) conducted a 6-week, lower body resistance exercise study on women with generalized anxiety disorder (DSM-IV criteria). They reported significant reductions in trait anxiety after week 4 (d = .36) and 6 (d = .52), respectively. This indicated over a ½ standard deviation improvement in trait anxiety after completing two resistance exercise sessions per week for 6 weeks. Similarly, Tsutsumi, Don, Zaichowsky, and Delizonna (1997) had 42 healthy, but sedentary, elderly adults randomly assigned to either a non-exercise control, a low-intensity/high volume (2 sets of 14–16 repetitions at 55–65% of 1 repetition maximum, RM), or a high-intensity/low volume (2 sets of 8–10 repetitions at 75–85% of 1-RM) resistance exercise group for a duration of 12 weeks. Exercise sessions occurred three times per week, and both high and low intensity resistance exercise was associated with marked improvements in anxiety scores (2.3 and 6.2 point decrease, respectively) as indexed by the STAI; the control group had a 3.2 point increase in anxiety (Tsutsumi et al. 1997). A similar follow-up resistance exercise study (same design and exercise interventions as previous study) with older women essentially showed the same results, with decreases in trait anxiety of 2.1 and 5.8 units for high and low intensity, respectively (Tsutsumi et al. 1998).

Cassilhas, Antunes, Tufik, and deMello (2010) randomly assigned elderly men (N = 20) to a resistance exercise group (6 different resistance exercises; 2 sets of 8 repetitions at 80% 1-RM for 1 hour 3 times·wk^{-1}) or to a control "exercise" group (N = 23; 1 hr·wk^{-1} of exercise with no overload) over 24 weeks, to assess changes in anxiety. Interestingly, participants in the resistance exercise group showed a significant reduction in trait anxiety (d = 1.20) from pre-post intervention, whereas the control group showed no change. While these findings may be limited to elderly men, this study provides evidence that, over time, high-intensity resistance exercise significantly reduces anxiety. This is important as previous research on acute bouts of resistance exercise have shown increases in anxiety following high-intensity resistance exercise.

Implications for the Health Sector and School

Based on the available literature, it is clear that aerobic exercise of both moderate and high intensity reliably reduces both state and trait anxiety. The magnitude of this effect remains unclear, but has been reported between low to moderate in meta-analytic reviews. This significant change could have important implications for the vast number of individuals currently living with an anxiety disorder. Also, well-designed studies assessing individuals with high baseline anxiety have found much larger effects of both acute and chronic (i.e. exercise interventions) exercise. Aerobic exercise has been recommended as a potential treatment/preventive measure with regards to anxiety.

Resistance exercise has shown more inconsistent results with respect to anxiety reduction. It is clear that many resistance exercise studies do show significant reductions in anxiety, but the literature is plagued by methodological issues that make systematic reviews difficult. Unlike the aerobic exercise literature, it appears that the most beneficial effects on state anxiety occur with low-to-moderate intensity resistance exercise, although there are exceptions. This is likely due to the temporary increase in anxiety following high-intensity resistance exercise that is most likely an artifact of increased activation/tension. Studies using high-intensity resistance exercise protocols (i.e. > 70% 1-RM) have shown elevated anxiety assessments immediately post exercise. However, high-intensity resistance exercise studies that assess anxiety in excess of 20–30 minutes post report anxiety returning to, and dropping below, baseline anxiety levels (Greene & Petruzzello 2014). Future research is necessary to understand if state anxiety following high-intensity exercise continues to decrease post exercise, or if it levels off after reaching pre-exercise levels; as well as to more carefully examine whether anxiety increases occur as a function of measurement issues (i.e. items reflecting effort/tension or activation/arousal being interpreted as increased anxiety).

Of particular interest to developing youth is the pattern for changes in severity and type of anxiety experienced throughout adolescence. Hale, Raaijmakers, Muris, Van Hoof, and Meeus (2008) assessed DSM-IV-TR anxiety disorders among junior high (*M* age = 12±0.57 yrs) and high school (*M* age = 16.6±0.65 yrs) students every year for 5 years. From year 1 to year 5, adolescent boys of both ages showed significant decreases in generalized anxiety. Interestingly, adolescent females in junior high displayed a significant increase in generalized anxiety throughout the 5-year prospective study while high school females showed no change in anxiety but did show significantly elevated anxiety at year 1 relative to all other groups. Previous literature has given evidence that females are at an increased risk of developing/being diagnosed with GAD relative to males. Hale and colleagues' findings provide evidence that this increased risk may stem from a critical developmental period as girls transition into high school. While more research is needed, it is clear that preventive measures (e.g. physical activity programming) should be implemented in schools to reduce the likelihood of developing anxiety disorders, and these measures should specifically target females.

Depression

Depression is often overlooked as a mental health disease. According to the Center for Disease Control, nearly 1 in 10 United States adults are currently living with depression (Lopez, Mathers, Ezzati, Jamison, & Murray 2006). However, unipolar depression is projected to be the second highest contributing factor to the global disease burden by 2030 (Mathers & Loncar 2006). In 2006, depression was the third highest leading cause of disease burden in high-income countries (Lopez, Mathers, Ezzati, Jamison, & Murray 2006), and is continuing to increase at an alarming rate. The number of adults who experienced a major depressive episode in the previous year significantly increased from 14.2 million in 2005 to 16.0 million in 2012 (Substance Abuse and Mental Health Services Administration 2013). In 2012, 11.6% of women aged 18–25 reported having a major depressive episode in the past year (Substance Abuse and Mental Health Services Administration 2013). Affecting around 9% of the population, depression is a major public health concern and is associated with high morbidity, co-morbidity, and mortality (Cassano & Fava 2002).

The challenge with depression lies within treatment and treatment options. Of those who received treatment, only 18–25% received adequate treatment (Ebmeier, Donaghey, & Steele 2006). While antidepressants are being prescribed at a much higher rate, over 50% of patients stop taking their antidepressants after one month (Cassano & Fava 2002). Antidepressants are also known for their serious side effects including withdrawal symptoms (Ebmeier et al. 2006); weight gain (Berken, Weinstein, & Stern 1984); dry mouth, nausea, insomnia, constipation, anxiety, and sexual dysfunction (Hsu & Shen 1995). It is clear that alternative treatment options need to be explored. Additionally, some antidepressant medication has been shown to increase depressive symptoms in children and adolescents (Hammad, Laughren, & Racoosin 2006). Physical exercise can be used as a treatment option for depression. Specifically, both resistance and aerobic exercise have been shown to improve risk factors, quality of life, and decrease morbidity from countless physical and psychological diseases, including depression (please see Chapter 15 in this volume for more detailed information about exercise and clinical depression).

Resistance Exercise and Depression

Exercise is an often sought after alternative treatment/management for a growing number of physical and mental health concerns. Resistance exercise appears to be an effective treatment option for reducing depression in both clinically depressed and non-clinically depressed participants.

Morvell and Belles (1993) assessed depressive symptoms in response to a 16-week circuit resistance training program in 43 males who were not clinically depressed. Participants met 3 times·wk^{-1} for a duration of 20 minutes in which resistance exercises were performed on 12 machines. Depression scores were compared to participants randomly assigned to a wait-list control group. Results indicated a significant reduction in depression scores following the resistance exercise intervention relative to the control group. Singh, Clements, and Fiatarone (1997) conducted a randomized controlled trial using either progressive resistance training (PRT) or an attention-control in older adults diagnosed with major or minor depression. Participants in the exercise intervention met for 45 minutes, 3 times·wk^{-1} for 10 weeks while participants in the control condition met for 60 minutes 2 times·wk^{-1} for 10 weeks. Results indicated a significant reduction in self-reported and therapist-rated depression following resistance exercise relative to control. Singh and colleagues also reported that 50% of the exercise group (compared to 26% of controls) had a clinically meaningful response to the treatment. At the end of the intervention, nearly 88% of the exercise participants no longer met diagnostic criteria for depression (compared to ~43% of controls). A similar trial by Singh and colleagues (2005) used 60 older adults with diagnosis of major or minor depression randomly assigned to high intensity PRT (3 sets of 8 repetitions at 80% 1-RM for 60 min 3 times·wk^{-1}), low intensity PRT (20% 1-RM), or standard care from a general practitioner for 8 weeks. As with the prior study, significant reductions were seen in both self-reported and therapist-rated depression, but these reductions were significantly larger in the high-intensity PRT group. These individuals also showed a sizable response to treatment (~55%) compared to the standard care group (~24%), with the low PRT group falling in between. Magnitude of change was d = ~1.8, ~1.3, and ~0.7 for the high PRT, low PRT, and standard care groups, respectively (slightly different depending on whether self-rated or therapist-rated). This work highlighted the dose-response nature of the effect of progressive resistance exercise on depression outcomes, but more work is needed to identify the optimal intensity, duration, and frequency of exercise needed.

Some studies have assessed changes in depression relative to both aerobic and non-aerobic exercise modes. Stein and Motta (1992) used a 7-week aerobic structured swimming, non-aerobic resistance exercise, or a general education control condition in a sample of undergraduates who were not clinically depressed. Following the intervention, there was a significant reduction in depressive symptoms (as measured by both the Beck Depression Inventory and the Depression Adjective Check-List), for both the aerobic and non-aerobic exercise conditions relative to control. Further, it was found that the resistance exercise condition was associated with an increase in self-concept, physical self, and social self that was not observed in the aerobic exercise condition. While there was a significant reduction in depression following both modes of exercise, there may be additional benefits to resistance exercise with regards to self-concept which is highly related to depression. Finally, Doyne et al. (1987) assessed changes in depression following 8 weeks of high intensity running (i.e. 80% HR_{max}), moderate intensity resistance exercise (i.e. 50–60% HR_{max}), or a no exercise control in female participants diagnosed with major or minor depression. A significant reduction in depression was shown with no differences observed between exercise modes.

Aerobic Exercise and Depression

There is a great deal of evidence to support the notion that aerobic activity helps aid the reduction of depressive symptoms. In a review by Danielsson, Noras, Waern, and Carlsson (2013), it was noted that most studies in this area have used aerobic exercise as a modality, performed at 60–80% HR_{max}, for 2 or 5 times·wk^{-1}, with an average duration of training lasting around 8 weeks. The weight of the evidence supports the use of aerobic exercise to treat some depressive symptoms. Danielsson et al. (2013) indicated that aerobic exercise is not more effective in the treatment of depression than other types of physical activity, but that these findings highlight the need for further investigations into the active components for the treatment of depression.

In spite of what appears to be effective symptom reduction with exercise, pharmacotherapy is often the first choice of health care providers. There have now been numerous studies done to determine the extent to which exercise compares to anti-depressant medication for reducing depressive symptoms in adults with MDD. Blumenthal and colleagues (1999) randomly assigned 156 participants (50–77 yrs old) with major depressive disorder (MDD; determined from Hamilton Rating Scale for Depression, HAM-D, scores > 13) to a 16 week intervention of either medication only, exercise only, or combination medication and exercise. Results showed no significant difference between the three treatment groups, with each showing a significant decrease in depressive symptoms (on both the HAM-D and Beck Depression Inventory). Further, 47.2% of the exercise group, 56.2% of the medication group, and 47.3% of the combination group were no longer classified as being clinically depressed (DSM-IV classified and HAM-D score < 6). While there was no difference between the three groups in terms of treatment for clinical depression, Blumenthal and colleagues (1999) also examined the rate at which depressive symptoms decreased. They showed differences across treatment groups where participants in the medication only group showed a quicker therapeutic response within the first few weeks) compared to exercise only and combination groups, but all three groups essentially ended up in the same place by the end of the intervention. These results demonstrate support for aerobic exercise as a viable and effective treatment for clinical depression in older individuals.

As a follow-up to Blumenthal and colleagues, Babyak et al. (2000) assessed presence and severity of depressive symptoms in the sample from the 1999 study 6 months post-treatment. At this post-treatment follow-up, individuals originally assigned to the exercise group had significantly lower relapse rates than those in the medication group, highlighting that continued exercise reduced the likelihood of depression recurring. Blumenthal et al. (2007) again compared the use of medication and exercise as a treatment for depression, this time in 202 participants with diagnosed MDD randomly assigned to either a medication, home-based exercise, supervised group exercise, or placebo control group. There was again no difference among exercise groups and medication or between home-based or group exercise interventions on depressive symptom reduction. All groups showed significant declines in HAM-D scores over the 16-week intervention period (home-based exercise = −7.1± 6.9, group exercise = −7.2±6.9, medication = −6.1±6.7, placebo = −6.1±7.3). These results again confirm the findings that aerobic exercise is comparable to the use of antidepressant medication in the treatment of MDD, also taking into account the potential for a placebo effect as well as other factors such as staff/group interaction and the expectation of reduced symptoms. This further supports the notion that exercise can be a cost-effective treatment for MDD, with little to no negative side effects. This is an important consideration in dealing with children and adolescents, where most antidepressants come with a warning that they have the ability to increase depressive symptoms in children and adolescents.

Regarding the use of exercise with children and adolescents, Åberg et al. (2012) found that low cardiovascular fitness at the age of 18 is predictive of greater risk of developing severe depression in adulthood. This may add to the value of physical activity for children and adolescents. Annesi (2004) conducted a study designed to examine relationships between changes in exercise self-efficacy scores and changes in Tension and Depression (as measured by the Profile of Mood States, POMS) in children involved in physical activity (PA) programs. A 12-week (3 times·wk^{-1} for 45 min) training program was implemented where participants (N = 54; 28 female) aged 9–12 years from low to middle SES engaged in PA. Resistance exercises were also performed 2 times·wk^{-1} on non-consecutive days where 5–6 exercises were performed (1–2 sets of 10–15 reps as well as 3–4 cardiovascular activities varying in intensity: low = 50–60% HR_{max}; moderate = 60–70% HR_{max}; high = 70–85% HR_{max}). On the non-exercise days, interactive sessions to increase self-efficacy were employed (included goal setting, positive self-talk, recruiting social support, and strategies for engaging in on-going physical activity; Annesi 2004). Following the completion of the 12-week program, there was a significant reduction in both tension (-1.2) and depression (-1.4; Annesi 2004). These findings suggest that well-designed PA programs can be implemented in children/adolescents and perhaps buffer against increasing depressive symptoms into adulthood.

Implications for the Health Sector and School

Based on the weight of the evidence to date, which includes numerous randomized controlled trials conducted with individuals having clinical diagnoses of depression, exercise (both aerobic and resistance) can be a valuable tool in the health care provider's toolbox in the treatment of depression and depressive symptoms (please refer to Chapter 15 in this volume for more detailed information about exercise and clinical depression). While there remain unanswered questions regarding the mechanisms underlying the anti-depressant effects of exercise, the fact that exercise has been shown to be effective is helpful for those providing care and treatment for individuals living with mild-to-severe forms of depression. There also remains work to be done to determine the optimal "dose" of exercise, but the evidence seems to indicate that "how much" and for "how long" is not quite as important as getting the depressed individual to simply engage in some form of exercise on a regular basis. The fact that a large percentage of depressed individuals get no relief from medication (Fournier et al. 2010) and that medication can often result in unwanted side effects makes the efficacy of exercise even more important for those living with depressive symptoms.

To say that there is a lack of information/evidence in children is an understatement when compared to what has been done regarding exercise and depression in adults. Children with mental health disorders like depression often become adults with mental health disorders, thus early detection and treatment is key in teaching coping and maintenance strategies when applicable. The National Institute of Mental Health estimates that 2.2 million US adolescents (9.1% of 12–17 yrs old) had at least one major depressive episode in the past year, and this is based on incomplete data (reasons for non-response included parental refusal (10%), youth refusal to participate (3%), respondent being unavailable (1%), and physical or mental incompetence (1%; SAMHSA, 2013)). Diagnosing depression in children and adolescents is difficult for these reasons.

As discussed previously, it is well known that physical activity offers multiple health benefits in addition to improving physical well-being. While the studies mentioned previously have demonstrated the potential for physical activity to improve the overall well-being of adults, Burdette and Whitaker (2005) suggest that play also has the ability to improve the physical, emotional, social, and cognitive well-being of children as well. Additionally, Burdette and Whitaker suggest using the term "play" rather than "exercise" or "physical activity" to increase movement in children. Both unstructured and outdoor play allows children the opportunity to learn to use their developing minds. Burdette and Whitaker (2005) suggested that adults have negative feelings toward sports/structured physical activities that began in childhood. Participating in unstructured play could potentially curb these negative feelings, decreasing depressive and health-related issues associated with negative overall well-being by increasing the positive associated with physical activity. Additionally, free play outdoors increases an individual's exposure to sunlight and absorption of vitamin D. Low levels of vitamin D have been associated with depression in adults. In a time where children are seemingly attached to technology (e.g. video games, cell phones, computers, etc.), free play is much more limited than it has been in the past. Burdette and Whitaker (2005) suggest that the benefits gained through play are those that will reinforce the important message of the benefits of physical activity. Burdette and Whitaker posit that, because physical activity has been shown to decrease depression and inactivity has been shown to increase the risk of developing depression, play in children will have the same effects on them as organized physical activity has on adults.

Cognitive Function

Exercise has long been considered as a stimulus to improve cognition or even as a preventive behavior to slow age-related cognitive decline. Since the turn of the century, researchers have been able to demonstrate positive effects of exercise on numerous aspects of cognition: learning, memory, creative thinking, executive functions (e.g. planning, scheduling, working memory (maintaining and manipulating information), and flexibly switching from task to task). These effects have been seen in response to acute (or short) bouts of exercise and chronic (long-term) exercise training. Generally, there is a consensus that exercise most often helps improve cognition and usually does not have any negative impacts on cognitive function (in addition to providing beneficial health outcomes such as improved cardiovascular health, body composition, and bone density; see also Chapter 7 in this volume). In children and young adults, high levels of cognitive function and academic performance are both desirable and extremely influential in the determination of social and financial standing later in life. Both age-related and disease-related cognitive decline negatively impact the quality of life of individuals experiencing these issues, as well as their caregivers. The capacity for exercise participation to induce beneficial changes in cognition provides an affordable, non-pharmaceutical intervention for enhancing brain function that also happens to provide numerous physiological health benefits (Hillman, Erickson, & Kramer 2008).

Acute Exercise and Cognitive Function

Reviews over the past 30 years have presented evidence that suggests an optimistic relationship between acute bouts of exercise and cognitive function (Etnier et al. 1997; Tomporowski 2003). When measuring cognitive performance after exercise, positive effects are seen when exercise duration is less than 60 minutes and intensity is moderate (Tomporowski 2003). However, there are studies providing evidence of either no effects or sometimes detrimental effects of exercise on cognitive performance (Dietrich & Sparling 2004). When detrimental effects are seen, many seem to be produced when higher-level cognitive tasks are assessed during the exercise bout itself, especially when the workload is of high intensity (review of studies from 1993–2002 by Brisswalter, Collardeau, & Arcelin 2002).

Tomporowski (2003) reviewed the effects of differing intensities of acute exercise on cognition. Short, high intensity bouts of exercise both enhanced and worsened cognitive performance, but any negative effects did not persist for long after completing the exercise bout. On average, reaction time (RT) latency appears to decrease (i.e. get faster) as intensity increases, occasionally demonstrating an inverted-U shape, and other times simply being faster relative to rest or low-intensity exercise. RT accuracy is either slightly improved or not changed on most tasks. Lambourne and Tomporowski (2010) reviewed papers from 1900–2008 which addressed the effect of exercise-induced arousal on cognitive performance both during and following acute bouts of exercise. Their review revealed a number of effects depending on: (a) the type of exercise (i.e. cycling seemed to enhance cognitive performance during and following exercise; treadmill running negatively affected performance during, with a small improvement following); (b) the timing of the cognitive assessment (cognitive performance generally impaired when assessed during exercise for first 20 min ($d = -0.14$) but enhanced if assessed after 20 min during exercise as well as following exercise ($d = 0.20$)); and the type of cognitive assessment being examined (i.e. speeded processing, memory storage, memory retrieval facilitated).

Upon examination of individual study findings, it becomes more obvious why the findings for the effects of acute exercise on cognition appear to be equivocal. Of the studies that have examined moderate intensity aerobic exercise, RTs on cognitive inhibition tasks (e.g. Flanker, Stroop tasks; Eriksen & Eriksen 1974; Stroop 1935) appear to be significantly faster following exercise compared to following a rest condition (Chen, Yan, Yin, Pan, & Chang 2014; DeBord & Deike 2015; Pontifex, Parks, Henning, & Kamijo 2015; Tonoli et al. 2014). Stroop color-word interference performance has also been shown to improve following high-intensity exercise (Hogervorst, Riedel, Jeukendrup, & Jolles 1996). However, Drollette et al. (2014) provided evidence that even in a case where RTs do not appear significantly different post-exercise compared to post-rest, cognitive processing speed (indicated by shorter P300 latency) was still faster following exercise than rest. This mimics findings of Kamijo et al. (2009) for both younger and older adults following moderate cycling exercise. For working memory tasks, studies have found RTs to be faster following exercise than rest or control (Gothe, Kramer, & McAuley 2014; Hogan, Mata, & Carstensen 2013), and even faster following yoga than aerobic exercise (Gothe et al. 2014). Speed of learning new vocabulary has been demonstrated to be faster in young adults immediately following a bout of intense sprinting, as compared to moderate aerobic running or resting (Winter et al. 2007). A newer area of interest has been addressing how attempting to learn while exercising might influence recall performance afterwards, both immediately and a few days later. One study demonstrated enhanced recall on a vocabulary test when participants memorized the list while exercising for 30 minutes compared to memorizing once the 30 minutes of either exercise or rest was completed (Schmidt-Kassow et al. 2013).

Cognitive performance has also been measured during exercise bouts. Random number generation (reflective of working memory) and cognitive inhibition (e.g. lower accuracy on incongruent flanker trials) have been reported as worsening during acute exercise bouts (McMorris, Collard, Corbett, Dicks, & Swain 2008; Pontifex & Hillman 2007). Cycling has been, by far, the most practiced mode of exercise in these cases, likely due to the relative ease of measurement in the cycling posture. Yet, Yagi, Coburn, Estes, and Arruda (1999) found oddball task performance to improve during exercise.

Scientists have concluded that acute aerobic exercise exerts its most prominent effects on aspects of *executive control,* that is "goal-directed cognitive processes underlying perception, memory, and action" (Khan & Hillman 2014, p. 140). However, others have documented no change in certain aspects of cognitive control following exercise. As Lambourne and Tomporowski (2010) demonstrated in their review, some of the inconsistencies could easily be due to the type of cognitive tasks examined, when the cognitive assessments are made, and the type of exercise that has been examined. This will be an area that continues to be investigated.

Chronic Exercise and Cognitive Function

Chronic aerobic exercise and aerobic fitness predominate the literature on chronic exercise and cognition. With adequate, repeated bouts of training comes improvements in aerobic fitness, which have been associated directly with the cognitive improvements seen from pre- to post-intervention. First, in older adults, exercise training that enhances aerobic fitness has been linked to improvements in cognitive function (Colcombe & Kramer 2003). Both greater levels of aerobic fitness and improvements in aerobic fitness through exercise training have been linked to higher brain activation levels in regions of the brain associated with inhibitory control, conflict monitoring, and spatial selection, while older adults complete cognitive tasks meant to tap into such functions (Colcombe et al. 2004). The demands of the 6-month training program used by Colcombe et al. were relatively minimal, starting with only 10–15 minutes of walking 3 times·wk^{-1} and increasing to around 45 minutes per session within the first 3 months (a control group completed a stretching and toning program, which led to minimal changes in fitness and is likely why improvements in cognitive performance were not seen). Higher cardiorespiratory fitness has also been associated with higher accuracy on task-switching tests, reflecting greater cognitive flexibility (Verstynen et al. 2012).

Diamond (2015) has suggested a move towards more dynamic exercise paradigms that explore movement patterns that are more complex than standard aerobic training protocols in order to provoke greater change in executive function over the intervention period. Alternative exercise training, such as yoga practice, has also been shown to benefit cognitive performance.

An 8-week hatha yoga training program (60 minutes, 3 times·wk^{-1}) with older adults (N = 118) resulted in improved working memory, memory span, and mental flexibility, when compared to individuals who spent the same time in a stretching and strengthening program (Gothe et al. 2014; it is unclear whether or not improvements in cognition were also seen in response to the stretching and strengthening program based on the results presented). In a review of strength training interventions for older adults, programs that involved moderate intensity training two times·wk^{-1} for around 6 months showed benefits for measures of attention, information processing, and some executive functions (Chang, Pan, Chen, Tsai, & Huang 2012).

Implications for the Health Sector and School

While it is sometimes difficult to see how basic scientific measures can be applied to broad, socially relevant health issues, there are instances where exercise is used as a tool for physical and mental wellness (e.g. "Exercise is Medicine" as a current platform espoused by the American College of Sports Medicine). As demonstrated above, both acute and chronic exercise have been shown to benefit cognitive performance and in many special populations. Positive outcomes of exercise training have been demonstrated in individuals with obesity, diabetes, brain injury, age-related cognitive decline, dementia, Alzheimer's disease (a severe form of dementia), Parkinson's disease, multiple sclerosis, and those with cancer or detrimental side effects of its aggressive treatment. There do not seem to be any studies involving exercise interventions aimed at improving cognitive performance in individuals with intellectual disabilities, though prevalence of physical activity participation appears to be low, and fitness modifiable, in this population (Chanias, Reid, & Hoover 1998).

Research has identified a relationship between obesity and cognitive dysfunction (see Miller & Spencer 2014) and exercise has demonstrated promise as an intervention for the cognitive declines that occur in conjunction with obesity and diabetes as people age (Chan, Yan, & Payne 2013). The majority of recent work has investigated the potential for exercise to enhance cognitive abilities of overweight children. Both academic achievement and executive function (i.e. "planning") showed significantly greater improvements in those who participated in ~13 wks of 40 min·wk^{-1} of aerobic exercise compared to a non-exercise control group, when baseline scores and other important demographic differences were accounted for (Davis et al. 2011).

Physical activity participation and greater aerobic fitness earlier in life have both been associated with the amelioration or delayed onset of cognitive decline (Sofi et al. 2011), dementia, and Alzheimer's Disease later in life (Larson et al. 2006; Lista & Sorrentino 2009; Zhu et al. 2014). Although aging itself cannot be stopped, fitness is a modifiable construct and exercise training has been shown to reverse some of the cognitive losses that occur with dementia (Colcombe & Kramer 2003; Heyn, Abreu, & Ottenbacher 2004). Recent evidence suggests that increasing regular physical activity participation, even later in life, can reduce risk for developing dementia, even in previously inactive individuals (Tolppanen et al. 2015). Greater fitness (relative to less fit patients) has also been associated with better cognitive performance and diminished risk for developing Parkinson's disease (Ahlskog 2011). A recent review of exercise effects on cognition in Parkinson's disease also supports the notion that it is beneficial to be physically active (Murray, Sacheli, Eng, & Stoessl 2014). A 2014 report of the influence of a 6-month aerobic training intervention (walking) in Parkinson's patients demonstrated significant improvements in executive control (i.e. inhibition), but not other aspects of cognition (e.g. language, verbal or visual memory, set shifting, visuospatial perception, or general cognition; Uc et al. 2014). Cancer and the side effects of chemotherapy treatment often involve some degree of cognitive impairment, and although some pharmacological methods are being attempted, exercise may provide another potential treatment (Cormie, Nowak, Chambers, Galvão, & Newton 2015). Animal models of cancer have shown reversal of cognitive impairments from chemotherapy when exercise was prescribed (Fardell, Vardy, Shah, & Johnston 2012). Cross-sectional reports of exercise behavior and cognitive function in chemotherapy-treated breast cancer patients compared to healthy adults also suggests a relationship between exercise behavior and visual memory (Crowgey et al. 2013).

As exercise participation (and then, fitness improvement) seems to exert a greater influence on those who have more room for improvement (e.g. aging and/or lower fit populations; individuals with dementia, cancer, Parkinson's disease), the need to examine the utility of exercise in individuals with cognitive impairment is important.

The use of exercise as a tool for enhancing cognitive performance (and potentially, academic achievement) in school-aged children could have extremely profound reach (e.g. ~95% of children in the United States attend school; Physical Activity Guidelines Advisory Committee 2008). The school setting provides a number of options for studying the effects of physical activity on cognitive performance: active transport to and from school, classroom-based physical activity, recess, and after-school programs. The available literature is largely cross-sectional, mainly reports academic achievement measures rather than specific cognitive-behavioral measures, and there is a general lack of follow-up data from exercise interventions in children (Lees & Hopkins 2013), but what is known suggests that physical activity is important in school settings.

Associations between physical activity and academic performance have been available for over 30 years (Gabbard & Barton 1979; Sibley & Etnier 2003). Generally, physical activity participation during the school day has not been shown to have any negative academic outcomes for students (Ahamed et al. 2007). Cross-sectional work has shown positive relationships between field tests of physical fitness and both standardized test scores and academic grades (Coe, Pivarnik, Womack, Reeves, & Malina 2012). Greater aerobic fitness has been linked to greater academic performance (Castelli, Hillman,

Buck, & Erwin 2007; Cooper Institute for Aerobics Research 1999), and cognitive function, including executive function (Hillman, Castelli, & Buck 2005; Tomporowski, Davis, Miller, & Naglieri 2008). It may be that aerobic fitness is influential in altering brain activation patterns and behavioral strategies chosen by children to process information and make decisions (Voss et al. 2011).

Can teachers do anything? Yes, and many are. One attempt has been the Physical Activity Across the Curriculum Project which addressed the addition of physically active live-classroom lessons and their potential influence on academic performance (Donnelly et al. 2009). Students who participated in 75 minutes of classroom-based physical activity each week improved their academic achievement scores by 6% as compared to controls who received no classroom-based physical activity. Students with teachers who were more physically active themselves, were also more physically active (Donnelly & Lambourne 2011). Other studies claim that participation in physical activity at recess is related to improved classroom behavior, perhaps because the physical activity leads to an improved learning environment (Jarrett et al. 1998; Mahar et al. 2006). Exercise may be impacting students in school settings in other ways as well: energy levels could be increasing and engagement in course material and "buy-in" to school could be occurring and resulting in improved learning and performance outcomes.

Overall, it is clear that attempts have been successfully made to increase physical activity for children at a national level in the United States and that these attempts have been motivated by promising research evidence suggesting that exercise may enhance cognition. What remains to be seen is whether or not these increases in physical activity participation consistently elicit the desired academic or cognitive benefits that have been shown in smaller-scale studies.

Psychological Well-Being

Defining psychological well-being is complex because it has multiple dimensions, meanings, and interpretations. *Psychological well-being* (PWB) could be defined as the absence of mental illness or disease or as the lack of mental distress. For the purposes of this chapter, PWB will be defined as a combination of positive affective states (e.g. happiness, satisfaction) relative to negative affective states (e.g. distress, unhappiness) coupled with the ability to function optimally and effectively in the world (Deci & Ryan 2008). Individuals who have higher degrees of PWB tend to live healthier, happier lives. This includes, but is not limited to, feelings of happiness, satisfaction, feeling capable, and well supported. Studies have shown that life longevity can be predicted by positive emotions in younger and older populations and are strongest for healthy adults (Winefield, Gill, Taylor, & Pilkington 2012). PWB has also been linked to reductions in chronic illness such as risk of coronary heart disease (Boehm, Peterson, Kivimaki, & Kubzansky 2011).

Psychological distress (i.e. lack of PWB) has been linked to negative mental and physical health outcomes. Anxiety, depression, irritability, and emotional instability are highly correlated with reductions in life span and overall quality of life along with higher prevalence of physical morbidity. The ability of exercise to improve PWB and improve overall quality of life, along with reducing side effects and risk factors that accompany poor PWB has become a focus of scientific inquiry (Scully, Kremer, Meade, Graham, & Dudgeon 1998). The next section will focus on the effects of both acute and chronic exercise on PWB.

Acute Exercise and Psychological Well-Being

Lack of energy or excessive fatigue is often cited as a reason for not exercising. However, research has shown that an acute bout of exercise can actually increase perceptions of energy and decrease the fatigue individuals experience in their daily lives. Lack of energy and a presence of fatigue are caused for many reasons, which can include poor PWB, presence of chronic illness, increased stress or worry, and many other reasons. These effects can be seen in both healthy populations as well as specific chronic populations.

College is a time of elevated stress and the student's environment is often changing and unstable. Herring and O'Connor (2009) conducted a study with sedentary college-aged women who reported persistent above average fatigue. The female volunteers completed three separate conditions on separate days: (a) no exercise control; (b) "placebo" condition, which involved lower body resistance exercise at 15% of 1-RM; and (c) lower body resistance exercise at 70% 1-RM. Self-report measures of vigor and fatigue were obtained before, during, and after each condition. Vigor scores were significantly higher following both exercise conditions when compared to the no exercise control condition, with no difference between placebo and 70% conditions. The placebo exercise condition resulted in significantly lower fatigue scores when compared to the no exercise control condition. Although the 70% 1-RM exercise condition did not result in significant reductions in fatigue, it did serve to reduce fatigue (standardized mean difference scores comparing each exercise condition to the no exercise control ranged from -0.09 to -0.35 for 70% 1-RM and -.23 to -0.63 for the placebo conditions). Herring and O'Connor concluded that acute moderate to high intensity resistance exercise increased feelings of energy and decreased feelings of fatigue (albeit more so in the light intensity/placebo condition) when compared to a no exercise control condition.

Fatigue and lack of energy are commonly reported as symptoms of many chronic illnesses. Symptoms like these can be a consequence of poor PWB as well as an antecedent of poor or decreased PWB. Individuals living with multiple sclerosis (MS) are at risk for, and often report high levels of, fatigue and low levels of energy. Petruzzello and Motl (2011) examined the effects of an acute bout of cycling exercise (20 min at 60% $VO_{2\,max}$) on fatigue in individuals with MS. Self-reported fatigue was obtained pre-exercise and at 5, 20, and 60 minutes post-exercise. The results revealed a reduction in fatigue, with effect sizes being small to moderate in nature. These results give physicians an alternate way to treat the debilitating side effect of fatigue in MS patients.

A relatively recent area of investigation has involved examination of "*green*" *exercise*, that is, engaging in physical activity in the presence of nature (Pretty, Peacock, Sellens, & Griffin 2005). Pretty and colleagues had participants exercise on a treadmill while being exposed to rural and urban pleasant and unpleasant scenes or to no scenes at all (exercise only). Reductions in blood pressure and improvements in mood and self-esteem were seen in the exercise only condition. The pleasant images (both rural and urban areas) had similar but stronger positive effects on mood and self-esteem, while the unpleasant urban and rural images resulted in decreases in mood and self-esteem (Pretty et al. 2005). In a synthesis of ten studies examining various facets of the green exercise dose-response, Barton and Pretty (2010) showed that the most positive impacts on mood and self-esteem occurred for brief bouts (5 min, d_s = ~0.6–0.7) of activity in natural environments, with the effects decreasing (although still positive) with longer exposures (weakest effects seen for "half a day" exposure, d_s = ~0.5) and for more vigorous intensities. Activity within the presence of water engendered the larger improvements, with similar improvements for men and women.

Chronic Exercise and Psychological Well-Being

The effects of chronic exercise on PWB are well documented and have been examined across a wide range of participant samples, including but not limited to: healthy individuals, cancer survivors, persons with spinal cord injuries, individuals living with mental health disorders (e.g. anxiety, depression), and those recovering from myocardial infarctions. Such effects include improvements in quality of life, reductions in mood disturbances, increases in energy, and decreases in fatigue.

A Finnish study examined physical exercise and PWB (Hassmen, Koivula, & Uutela 2000) in adults (ages 25–64 yrs). Participants completed questionnaires assessing their exercise habits, perceived health and fitness, depression (Beck Depression Inventory), anger (State-Trait Anger Scale), and measures of distrust and coherence (Cynical Distrust Scale, Sense of Coherence Inventory) as part of the Finnish cardiovascular risk factor survey. Those who reported exercising at least 2–3 times·wk^{-1} were significantly less likely to experience or report feelings of depression, cynical distrust, and stress than those who exercised less often or not at all. Regular exercisers reported higher levels of coherence and stronger feelings of social interactions/integration when compared to those who were less active. Such findings provide support for the idea that greater levels of exercise/physical activity are associated with enhanced PWB.

As mentioned in the previous section, chronic illness/disease is often accompanied by a decrease in PWB. Cancer survivors represent a particular population whose PWB is highly affected by the side effects of their disease. Such side effects include, but are not limited to, fatigue, reductions in energy, and reduced overall quality of life (mental and physical). Burnham and Wilcox (2002) examined the effects of exercise on a variety of both physiological (e.g. body composition, flexibility) and psychological (e.g. self-rated energy) outcomes. Cancer survivors (breast and colon cancer; 40–65 yrs old, $N = 18$) were randomly assigned to either a no-exercise control group, a low intensity exercise group (25–35% of $HR_{Reserve}$), or a moderate intensity exercise group (40–50% of $HR_{Reserve}$). Both exercise interventions resulted in significant improvements in aerobic capacity, flexibility, and body composition along with improved quality of life and increased energy. Once again, findings like these (and numerous other examples with other types of chronic diseases and illnesses) provide physicians and health practitioners with further evidence that exercise can be an effective alternative to traditional treatment approaches in managing and treating the side effects often accompanying such illnesses/diseases.

Exercise interventions have also been examined for school-aged children (7–8 yrs old). Hind et al. (2014) developed a grade school-based exercise intervention which varied the duration and type of activity: Week 1 consisted of 5-minute activity breaks done at four times during the school day (classroom based and involved music); Week 2 included 10-minute activity breaks done twice per day (carried out in the hall or outside and involving jumps and circuit training to music); and Week 3, 20 minutes per day which included circuit training to music done outside or in the hall. Interviews with the teachers revealed positive opinions for each of the durations/types of activity, although slightly less so for the full 20-minute bout. Teachers commented that the students seemed to concentrate better, enjoyed the activities, and they fit fairly well within the school day. Interviews with the children revealed that they uniformly enjoyed the activities, that the 20-minute exercise sessions were rated as more physically demanding, particularly by girls, but all of the children reported feeling better after the exercises. These results provide evidence that even with varied activity duration, time, and intensity, positive impacts can be seen in the school setting.

Implications for the Health Sector and School

In addition to the effects previously mentioned, a growing body of literature continues to demonstrate the positive effects that exercise and physical activity can have on the population. It appears somewhat obvious that exercise, both aerobic and resistance, can provide an effective treatment option for the health practitioner. This section shows that, in addition to the reductions in things like stress, anxiety, and depression, exercise can also serve to increase feelings of energy and reduce fatigue. Such symptoms accompany many of the chronic diseases that have become increasingly prevalent. Such positive changes are also accompanied by an improved quality of life, providing the affected individual with a self-managed option for improving their life.

A disturbing trend of late has been the removal of physical education programs and recess from the primary and secondary school curricula. Justification for the removal of such activities often cites the need to increase time in the classroom and attention to academics. As noted previously, physical activity increases PWB in adult populations and should offer great promise in doing the same for children and adolescents.

As shown in the work of Pretty and colleagues, "green exercise" improves health and PWB (i.e. improved mood and self-esteem; Barton & Pretty 2010; Pretty et al. 2005; Pretty et al. 2007). While the majority of this work has been done in adult samples, it is interesting to speculate that similar effects might be seen in children and adolescents. This is particularly worth investigating because of the physical changes occurring during this growth period and the ever-present peer pressure. Varying the type and intensity of physical activity has been shown to result in enjoyment (Hind et al. 2014) and outdoor physical activity (e.g. recess) could be a very important and useful strategy for improving a child's self-esteem and/or managing mood disturbances.

Conclusion

The overwhelming majority of work published to date has examined the effects of aerobic exercise and changes in aerobic fitness on the various aspects of mental health reviewed here (i.e. stress, anxiety, depression, cognitive function, psychological well-being), rather than resistance or alternative training modes. There is relatively more research available on exercise training and aerobic fitness in adults and older adults (Colcombe & Kramer 2003; Kamijo et al. 2009), with growing levels of information on the effects of exercise in children (although the acute exercise literature has tended to examine young adult populations (i.e. convenience samples of college students). The general consensus is that cross-sectional evidence is strong for a relationship between better mental health and higher aerobic fitness and that some aspects of mental health are selectively enhanced by changes in fitness (Hillman et al. 2008; Kramer, Hahn, & McAuley 2000).

Exercise can come in many forms: aerobic exercise, resistance training, yoga, other alternative physical activities, and various combinations of those listed, along with a continuum of intensities (e.g. low, moderate, high). The infinite combinations of exercise duration, mode, and intensity coupled with individual biologies, other lifestyle factors, and methodological differences reported in the literature, allow for any number of possible reactions to exercise. Just as eating an apple a day will not prevent everyone from ever needing to visit the doctor, exercise cannot and will not have the same positive effect on all individuals. More replication studies and randomized controlled trials need to be completed before generalizations can be made about specific effects of exercise on the variety of distinct aspects of mental health. Research can only uncover strong patterns that will lead us to finding greater probabilities for successful mental health enhancement in distinct populations under particular circumstances. Generally, the mental health benefits discussed require thorough investigation across all populations so that we can better understand how individuals might respond to an exercise "treatment."

Note

1 These reviews will not, by design, be as comprehensive as what appears elsewhere in this volume for several of the outcomes.

References

Åberg, M. A. I., M. Waern, J. Nyberg, N. L. Pedersen, Y. Bergh, N. D. Åberg,...K. Torén. 2012. "Cardiovascular fitness in males at age 18 and risk of serious depression in adulthood: Swedish prospective population-based study." *Br J Psychiatry 201* (5): 352–9.

Ahamed, Y., H. Macdonald, K. Reed, P.J. Naylor, T. Liu-Ambrose, & H. McKay. 2007. "School-based physical activity does not compromise children's academic performance." *Med Sci Sports Exerc 39*: 371–6.

Ahlskog, J. E. 2011. "Does vigorous exercise have a neuroprotective effect in Parkinson disease?" *Neurol 77* (3): 288–94.

American Psychological Association. *Stress in America™*. 2014. Full text of report available at www.apa.org/news/press/releases/stress/2013/stress-report.pdf

Annesi, J. 2004. "Relationship between self-efficacy and changes in rated tension and depression for 9 to 12 year-old children enrolled in a 12-week after school physical activity program." *Percept Mot Skills 99*: 191–4.

Arent, S. M., B. L. Alderman, E. J. Short, & D. M. Landers. 2007. "The impact of the testing environment on affective changes following acute resistance exercise." *J Appl Sport Psychol 19*: 364–78.

Asmundson, G., M. Fetzner, L. DeBoer, M. Powers, M. Otto, & J. Smits. 2013. "Let's get physical: A contemporary review of the anxiolytic effects of exercise for anxiety and its disorders." *Depress Anxiety 30*: 362–73.

Babyak, M., J. A. Blumenthal, S. Herman, P. Khatri, M. Doraiswamy, K. Moore,...K. R. Krishnan. 2000. "Exercise treatment for major depression: Maintenance of therapeutic benefit at 10 months." *Psychosom Med 62*: 633–8.

Bartholomew, J. B. 1999. "The effect of resistance exercise on manipulated preexercise mood states for male exercisers." *J Sport Exerc Psychol 21*: 39–51.

Bartholomew, J. B., & D. E. Linder. 1998. "State anxiety following resistance exercise: The role of gender and exercise intensity." *J Behav Med 21*(2): 205–19.

Barton, J., & J. Pretty. 2010. "What is the best dose of nature and green exercise for improving mental health? A multi-study analysis." *Environ Sci Technol 44*: 3947–55.

Berken, G. H., D. O. Weinstein, & W. C. Stern. 1984. "Weight gain: A side-effect of tricyclic antidepressants." *J Affect Disord 7* (2): 133–8.

Blumenthal, J. A., M. A. Babyak, M. Doraiswamy, L. Watkins, B. M. Hoffman, K. A. Barbour,...A. Sherwood. 2007. "Exercise and pharmacotherapy in the treatment of major depressive disorder." *Psychosom Med 69*: 587–96.

Blumenthal, J. A., M. A. Babyak, K. A. Moore, W. E. Craighead, S. Herman, P. Khatri,...K. R. Krishna. 1999. "Effects of exercise training on older patients with major depression." *Arch Intern Med 159*: 2349–56.

Boehm, J. K., C. Peterson, M. Kivimaki, & L. Kubzansky. 2011. "A prospective study of positive psychological well-being and coronary heart disease." *Health Psychol 30* (3): 259–67.

Brisswalter, J., M. Collardeau, & R. Arcelin. 2002. "Effects of acute physical exercise on cognitive performance." *Sports Med 32*: 555–66.

Broman-Fulks, J., & K. Storey. 2008. "Evaluation of a brief aerobic exercise intervention for high anxiety sensitivity." *Anxiety Stress Coping 21* (2): 117–28.

Burdette, H. & R. Whitaker. 2005. "Resurrecting free play in children: Looking beyond fitness and fatness to attention, affiliation, and affect." *Arch Pediatr Adolesc Med 159*: 46–50.

Burnham, T. R., & A. Wilcox. 2002. "Effects of exercise on physiological and psychological variables in cancer survivors." *Med Sci Sports Exerc 34* (12): 1863–7.

Cassano, P., & M. Fava. 2002. "Depression and public health: An overview." *J Psychosom Res 53* (4): 849–57.

Cassilhas, R. C., H. K. M., Antunes, S. Tufik, & M. T. deMello. 2010. "Mood, anxiety, and serum IGF-1 in elderly men given 24 weeks of high resistance exercise." *Percept Mot Skills 110* (1): 265–76.

Castelli, D. M., C. H. Hillman, S. M. Buck, & H. E. Erwin. 2007. "Physical fitness and academic achievement in third- and fifth-grade students." *J Sport Exerc Psychol 29*: 239–52.

Chan, J. S. Y., J. H. Yan, & V. G. Payne. 2013. "The impact of obesity and exercise on cognitive aging." *Front Aging Neurosci 5*: 97.

Chang, Y. K., C. Y. Pan, F. T. Chen, C. L. Tsai, & C. C. Huang. 2012. "Effect of resistance-exercise training on cognitive function in healthy older adults: A review." *Journal of Aging and Physical Activity 20* (4): 497–517.

Chanias, A. K., G. Reid, & M. L. Hoover. 1998. "Exercise effects on health-related physical fitness of individuals with an intellectual disability: A meta-analysis." *Adapt Phys Activ Q 15* (2): 119–40.

Chen, A., J. Yan, H. Yin, C. Pan, & Y. Chang. 2014. "Effects of acute aerobic exercise on multiple aspects of executive function in preadolescent children." *Psychol Sport Exerc 15* (6): 627–36.

Chrousos, G. P., & P. W. Gold. 1992. "The concepts of stress and stress system disorders: Overview of physical and behavioral homeostasis." *JAMA 267* (9): 1244–52.

Coe, D. P., J. M. Pivarnik, C. J. Womack, M. J. Reeves, & R. M. Malina. 2012. "Health-related fitness and academic achievement in middle school students." *J Sports Med Phys Fitness 52* (6): 654–60.

Colcombe, S. J., & A. F. Kramer. 2003. "Fitness effects on the cognitive function of older adults: A meta-analytic study." *Psychol Sci 14*: 125–30.

Colcombe, S. J., A. F. Kramer, K. I. Erickson, P. Scalf, E. McAuley, N. J. Cohen,…S. Elavsky. 2004. "Cardiovascular fitness, cortical plasticity, and aging." *Proceedings of the Natl Acad Sci 101* (9): 3316–21.

Cooper Institute for Aerobics Research. 1999. *Fitnessgram: Test Administration Manual* Champaign, IL: Human Kinetics.

Cormie, P., A. K. Nowak, S. K. Chambers, D. A. Galvão, & R. U. Newton. 2015. "The potential role of exercise in neuro-oncology." *Front Oncol 5*: 85.

Crews, D. J., & D. M. Landers. 1987. "A meta-analytic review of aerobic fitness and reactivity to psychosocial stressors." *Med Sc Sports Exerc 19*: 114–20.

Crowgey, T., K. B. Peters, W. E. Hornsby, A. Lane, F. McSherry, J. E. Herndon,…L.W. Jones. 2013. "Relationship between exercise behavior, cardiorespiratory fitness, and cognitive function in early breast cancer patients treated with doxorubicin-containing chemotherapy: A pilot study." *Appl Physiol Nutr Metab 39* (6): 724–9.

Danielsson, L., A. M. Noras, M. Waern, & J. Carlsson. 2013. "Exercise in the threatment of major depression: A systematic review grading the quality of evidence." *Physiother Theory Pract 29* (8): 573–85.

Davis, C. L., P. D. Tomporowski, J. E. McDowell, B. P. Austin, P. H. Miller, N. E. Yanasak,...J.A. Naglieri. 2011. "Exercise improves executive function and achievement and alters brain activation in overweight children: A randomized, controlled trial." *Health Psychol 30* (1): 91–8.

DeBord, K., & E. Deike. 2015. "Effects of acute exercise on executive functions of cognition. *Intl J Exerc Sci (Conference Proceedings) 2* (7): Article 26.

Deci E. L., & R. M. Ryan. 2008. "Hedonia, eudaimonia, and well-being: An introduction." *J Happiness Stud 9*: 1–11.

Diamond, A. 2015. "Effects of physical exercise on executive functions: Going beyond simply moving to moving with thought." *Ann Sports Med Res 2* (1): 1011.

Dietrich, A., & P. B. Sparling. 2004. "Endurance exercise selectively impairs prefrontal-dependent cognition." *Brain Cogn 55* (3): 516–24.

Donnelly, J. E., J. L. Greene, C. A. Gibson, B. K. Smith, R. A. Washburn, D. K. Sullivan,...S. L. Williams. 2009. "Physical Activity Across the Curriculum (PAAC): A randomized controlled trial to promote physical activity and diminish overweight and obesity in elementary school children." *Prev Med 49*: 336–41.

Donnelly, J. E., & K. Lambourne. 2011. "Classroom-based physical activity, cognition, and academic achievement." *Prev Med 52*(Suppl1): S36–42.

Doyne, E. J., D. J. Ossip-Klein, E. D. Bowman, K. M. Osborn, I. B. McDougall-Wilson, & R. A. Neimeyer. 1987. "Running versus weight lifting in the treatment of depression." *J Consult Clin Psychol 55* (5): 748–54.

Drollette, E. S., M. R. Scudder, L. B. Raine, R. D. Moore, B. J. Saliba, M. B. Pontifex, & C. H. Hillman. 2014. "Acute exercise facilitates brain function and cognition in children who need it most: An ERP study of individual differences in inhibitory control capacity. *Dev Cogn Neurosci 7*: 53–64.

Ebmeier, K. P., C. Donaghey, & J. D. Steele. 2006. "Recent developments and current controversies in depression." *Lancet 367* (9505): 153–67.

Ensari, I., T. A. Greenlee, R. W. Motl, & S. J. Petruzzello. 2015. "Meta-analysis of acute exercise effects on state anxiety: An update of randomized controlled trials over the past 25 years." *Depress Anxiety 32*: 624–34.

Eriksen, C. W., & B. A. Eriksen. 1974. "Effects of noise letters upon the identification letter in a non-search task." *Perceptual Psychophysiol 16*: 143–9.

Etnier, J. L., W. Salazar, D. M. Landers, S. J. Petruzzello, M. Han, & P. Nowell. 1997. "The influence of physical fitness and exercise upon cognitive functioning: A meta-analysis." *J Sport Exerc Psychol 19*: 249–77.

Fardell, J. E., J. Vardy, J. D. Shah, & I. N. Johnston. 2012. "Cognitive impairments caused by oxaliplatin and 5-fluorouracil chemotherapy are ameliorated by physical activity." *Psychopharmacol 220* (1): 183–93.

Fleshner, M. 2005. "Physical activity and stress resistance: Sympathetic nervous system adaptations prevent stress-induced immunosuppression." *Exerc Sport Sci Rev 33*: 120–6.

Forcier, K., L. R. Stroud, G. D. Papandonatos, B. Hitsman, M. Reiches, J. Krishnamoorthy, & R. Niaura. 2006. "Links between physical fitness and cardiovascular reactivity and recovery to psychological stressors: A meta-analysis." *Health Psychol 25* (6): 723–39.

Fournier, J. C., R. J. DeRubeis, S. D. Hollon, S. Dimidjian, J. D. Amsterdam, R. C. Shelton, & J. Fawcett. 2010. "Antidepressant drug effects and depression severity: A patient-level meta-analysis." *JAMA 303* (1): 47–53.

Gabbard, C., & J. Barton. 1979. "Effects of physical activity on mathematical computation among young children." *J Psychol 103*: 287–8.

Gothe, N. P., A. F. Kramer, & E. McAuley. 2014. "The effects of an 8-week Hatha yoga intervention on executive function in older adults." *J Gerontol A Biol Sci Med Sci 69* (9): 1109–16.

Greene, D. R., & S. J. Petruzzello. 2014. "More isn't necessarily better: Examining the intensity-affect-enjoyment relationship in the context of resistance exercise." *Sport Exerc Perform Psychol 4* (2): 75–87.

Gum, A. M., B. King-Kallimanis, & R. Kohn. 2009. "Prevalence of mood, anxiety, and substance-abuse disorders for older Americans in the National Comorbidity Survey-replication." *Am J Geriatr Psychiatry 17* (9): 769–81.

Guszkowska, M., & S. Sionek. 2009. "Changes in mood states and selected personality traits in women participating in a 12-week exercise program." *Hum Mov 10* (2): 163–9.

Hale, W. W., Q. Raaijmakers, P. Muris, A. Van Hoof, & W. Meeus. 2008. "Developmental trajectories of adolescent anxiety disorders symptoms: A 5-year prospective community study." *J Am Acad Child Adolesc Psychiatry 47* (5): 556–64.

Hamer, M., & A. Steptoe. 2013. "Physical activity, stress reactivity, and stress-mediated pathophysiology." In P. Ekkekakis (Ed.), ***Routledge handbook of physical activity and mental health*** (pp. 303–15). New York: Routledge.

Hamer, M., A. Taylor, & A. Steptoe. 2006. "The effect of acute aerobic exercise on stress related blood pressure responses: A systematic review and meta-analysis." *Biol Psychol 71*: 183–90.

Hammad, T. A., T. Laughren, & J. Racoosin. 2006. "Suicidality in pediatric patients treated with antidepressant drugs." *Arch Gen Psychiatry 63*: 332–9.

Hassmen, P., N. Koivula, & A. Uutela. 2000. "Physical exercise and psychological well-being: A population study in Finland." *Prev Med 30*: 17–25.

Herring, M., M. Jacob, C. Suveg, & P. O'Connor. 2011. "Effects of short-term exercise training on signs and symptoms of generalized anxiety disorder." *Ment Health Phys Act 4*: 71–7.

Herring, M. P., & P. J. O'Connor. 2009. "The effect of acute exercise on feelings of energy and fatigue." *J Sports Sci 27* (7): 701–9.

Heydari, M., Y. N. Boutcher, & S. H. Boutcher. 2013. "The effects of high-intensity intermittent exercise training on cardiovascular response to mental and physical challenge." *Int J Psychophysiol 87*: 141–6.

Heyn, P., B. C. Abreu, & K. J. Ottenbacher. 2004. "The effects of exercise training on elderly persons with cognitive impairment and dementia: A meta-analysis." *Arch Phys Med Rehabil 85* (10): 1694–1704.

Hillman, C. H., D. M. Castelli, & S. M. Buck. 2005. "Aerobic fitness and neurocognitive function in healthy preadolescent children." *Med Sci Sports Exerc 37*: 1967–74.

Hillman, C. H., K. I. Erickson, & A. F. Kramer. 2008. "Be smart, exercise your heart: Exercise effects on brain and cognition." *Nat Rev Neurosci 9*: 58–65.

Hind, K., D. Torgerson, J. Mckenna, R. Ashby, A. Daly-Smith, & A. Jennings. 2014. "Developing interventions for children's exercise (DICE): A pilot evaluation of school-based exercise interventions for primary school children aged 7 to 8 years." *J Phys Act Health 11*: 699–704.

Hogan, C. L., J. Mata, & L. L. Carstensen. 2013. "Exercise holds immediate benefits for affect and cognition in younger and older adults." *Psychol Aging 28* (2): 587–94.

Hogervorst, E., W. Riedel, A. Jeukendrup, & J. Jolles. 1996. "Cognitive performance after strenuous physical exercise." *Percept Mot Skills 83*: 479–88.

Hsu, J. H., & W. W. Shen. 1995. "Male sexual side effects associated with antidepressants: A descriptive clinical study of 32 patients." *Int J Psychiatry Med 25* (2): 191–201.

Jackson, E. M., & R. K. Dishman. 2006. "Cardiorespiratory fitness and laboratory stress: A meta-regression analysis." *Psychophysiol 43*: 57–72.

Jarrett, O. S., D. M. Maxwell, C. Dickerson, P. Hoge, G. Davies, & A. Yetley. 1998. "Impact of recess on classroom behavior: Group effects and individual differences." *J Educ Res 92*: 121–6.

Jayakody, K., S. Gunadasa, & C. Hosker. 2014. "Exercise for anxiety disorders: Systematic review." *Br J Sports Med 48*: 187–96.

Kamijo, K., Y. Hayashi, T. Sakai, T. Yahiro, K. Tanaka, & Y. Nishihira. 2009. "Acute effects of aerobic exercise on cognitive function in older adults." *J Gerontol B Psychol Sci Soc Sci 64*: 356–63.

Kessler, R. C., W. T. Chiu, O. Demler, & E. E. Walters. 2005. "Prevalence, severity, and comorbidity of 12-month DSM-IV disorders in the National Comorbidity Survey Replication." *Arch Gen Psychiatry 62*: 617–27.

Kessler, R. C., K. A. McGonagle, S. Zhao, C. B. Nelson, M. Hughes, S. Eshleman,… K. S. Kendler. 1994. "Lifetime and 12-month prevalence of DSM-III-R psychiatric disorders in the United States: Results from the National Comorbidity Survey." *Arch Gen Psychiatry 51*: 8–18.

Khan, N. A., & C. H. Hillman. 2014. "The relation of childhood physical activity and aerobic fitness to brain function and cognition: A review." *Ped Exerc Sci 26* (2): 138–46.

King, A. C., K. Baumann, P. O'Sullivan, S. Wilcox, & C. Castro. 2002. "Effects of moderate-intensity exercise on physiological, behavioral, and emotional responses to family caregiving: A randomized controlled trial." *J Gerontol A Biol Sci Med Sci 57*: M26–36.

Kramer, A. F., S. Hahn, E. McAuley. 2000. "Influence of aerobic fitness on the neurocognitive function of older adults." *J Aging Phys Activ 8*: 379–85.

Lambourne, K., & P. Tomporowski. 2010. "The effect of exercise-induced arousal on cognitive task performance: A meta-regression analysis." *Brain Res 1341*: 12–24.

Larson, E. B., L. Wang, J. D. Bowen, W. C. McCormick, L. Teri, P. Crane, & W. Kukull. 2006. "Exercise is associated with reduced risk for incident dementia among persons 65 years of age and older." *Ann Intern Med 144*: 73–81.

Lees, C., & J. Hopkins. 2013. "Effect of aerobic exercise on cognition, academic achievement, and psychosocial function in children: A systematic review of randomized control trials." *Prev Chronic Dis 10*: E174.

Leith, L.M. 1994. *Foundations of exercise and mental health*. Morgantown, WV: Fitness Information Technology.

Lista, I., & G. Sorrentino. 2009. "Biological mechanisms of physical activity in preventing cognitive decline." *Cell Mol Neurobiol 30* (4): 493–503.

Longo, L. P., & B. Johnson. 2000. "Addiction: Part I. Benzodiazepines-side effects, abuse risk and alternatives." *Am Fam Physician 61* (7): 2121–8.

Lopez, D., C. D. Mathers, M. Ezzati, D. T. Jamison, & C. J. L. Murray. 2006. "Global and regional burden of disease and risk factors, 2001: Systematic analysis of population health data." *Lancet 367* (9524): 1747–57.

Lupien, S. J., B. S. McEwen, M. R. Gunnar, & C. Heim. 2009. "Effects of stress throughout the lifespan on the brain, behavior and cognition." *Nat Rev Neurosci 10*: 434–45.

Lupien, S. J., F. Maheu, M. Tu, A. Fiocco, & T. E. Schramek. 2007. "The effects of stress and stress hormones on human cognition: Implications for the field of brain and cognition." *Brain Cogn 65*: 209–37.

McMorris, T., K. Collard, J. Corbett, M. Dicks, & J. P. Swain. 2008. "A test of the catecholamines hypothesis for an acute exercise-cognition interaction." *Pharmacol Biochem Behav 89*: 106–15.

Mahar, M. T., S. K. Murphy, D. A. Rowe, J. Golden, A. T. Shields, & T. D. Raedeke. 2006. "Effects of a classroom-based program on physical activity and on-task behavior." *Med Sci Sports Exerc 38*: 2086–94.

Mathers, C. D., & D. Loncar. 2006. "Projections of global mortality and burden of disease from 2002 to 2030." *PLoS Med 3* (11): e442.

Miller, A. A., & S. J. Spencer. 2014. "Obesity and neuroinflammation: A pathway to cognitive impairment." *Brain Behav Immun 42*: 10–21.

Morvell, N., & D. Belles. 1993. "Psychological and physical benefits of circuit weight training in law enforcement personnel." *J Consult Clin Psychol 61* (3): 520–7.

Motl, R., & R. K. Dishman. 2004. "Effects of acute exercise on the soleus H-reflex and self-reported anxiety after caffeine ingestion." *Physiol Behav 80*: 577–85.

Murray, D. K., M. A. Sacheli, J. J. Eng, & A. J. Stoessl. 2014. "The effects of exercise on cognition in Parkinson's disease: A systematic review." *Transl Neurodegener 3*: 5.

O'Connor, P. J., C. X. Bryant, J. P. Veltri, & S. H. Gebhardt. 1993. "State anxiety and ambulatory blood pressure following resistance exercise in females." *Med Sci Sports Exerc 25*: 516–21.

Petruzzello, S. J., D. M. Landers, B. D. Hatfield, K. A. Kubitz, & W. Salazar. 1991. "A meta-analysis on the anxiety-reducing effects of acute and chronic exercise." *Sports Med 11* (3): 143–82.

Petruzzello, S. J., & R. W. Motl. 2011. "Acute moderate-intensity cycling exercise is associated with reduced fatigue in person with multiple sclerosis." *Ment Health Phys Activ 4*: 1–4.

Petruzzello, S. J., E. M. Snook, R. C. Gliottoni, & R. W. Motl. 2009. "Anxiety and mood changes associated with acute cycling in persons with multiple sclerosis." *Anxiety Stress Coping 22* (3): 297–307.

Physical Activity Guidelines Advisory Committee. 2008. *Physical Activity Guidelines Advisory Committee Report, 2008*. Washington, DC: US Department of Health and Human Services, A1-H14.

Pontifex, M. B., & C. H. Hillman. 2007. "Neuroelectric and behavioral indices of interference control during acute cycling." *Clin Neurophysiol 118*: 570–80.

Pontifex, M. B., A. C. Parks, D. A. Henning, & K. Kamijo. 2015. "Single bouts of exercise selectively sustain attentional processes." *Psychophysiol 52* (5): 618–25.

Pretty, J., J. Peacock, R. Hine, M. Sellens, N. South, & M. Griffin. 2007. "Green exercise in the UK countryside: Effects on health and psychological well-being, and implications for policy and planning." *J Environ Plan Manage 50* (2): 211–31.

Pretty, J., J. Peacock, M. Sellens, & M. Griffin. 2005. "The mental and physical health outcomes of green exercise." *International J Environ Health Res 15* (5): 319–37.

Rejeski, W. J., C. J. Hardy, & J. Shaw. 1991. "Psychometric confounds of assessing state anxiety in conjunction with acute bouts of vigorous exercise." *J Sport Exerc Psychol 13*: 65–74.

Rejeski, W. J., A. Thompson, P. H. Brubaker, & H. S. Miller. 1992. "Acute exercise: Buffering psychosocial stress in women." *Health Psychol 11*: 355–62.

Rimmele, U., R. Seiler, B. Marti, P. H. Wirtz, U. Ehlert, & M. Heinrichs. 2009. "The level of physical activity affects adrenal and cardiovascular reactivity to psychosocial stress." *Psychoneuroendocrinology 34*: 190–8.

Rimmele, U., B. C. Zellweger, B. Marti, R. Seiler, C. Mohiyeddini, U. Ehlert, & M. Heinrichs. 2007. "Trained men show lower cortisol, heart rate and psychological responses to psychosocial stress compared with untrained men." *Psychoneuroendocrinology 32*: 627–35.

Roemmich, J. N., M. Lambaise, S. J. Salvy, & P. J. Horvath. 2009. "Protective effect of interval exercise on psychophysiological stress reactivity in children." *Psychophysiol 46*: 852–61.

Sapolsky, R. 2003. "Taming stress." *Sci Am 289* (3): 88–95.

Schmidt-Kassow, M., M. Deusser, C. Thiel, S. Otterbein, C. Montag, M. Reuter,...J. Kaiser. 2013. "Physical exercise during encoding improves vocabulary learning in young female adults: A neuroendocrinological study." *PLoS One 8* (5): e64172.

Scully, D., J. Kremer, M. M. Meade, R. Graham, & K. Dudgeon. 1998. "Physical exercise and psychological well-being: A critical review." *Br J Sport Med 32*: 111–20.

Sibley, B. A., & J. L. Etnier. 2003. "The relationship between physical activity and cognition in children: A meta-analysis." *Ped Exerc Sci 15*: 243–56.

Singh, N. A., K. M. Clements, & M. A. Fiatarone. 1997. "A randomized controlled trial of progressive resistance training in depressed elders." *J Gerontol A Bio Sci Med Sci 52*: M27–35.

Singh, N. A., T. M. Stavrinos, Y. Scarbeck, G. Galambos, C. Liber, & M. A. Fiatarone Singh. 2005. "A randomized controlled trial of high versus low intensity weight training versus general practitioner care for clinical depression in older adults." *J Gerontol A Bio Sci Med Sci 60*: 768–76.

Sofi, F., D. Valecchi, D. Bacci, R. Abbate, G. F. Gensini, A. Casini, & C. Macchi. 2011. "Physical activity and risk of cognitive decline: A meta-analysis of prospective studies." *J Intern Med 269* (1): 107–17.

Sothmann, M. S., J. Buckworth, R. P. Claytor, R. H. Cox, J. E. White-Welkley, & R. K. Dishman. 1996. "Exercise training and the cross-stressor adaptation hypothesis." *Exerc Sport Sci Rev 24*: 267–87.

Stein, P. N. & R. W. Motta. 1992. "Effects of aerobic and nonaerobic exercise on depression and self-concept." *Percept Mot Skills 74*: 79–89.

Strickland, J. C., & M. A. Smith. 2014. "The anxiolytic effects of resistance exercise." *Front Psychol Mov Sci Sport Psychol 5*: 1–6.

Stroop, J. R. 1935. "Studies of interference in serial verbal reactions." *J Exp Psychol 18* (6): 643–62.

Substance Abuse and Mental Health Services Administration. 2013. *Results from the 2012 National Survey on Drug Use and Health: Mental Health Findings*, NSDUH Series H-47, HHS Publication No. (SMA) 13–4805. Rockville, MD: Substance Abuse and Mental Health Services Administration.

Taylor, S. E., L. C. Klein, B. P. Lewis, T. L. Gruenewald, R. A. Gurung, & J. A. Updegraff. 2000. "Biobehavioral responses to stress in females: Tend-and-befriend, not fight-or-flight." *Psychol Rev 107* (3): 411–29.

Tolppanen, A. M., A. Solomon, J. Kulmala, I. Kåreholt, T. Ngandu, M. Rusanen,...M. Kivipelto. 2015. "Leisure-time physical activity from mid- to late life, body mass index, and risk of dementia." *Alzheimers Dement 11*: 434–43.

Tomporowski, P. D. 2003. "Effects of acute bouts of exercise on cognition." *Acta Psychol 112*: 297–324.

Tomporowski, P. D., C. L. Davis, P. H. Miller, & J. A. Naglieri. 2008. "Exercise and children's intelligence, cognition, and academic achievement." *Educ Psychol Rev 20 (2)*: 111–31.

Tonoli, C., E. Heyman, B. Roelands, N. Pattyn, L. Buyse, M. F. Piacentini,...R. Meeusen. 2014. "Type 1 diabetes-associated cognitive decline: A meta-analysis and update of the current literature." *J Diabetes 6*: 499–513.

Traustadottir, T., P. R. Bosch, & K. S. Matt. 2005. "The HPA axis response to stress in women: Effects of aging and fitness." *Psychoneuroendocrinology 30*: 392–402.

Tsatsoulis, A., & S. Fountoulakis. 2006. "The protective role of exercise on stress system dysregulation and comorbidities." *Ann NY Acad Sci 1083*: 196–213.

Tsutsumi, T., B. M. Don, L. D. Zaichowsky, & L. L. Delizonna. 1997. "Physical fitness and psychological benefits of strength training in community dwelling older adults." *Appl Human Sci 16*: 257–66.

Tsutsumi, T., B. M. Don, L. D. Zaichowsky, K. Takenaka, K. Oka, & T. Ohno. 1998. "Comparison of high and moderate intensity of strength training on mood and anxiety in older adults." *Percept Mot Skills 87*: 1003–11.

Uc, E. Y., K. C. Doerschug, V. Magnotta, J. D. Dawson, T. R. Thomsen, J. N. Kline,... W.G. Darling. 2014. "Phase I/II randomized trial of aerobic exercise in Parkinson disease in a community setting." *Neurology 83* (5): 413–25.

U.S. Department of Health and Human Services. 1999. *Mental Health: A Report of the Surgeon General*. Rockville, MD: National Institute of Mental Health.

Verstynen, T. D., B. Lynch, D. L. Miller, M. W. Voss, R. S. Prakash, L. Chaddock, & K. I. Erickson. 2012. "Caudate nucleus volume mediates the link between cardiorespiratory fitness and cognitive flexibility in older adults." *J Aging Res 2012*: 939285.

Voss, M. W., L. Chaddock, J. S. Kim, M. Vanpatter, M. B. Pontifex, L. B. Raine,...A.F. Kramer. 2011. "Aerobic fitness is associated with greater efficiency of the network underlying cognitive control in preadolescent children." *Neurosci 199*: 166–76.

Weinberg, R., A. Jackson, & K. Kolodny. 1987. "The relationship of massage and exercise to mood enhancement." *Sport Psychol 2*: 202–11.

Winefield, H. R., T. K. Gill, A. W. Taylor, & R. M. Pilkington. 2012. "Psychological well-being and psychological distress: Is it necessary to measure both?" *Psychol Well Being 2*: 3.

Winter, B., C. Breitenstein, F. C. Mooren, K. Voelker, M. Fobker, A. Lechtermann,... S. Knecht. 2007. "High impact running improves learning." *Neurobiol Learn Mem 87* (4): 597–609.

Wipfli, B. M., C. D. Rethorst, & D. M. Landers. 2008. "The anxiolytic effects of exercise: A meta-analysis of randomized trials and dose-response analysis. *J Sport Exerc Psychol 30*: 392–410.

Yagi, Y., K. L. Coburn, K. M. Estes, & J. E. Arruda. 1999. "Effects of aerobic exercise and gender on visual and auditory P300, reaction time, and accuracy." *Eur J Appl Physiol Occup Physiol 80* (5): 402–8.

Zhu, N., D. Jacobs, P. Schreiner, K. Yaffe, N. Bryan, L. Launer,...B. Sternfeld. 2014. "Cardiorespiratory fitness and cognitive function in middle age: The CARDIA Study." *Neurology 82* (15): 1339–46.

18 Can Physical Activity Prevent Mental Illness?

Viviane Grassmann, George Mammen, and Guy Faulkner

Introduction

Mental disorders are primary contributors to the global burden of disease and they account for approximately 7.4% of disease burden worldwide (Whiteford et al. 2013). Mental illness and poor mental health is a tremendous burden on quality of life for individuals and their families, and the health care system. In any given year, one in five people in Canada experiences a mental health problem or illness, with a cost to the economy estimated to be in excess of $50 billion annually (Smetanin et al. 2011). In the province of Ontario, Canada, the burden of mental illness and addictions in Ontario is more than 1.5 times that of all cancers and more than seven times that of infectious disease (Ratnasingham, Cairney, Rehm, Manson, & Kurdyak 2012).

Treatment remains a priority and improving access to treatment doubly so. However, improvement in population health is only possible if the prevention of mental and substance use disorders turns into a public health priority (Whiteford et al. 2013). Accordingly, the focus of this chapter is to explore whether physical activity may also serve a role in preventing mental disorders – in particular dementia and depression, both are highly prevalent now and expected to rise in the future (Prince et al. 2013; World Health Organization 2012).

Does Physical Activity Prevent Dementia?

Dementia is an overall term for a set of symptoms reflecting significant cognitive decline in one or more domains (such as: attention, executive function, learning and memory, language, social cognition, and perceptual motor) that interferes with the ability to engage in activities of daily living. In 2014, the DSM-V (Diagnostic and Statistical Manual of Mental Disorders, fifth edition) started to use the term *Major Neurocognitive Disorder* as a substitute for dementia, although dementia is often used synonymously (American Psychiatric Association 2013). The most common types of dementia are Alzheimer's disease and vascular dementia. *Alzheimer's disease* shows a progressive and gradual decline in cognitive function that is usually associated with a genetic mutation. This cognitive decline is mainly associated with a decrease in memory and learning performance, but can also be associated with executive dysfunction. People with *vascular dementia* show attention and executive function decline after one or more cerebral vascular events (American Psychiatric Association 2013).

The worldwide prevalence of dementia in the population aged ≥ 60 years is between 5%–7% in most regions of the world. It was estimated that, in 2010, more than 35 million people had dementia, but those numbers are expected to increase over the next 20 years to more than 65 million in 2030 and 115.4 million in 2050 (Prince et al. 2013). This is due to increased life expectancy, as the prevalence of dementia increases with increasing age, doubling every five years after age 65 (Hugo & Ganguli 2014).

Dementia has tremendous consequences for individuals, the healthcare system, the economy, and to families and caregivers (that can experience increased emotional stress, depression, and other health problems) (Hugo & Ganguli 2014). This also affects the quality of life in these patients, by affecting their relationship with other people, decreasing their agency in life (autonomy and independence), wellness and sense of place (disorientation), which is associated with sadness (O'Rourke, Duggleby, Fraser, & Jerke 2015).

Although there are different types of pharmacological treatment for dementia (e.g. cholinesterase inhibitors, NMDA receptor antagonist, dopamine blocking agents, serotonergic agents), these can only treat the symptoms and not the cause of the illness (Hugo & Ganguli 2014). Thus, prevention is the best form of treatment. Given also the impact of dementia on quality of life and health care costs, a focus on prevention is clearly warranted. Although there are many non-modifiable variables related to this disease (e.g. age, genetic predisposition, and sex), there is suggestive evidence that lifestyle factors, including PA, may protect against cognitive decline and dementia (Plassman, Williams, Jr., Burke, Holsinger, & Benjamin 2010). Blondell, Hammersley-Mather, and Veerman (2014) conducted the most recent systematic review analysing the relationship between dementia and PA. Based on 21 studies, the authors put forward the case for a causal relationship between PA and reduced risk of dementia. This relationship was most evident when participation in higher levels of PA (types, intensity, and/or frequency) was compared to a lower level of participation (RR 0.86, 95% CI 0.76–0.97).

It is important to note that other factors, such as genetic predisposition, could be moderating the relationship between PA and the reduced risk of dementia. The best-established genetic factor associated with dementia is the APOE (apolipoprotein E) polymorphism on the Chromosome 19 (Hugo & Ganguli 2014). The APOE is a prevalent brain lipoprotein that is expressed in three common forms (APOE ε2; APOE ε3; APOE ε4) that differs by one or two amino acids. In people that carry APOE ε4 alleles, this lipoprotein is deposited in neurofibrillary tangles and neuritic plaques, and this contributes to the neurodegenerative process of Alzheimer's disease (Michaelson 2014). Although APOE ε4 is a genetic risk factor that is mainly related to Alzheimer's Disease (Michaelson 2014), the presence of APOE ε4 alleles is also associated with the risk for vascular dementia (Rohn 2014). The presence of APOE ε4 alleles can modify a wide variety of responses to therapeutic treatment in this population, thus the presence of APOE ε4 alleles should be considered when studies evaluate dementia risk (Hanson, Craft, & Banks 2015).

Therefore the purpose of this section is to update the review conducted by Blondell and colleagues (2014) examining the association between PA and reduced risk of dementia and explore the level of PA required to prevent dementia. We extend Blondell and colleagues' (2014) review by explicitly evaluating the possibility of genetic influence on the relationship. An overview of possible mechanisms and methodological issues and future research needs is then provided.

Search Strategy

We replicated the search strategy described by Blondell and colleagues (2014) in identifying relevant published articles between January 2014 and June

2015 on PubMed and PsyInfo databases. The search terms used were "physical activity" OR "exercise" and "cognitive decline" OR "dementia" OR "cognitive impairment" OR "Alzheimer's disease" OR "cognition."

Inclusion/Exclusion Criteria

The studies were chosen based on the following criteria: (1) prospective design; (2) population-based sample; (3) with a definition of what constitutes dementia; (4) with a description of the methods used to assess the disease; (5) with physical activity assessed at baseline; (6) reports estimates of association between physical activity subgroups and dementia; and (7) with participants aged ≥ 40 years.

Study Identification and Selection

The searches yielded 1242 titles, 28 of which were identified as a potentially relevant study after reviewing the title/abstract. After a full text reading, three new studies were included and analysed with the 21 studies identified in Blondell and colleagues' systematic review (2014) (for a summary of included studies see Table 18.1). From the 25 excluded studies, five had an inappropriate design; 12 were not specific to dementia; five did not have a definition for dementia; two did not exclude individuals with dementia at baseline; and one did not assess PA.

Study Characteristics

The majority of studies were conducted in North America (n=10; Bowen 2012; Buchman et al. 2012; Larson et al. 2006; Laurin, Verreault, Lindsay, MacPherson, & Rockwood 2001; Podewils et al. 2005; Scarmeas et al. 2009; Taaffe et al. 2008; Verghese et al. 2003; Wang, Luo, Barnes, Sano, & Yaffe 2014; Wilson et al. 2002) and Europe (n=8; Chang et al. 2010; de Bruijn et al. 2013; Fabrigoule et al. 1995; Luck et al. 2014; Morgan et al. 2012; Ravaglia et al. 2008; Rovio et al. 2005; Rovio et al. 2007). In all studies the subjects did not have any type of dementia at baseline. Although most of the studies included participants of both sexes, three studies (Abbott et al. 2004; Morgan et al. 2012; Taaffe et al. 2008) were conducted only with males and one study included only females (Wang et al. 2014). Conventionally, people aged 65 years old and older are considered elderly (Orimo et al. 2006). The youngest mean of age of participants at baseline was 51 years old, although almost half of the studies (n=11) were conducted with participants 65 years old and older. The length of follow-up from baseline varied from six months to 26 years, and half of the studies varied from four to seven years. The number of participants that completed a follow-up assessment varied between 469 and 4761.

Most studies evaluated people that developed more than one type of dementia (n=21), but three studies evaluated only people with Alzheimer's disease. All studies used well-validated methods such as DSM (Diagnostic and Statistical Manual of Mental Disorders-III, III-R or IV), National Institute of Neurological Disorders and Stroke (NINDS), Alzheimer's Disease and Related Disorders Association (ADRDA) and/or Association *Internationale pour la Recherche et l'Enseignement en Neurosciences* (AIREN), International Statistical Classification of Diseases and Related Health Problems (ICD-10) and/or Cognitive Abilities Screening Instrument (CASI) to diagnose dementia. Regarding PA, all but one study used a self-report measure. Only eight studies (Buchman et al. 2012; de Bruijn et al. 2013; Gureje, Ogunniyi, Kola, & Abiona 2011; Laurin et al. 2001; Luck et al. 2014; Morgan et al. 2012; Podewils et al. 2005; Scarmeas et al. 2009) used valid and reliable questionnaires; however, those studies were heterogeneous, as none of them used the same measure. Podewils et al. (2005) and Morgan et al. (2012), used similar measures (the Minnesota Leisure Time Activity Questionnaire), although Podewils et al. (2005) used a modified version that included more activities.

In contrast, Buchman et al. (2012) evaluated total daily PA of participants for up to 10 days (24 hours a day) using actigraphy on the non-dominant wrist. The total daily PA was considered as the average sum of all recorded activity in the day. They also used a self-report measure asking participants the time they spent in five different types of PA during the week.

Another important point is the variation in the number of confounders controlled for. Although most articles found a preventive effect of PA after controlling for age, sex, years of education, smoking/alcohol habits, and the presence of the APOE ε4 gene, one study controlled only for age (Fabrigoule et al. 1995); and one did not include any other variable in the model (Wei et al. 2014) when the preventive effect of PA was evaluated in any type of dementia. For Alzheimer's disease two studies only controlled for age (Fabrigoule et al. 1995; Taaffe et al. 2008) and one study only included age and sex (Larson et al. 2006). For vascular disease, one study included only age in the analysis (Morgan et al. 2012).

Does Being Physically Active Prevent Dementia?

The majority (n=19) of the studies reported a significant relationship between PA at baseline and a decrease in the risk of developing at least one type of dementia later in life (Abbott et al. 2004; Bowen 2012; Buchman et al. 2012; Chang et al. 2010; de Bruijn et al. 2013; Fabrigoule et al. 1995; Kim et al. 2011; Larson et al. 2006; Laurin et al. 2001; Luck et al. 2014; McCallum, Simons, Simons, & Friedlander 2007; Podewils et al. 2005; Ravaglia et al. 2008; Rovio et al. 2005; Scarmeas et al. 2009; Taaffe et al. 2008; Wang et al. 2014; Wei et al. 2014; Yoshitake et al. 1995) with a range of reduced risk between RR 0.18 (CI 0.06–0.61) (Yoshitake et al. 1995) and HR 0.84 (CI 0.73–0.97) (De Brujin et al. 2013). This suggests a preventive effect of PA on dementia.

Most of the studies evaluated more than one type of dementia (n=14). From the 20 studies that evaluated the relationship between PA and any type of dementia (Abbott et al. 2004; Bowen 2012; Chang et al. 2010; de Bruijn et al. 2013; Fabrigoule et al. 1995; Gureje et al. 2011; Kim et al. 2011; Laurin et al. 2001; Larson et al. 2006; Luck et al. 2014; McCallum et al. 2007; Morgan et al. 2012; Podewils et al. 2005; Ravaglia et al. 2008; Rovio et al. 2005; Rovio et al. 2007; Taaffe et al. 2008; Verghese et al. 2003; Wang et al. 2014; Wei et al. 2014), 16 (80%) reported a protective effect of PA (Abbott et al. 2004; Bowen 2012; Chang et al. 2010; de Bruijn et al. 2013; Fabrigoule et al. 1995; Kim et al. 2011; Laurin et al. 2001; Larson et al. 2006; Luck et al. 2014; McCallum et al. 2007; Podewils et al. 2005; Ravaglia et al. 2008; Rovio et al. 2005; Taaffe et al. 2008; Wang et al. 2014; Wei et al. 2014).

Considering only Alzheimer's disease, from the 16 studies that evaluated this type of dementia (Abbott et al. 2004; Buchman et al. 2012; de Bruijn et al. 2013; Fabrigoule et al. 1995; Larson et al. 2006; Laurin et al. 2001; Luck et al. 2014; Podewils et al. 2005; Ravaglia et al. 2008; Rovio et al. 2005; Rovio et al. 2007; Scarmeas et al. 2009; Taaffe et al. 2008; Verghese et al. 2003; Wilson et al. 2002; Yoshitake et al. 1995), 11 (69%) found a protective effect of PA (Abbott et al. 2004; Buchman et al. 2012; Fabrigoule et al. 1995; Larson et al. 2006; Laurin et al. 2001; Luck et al. 2014; Podewils et al. 2005; Rovio et al. 2005; Scarmeas et al. 2009; Taaffe et al. 2008; Yoshitake et al. 1995). While for vascular dementia, from the eight studies (Abbott et al. 2004; Laurin et al. 2001; Morgan et al. 2012; Podewils et al. 2005; Ravaglia et al. 2008; Taaffe et al. 2008; Verghese et al. 2003; Yoshitake et al. 1995) only two (25%) observed a protective effect of PA (Morgan et al. 2012; Ravaglia et al. 2008).

Among the five studies that did not show a preventive effect of PA, one evaluated this effect only in Alzheimer's Disease regardless of other types of dementia (Wilson et al. 2002), while the other four studies included different types of dementia (Gureje et al. 2011; Morgan et al. 2012; Rovio et al. 2007; Verghese et al. 2003). There were also limitations in how PA was assessed. No information was presented regarding the validity of the PA measure used in three studies (Rovio et al. 2007; Verghese et al. 2003; Wilson et al. 2002), nor reliability in two studies (Rovio et al. 2007; Wilson et al. 2002). Another important point is that, unlike other studies, only Rovio et al. (2007) and Morgan et al. (2012) evaluated work-related PA. While Rovio et al. (2007) did not find any effect of occupational PA, Morgan et al. (2012) showed that increasing work-related activity results in an increase in the odds of developing vascular and non-vascular disease. Nonetheless, after full adjustment in the model, including socioeconomic status, this relationship was no longer present.

Only one study directly compared differences between sexes. Laurin et al. (2001) showed that when both men and women were evaluated together there was a preventive effect of PA on any type of dementia and Alzheimer's disease. However, when they were evaluated separately there was only a preventive effect for women. In contrast, two other studies evaluated only men and showed a preventive effect of PA on the same types of dementia (Abbott et al. 2004; Taaffe et al. 2008).

Another important aspect is the length of follow-up, since it is possible that the effect of PA could decrease after a long period of time from the baseline PA measure. The time of follow-up also varied across the studies, but it can be noticed that in a small period of follow-up (less than 3 years) all studies were able to show the preventive effect of PA on dementia (any type and Alzheimer's) (Fabrigoule et al. 1995; Kim et al. 2011; Wei et al. 2014), while after a period between 3 and 9 years of follow-up, most of them showed this preventive effect on dementia (any type) and Alzheimer's Disease (Abbott et al. 2004; Bowen 2012; Buchman et al. 2012; de Bruijn et al. 2013; Larson et al. 2006; Laurin et al. 2001; Luck et al. 2014; Podewils et al. 2005; Ravaglia et al. 2008, Scarmeas et al. 2009; Taaffe et al. 2008; Wang et al. 2014; Yoshitake et al. 1995), but not for vascular disease (Abbott et al. 2004; Laurin et al. 2001; Taaffe et al. 2008; Verghese et al. 2003; Yoshitake et al. 1995). However, after a long-term period (more than 10 years), almost half of studies reported this association (Chang et al. 2010; McCallum et al. 2007; Rovio et al. 2005). Thus, the preventive effect of PA on dementia (any type) and Alzheimer's Disease seems to weaken with longer years of follow-up. It is important to note that those studies with a long-term period of follow-up only assessed PA once at baseline. Given PA declines with age (Sallis 2000), it is possible that as people become less active over time the protective effect weakens. Currently, there are no studies examining differences in dementia incidence amongst those who engage in PA versus those who do not at various follow-up periods. It is also plausible that other modifiable and/or non-modifiable risk factors may have a stronger influence on dementia incidence above that of PA later in life.

How Much Physical Activity Is Needed to Prevent Dementia?

Given the heterogeneity in how PA was assessed across studies in terms of dimensions such as type, intensity, and frequency, it is difficult to identify an optimal dose of PA needed to prevent dementia. However, most of the studies (n=13) showed that a higher level (many or several times per week and/or moderate to vigorous intensity) of PA has a more protective effect against Alzheimer's disease, vascular dementia or any type of dementia when compared to a lower level (Abbott et al. 2004; Buchman et al. 2012; Kim et al. 2011; Laurin et al. 2001; Larson et al. 2006; Luck et al. 2014; McCallum et al. 2007; Podewils et al. 2005; Ravaglia et al. 2008; Rovio et al. 2005; Scarmeas et al. 2009; Taaffe et al. 2008; Wang et al. 2014). Contradicting this evidence, Chang et al. (2010) showed that there was no preventive effect of PA for people who participate in more than 5 hours of PA per week during midlife when compared to those who never exercised (at midlife). Nonetheless, a preventive effect was found in those who practice less than 5 hours of PA per week when compared to the same population. They further explained that a moderate amount of PA was better than high or no PA at all. On the other hand, Rovio et al. (2007) presented the opposite result, demonstrating that an active commute to work (more than 60 minutes per day) or being sedentary during the commute to work reduces the odds of developing dementia or Alzheimer's disease when compared to people who have a moderate length of active commute (<59 minutes per day). It is important to note that those two studies have some bias, as they did not use valid and reliable measures of PA.

Although many studies used a valid and/or reliable measure for PA, in fact, only one study used an objective measure (10 days of actigraphy). This study reported that people with low total daily PA (10th percentile of the average sum) had 2.3 times more risk of developing Alzheimer's disease when compared to people that had a high total daily PA (90th percentile of the average sum) (Buchman et al. 2012).

Most national guidelines for PA recommend 150 minutes of moderate to vigorous PA in order to improve health (e.g. increasing bone health, decreasing chronic diseases and controlling weight). Despite the heterogeneity of the PA evaluation across the studies, it was evident that even less than 150 minutes per week of PA may prevent dementia (Chang et al. 2010; Larson et al. 2006; Rovio et al. 2005; Scarmeas et al. 2009). As reported by Blondell and colleagues (2014), there was no evidence of a linear dose-response relationship at this stage of the evidence, although studies have generally not tested this.

Gene–Environment Interaction to Prevent Dementia

One remaining question is for whom does PA have the greater preventive effect: the APOE ε4 carriers or the non-carriers? In order to answer that, five studies compared the preventive effect of PA in a group of APOE ε4 carriers with non-carriers. Three studies reported no interaction effect, showing that PA was effective in preventing Alzheimer's disease or any type of dementia among both APOE ε4 carriers or non-carriers (Kim et al. 2011; Taaffe et al. 2008; Rovio et al. 2005). Podewils et al. (2005) showed a preventive effect of PA only for persons that were APOE ε4 non-carriers, while Luck et al. (2014) reported that the APOE ε4 carriers that participated in one or more type of PA (at least several times per week) and the APOE ε4 non-carriers that participated in less than one activity had a greater risk in developing Alzheimer's and any type of dementia when compared to individuals who were APOE ε4 non-carriers and also participated in more than two types of PA (at least several times per week). Also, those who were APOE ε4 carriers and participated in less than one activity had the greatest risk of developing those diseases when compared to all others. Therefore, most studies show that regardless of being an APOE ε4 carrier or not, being physically active seems to be beneficial in preventing dementia.

Methodological Issues and Future Research

The purpose of this review was to provide an update on research examining the association between PA and dementia. This review extended the Blondell et al. (2014) review by adding three recent studies and by focusing on studies that have considered genetic predisposition for dementia. Overall, a consistent pattern is emerging that there is a preventive effect of PA on dementia, particularly in terms of Alzheimer's disease. However, the role of exercise for the prevention of vascular dementia is still unclear, as only a few studies have reported a preventive effect.

Caution is required in making definitive conclusions given a range of methodological limitations in this literature. One example is the length of the follow-up. A preventive function of PA appears to weaken over time. PA was not assessed during the course of studies so it is not possible to confirm whether individuals maintained PA levels reported at baseline. PA may only protect when sustained over time. Ideally future studies should track PA at multiple time points.

Another limitation is how PA is assessed as it is critical to understanding this relationship. The majority of studies have used self-report measures and these are prone to reporting bias. There is only modest agreement between self-reported and objectively measured PA (Steene-Johannessen et al. 2015). Notably, the one study that used an objective measure of PA, via accelerometers, reported that higher levels of objectively measured daily PA was significantly associated with reduced risk of Alzheimer's disease (Buchman et al. 2012).

Assessing the type of PA is also an important issue. Although many studies included a question about the type of PA participated in through self-report, no study assessed if the relationship between PA and dementia was moderated by type of PA. It is possible that variation in the type of PA could improve cognitive function by enhancing the environment or by requiring more cognitive abilities in different ways. Angevaren et al. (2007) studied the effect of the variation of PA on cognitive function in people between 45–70 years of age. Variation in this study was considered as the sum of all types of PA (walking, cycling, housekeeping, odds jobs, gardening, and sports). The authors noted that the variation in PA was positively associated with some aspects of cognitive function (i.e. processing speed, memory, and mental flexibility). Speculatively, it is not just the volume of PA that is important but the nature of PA participation. Taking part in PA that includes a cognitive element (e.g. strategizing in a game of bowling or golf) may be more beneficial than walking on a treadmill. Thus, future studies that evaluate the potential preventive effect of PA on dementia should assess the type of PA in addition to just volume. Other dimensions such as frequency and intensity are also likely important.

A final limitation is the possibility that other factors explain the relationship between PA and reduced risk of dementia. For example, physically active individuals tend to be more educated and less likely to smoke (Lim & Taylor 2005; Yaffe, Barnes, Nevitt, Lui, & Covinsky 2001). Thus, other demographic or lifestyle factors may be contributing to reduced risk. However, most of the studies did control for some confounders such as age, sex, years of education, smoking/alcohol habits, and presence of the APOE ε4 gene.

The three most common confounders addressed were age, sex, and presence of the APOE ε4 gene. Age is important as the prevalence of dementia increases with the increasing of age (Hugo & Ganguli 2014). Sex is also likely to be important as many women take hormonal replacement. PA seems to have an independent and complementary effect on postmenopausal women who take hormonal replacement. Also, PA could show a more preventive effect in women since they are less likely to consume alcohol (York & Welte 1994) and nicotine (Hitchman & Fong 2011) compared to men. In one study, a protective effect against Alzheimer's and any type of dementia was found only for women (Laurin et al. 2001). However, two studies have shown a preventive effect for men (Abbott et al. 2004; Taaffe et al. 2008). More studies are needed to state if there are differences between sexes.

The last aspect is related to the genetic influence. Considering this, five studies compared APOE ε4 carriers with non-carriers. Most of them showed a genetic independent effect, as there was a preventive effect among both APOE ε4 carriers and non-carriers (Kim et al. 2011; Rovio et al. 2005; Taaffe et al. 2008). Nevertheless, one study reported that being an APOE ε4 carrier and sedentary increased the risk of developing dementia (any type and Alzheimer's) (Luck et al. 2014). Even in the presence of a genetic predisposition for dementia, PA appears to have a protective role to play.

What Mechanisms Could Explain the Preventive Effect of Physical Activity on Dementia?

Although this effect seems to be genetic independent, some biological hypotheses have been proposed that may explain the potential protective effect of PA. One hypothesis is based on the fact that PA decreases cardiovascular risk (i.e. hypertension, dyslipidemia, diabetes and others); however, it is still unclear how this improvement could serve as a mediator of improved cognitive performance (Kirk-Sanchez & McGough 2014). Also, vascular dementia does not seem to be influenced by physical activity, as most of the studies that evaluated this type of dementia failed to find a preventive effect.

The formation of amyloid-β plaques is one characteristic of Alzheimer's disease. Thus, another hypothesis is that PA could reduce those plaques. Currently, it is unclear if PA could reduce the amyloid-β (Brown, Peiffer, & Martins 2013). Another perspective is related to neurotrophic factors. PA also seems to enhance some neurotrophins such as IGF-1 (Insulin-like Growth Factor-1) and BDNF (Brain-Derived Neurotrophic Factor). The IGF-1 promotes neuronal survival, differentiation, and growth (Cotman & Berchtold 2002). It has been reported that exercise can increase IGF-1 even in healthy older adults after 6 months of moderate and high level of training (Cassilhas et al. 2007). Another possible important function of the IGF-1 is to be an upstream mediator of the BDNF gene regulation. The BDNF facilitates plasticity and enhances neurovascularity in the brain (including the hippocampus) (Cotman & Berchtold 2002; Vaynman & Gomez-Pinilla 2005). While a number of plausible mechanisms have been proposed, there remains no consensus as to what may explain the protective effects of PA on dementia.

Conclusion

There is consistent evidence that PA is associated with reduced risk of dementia, primarily in terms of Alzheimer's disease. Some caution is required given a range of methodological limitations. However, the balance of evidence suggests public health messaging that PA may protect against dementia is warranted. There is no clear evidence of an optimal dose but national guidelines for PA appear an appropriate recommendation (see Chapter 6 in this volume for more information about physical activity and mental health in older adults).

Table 18.1 Physical activity and the prevention of depression and dementia – summary of included studies

	Depression	*Dementia*
Studies included in analysis	36	24
Studies showing a preventive effect	30	19
Range of follow-up	1–27 years	0.5–26 years
Range in age	11–100 years	51 years or older
Range of sample	497–1, 117, 294	469–4, 761
Minimum amount of PA dosage	10 min of daily walking	48 minutes of PA per week
Most used PA measure	Self-report	Self-report
Most used Depression/ Dementia measure	Center for Epidemiologic Studies Depression Scale (CESD)	NINDS-ADRDA; NINDS-AIREN; ICD-10; CASI
Proposed mechanisms	Physiological: - increased body temperature and endorphins; stimulated endocannabinoid system Psychological - distraction hypothesis - increased self-efficacy	- Decrease in some cardiovascular risks - Reduction in amyloid-β plaque - Enhancement of some neurotrophins (i.e. IGF-1 and BDNF)

Note: BDNF: Brain-Derived Neurotrophic Factor; CASI: Cognitive Abilities Screening Instrument; ICD-10: International Statistical Classification of Diseases and Related Health Problems; IGF-1: Insulin-like growth factor-1; PA: Physical Activity; NINDS- ADRDA: National Institute of Neurological Disorders and Stroke- Alzheimer's Disease and Related Disorders Association; NINDS-AIREN: National Institute of Neurological Disorders and Stroke- Association *Internationale pour la Recherche et l'Enseignement en Neurosciences.*

Physical Activity and the Prevention of Depression

Depression is a common mental illness that is prevalent globally. It is estimated that 350 million people are currently affected by depression and by the year 2020 it is predicted to be the leading cause of disability worldwide (Ferrari et al. 2013; World Health Organization 2012). This projection is worrisome given depression's detrimental impact to overall health and well-being. Specifically, depression is commonly associated with a multitude of symptoms including: low energy, interest, self-worth, and concentration; poor sleeping patterns and appetite; and overwhelming feelings of sadness, stress, and anxiety (World Health Organization 2012). Kessler and Bromet (2013) reviewed international data summarizing other psychosocial and secondary disorders linked with depression such as reduced role functioning (e.g. low marital and work performance quality) and an elevated risk of chronic secondary disorders (e.g. cancer, cardiovascular disease, diabetes). Furthermore, depression has been identified as being a recurrent and chronic condition (Hardeveld, Spijker, De Graaf, Nolen, & Beekman 2010) that can ultimately lead to suicidal tendencies and acts.

The adverse health effects of depression extend beyond the individual. As a result of lost work productivity and health care costs, the incremental economic burden of depression has risen from $83.1 billion to 210.5 billion between the years 2000 and 2010 (Greenburg, Fournier, Sisitsky, Pike, & Kessler 2015). With the high prevalence of depression and its burden on health and the economy, there is an urgent need to halt these growing trends. Hence, in 2012 the World Health Organization declared that preventing depression is an area that warrants greater attention. Of interest to this chapter, one modifiable behavior shown to prevent the onset of depression is PA.

PA has long been known for its physical and mental health benefits. Pertaining to depression, reviews have shown PA to help treat depression (Nyström, Neely, Hassmén, & Carlbring 2015; Stanton & Reaburn 2014). In 2013, we provided the first systematic review to determine if PA (i.e. aerobic) can prevent depression from ever developing in those not depressed at baseline (Mammen & Faulkner 2013). This review, which searched multiple databases from 1976–2012, included 30 prospective studies and found the majority (i.e. 25) to show a preventive effect of PA on depression. Since the search criteria cut-off date (i.e. December 2012), a number of longitudinal studies examining the links between baseline PA and follow-up depression have been published. Thus, a primary objective of this section is to update this systematic review and address three questions: Does PA help prevent depression? How much PA is needed to help prevent depression? Can fluctuating levels of PA over time change outcomes in depression? To conclude, an overview of possible mechanisms, and methodological issues and future research needs is provided.

Search Strategy

The following databases were initially searched to locate studies examining the longitudinal relationship between PA and depression from January 1976 to December 2012: Medline, Embase, PubMed, PsycINFO, SPORTDiscus, and Cochrane. From January 2013 to August 2015, an updated search was conducted in Medline only. Search terms related to four concepts in physical activity (e.g. exercise, physical fitness), depression (e.g. depression disorder, major depressive episodes,) study design (e.g. cohort studies, longitudinal studies), and association (e.g. risk factor).

Inclusion/Exclusion Criteria

The studies were chosen based on the following criteria: (1) employed a prospective design; (2) assessed PA and depression over at least two time intervals; (3) identified PA as the exposure variable and depression as the outcome variable; (4) defined depression using depression threshold cut-off scores on self-report scales or using more-direct measures of depression such as a physician's diagnosis or hospital discharge records.

Study Identification and Selection

The initial search yielded a total of 6363 citations: 2111 in Embase, 1762 in MEDLINE, 1004 in PubMed, 693 in Cochrane, 684 in PsycINFO, and 109 in SPORTDiscus. All abstracts were screened, and 30 studies were included for analyses. The updated search, which used MEDLINE only, generated 1468 hits, and found an additional six studies meeting inclusion criteria. Thus, 36 studies were included for analyses (Aberg et al. 2012; Augestad, Slettemoen, & Flanders 2008; Backmand, Kaprio, Kujala, & Sarna 2003; Ball, Burton, & Brown 2009; Bernaards et al. 2006; Brown, Ford, Burton, Marshall, & Dobson 2005; Camacho, Roberts, Lazarus, Kaplan & Cohen 1991; Carroll, Blanck, Serdula, & Brown 2010; Cooper-Patrick, Ford, Mead, Chang, & Klag 1997; Farmer et al. 1988; Gallegos-Carrillo et al. 2013; Hamer, Molloy, Oliveria, & Demakakos 2009; Jerstad, Boutelle, Ness, & Stice 2010; Jonsdottir, Rödjerc, Hadzibajramovica, Börjesson, & Ahlborg 2010; Kritz-Silverstein, Barrett-Connor, & Corbeau 2001; Ku, Fox, & Chen 2009; Lampinen, Heikkinen, & Ruoppila 2000; Lucas et al. 2011; McKercher et al. 2014; Mikkelsen et al. 2010; Mobily, Rubinstein, Lemke, O'Hara, & Wallace 1996; Morgan & Bath 1998; Paffenbarger, Lee, & Leung 1994; Rothon et al. 2010; Smith et al. 2010; Stavrakakis et al. 2013; Strawbridge, Deleger, Roberts, & Kaplan 2002; Sund, Larsson, & Wichstrom 2011; Tsai, Chi, & Wang 2013; Uebelacker et al. 2013; van Gool et al. 2006; van Uffelen et al. 2013; Wang et al. 2011; Weyerer 1992; Wise, Adams-Campbell, Palmer, & Rosenberg 2006) (for a summary of included studies see Table 18.1). A total of 70 studies were excluded (experimental or cross-sectional in design [n=18]; not specific to depression [n=14]; had no cut-off criteria score to assess depression [n=26]; were review articles [n=7]; did not exclude individuals with depression at baseline [n=5]).

Study Characteristics

The majority of studies were conducted in North America (n=15) and Europe (n=16).

All studies contained nonclinical community samples of males and females ranging in age from 11 to 100 years. Twenty five studies examined associations for both genders; seven were female only; four were male only. Follow- up assessments ranged between 1–27 years. Relative to those using subjective PA measures of aerobic activity, only one study objectively measured PA via ergometer cycling (Aberg et al. 2012). The majority of studies measured depression through reliable and valid instruments such as the Center for Epidemiologic Studies Depression Scale (CESD). Six studies measured depression more directly, via physician diagnosis (n=3); hospital discharge register (n=2); or use of antidepressants (n=1).

Can Physical Activity Help Prevent Depression?

Among the 36 included studies, 30 observed a significant inverse relationship between baseline PA and follow-up depression. The large majority of these studies maintained high methodological rigour in relation to limited selection bias, using valid instruments to assess PA and depression, and accounting for confounding variables known to correlate with both PA and depression. For instance, most studies controlled for age, sex, comorbidities, occupational and socioeconomic status, alcohol and tobacco use, and marital status, all of which have shown to influence levels PA and the degree of depressive symptoms. Taken together, the scientific evidence provides consistent evidence that PA can prevent depression.

How Much PA Is Needed to Help Prevent Depression?

Determining the quantity of PA required to reduce the risk of depression is often challenging to address due to the methodological heterogeneity in PA measurement across studies. However, the following provides a summary of findings in which PA was assessed in terms of minutes per week, daily PA, and walking.

Ten studies measured PA in terms of minutes per week (Ball et al. 2009; Brown et al. 2005; Jonsdottir et al. 2010; McKercher et al. 2014; Mickkelsen et al. 2010; Paffenbarger et al. 1994; Strawbridge et al. 2002; Tsai et al. 2013; van Uffelen et al. 2013; Wise et al. 2006). In relation to common national PA guidelines for adults (e.g. 150 minutes of moderate-vigorous PA/week), four found that engaging in this amount or less could protect against depression (Ball et al. 2009; Brown et al. 2005; Jonsdottir et al. 2010; Van Uffelen et al. 2013). For example, in Van Uffelen and colleagues' 9-year study, women (n=8950) aged between 50–55 years were more likely to develop depression when not meeting PA guidelines compared to those who did (OR=1.26). Jonsdottir and colleagues (2010) reported that subjects engaging in even low-intensity PA (i.e. gardening/walking) for 120 minutes per week were at a 63% reduced risk of developing future depression relative to those who were sedentary. Although the other two studies found a smaller effect (Ball et al. 2009; Brown et al. 2005), it is still promising that even low amounts of PA can help prevent depression.

Aligned with these findings, three studies found that walking was linked with a decreased risk of depression (Augestad et al. 2008; Bernaards et al. 2006; Lucas et al. 2011). Two of these studies revealed that even low levels of daily walking protected against depression by up to 40% (Augestad et al. 2008; Bernaards et al. 2006). Lucas and colleagues (2011) showed that walking less than 20 minutes per day could decrease risk of depression by 6%. Their results further revealed a dose-response relationship with higher levels of daily PA (60–90 minutes/day, RR=0.84; 490 minutes/day, RR=0.80) decreasing the risk of depression.

Five other studies in this review did find that greater intensity (Jonsdottir et al. 2010; Paffenbarger et al. 1994), frequency (Gallegos-Carrillo et al. 2013; Ku et al. 2009), or volume (Sund et al. 2011) is associated with reduced odds in developing depression. For instance, in their 6-year prospective study of Mexican adults, Gallegos-Carrillo et al. found "higher levels of PA patterns" (OR=0.46) had a stronger protective effect against depression relative to those reporting "moderately active PA patterns" (OR=0.57). However, due to the various measures of PA, a clear dose-response relationship between PA and reduced depression is not readily apparent across studies.

Can Fluctuating Levels of Physical Activity Over Time Influence Outcomes in Depression?

Consistent evidence from this review shows that individuals who are active at baseline are less likely to develop depression at follow-up. But can increasing PA levels between baseline and follow-up periods reduce the onset of depression?

Twelve of the 36 studies examined changes in PA over time and its association with depression (Ball et al. 2009; Brown et al. 2005; Camacho et al. 1991; Carroll et al. 2010; Hamer et al. 2009; Ku et al. 2009; Lampinen et al. 2000; McKercher et al. 2014; Rothon et al. 2010; Van Gool et al. 2006; Wang et al. 2011; Wise et al. 2006). Only two of these studies showed no significant association between changes in PA and depression (Ku et al. 2009; Rothon et al. 2010). Four studies concluded that reducing PA over time increased the risk of developing depression relative to remaining active or increasing activity level (Camacho et al. 1991; Carroll et al. 2010; Hamer et al. 2009; Wise et al.

2006). One study found a large effect, noting that individuals who reduced PA over time were more than ten times more likely to develop depression (Lampinen et al. 2000).

Conversely, four studies revealed that subjects who increased PA over time were at a reduced risk of subsequent depression (Ball et al. 2009; Brown et al. 2005; McKercher et al. 2014; Wise et al. 2006), and three showed that subjects who maintained their levels of PA were at a lower risk of depression relative to those who were inactive throughout (McKercher et al. 2014; Van Gool et al. 2006; Wise et al. 2006). Using a sample of 1630 males and females, McKercher and colleagues (2014) found that over 20 years, individuals who were "increasingly" and "persistently active" had a 51%–69% reduced risk of subsequent depression relative to those "persistently inactive." Overall, these studies suggest that in order to further help prevent depression, individuals should sustain their activity levels over time and that despite an individual's history of inactivity, the initiation of PA may prevent depression.

What Mechanisms Explain Physical Activity's Preventive Effect on Depression?

This review indicates that PA can reduce the risk of developing depression. An often-contemplated question regarding this finding relates to the mechanisms responsible for this protective effect. Though the study findings from this review are consistent in terms of PA's effect on depression, the underlying mechanisms supporting this relationship remain unclear. Nonetheless, the following will provide a brief summary of the plausible physiological and psychological mechanisms that are often cited in the literature.

Physiologically, mechanisms that are often discussed include the thermogenic, endorphin, and monoamine hypothesis. The ***thermogenic hypothesis*** postulates PA to facilitate an increase in body temperature. Consequently, the increased temperature in specific brain regions, such as the brain stem, can lead to a more relaxed state (deVries 1981). The *endorphin hypothesis* suggests PA to increase levels of endorphins, which results in elevated mood and may help explain the "runner's high" euphoria experienced after a bout of PA (Fox 1999). The *monoamine hypothesis* states that PA replenishes the diminished brain neurotransmitters in those affected by depression, such as norepinephrine and serotonin (Craft & Perna 2004).

More recently, the *endocannabinoid system* has received attention as a plausible mechanism. This system is a neuromodulatory system known to regulate emotional and cognitive processes. It is comprised of cannabinoid 1 and 2 (CB1 and CB2) receptors that are expressed at high density in the brain and periphery (Piomelli 2003). Interestingly, a review conducted by Hill and colleagues (2009) provides strong evidence indicating a deficiency in endocannabinoids (e.g. CB1, CB2 receptors) in the etiology of depression. The review contains supporting studies to show that deficient endocannabinoid signaling can trigger depression and that impairments in this system were evident among those depressed. As such, suggestions have been made by researchers to target the endocannabinoid system to help with depression. PA has been identified as one modifiable behavior that can help facilitate the endocannabinoid system, which may mediate PA's beneficial impact on cognition and mood (Tantimonaco et al. 2014).

Psychologically, the distraction and self-efficacy hypotheses have been proposed as mechanisms explaining PA's preventive role with depression. The distraction hypothesis proposes that PA acts as a distraction from the cognitive symptoms associated with depression such as worry (Searle et al. 2011). The *self-efficacy hypothesis* suggests an individual's confidence in executing various tasks can be strengthened as a result of PA. For instance, individuals who suffer from depression are known to negatively self-evaluate and ruminate. This hypothesis suggests PA provides a channel where self-efficacy can be enhanced based on individuals executing meaningful mastery experiences (Fox 1999).

Overall, while these mechanisms are plausible, there remains limited evidence to support one definitive hypothesis. Faulkner and Carless (2006) rather suggest a concurrent effect of these various PA mechanisms (e.g. physiological, biochemical or psychosocial) in helping prevent depression. However, as Craft and Perna (2004) explain, more research using the growing innovative neuroimaging techniques will help better determine which mechanisms, or combinations thereof, are responsible for PA's preventive effect on depression.

Methodological Issues and Future Research

Despite the consistent evidence regarding PA's protective effect on depression, some caution is required since key covariates, such as genetic variations, may not have been fully accounted for. For instance, perhaps those individuals with a family history of depression are more likely to develop depression (Sullivan, Neale, & Kendler 2000) irrespective of their PA levels. Hence, future investigation in this area should aim to capture genetic information regarding family history of depression.

The use of subjective (i.e. self-report) measures for PA and depression is another methodological weakness of the included studies. Although it may not be feasible to objectively measure PA (e.g. accelerometers) and depression (e.g. diagnostic interviews) in large cohorts, self-report assessments are prone to bias (e.g. recall, social desirability), and are more focused on PA behavior as opposed to energy expenditure. This bias may exaggerate or conceal the true associations between PA and depression. Furthermore, the inconsistent use of PA self-report measures between studies limits the ability to examine dose-response relationships between PA and depression. Future measures of PA ought to capture various components of PA (e.g. frequency, duration, intensity) for cross-study comparability. Researchers could also explore the relationships between types (e.g. aerobic versus strength training) and domains (e.g. occupational; household) of PA and the prevention of depression.

Similarly, the use of various subjective and direct measures of depression limits comparability between studies due to different criteria for defining depression. For example, the cut-off score for "significant" depressive symptomatology using the CESD is typically 16 (Eaton et al. 2004). Although some studies used this cut-off and found a protective effect, other studies using the same measure, and sometimes different versions of the CESD, utilized lower threshold cut-offs. This inconsistency in classifying depression has implications for inter-study comparisons, as well as estimating whether symptoms are clinically meaningful. Hence, more investigations need to elucidate the relationship between PA and *clinical* depression by using more direct measures (e.g. use of antidepressants or hospital discharge registers).

Conclusion

There is sufficient evidence to conclude that PA can help reduce the risk of developing depression. Thirty of the 36 studies found baseline PA to be inversely related to follow-up depression. The inclusion of only longitudinal studies adds weight to a causal relationship between PA and reductions in depression (Hill 1965). Further, everyday activities such as walking appear to confer a benefit. Meeting the recommended levels of PA, established for physical health benefits, appear equally appropriate for preventing depression. Data from the studies also suggest for individuals who are currently active to sustain their PA habits and those who are inactive to initiate a physically active lifestyle to help reduce odds of developing depression. From a health promotion perspective, this review suggests that promoting any level of PA could be an important strategy for the prevention of future depression in addition to the wealth of other physical and mental health benefits accrued through a physically active lifestyle (see Chapter 15 in this volume for more information on physical activity and depression).

A Final Overview

There are a number of consistent issues regarding the evidence for both disorders and these also have implications for future research. First, it is not possible to identify an optimal dosage of PA required to prevent dementia (primarily in terms of Alzheimer's disease) or depression. Certainly, recommending attaining national PA guidelines appears as a reasonable starting point. Yet, reflecting evidence that the dose-response relationship between PA and many health benefits is likely curvilinear in nature (Warburton, Charlesworth, Ivey, Nettlefold, & Bredin 2010), some PA appears better than none in preventing depression and dementia. The most effective dose of PA is likely the one that individuals enjoy and find pleasant. For both disorders, it is also evident that PA must be maintained to have a reliable preventive effect. Future research ideally would include objective measures of PA over multiple time points to avoid problems of self-report bias and to explore the impact of variations in PA over time. Self-report is still needed but should focus more on what people are doing (e.g. type of physical activity) and the context in which they are doing it (e.g. individual or group setting).

Second, despite consistency in the literature regarding a protective function of physical activity, some caution is required given that there may be a number of other factors, such as genetic variations (De Moor, Boomsma, Stubbe, Willemsen & de Geus 2008), that predict both PA and reduced risk of depression or dementia. Essentially, there may be unexplained, confounding factors that explain the preventive function being attributed to PA. Future research should carefully control for such factors and explore theoretically informed investigations of specific confounders.

Third, the precise mechanisms through which PA may prevent dementia or depression are not known. Most population-based, prospective studies reviewed here have not been established to specifically answer the question – does physical activity prevent mental disorder? Consequently, studies have not been designed to explore such mechanistic questions. However, it is unlikely one specific mechanism can explain the preventive benefits of PA given the large number of potential influences that may be experienced through PA (Faulkner & Carless 2006). Rather, several mechanisms are most likely operating in concert, with the precise combination being highly individual-specific (Fox 1999). Although not without challenges, future research should explore potential mechanisms where possible. In turn, this may inform health messaging about the optimal doses of PA.

PA is recognized as one of the "Best Buy" interventions for non-communicable disease prevention and control (Bloom et al. 2011). Overall, the presented evidence (see Table 18.1) supports the possibility that PA promotion might also be considered a strategy for lowering the risk of mental disorders such as depression and dementia. This provides yet further justification for investing in PA promotion at a population level.

References

Abbott, R. D., L. R. White, G. W. Ross, K. H. Masaki, J. D. Curb, & H. Petrovitch. 2004. "Walking and dementia in physically capable elderly men". *JAMA 292* (12): 1447–53. doi:10.1001/jama.292.12.1447

Aberg, M., M. Waern, J. Nyberg, N. L. Pedersen, Y. Bergh, N. D. Aberg,...K. Torén. 2012. "Cardiovascular fitness in males at age 18 and risk of serious depression in adulthood: Swedish prospective population-based study." *BJ Psych 10*: 1–8. doi:10.1192/bjp.bp.111.103416

American Psychiatric Association. 2013. Diagnostic and statistical manual of mental disorders. 5th ed. Washington, DC.

Angevaren, M., L. Vanhees, W. Wendel-Vos, H. J. J. Verhaar, G. Aufdemkampe, A. Aleman, & V. M. M. Verschuren. 2007. "Intensity, but not duration, of physical activities is related to cognitive function." *Eur J Cardiovasc Prev Rehabil 14* (6): 825–30. doi:10.1097/HJR.0b013e3282ef995b

Augestad, L. B., R. P. Slettemoen, & W. D. Flanders. 2008. "Physical activity and depressive symptoms among Norwegian adults aged 20–50." *Public Health Nurs 25*: 536–45. doi:10.1111/j.1525-1446.2008.00740.x

Backmand, H., J. Kaprio, U. Kujala, & S. Sarna. 2003. "Influence of physical activity on depression and anxiety of former elite athletes." *Int J Sports Med 24* (8): 609–19. doi:10.1055/s-2003-43271

Ball, K., N. Burton, & W. Brown. 2009. "A prospective study of overweight, physical activity, and depressive symptoms in young women." *Obesity 17*: 66–71. doi:10.1038/oby.2008.497

Bernaards, C. M., M. P. Jans, S. G. van den Heuvel, I. J. Hendriksen, I. L. Houtman, & P. M. Bongers. 2006. "Can strenuous leisure time physical activity prevent psychological complaints in a working population?" Occup Environ Med *63*: 10–16. doi:10.1136/oem.2004.017541

Blondell, S. J., R. Hammersley-Mather, & J. L. Veerman. 2014. "Does physical activity prevent cognitive decline and dementia? A systematic review and meta-analysis of longitudinal studies." *BMC Public Health 14*: 510. doi:10.1186/1471-2458-14-510

Bloom D. E., E. T. Cafiero, E. Jané-Llopis, S. Abrahams-Gessel, L. R. Bloom, S. Fathima,...C. Weinstein. 2011. *The global economic burden of noncommunicable diseases.* Geneva: World Economic Forum.

Bowen, M. E. 2012. "A prospective examination of the relationship between physical activity and dementia risk in later life." *Am J Health Promot 26* (6): 333–40. doi:10.4278/ajhp.110311-QUAN-115

Brown, B. M., J. J. Peiffer, & R. N. Martins. 2013. "Multiple effects of physical activity on molecular and cognitive signs of brain aging: Can exercise slow neurodegeneration and delay Alzheimer's disease?" *Mol Psychiatry 18* (8): 864–74. doi:10.1038/mp.2012.162

Brown, W. J., J. H. Ford, N. W. Burton, A. L. Marshall, & A. J. Dobson. 2005. "Prospective study of physical activity and depressive symptoms in middle-aged women." *Am J Prev Med 29*: 265–72.

Buchman, A. S., P. A. Boyle, L. Yu, R. C. Shah, R. S. Wilson, & D. A. Bennett. 2012. "Total daily physical activity and the risk of AD and cognitive decline in older adults." *Neurology 78* (17): 1323–9. doi:10.1212/WNL.0b013e3182535d35

Camacho, T. C., R. E. Roberts, N. B. Lazarus, G. A. Kaplan, & R. D. Cohen. 1991. "Physical activity and depression: Evidence from the Alamada County study." *Am J Epidiomiol 134*: 220–31.

Carroll, D. D., H. M. Blanck, M. K. Serdula, & D. R. Brown. 2010. "Obesity, physical activity, and depressive symptoms in a cohort of adults aged 51 to 61." *J Aging Health 22*: 384–98.

Cassilhas, R. C., V. A. Viana, V. Grassmann, R. T. Santos, R. F. Santos, S. Tufik, & M. T. Mello. 2007. "The impact of resistance exercise on the cognitive function of the elderly." *Med Sci Sports Exerc 39* (8): 1401–7. doi:10.1249/mss.0b013e318060111f

Chang, M., P. V. Jonsson, J. Snaedal, S. Bjornsson, J. S. Saczynski, T. Aspelund,...L. J. Launer. 2010. "The effect of midlife physical activity on cognitive function among older adults: AGES–Reykjavik Study." *J Gerontol A Biol Sci Med Sci 65* (12): 1369–74. doi:10.1093/gerona/glq152

Cooper-Patrick, L., D. E. Ford, L. A. Mead, P. P. Chang, & M. J. Klag. 1997. "Exercise and depression in midlife: A prospective study." *Am J Public Health 87*: 670–3.

Cotman, C. W., & N. C. Berchtold. 2002. "Exercise: A behavioral intervention to enhance brain health and plasticity." *Trends Neurosci 25* (6): 295–301.

Craft, L. L., & F. M. Perna. 2004. "The benefits of exercise for the clinically depressed." *Primary Care Companion to the Journal of Clinical Psychiatry 6*: 104.

de Bruijn, R. F., E. M. Schrijvers, K. A. de Groot, J. C. Witteman, A. Hofman, O. H. Franco,...M.A. Ikram. 2013. "The association between physical activity and dementia in an elderly population: The Rotterdam Study." *Eur J Epidemiol 28* (3): 277–83. doi:10.1007/s10654-013-9773-3

De Moor M. H. M., D. I. Boomsma, J. H. Stubbe, G. Willemsen, & E. J. C. de Geus. 2008. "Testing causality in the association between regular exercise and symptoms of anxiety and depression." *Archives of General Psychiatry 65* (8): 897–905.

deVries, H. A. 1981. "Tranquilizer effects of exercise: A critical review." *Phys Sportsmed 9*: 46–55.

Eaton, W. W., C. Muntaner, C. Smith, M. Ybarra, C. Muntaner, A. Tien, & M. E. Maruish. 2004. "Center for Epidemiologic Studies Depression Scale: Review and revision (CESD and CESD-R)." In M. E. Maruish (Ed.), *The use of psychological testing for treatment planning and outcomes assessment, 3rd Edition* (pp. 363–77). Mahwah, NJ: Lawrence Erlbaum.

Fabrigoule, C., L. Letenneur, J. F. Dartigues, M. Zarrouk, D. Commenges, & P. Barberger-Gateau. 1995. "Social and leisure activities and risk of dementia: A prospective longitudinal study." *J Am Geriatr Soc 43* (5): 485–90.

Farmer, M. E., B. Z. Locke, E. K. Mosciki, A. L. Danmenberg, D. B. Larson, & L. S. Radloff. 1988. "Physical activity and depressive symptoms: The NHANES I epidemiologic follow-up study." *Am J Epidemiol 128*: 1340–51.

Faulkner, G., & D. Carless. 2006. "Physical activity and the process of psychiatric rehabilitation: Theoretical and methodological issues." *Psych Rehab J 29*: 258–66.

Ferrari, A. J., F. J. Charlson, R. E. Norman, S. B. Patten, G. Freedman, C. J. Murray,... H. A. Whiteford. 2013. "Burden of depressive disorders by country, sex, age, and year: Findings from the global burden of disease study 2010." *PLos Med 10*(11): e1001547.

Fox, K. R. 1999. "The influence of physical activity on mental well-being." *Pub Health Nutr 2*: 411–18.

Gallegos-Carrillo, K., Y. N. Flores, E. Denova-Gutiérrez, P. Méndez-Hernández, L. D. Dosamantes-Carrasco, S. Henao-Morán,...J. Salmerón. 2013. "Physical activity and reduced risk of depression: Results of a longitudinal study of Mexican adults." *Health Psychol 32* (6): 609–15. doi:10.1037/a0029276

Greenberg, P. E., A. A. Fournier, T. Sisitsky, C. T. Pike, & R. C. Kessler. 2015. "The economic burden of adults with major depressive disorder in the United States (2005 and 2010)." *The Journal of Clinical Psychiatry 76*: 155–62.

Gureje, O., A. Ogunniyi, L. Kola, & T. Abiona. 2011. "Incidence of and risk factors for dementia in the Ibadan study of aging." *J Am Geriatr Soc 59* (5): 869–74. doi:10.1111/j.1532-5415.2011.03374.x

Hamer, M., G. Molloy, C. Oliveria, & P. Demakakos. 2009. "Leisure time physical activity, risk of depressive symptoms, and inflammatory mediators: The English longitudinal study of ageing." *Psychoneuroendocrino 34*: 1050–5.

Hanson, A. J., S. Craft, & W. A. Banks. 2015. "The APOE genotype: Modification of therapeutic responses in Alzheimer's disease." *Curr Pharm Des 21* (1): 114–20.

Hardeveld F., J. Spijker, R. De Graaf, W. A. Nolen, & A. T. Beekman. 2010. "Prevalence and predictors of recurrence of major depressive disorder in the adult population." *Acta Psychiatr Scand 122*: 184–91.

Hill, A. B. 1965. "The environment and disease: Association or causation?" *Proceedings of the Royal Society of Medicine 58*: 295.

Hill, M. N., C. J. Hillard, F. R. Bambico, S. Patel, B. B. Gorzalka, & G. Gobbi. 2009. "The therapeutic potential of the endocannabinoid system for the development of a novel class of antidepressants." *Trends in Pharmacological Sciences 30*: 484–93. doi:10.1016/j.tips.2009.06.006

Hitchman S. C., & G. T. Fong. 2011. "Gender empowerment and female-to-male smoking prevalence ratios." *Bull World Health Organ* Mar 1; *89*(3): 195–202. doi:10.2471/BLT.10.079905

Hugo, J., & M. Ganguli. 2014. "Dementia and cognitive impairment: Epidemiology, diagnosis, and treatment." *Clin Geriatr Med 30* (3): 421–42. doi:10.1016/j.cger.2014.04.001

Jerstad, S. J., K. N. Boutelle, K. K. Ness, & E. Stice. 2010. "Prospective reciprocal relations between physical activity and depression in female adolescents." *J Consult Clin Psychol 78*: 268–72.

Jonsdottir, I. H., L. Rödjerc, E. Hadzibajramovica, M. Börjesson, & G. J. Ahlborg. 2010. "A prospective study of leisure time physical activity and mental health in Swedish health care workers and social insurance officers." *Prev Med 51*: 373–7.

Kessler, R. C., & E. J. Bromet. 2013. "The epidemiology of depression across cultures." *Annual Review of Public Health 34*: 119.

Kim, J. M., R. Stewart, K. Y. Bae, S. W. Kim, S. J. Yang, K. H. Park,...J. S. Yoon. 2011. "Role of BDNF val66met polymorphism on the association between physical activity and incident dementia." *Neurobiol Aging 32* (3): 551 e5-12. doi:10.1016/j.neurobiolaging.2010.01.018

Kirk-Sanchez, N. J., & E. L. McGough. 2014. "Physical exercise and cognitive performance in the elderly: Current perspectives." *Clin Interv Aging 9*: 51–62. doi:10.2147/CIA.S39506

Kritz-Silverstein, D., E. Barrett-Connor, & C. Corbeau. 2001. "Cross-sectional and prospective study of exercise and depressed mood in the elderly: The Rancho Bernardo study." *Am J* Epidemiol *153*: 596–603.

Ku, P. W., K. R. Fox, & L. J. Chen. 2009. "Physical activity and depressive symptoms in Taiwanese older adults: A seven-year follow-up study." *Prev Med 48*: 250–5.

Lampinen, P., R. L. Heikkinen, & I. Ruoppila. 2000. "Changes in intensity of physical exercise as predictors of depressive symptoms among older adults: An eight-year follow-up study." *Prev Med 12*:113–80.

Larson, E. B., L. Wang, J. D. Bowen, W. C. McCormick, L. Teri, P. Crane, & W. Kukull. 2006. "Exercise is associated with reduced risk for incident dementia among persons 65 years of age and older." *Ann Intern Med 144* (2): 73–81.

Laurin, D., R. Verreault, J. Lindsay, K. MacPherson, & K. Rockwood. 2001. "Physical activity and risk of cognitive impairment and dementia in elderly persons." *Arch Neurol 58* (3): 498–504.

Lim, K., & L. Taylor. 2005. "Factors associated with physical activity among older people – a population-based study." *Prev Med 40* (1): 33–40.

Lucas, M., R. Mekary, A. Pan, F. Mirzaei, E. J. O'Reilly, W. C. Willett,...A. Ascherio. 2011. "Relation between clinical depression risk and physical activity and time spent watching television among older women: A 10-year prospective follow-up study." *Am J Epidemiol 174*: 1017–27.

Luck, T., S. G. Riedel-Heller, M. Luppa, B. Wiese, M. Köhler, F. Jessen,...W. Maier. 2014. "Apolipoprotein E epsilon 4 genotype and a physically active lifestyle in late life: Analysis of gene-environment interaction for the risk of dementia and Alzheimer's disease dementia." *Psychol Med 44* (6): 1319–29. doi:10.1017/S0033291713001918

McCallum, J., L. A. Simons, J. Simons, & Y. Friedlander. 2007. "Delaying dementia and nursing home placement: The Dubbo study of elderly Australians over a 14-year follow-up." *Ann N Y Acad Sci 1114*: 121–9. doi:10.1196/annals.1396.049

McKercher, C., K. Sanderson, M. D. Schmidt, P. Otahal, G. C. Patton, T. Dwyer, & A. J. Venn. 2014. "Physical activity patterns and risk of depression in young adulthood: A 20-year cohort study since childhood." *Soc Psych Epi 49*: 1823–34.

Mammen, G., & G. Faulkner. 2013. "Physical activity and the prevention of depression: A systematic review of prospective studies." *AJPM 45*: 649–57.

Michaelson, D. M. 2014. "APOE epsilon4: The most prevalent yet understudied risk factor for Alzheimer's disease." *Alzheimers Dement 10* (6): 861–8. doi:10.1016/j.jalz.2014.06.015

Mikkelsen, S. S., J. S. Tolstrup, E. M. Flachs, E. L. Mortensen, P. Schnohr, & T. Flensborg-Madsen. 2010. "A cohort study of leisure time physical activity and depression." *Prev Med 51*: 471–5.

Mobily, K. E., L. M. Rubinstein, J. H. Lemke, M. W. O'Hara, & R. B. Wallace. 1996. "Walking and depression in a cohort of older adults: The Iowa 65+ Rural Health Study." *JAPA 4*: 119–35.

Morgan, G. S., J. Gallacher, A. Bayer, M. Fish, S. Ebrahim, & Y. Ben-Shlomo. 2012. "Physical activity in middle-age and dementia in later life: Findings from a prospective cohort of men in Caerphilly, South Wales and a meta-analysis." *J Alzheimers Dis 31* (3): 569–80. doi:10.3233/JAD-2012–112171

Morgan, K., & P. A. Bath. 1998. "Customary physical activity and psychological well-being: A longitudinal study." *Ageing 27*: 35–40.

Nyström, M. B., G. Neely, P. Hassmén, & P. Carlbring. 2015. "Treating major depression with physical activity: A systematic overview with recommendations." *Cognitive Behaviour Therapy* 1–12.

Orimo, H., H. Ito, T. Suzuki, A. Araki, T. Hosoi, & M. Sawabe. 2006. "Reviewing the definition of 'elderly.'" *Geriatrics and Gerontology International 6* (3): 149–58. doi:10.1111/j.1447-0594.2006.00341.x

O'Rourke, H. M., W. Duggleby, K. D. Fraser, & L. Jerke. 2015. "Factors that affect quality of life from the perspective of people with dementia: A metasynthesis." *J Am Geriatr Soc 63* (1): 24–38. doi:10.1111/jgs.13178

Paffenbarger, R. S., I. M. Lee, & R. Leung. 1994. "Physical activity and personal characteristics associated with depression and suicide in American college men." *Acta Psychiatr Scan* Suppl *377*: 16–22.

Piomelli, D. 2003. "The molecular logic of endocannabinoid signaling." *Nature Reviews Neuroscience 4*: 873–84.

Plassman, B. L., J. W. Williams, Jr., J. R. Burke, T. Holsinger, & S. Benjamin. 2010. "Systematic review: Factors associated with risk for and possible prevention of cognitive decline in later life." *Ann Intern Med 153* (3): 182–93. doi:10.7326/0003-4819-153-3-201008030-00258

Podewils, L. J., E. Guallar, L. H. Kuller, L. P. Fried, O. L. Lopez, M. Carlson, & C. G. Lyketsos. 2005. "Physical activity, APOE genotype, and dementia risk: Findings from the Cardiovascular Health Cognition Study." *Am J Epidemiol 161* (7): 639–51. doi:10.1093/aje/kwi092

Prince, M., R. Bryce, E. Albanese, A. Wimo, W. Ribeiro, & C. P. Ferri. 2013. "The global prevalence of dementia: A systematic review and metaanalysis." *Alzheimers Dement 9* (1): 63–75 e2. doi:10.1016/j.jalz.2012.11.007

Ratnasingham S., J. Cairney, J. Rehm, H. Manson, & P. A. Kurdyak. 2012. *Opening Eyes, Opening Minds: The Ontario Burden of Mental Illness and Addictions Report.*

An ICES/PHO Report. Toronto: Institute for Clinical Evaluative Sciences and Public Health Ontario.

Ravaglia, G., P. Forti, A. Lucicesare, N. Pisacane, E. Rietti, M. Bianchin, & E. Dalmonte. 2008. "Physical activity and dementia risk in the elderly: Findings from a prospective Italian study." *Neurology 70* (19 Pt 2): 1786–94. doi:10.1212/01.wnl.0000296276.50595.86

Rohn, T. T. 2014. "Is apolipoprotein E4 an important risk factor for vascular dementia?" *Int J Clin Exp Pathol 7* (7): 3504–11.

Rothon, C., P. Edwards, K. Bhui, R. M. Vine, S. Taylor, & S. A. Stansfeld. 2010. "Physical activity and depressive symptoms in adolescents: A prospective study." *BMC Med 8*: 32.

Rovio, S., I. Kåreholt, E. L. Helkala, M. Viitanen, B. Winblad, J. Tuomilehto,...M. Kivipelto. 2005. "Leisure-time physical activity at midlife and the risk of dementia and Alzheimer's disease." *Lancet Neurology 4* (11): 705–11. doi:10.1016/s1474-4422(05)70198-8

Rovio, S., I. Kareholt, M. Viitanen, B. Winblad, J. Tuomilehto, H. Soininen,...M. Kivipelto. 2007. "Work-related physical activity and the risk of dementia and Alzheimer's disease." *Int J Geriatr Psychiatry 22* (9): 874–82. doi:10.1002/gps.1755

Sallis, J. F. 2000. "Age-related decline in physical activity: A synthesis of human and animal studies." *Med Sci Sports Exerc 32* (9): 1598–600.

Scarmeas, N., J. A. Luchsinger, N. Schupf, A. M. Brickman, S. Cosentino, M. X. Tang, & Y. Stern. 2009. "Physical activity, diet, and risk of Alzheimer disease." *JAMA 302* (6): 627–37. doi:10.1001/jama.2009.1144

Searle, A., M. Calnan, G. Lewis, J. Campbell, A. Taylor, & K. Turner. 2011. "Patients' views of physical activity as treatment for depression: A qualitative study." *Br J Gen Pract 61*: 149–56.

Smetanin, P., D. Stiff, C. Briante, C. E. Adair, S. Ahmad, & M. Khan. 2011. *The Life and Economic Impact of Major Mental Illnesses in Canada: 2011 to 2041*. RiskAnalytica, on behalf of the Mental Health Commission of Canada.

Smith, T., K. Masaki, K. Fong, R. D. Abbott, G. W. Ross, H. Petrovitch,...L.R. White. 2010. "Effect of walking distance on 8-year incident depressive symptoms in elderly men with and without chronic disease: The Honolulu-Asia aging study." *J Am Geriatr Soc 58*: 1447–52.

Stanton, R., & P. Reaburn. 2014. "Exercise and the treatment of depression: A review of the exercise program variables." *Journal of Science and Medicine in Sport 17*: 177–82.

Stavrakakis, N., A. M. Roest, F. Verhulst, J. Ormel, P. de Jonge, & A.J. Oldehinkel. 2013. "Physical activity and onset of depression in adolescents: A prospective study in the general population cohort TRAILS." *J Psych Res 47*: 1304–8.

Steene-Johannessen, J., S. A. Anderssen, H. P. van der Ploeg, I. J. Hendriksen, A.E. Donnelly, S. Brage, & U. Ekelund. 2015. "Are self-report measures able to define individuals as physically active or inactive?" *Med Sci Sports Exerc* doi:10.1249/MSS.0000000000000760

Strawbridge, W. J., S. Deleger, R. E. Roberts, & G. A. Kaplan. 2002. "Physical activity reduces the risk of subsequent depression for older adults." *Am J Epidemiol 156*: 328–34.

Sullivan, P. F., M. C. Neale, & K.S. Kendler. 2000. "Genetic epidemiology of major depression: Review and meta-analysis." *Am J Psych 157*: 1552–62.

Sund, A. M., B. Larsson, & L. Wichstrom. 2011. "Role of physical and sedentary activities in the development of depressive symptoms in early adolescence." *Soc Psychiatry Psychiatr Epidemiol 46*: 431–41.

Taaffe, D. R., F. Irie, K. H. Masaki, R. D. Abbott, H. Petrovitch, G. W. Ross, & L. R. White. 2008. "Physical activity, physical function, and incident dementia in elderly men: The Honolulu-Asia Aging Study." *J Gerontol A Biol Sci Med Sci 63* (5): 529–35.

Tantimonaco, M., R. Ceci, S. Sabatini, M. V. Catani, A. Rossi, V. Gasperi, & M. Maccarrone. 2014. "Physical activity and the endocannabinoid system: An overview." *Cell Mol Life Sci 71*: 2681–98.

Tsai, A. C., S. H. Chi, & J. Y. Wang. 2013. "Cross-sectional and longitudinal associations of lifestyle factors with depressive symptoms in≥ 53-year old Taiwanese-results of an 8-year cohort study." *Prev Med 57*: 92–7.

Uebelacker, L. A., C. B. Eaton, R. Weisberg, M. Sands, C. Williams, D. Calhoun,... T. Taylor. 2013. "Social support and physical activity as moderators of life stress in predicting baseline depression and change in depression over time in the Women's Health Initiative." *Soc Psych Epi 48*: 1971–82.

van Gool, C. H., G. I. Kempen, H. Bosma, M. P. van Boxtel, J. Jolles, & J. T. van Eijk. 2006. "Associations between lifestyle and depressed mood: Longitudinal results from the Maastricht Aging Study." *Am J Public Health 97*: 887–94.

Van Uffelen, J. G., Y. R. van Gellecum, N.W. Burton, G. Peeters, K. C. Heesch, & W. J. Brown. 2013. "Sitting-time, physical activity, and depressive symptoms in mid-aged women." *Am J Prev Med 45*: 276–81.

Vaynman, S., & F. Gomez-Pinilla. 2005. "License to run: Exercise impacts functional plasticity in the intact and injured central nervous system by using neurotrophins." *Neurorehabil Neural Repair 19* (4): 283–95. doi:10.1177/1545968305280753

Verghese, J., R. B. Lipton, M. J. Katz, C. B. Hall, C. A. Derby, G. Kuslansky,...H. Buschke. 2003. "Leisure activities and the risk of dementia in the elderly." *N Engl J Med 348* (25): 2508–16. doi:10.1056/NEJMoa022252

Wang, F., M. DesMeules, W. Luo, S. Dai, C. Lagace, & H. Morrison. 2011. "Leisure time physical activity and marital status in relation to depression between men and women: A prospective study." *Health Psychol 30*: 204–11. doi:10.1037/a0022434

Wang, S., X. Luo, D. Barnes, M. Sano, & K. Yaffe. 2014. "Physical activity and risk of cognitive impairment among oldest-old women." *Am J Geriatr Psychiatry 22* (11): 1149–57. doi:10.1016/j.jagp.2013.03.002

Warburton D. E. R., S. Charlesworth, A. Ivey, L. Nettlefold, & S. S. Bredin. 2010. "A systematic review of the evidence for Canada's Physical Activity Guidelines for Adults." *Int J Behav Nutr Phys Act 11* (7): 39. doi:10.1186/1479-5868-7-39

Wei, C. J., Y. Cheng, Y. Zhang, F. Sun, W. S. Zhang, & M. Y. Zhang. 2014. "Risk factors for dementia in highly educated elderly people in Tianjin, China." *Clin Neurol Neurosurg 122*: 4–8. doi:10.1016/j.clineuro.2014.04.004

Weyerer, S. 1992. "Physical inactivity and depression in the community: Evidence from the Upper Bavarian field study." *Int J Sports Med 13*: 492–6.

Whiteford H. A., L. Degenhardt, J. Rehm, A. J. Baxter, A. J. Ferrari, H. E. Erskine,... T. Vos. 2013. "Global burden of disease attributable to mental and substance use disorders: Findings from the Global Burden of Disease Study 2010." *Lancet 382* (9904): 1575–86.

Wilson, R. S., D. A. Bennett, J. L. Bienias, N. T. Aggarwal, C. F. Mendes De Leon, M. C. Morris,...D.A. Evans. 2002. "Cognitive activity and incident AD in a population-based sample of older persons." Neurology *59* (12): 1910–4.

Wise, L. A., L. L. Adams-Campbell, J. R. Palmer, & L. Rosenberg. 2006. "Leisure time physical activity in relation to depressive symptoms in the Black Women's Health Study." *Ann Behav Med 32*: 68–76.

World Health Organization. 2012. "Depression: A global crisis." www.who.int/mental_health/management/depression/wfmh_paper_depression_wmhd_2012.pdf.

Yaffe, K, D. Barnes, M. Nevitt, L. Y. Lui, & K. Covinsky. 2001. "A prospective study of physical activity and cognitive decline in elderly women: Women who walk." *Arch Intern Med.* Jul 23; *161* (14): 1703–8.

York J. L., & J. W. Welte. 1994. "Gender comparisons of alcohol consumption in alcoholic and nonalcoholic populations." *J Stud Alcohol 55*: 743–750.

Yoshitake, T., Y. Kiyohara, I. Kato, T. Ohmura, H. Iwamoto, K. Nakayama,...K. Ueda. 1995. "Incidence and risk factors of vascular dementia and Alzheimer's disease in a defined elderly Japanese population: The Hisayama Study." Neurology *45* (6): 1161–8.

Index

Note: Page numbers for tables are identified by **bold** type, and figures by *italic* type.

Printed in the United States
by Baker & Taylor Publisher Services

Printed in the United States
by Baker & Taylor Publisher Services